.

Methods in Molecular Genetics

Volume 6

Microbial Gene Techniques

Methods in Molecular Genetics

Edited by

Kenneth W. Adolph

Department of Biochemistry
University of Minnesota Medical School
Minneapolis, Minnesota

Volume 6

Microbial Gene Techniques

ACADEMIC PRESS
San Diego   New York   Boston   London   Sydney   Tokyo   Toronto

Front cover photograph: Transcription Factor–Nucleosome Interactions: The illustration indicates functional steps in chromatin remodeling by sequence-specific transcription factors in eukaryotic cells. *Top*: Eukaryotic DNA (thick black line) is packaged in the nucleus by wrapping around histone octamers which are composed of one H3/H4 tetramer (blue ovals) and two H2A/H2B dimers (green ovals). *Center*: Multiple mechanisms stimulate the binding of transcription factors (red) to their recognition sites in nucleosomal DNA. This results in the formation of ternary complexes which contain bound factors, histones, and DNA. Importantly, factor binding destabilizes nucleosome structure. *Bottom*: Destabilized histones in transcription factor–nucleosome ternary complexes can be transferred onto other histone-binding molecules. This results in complete nucleosome displacement and provides the opportunity for additional factor binding. For details of the analysis of transcription factor–nucleosome interactions see the chapters by Côté *et al.* and Vettese-Dadey *et al.* in this volume. Courtesy of Dr. Jerry L. Workman, Department of Biochemistry and Molecular Biology and The Center for Gene Regulation, The Pennsylvania State University, University Park, Pennsylvania.

Academic Press, Inc.
A Division of Harcourt Brace & Company
525 B Street, Suite 1900, San Diego, California 92101-4495

United Kingdom Edition published by
Academic Press Limited
24-28 Oval Road, London NW1 7DX

International Standard Serial Number: 1067-2389

International Standard Book Number: 0-12-044308-2

PRINTED IN THE UNITED STATES OF AMERICA
95 96 97 98 99 00 EB 9 8 7 6 5 4 3 2 1

Table of Contents

Section III Bacterial Gene Structure and Regulation

Section IV Plasmids and Phages: Replication, Transcription, Assembly

Contributors to Volume 6

Numbers in parentheses indicate the pages on which the authors' contributions begin.

CHRISTOPHER C. ADAMS (129), Department of Biochemistry and Molecular Biology, and Center for Gene Regulation, The Pennsylvania State University, University Park, Pennsylvania 16802

JAMES W. AJIOKA (30), Laboratory for Parasite Genome Analysis, Department of Pathology, University of Cambridge, Cambridge CB2 1QP, United Kingdom

VASCO AZEVEDO (323), Ministério de Educação, Universidade Federal de Ouro Preto, Laboratório de Bioquímico e Fisiológico de Microorganismos, 35,400-000-Ouro, Preto-MG, Brazil

DENNIS H. BAMFORD (455), Institute of Biotechnology and Department of Biosciences, Division of Genetics, University of Helsinki, FIN-00014 Helsinki, Finland

JAANA K. H. BAMFORD (455), Department of Biosciences, Division of Genetics, University of Helsinki, FIN-00014 Helsinki, Finland

MICHAEL BLACK (3), Department of Microbiology and Immunology, Stanford University School of Medicine, Stanford, California 94305

JOHN C. BOOTHROYD (3), Department of Microbiology and Immunology, Stanford University School of Medicine, Stanford, California 94305

ROLAND BROUSSEAU (186), Biotechnology Research Institute, Montreal, Quebec, Canada Y4P 2R2

MEMMO BUTTINELLI (168), Fondazione "Istituto Pasteur-Fondazione Cenci Bolognetti", Dipartimento di Genetica e Biologia Molecolare, Università di Roma "La Sapienza", 00185 Rome, Italy

GIORGIO CAMILLONI (168), Centro di Studio per gli Acidi Nucleici, CNR, Dipartimento di Genetica e Biologia Molecolare, Università di Roma "La Sapienza", 00185 Rome, Italy

VINCENT J. CANNISTRARO (280), Department of Molecular Microbiology, Washington University School of Medicine, St. Louis, Missouri 63110

DHRUBA K. CHATTORAJ (400), Laboratory of Biochemistry, National Cancer Institute, National Institutes of Health, Bethesda, Maryland 20892

HYON E. CHOY (389), Developmental Genetics Section, Laboratory of Molecular Biology, National Cancer Institute, National Institutes of Health, Bethesda, Maryland 20892

ANTHONY J. CLARKE (301), Department of Microbiology, University of Guelph, Guelph, Ontario, Canada N1G 2W1

GIOVANNA COSTANZO (168), Centro di Studio per gli Acidi Nucleici, CNR, Dipartimento di Genetica e Biologia Molecolare, Università di Roma "La Sapienza", 00185 Rome, Italy

JACQUES CÔTÉ (108, 129), Department of Biochemistry and Molecular Biology, and Center for Gene Regulation, The Pennsylvania State University, University Park, Pennsylvania 16802

RICHARD M. R. COULSON (30), Laboratory for Parasite Genome Analysis, Department of Pathology, University of Cambridge, Cambridge CB2 1QP, United Kingdom

SIMON M. CUTTING (323), Department of Microbiology, University of Pennsylvania School of Medicine, Philadelphia, Pennsylvania 19104

JOHN DAVEY (247), School of Biochemistry, The University of Birmingham, Birmingham B15 2TT, England

ERNESTO DI MAURO (168), Fondazione "Istituto Pasteur-Fondazione Cenci Bolognetti", Dipartimento di Genetica e Biologia Molecolare, Università di Roma "La Sapienza", 00185 Rome, Italy

JUSTIN A. DIBBENS (400), Laboratory of Biochemistry, National Cancer Institute, National Institutes of Health, Bethesda, Maryland 20892

BERNARD DUJON (81), Unité de Génétique Moléculaire des Levures, Département de Biologie Moléculaire, Institut Pasteur, Université Pierre et Marie Curie, F-75724 Paris, France

RICHARD EGEL (247), Department of Genetics, Institute of Molecular Biology, University of Copenhagen, DK-1353 Copenhagen K, Denmark

JOSE R. ESPINOZA, (30), Laboratory for Parasite Genome Analysis, Department of Pathology, University of Cambridge, Cambridge CB2 1QP, United Kingdom

CÉCILE FAIRHEAD (81), Unité de Génétique Moléculaire des Levures, Département de Biologie Moléculaire, Institut Pasteur, F-75724 Paris, France

MIKKO FRILANDER (455), Department of Biosciences, Division of Genetics, University of Helsinki, FIN-00014 Helsinki, Finland

SUSAN GARGES (389), Laboratory of Molecular Biology, National Cancer Institute, National Institutes of Health, Bethesda, Maryland 20892

GUSTAVO H. GOLDMAN (48), Department of Pharmacology, University of Medicine and Dentistry of New Jersey, Robert Wood Johnson Medical School, Piscataway, New Jersey 08854

JAMES E. HABER (204), Rosenstiel Basic Medical Sciences Research Center, Department of Biology, Brandeis University, Waltham, Massachusetts 02254

WOLFRAM HÖRZ (153), Institut für Physiologische Chemie, Universität München, 80336 München, Germany

PADMAN JAYARATNE (301), Microbiology Section, Department of Laboratory Medicine, Hamilton General Hospital, Hamilton, Ontario, Canada L8L 2X2

WENDY J. KEENLEYSIDE (301), Department of Microbiology, University of Guelph, Guelph, Ontario, Canada N1G 2W1

DAVID KENNELL (280), Department of Molecular Microbiology, Washington University School of Medicine, St. Louis, Missouri 63110

SHAHID M. KHAN (30), Laboratory for Parasite Genome Analysis, Department of Pathology, University of Cambridge, Cambridge CB2 1QP, United Kingdom

KAMI KIM (3), Department of Microbiology and Immunology, Stanford University School of Medicine, Stanford, California 94305

DOMINIQUE LALO (227), Service de Biochimie et de Génétique Moléculaire, Département de Biologie Cellulaire et Moléculaire, Centre d'Etudes de Saclay, F-91191 Gif-sur-Yvette, France

P. RONALD MACLACHLAN (301), Veterinary Infectious Diseases Organization, University of Saskatchewan, Saskatoon, Saskatchewan, Canada S7N 0W0

M. G. MARINUS (267), Department of Pharmacology, University of Massachusetts Medical School, Worcester, Massachusetts 01655

GEORGE A. MARZLUF (66), Department of Biochemistry, The Ohio State University, Columbus, Ohio 43210

SARA E. MELVILLE (30), Laboratory for Parasite Genome Analysis, Department of Pathology, University of Cambridge, Cambridge CB2 1QP, United Kingdom

N. RONALD MORRIS (48), Department of Pharmacology, University of Medicine and Dentistry of New Jersey, Robert Wood Johnson Medical School, Piscataway, New Jersey 08854

GUARANGA MUKHOPADHYAY (400), Laboratory of Biochemistry, National Cancer Institute, National Institutes of Health, Bethesda, Maryland 20892

VINCENT MULHOLLAND (439), Scottish Agricultural Science Agency, Edinburgh EH12 8NJ, Scotland

RODOLFO NEGRI (168), Centro di Studio per gli Acidi Nucleici, CNR, Dipartimento di Genetica e Biologia Molecolare, Università di Roma "La Sapienza", 00185 Rome, Italy

OLAF NIELSEN (247), Department of Genetics, Institute of Molecular Biology, University of Copenhagen, DK-1353 Copenhagen K, Denmark

PÄIVI M. OJALA (455), Department of Cell Biology, Yale University School of Medicine, New Haven, Connecticut 06520

GEORGE W. ORDAL (339), Department of Biochemistry, University of Illinois at Urbana-Champaign, Urbana, Illinois 61801

GUOHUA PAN (186), Department of Molecular and Medical Genetics, University of Toronto, Toronto, Ontario, Canada M5S 1A8

ARNAUD PERRIN (81), Unité de Génétique Moléculaire des Levures, Département de Biologie Moléculaire, Institut Pasteur, F-75724 Paris, France

ELMER R. PFEFFERKORN (3), Department of Microbiology, Dartmouth Medical School, Hanover, New Hampshire 63755

LENE JUEL RASMUSSEN (267), Department of Molecular and Cellular Toxicology, Harvard School of Public Health, Boston, Massachusetts 02115

JONATHAN REIZER (375), Department of Biology, University of California at San Diego, La Jolla, California 92093

FERNANDO ROJO (421), Centro de Biologia Molecular "Severo Ochoa" (CSIC-UAM), Universidad Autónoma, Canto Blanco, 28049 Madrid, Spain

MIA MAE L. ROSARIO (339), Department of Biochemistry, University of Illinois at Urbana-Champaign, Urbana, Illinois 61801

SANGRYEOL RYU (389), Developmental Genetics Section, Laboratory of Molecular Biology, National Cancer Institute, National Institutes of Health, Bethesda, Maryland 20892

PAUL D. SADOWSKI (186), Department of Molecular and Medical Genetics, University of Toronto, Toronto, Ontario, Canada M5S 1A8

MILTON H. SAIER, JR. (375), Department of Biology, University of California at San Diego, La Jolla, California 92093

MARGARITA SALAS (421), Centro de Biologia Molecular "Severo Ochoa" (CSIC-UAM), Universidad Autónoma, Canto Blanco, 28049 Madrid, Spain

GEORGE P. C. SALMOND (439), Department of Biological Sciences, University of Warwick, Coventry CV4 7AL, United Kingdom

FRANK SEEBER (3), Department of Microbiology and Immunology, Stanford University School of Medicine, Stanford, California 94305

DAVID SIBLEY (3), Department of Molecular Microbiology, Washington University School of Medicine, St. Louis, Missouri 63110

DOMINIQUE SOLDATI (3), Department of Microbiology and Immunology, Stanford University School of Medicine, Stanford, California 94305

SOPHIE STETTLER (227), Service de Biochimie et de Génétique Moléculaire, Département de Biologie Cellulaire et Moléculaire, Centre d'Etudes de Saclay, F-91191 Gif-sur-Yvette, France

NEAL SUGAWARA (204), Rosenstiel Basic Medical Sciences Research Center, Department of Biology, Brandeis University, Waltham, Massachusetts 02254

JOHN SVAREN (153), Institut für Physiologische Chemie, Universität München, 80336 München, Germany

HERVÉ TETTELIN (81), Unité de Génétique Moléculaire des Levures, Département de Biologie Moléculaire, Institut Pasteur, F-75724 Paris, France

AGNÈS THIERRY (81), Unité de Génétique Moléculaire des Levures, Département de Biologie Moléculaire, Institut Pasteur, F-75724 Paris, France

PIERRE THURIAUX (227), Service de Biochimie et de Génétique Moléculaire, Département de Biologie Cellulaire et Moléculaire, Centre d'Etudes de Saclay, F-91191 Gif-sur-Yvette, France

RHEA T. UTLEY (108), Department of Biochemistry and Molecular Biology, and Center for Gene Regulation, The Pennsylvania State University, University Park, Pennsylvania 16802

SABRINA VENDITTI (168), Fondazione "Istituto Pasteur-Fondazione Cenci Bolognetti", Dipartimento di Genetica e Biologia Molecolare, Università di Roma "La Sapienza", 00185 Rome, Italy

PATRIZIA VENDITTI (168), Centro di Studio per gli Acidi Nucleici, CNR, Dipartimento di Genetica e Biologia Molecolare, Università di Roma "La Sapienza", 00185 Rome, Italy

ULRIKE VENTER (153), Institut für Physiologische Chemie, Universität München, 80336 München, Germany

MICHELLE VETTESE-DADEY (129), Department of Biochemistry and Molecular Biology, and Center for Gene Regulation, The Pennsylvania State University, University Park, Pennsylvania 16802

LAURA WALIN (455), Department of Biosciences, Division of Genetics, University of Helsinki, FIN-00014 Helsinki, Finland

PHILLIP WALTER (129), Department of Biochemistry and Molecular Biology and Center for Gene Regulation, The Pennsylvania State University, University Park, Pennsylvania 16802

KIEW-LIAN WAN (30), Laboratory for Parasite Genome Analysis, Department of Pathology, University of Cambridge, Cambridge CB2 1QP, United Kingdom

MICHEL WERNER (227), Service de Biochimie et de Génétique Moléculaire, Département de Biologie Cellulaire et Moléculaire, Centre d'Etudes de Saclay, F-91191 Gif-sur-Yvette, France

CHRIS WHITFIELD (301), Department of Microbiology, University of Guelph, Guelph, Ontario, Canada N1G 2W1

DAVID B. WILSON (367), Section of Biochemistry, Molecular and Cell Biology, Cornell University, Ithaca, New York 14853

JERRY L. WORKMAN (108, 129), Department of Biochemistry and Molecular Biology and Center for Gene Regulation, The Pennsylvania State University, University Park, Pennsylvania 16802

Preface

The new series *Methods in Molecular Genetics* provides practical experimental procedures for use in the laboratory. Because the introduction of molecular genetic techniques and related methodology has revolutionized biological research, a wide range of methods are covered. The power and applicability of these techniques have led to detailed molecular answers to important biological questions and have changed the emphasis of biological research, including medical research, from the isolation and characterization of cellular material to studies of genes and their protein products.

Molecular genetics and related fields are concerned with genes: DNA sequences of genes, regulation of gene expression, and the proteins encoded by genes. The consequences of gene activity at the cellular and developmental levels are also investigated. In medical research, knowledge of the causes of human disease is reaching an increasingly sophisticated level now that disease genes and their products can be studied. The techniques of molecular genetics are also being widely applied to other biological systems, including viruses, bacteria, and plants, and the utilization of gene cloning methodology for the commercial production of proteins for medicine and industry is the foundation of biotechnology. The revolution in biology that began with the introduction of DNA sequencing and cloning techniques will continue as new procedures of increasing usefulness and convenience are developed.

In addition to the basic DNA methods, instrumentation and cell biology innovations are contributing to the advances in molecular genetics. Important examples include gel electrophoresis and DNA sequencing instrumentation, *in situ* hybridization, and transgenic animal technology. Such related methodology must be considered along with the DNA procedures.

Microbial Gene Techniques consists of twenty-five chapters in four sections: Molecular Genetics of Eukaryotic Microbes; Yeast Chromosomes: Transcription, Recombination, Replication; Bacterial Gene Structure and Regulation; Plasmids and Phages: Replication, Transcription, Assembly. Methods are presented for investigating microbial gene and genome structure, as well as genome functioning in transcription, replication, and recombination. The experimental systems discussed include protozoa, filamentous fungi, yeast, bacteria (*Escherichia coli, Bacillus subtilis*), bacterial plasmids, and bacteriophages.

Methods in Molecular Genetics will be of value to researchers, as well as to students and technicians, in a number of biological disciplines because of the wide applicability of the procedures and the range of topics covered.

KENNETH W. ADOLPH

xv

Methods in Molecular Genetics

Section I

Molecular Genetics of Eukaryotic Microbes

[1] Forward and Reverse Genetics in the Study of the Obligate, Intracellular Parasite *Toxoplasma gondii*

John C. Boothroyd, Michael Black, Kami Kim,
Elmer R. Pfefferkorn, Frank Seeber, David Sibley,
and Dominique Soldati

Overview

The study of parasitic protozoa has benefited relatively little from the application of genetics. In part this is because most of the organisms that have been models for biochemical and molecular analyses do not lend themselves to such approaches. In many cases, in fact, no sexual cycle is known, and culture methods are so cumbersome that *in vitro* manipulations are extremely difficult.

One exception to both these limitations is the apicomplexan protozoan *Toxoplasma gondii* (1, 2). It has long been known that toxoplasma is capable of unlimited asexual growth as a haploid form in almost any warm-blooded vertebrate. In the early 1970s it was discovered that it can also undergo a sexual cycle in felines that culminates in the generation of many robust oocysts (this aspect of the life cycle is described further below). Particularly exciting to the geneticist is the fact that the oocysts apparently contain all the progeny of the meiotic process, analogous to fungal asci. Thus the requirement of a well-described, readily manipulable sexual cycle is met by this parasite.

The second limitation often encountered with parasites is *in vitro* growth. Although it is an obligate intracellular parasite, *T. gondii* has the ability to invade and proliferate in almost any animal cell, which has made its adaptation to *in vitro* culture in a variety of host cell lines relatively simple. Indeed, lytic properties of *T. gondii* make possible its quantitation by plaque assays (3).

In addition to its use as a model organism for the study of invasion and intracellular survival, *T. gondii* is of great concern as a potent opportunistic pathogen. Today, its role as the etiologic agent of congenital toxoplasmosis (blindness, retardation, hydrocephaly, etc.) is being eclipsed by the scale of disease it causes in patients with AIDS (acquired immunodeficiency syndrome) (4). After infection with human immunodeficiency virus (HIV) itself, toxoplasmic encephalitis is the major disease of the central nervous system in such patients. This potentially fatal condition is found in up to one-half of all AIDS patients (depending on the country).

Efforts to develop the potential of this system for genetic analysis have accelerated. The point has now been reached where the full power of genetics can be applied to detailed investigations on all aspects of this important organism. In this chapter, we describe methodologies for both forward genetics (i.e., conventional, transmission genetics) and reverse genetics (transfection/transformation).

Forward Genetics

The life cycle of toxoplasma can be thought of as having two discrete components, one sexual and the other asexual. The asexual cycle relies on carnivorism and scavenging to effect transmission between one warm-blooded vertebrate and another. Within the hosts, the parasite exists in two forms, the rapidly growing tachyzoite and the slow-growing, encysted form, the bradyzoite. It is the latter form which normally transmits the infection to other animals.

The sexual cycle, which occurs only in felines, begins when a cat feeds on infected prey. The tissue cysts within the prey are ingested, and the bradyzoites within are released in the digestive tract; an infection then ensues in the cat. Responding to some unknown stimulus, a portion of the parasites infecting the epithelium of the intestine enter gametogenesis. Male and female gametes result which on fusion produce a zygote. The zygote begins the process of generating a cyst wall, after which it enters the lumen of the intestine and is excreted in the feces as an immature oocyst.

It is believed that the immature oocyst is diploid, that is, no replication of the DNA occurs prior to maturation. After fecal excretion, maturation begins, involving full meiosis with, presumably, an extra mitotic division at the end. What is sure is that eight haploid sporozoites are ultimately found in the mature oocyst. These eight are subcompartmentalized as two groups of four, each group comprising a sporocyst. The mature oocyst is highly stable in the environment, being capable of surviving many months in the soil (cool, damp conditions being optimal).

Transmission occurs when the mature oocysts are ingested by animals. They rupture in the intestine of the host, and the sporozoites within rapidly initiate a new infection. Initially, this new infection is dominated by the tachyzoite form but with time, tissue cysts with their constituent bradyzoites appear.

Sporozoites can be released in the laboratory through mechanical and enzymatic means that mimic natural ingestion. The sporozoites are capable of invading the usual experimental host cells and transforming themselves into tachyzoites, and so further analysis can proceed entirely *in vitro,* if appropriate or desired. It may also be possible to microdissect the oocyst (D. Roos, personal communication), which may ultimately permit "tetrad analyses" of the sort that have proved so powerful in fungal genetics.

The ability to generate mutants and do crosses is a powerful tool to the biologist. It is far more useful, however, if a map exists so that the mutations responsible for a given phenotype can be positioned relative to other genes. This gives information on the number of genes involved in a phenotype and provides clues to the identity of the genes. The ultimate goal, however, is to be able to use the map to clone the gene(s). As detailed below, a genetic map has been generated (5) with 64 markers localized relative to one another. Although of low resolution, the map appears sufficient to allow the rapid mapping of any mutant nuclear gene.

Conduction of Genetic Crosses

Detailed protocols for the methods used in performing genetic crosses are presented. First, however, some comments on laboratory safety are critical. *Toxoplasma gondii* is a significant human pathogen. Most laboratory infections in other institutions have been traced to ingested oocysts or to needle-stick accidents with syringes. Our laboratories function at the standards of a P-2 facility. All manipulations are carried out in regularly certified and inspected class-2 biosafety hoods. All centrifuging is done in closed tubes. Everyone wears gloves when working with *T. gondii*. No mouth pipetting is permitted. Infected kittens excreting oocysts are housed in a separate containment facility to which only experienced investigators have access. Feces are collected daily and the cages are cleaned daily. This offers excellent protection because freshly excreted oocysts do not become infectious (sporulated) for at least 36 hr. The purification and opening of oocysts are done only by experienced investigators with the additional precautions of wearing a face mask and immediately sterilizing all materials. We measure antitoxoplasma antibody titers of all laboratory personnel potentially exposed to the oocysts at regular intervals and have never observed a seroconversion.

Cultured Host Cells

We use cultured cells for the selection of mutants and for the analysis of genetic crosses. The host cells are human fibroblasts that have been established from a foreskin. Suitable cultures can be purchased from the American Type Culture Collection (Rockville, MD). Occasionally, other cell lines can be used as the parasite is relatively nonfastidious in this regard. Among those that work well are Vero, macrophages, and T cells. HeLa cells can also be used but tend to be less efficient as far as yield of parasites obtained.

Parasites

We use three cloned strains of *T. gondii*, all of which make plaques in human fibroblasts. Each strain has advantages and disadvantages in different experimental contexts, and thus we attempt to choose the most appropriate strain for particular experiments. The RH strain (6) is the easiest to grow both in cultured cells and in animals. The yield of parasites in cultured cells is high, and macroscopic plaques are produced in 5 days. The principal disadvantage of the RH strain is that, perhaps because of prolonged passage in mice, it is not able to make gametes in the feline host. Thus, the RH strain cannot be used in genetic crosses. We use the RH strain primarily for biochemical experiments, particularly in the characterization of drug-resistant mutants. This aspect of our work is facilitated by the ability to produce relatively large quantities of purified parasites. The population structure of toxoplasma is unusual, comprising three main clonal lineages (7) (D. Howe and L. D. Sibley unpublished). This clonality has important implication for experimental genetics as well as biochemical studies on drug resistance, virulence, etc. The RH strain is similar to many naturally occurring isolates from animals and human infections that share a similar genotype and have a virulent phenotype in mice. The RH strain has also proved to be a good model for studying the biochemical basis of drug-resistant mutants in other, less related strains.

Two other common laboratory strains of toxoplasma, CEP and PLK, were originally cloned at Dartmouth Medical School and provide characterized type strains for the remaining two clonal lineages. Strain CEP came from an infected cat (8); strain PLK (9) is a subclone of the ME49 strain (10) which originally came from an infected sheep. Both strains are capable of completing the sexual cycle in the feline host.

We have done most of our genetic crosses with mutants of the CEP strain (11, 12). As a by-product of these crosses, we have frozen stocks of the strain bearing various combinations of the drug resistance markers adenine arabinoside (ARA^R), fluorodeoxyuridine (Fud^R), and sinefungin (Snf^R). The ME49 strain is valuable in genetic studies because a large number of restriction fragment length polymorphism (RFLP) markers have been identified that distinguish it from the CEP strain. Analysis of the recombinant progeny of a cross between ME49 and CEP parasites has allowed the detection of recombination of RFLP markers and the assignment of these markers to individual chromosomes resolved by pulsed-field electrophoresis (5).

Measurement of Parasite Infectivity and Growth

We measure the infectivity of *T. gondii* with a simple plaque assay in 25-cm^2 flasks with a confluent monolayer of human fibroblasts (3). We also use plaque formation in the cloning of the parasite. Suitable dilutions of *T. gondii* are allowed to make plaques in half-sized (16mm^2) 96-well trays (Corning, Corning, NY). Because of the small size, the wells can be rapidly inspected under low-power magnification to identify those that unequivocally contain a single plaque. Subculture of the parasites from wells in a single plaque yields a clone. To assay the extent of drug resistance in the characterization of a mutant, we use 24-well trays with quadruplicate wells at all drug concentrations, which are generally 2-fold apart. Several days after infection and drug treatment, parasite growth is measured by assaying the incorporation of [^{3}H]uracil or [^{3}H]xanthine into acid-precipitable form during a 4-hr pulse. This precursor is highly specific for the parasite because it is not incorporated by the host cell (8).

Mutagenesis and Mutant Selection

Although *T. gondii* is haploid, efficient chemical mutagenesis is critical for the induction of drug-resistant mutants. This is particularly true in those cases in which resistance can arise only when nucleotide sequence changes occur within a short region of the gene. In our early published experiments (3) we aimed at 95% killing of the parasites because at that level of killing the desired mutation was relatively common in the survivors. However, the lethal action of the mutagen is an exponential function of the mutagen concentration, whereas the production of mutations in any one gene is a linear function. Thus with more severe mutagenesis, the advantage of the increased fraction of the survivors that were mutants is far outweighed by the more extensive decrease in the number of survivors. Theoretical calculations show that the total yield of mutants is maximal at about 37% (actually $1/e$) survival. Unpublished experiments have confirmed this conclusion.

We find that ethylnitrosourea is the most efficient mutagen for *T. gondii*. The treated parasites are grown for about 10–20 divisions to allow phenotypic expression of induced mutations and then challenged with a predetermined minimally inhibitory concentration of the drug for which resistant mutants are being sought. During serial subculture in the presence of the selective drug, resistant mutants emerge and are cloned as described above.

Preparation of Chronically Infected Mice as Source of Encysted Bradyzoites

We infect mice by the intraperitoneal injection of 6×10^4 tachyzoites grown in cultured cells. At the first sign of illness (ruffled fur), the mice are treated with 500 μg/ml sulfadiazine in the drinking water for 10 days. To prepare the encysted parasites for consumption by cats, the mice are first sacrificed by ether overdose. The brains are removed aseptically and individually homogenized in a small volume of saline. The number of cysts present is estimated microscopically (13).

Genetic Crosses in Cats

The crosses are carried out in 8- to 12-week-old cats. The cats are maintained in the animal facility for 10 days after they are purchased in order to accustom them to the unfamiliar environment and food. This period is essential because the entire expensive experiment can be compromised if the animals eat only a small portion of the brain cyst-laden meal. During this interval, the seronegative status of the animals is confirmed. At noon on the day before the infection, food is withdrawn from the cats to increase the likelihood that they will eat all of the brain cysts. On the day of infection, a small portion of canned cat food is thoroughly mixed with homogenized brains from mice chronically infected with the parasite mutants to be used in the crosses. Each cat receives approximately 1500 cysts from each of the two strains of *T. gondii* to be crossed. Beginning 2 days after infection, total fecal specimens are collected daily and immediate processed as described below to purify the oocysts. The experiment is terminated when the daily fecal output of oocysts falls to 10% of the maximal observed output.

Purification of Oocysts and Release of Sporozoites

Fecal specimens are soaked in about 2 volumes of water for 1 hr and then stirred into an even suspension. The suspension is filtered through several layers of gauze to remove the kitty litter and other debris and then adjusted to 1.1 *M* sucrose by adding an equal volume of 2.2 *M* sucrose. The fecal suspension is placed in a (250 ml) centrifuge tube and carefully overlaid successively with 40 ml of 0.9 *M* sucrose and 40 ml of 0.7 *M* sucrose. After centrifugation at 1100 *g* for 20 min at room temperature, the oocysts are found at the top of the 0.7 *M* sucrose and are skimmed off with a pipette in a total volume of 12 ml, diluted with 36 ml of 0.36 *M* H_2SO_4, and collected by centrifugation. The pellet is resuspended in 10 ml of 0.9 *M* sucrose and then successively overlaid with 0.7 and 0.3 *M* sucrose. After centrifugation at 1100 *g* for 20 min, a white band of highly purified oocysts is found at the 0.7–0.3 *M* interface. This band is collected, diluted with 3 volumes of 0.36 *M* H_2SO_4, and centrifuged. The resulting pellet is suspended in 5 ml of 0.36 *M* H_2SO_4 and incubated as a thin layer in a tightly closed bottle for 3 days at room temperature. After microscopic examination for completeness of sporulation, the sporulated oocysts are centrifuged, resuspended in phosphate-buffered saline (PBS) containing 1 mg/ml glucose, and stored at 4°C. They remain viable for at least 1 year.

The sporulated oocysts are opened through a simple procedure. Oocysts are transferred to Hanks' balanced salt solution by centrifugation. A concentrated suspension in a volume of about 0.5 ml is homogenized for 30 sec with a motor-driven Potter–Elvehjem homogenizer. The resulting suspension is collected by centrifugation (1000 *g* for 10 min) and incubated for 10 min at 37°C in Hanks' balanced salt solution containing 5% sodium taurodeoxycholate (w/v). After several centrifugations to remove the taurodeoxycholate, the pellet (a mixture of oocysts, sporocysts, and a few free sporozoites) is resuspended and used to infect a culture of human fibroblasts. In the culture, additional sporozoites exit the sporocysts, infect the host cells, and multiply as tachyzoites. We find roughly 0.5 infectious sporozoites per sporulated oocyst. The theoretical value of 8 (the number of sporozoites per oocyst) is not achieved because about half of the oocysts are not opened and not all of the released sporozoites are infectious. All of the results from previous crosses suggest that the infectious sporozoites are representative of the entire population.

Characterization of Physical Linkage Groups

The first step in the establishment of genetics in any system is to determine the number of linkage groups (14). Traditionally, this has been done in either of two ways: cytogenetic examination of the metaphase chromosomes or analysis of a large number of markers. Pulsed-field gel electrophoresis, however, has provided a third alternative for organisms where a large number of auxotrophic markers did not exist and where the chromosomes did not condense or were too small to see through the microscope.

Transverse-alternating field electrophoresis [TAFE I (15)] was used to separate toxoplasma chromosomes into 9 distinct bands ranging from 2 to greater than 10 Mb.

A single 4–6 day run was sufficient to separate the entire karyotype, and bands were sufficiently sharp for blotting and hybridization studies (14). Attempts to resolve toxoplasma chromosomes using contour-clamped homogeneous electric field electrophoresis [CHEF (16)] resulted in separation of only six discrete bands using a fixed angle of 120° between the two opposing electric fields. Newer designs of the CHEF apparatus allow for smaller angles of separation (i.e., 105°) that greatly improve the resolution of larger chromosomes. On variable angle CHEF systems, it has also been possible to separate all nine bands of the toxoplasma karyotype (M. Messina and L. D. Sibley, unpublished).

A total of 57 probes representing cloned segments of the toxoplasma genome were hybridized to blots of the pulse-field gels (the probes are described in further detail below). Initial work focused on the RH strain. In the vast majority of cases, the probes hybridized to a single chromosomal band. This was expected given the fact that the probes were believed to represent unique sequences in the toxoplasma genome, judging from conventional Southern blot analysis with digested DNA where the size and number of bands observed was commensurate with the size of the probe. In the rare cases where hybridization to two bands was observed, this was noted for comparison with the genetic analysis.

The chromosome analysis was repeated for two other strains that form the basis of our genetic mapping, CEP and PLK. Although some minor variation was seen, the approximate size and number of the chromosomes were similar in all three strains. Such "minor variation" may not significantly impact the molecular karyotype, but the large size of the toxoplasma chromosomes (compared with, say, plasmodium and leishmania) means that a small shift in relative mobility can represent a difference of a few hundred kilobase pairs. This raised the possibility that crossing different strains might give a complicated result because essential genes might be on different linkage groups in the two strains. We found, however, that none of the probes mapped to different chromosomes in the two strains. The differences, therefore, may be in repetitive DNA and thus may not represent a problem in the genetic cross (as eventually borne out when a cross was attempted, as described below).

Creation of a Low Resolution Genetic Map

The first stage in deriving a genetic map is to identify RFLPs for the gene probes used in the characterization of the physical linkage groups (5). We employed probes of two types. Initially, we used small DNA fragments, typically cDNA clones in the 1- to 2-kb range and often of known coding function. However, we empirically find that such small probes necessitate looking at a great many restriction digests (typically 20–40) to find an RFLP for each probe.

Subsequently, we selected use of cosmids containing large segments (~30–40 kb) of toxoplasma DNA as probes. This has the inherent risk of including repetitive DNA (as is often observed for cosmid clones of human DNA), but we found that the initial cosmids chosen for toxoplasma RFLP analysis apparently contained only unique DNA. Hence, the pattern obtained on Southern blotting of toxoplasma DNA digested with a hexanucleotide-specific restriction endonuclease (e.g., *Eco*RI) and probed with cosmid clone was relatively simple, consisting of about 5–15 bands. We did our first such analyses with cosmids generated in association with other work in the laboratory (e.g., those containing surface antigen genes). Subsequently, we examined about 60 cosmid clones isolated at random from the library. Of the cosmids, only two have been found to apparently contain repetitive DNA, and these have not been further pursued for RFLP analysis. The majority of the remainder yield a usable RFLP after comparison of typically only four to six different enzymes. One advantage of this is that enzymes can be selected because of their price and efficiency. More importantly, however, a given blot can be successively (or even simultaneously) hybridized to several probes, saving enormously on manipulations and materials needed to perform a complete analysis.

In all, 64 probes were generated where there was a RFLP that distinguished PLK from CEP (5). These were used in the next phase, namely, determining through genetic crosses the relative linkage of the loci represented by these probes. Kittens were fed with pooled, freshly removed brains from mice individually infected with ME49 (the parent of PLK, see above) and CEP strains. The CEP strain carries unlinked resistance genes for two drugs, sinefungin (Snf^R) and adenine arabinoside (Ara^R). Oocysts were harvested, allowed to mature, and opened, and the resulting sporozoites were used to initiate *in vitro* cultures (as described above). These cultures were expanded and the resulting tachyzoites cloned. Recombinants are preliminarily identified based on resistance to either Snf or Ara but not both [reversion to drug sensitivity is known to be low for the markers (12)]. Using this approach, 19 "recombinants" were readily identified. This indicates that the two strains are indeed capable of productive cross-fertilization.

The results of the linkage analysis are presented elsewhere (5). Briefly, all probes fell into 1 of 11 linkage groups, corresponding to the 10 bands (one of which is a doublet) seen in the physical analysis, described above. Based on segregation frequencies, we were able to construct a low resolution genetic map for all 11 linkage groups. The Ara^R and Snf^R genes mapped to chromosomes V and IX, respectively.

In addition to the generation of the genetic map, several other conclusions came out of the analysis. (1) Repetitive DNA is not so abundant or so dispersed as to pose a problem in analysis of the toxoplasma genome (unlike, e.g., the case with the human genome where repetitive elements such as Alu repeats preclude the use of cosmids which almost invariably possess at least one such repeat). (2) The recombination frequency is of the order of 150–300 kb per centimorgan, which is low enough that with the exception of the biggest linkage group, X, apparently all loci on a given chromosome are linked. Consequently, only about one marker per chromosome is needed to determine the linkage group to which a given mutation maps. (3) Recombination does not appear to be an obligatory event for any given chromosome in any given meiosis. This is an unusual situation not seen in most other organisms. (4) Independent strains of *T. gondii* are capable of crossing in a productive way. Although expected, such a cross had never been described before; it was possible that the parasite populations are in the process of speciating and that two random strains might not cross.

All of the above refers, of course, only to the nuclear genome. In addition to this there are believed to be at least two nonnuclear genomes (17, 18) originating, respectively, from the mitochondrion and an as yet uncharacterized organelle (that may be plastid-like). Once these genomes are better characterized, it will be possible to generate markers that will allow mutations within them to be mapped.

Mapping of Unknown Mutations

What follows is a suggested protocol for mapping unknown genes based on the above experience. First, a biological or biochemical property of interest must be defined. Next, a mutant phenotype that can be readily screened or, preferably, selected is chosen. Mutagenesis of a suitable parent (e.g., Fud^R ME49) is performed, if necessary, and the selection carried out. A cross is then performed with another suitable parent (e.g., Snf^R CEP) and recombinant progeny isolated based on their double resistance (i.e., Snf^R Fud^R).

Given the existence of a rudimentary but apparently complete linkage map for the nuclear genome, the next task is to localize the mutant gene relative to a subset of the known markers. This is where the relatively low recombination frequency in toxoplasma becomes a distinct advantage: it is necessary to use only about 15 markers to determine the chromosome on which the mutation lies.

Twenty F_1 recombinants should be selected that possess a nonparental drug resistance phenotype; that is, progeny should be selected in the presence of the two drugs to which the parents were individually resistant. These progeny must then be typed with respect to the mutant phenotype of interest. Unless the responsible mutation is closely linked to one or other of the drug resistance markers (possible but unlikely), and assuming there is no bias for or against the mutant phenotype, one should obtain about a 1:1 ratio of mutant to wild type. If there is linkage to one or other drug resistance marker, a skewed ratio will be observed. Analysis of the recombinants with the RFLP markers (see below), including one that is tightly linked to the drug resistance phenotype in question, will quickly confirm or eliminate any tentative linkage.

The 20 progeny should then be grown to sufficient levels to allow genomic DNA to be prepared [in the future a complete set of polymerase chain reaction (PCR) RFLP markers should be available that allows analysis without the need for expanding each clone; D. Howe and L. D. Sibley, unpublished; J. Ajioka and J. Blackwell, personal communication]. Southern blots are then prepared where all 20 progeny are digested with each of the necessary enzymes. The blots are then probed with each of the appropriate RFLP markers. Because many of the enzymes that show polymorphism are the same for the different markers, the number of actual blots that need be performed is considerably less than 15. A table can then be produced that shows which allele is present at leach locus as well as the phenotype of interest. In a sample of 20 progeny, statistically significant linkage exists if two loci cosegregate in 16 or more of the progeny ($P < 0.01$). Assuming that a single mutation was responsible for the mutant phenotype, linkage to only a single region should be observed. If linkage is seen to two or more regions, this may indicate that two mutations are necessary for the phenotype or that one of the linkages is untrue. The simplest way to discriminate between the two choices is to analyze a further 10 recombinant progeny with the two probes in question to see if the linkage holds up.

Refinement of the initial mapping is done through analysis of further probes that span the identified linkage group. Initially, this need be done with only the original set of 20 recombinants. In this way, the mutation should be mappable relative to the ordered probes.

To date, this is as far as we have actually carried the genetic mapping of any mutant. Further refinement requires the analysis of more progeny with respect to markers that are determined to lie nearest to the mutation. This can be done either by conventional Southern blotting or, more likely, through PCR. The latter approach requires a small investment by way of sequencing the alleles present in the two parents for these two markers to find an RFLP that can be detected in a PCR product. Ultimately, however, such an investment is quickly repaid with an accurate map position.

In the future, it should be relatively straightforward to merge this technology with what is described below to move rapidly from the map to having the cloned gene in hand. This will be done through complementation using defined gene libraries. That is, once the map position is known, cloned DNA molecules corresponding to this general region from a wild-type parasite can be introduced into the mutant and assessed for the ability to confer a wild-type phenotype (i.e., to complement the mutation).

Reverse Genetics

As already mentioned, the ability to generate mutants, map the mutations, and isolate the genes involved is the first step in establishing structure–function relationships. Further refinement in any but the simplest of systems (such as phage) requires the ability to manipulate the gene *in vitro,* reintroduce the gene into the organism, and assay the function of its altered product. Because one starts with a known mutation and then determines the phenotype, rather than the other way around, this process has come to be known as reverse genetics.

On the basis of early studies with bacteria, the introduction of DNA into a cell such that its phenotype is altered was termed transformation. Although there is the obvious potential for confusion with processes that lead to conversion of mammalian cells to the immortal or "transformed" state, this original term for conversion of a cell from one stable phenotype to another through introduction of DNA has been maintained. Transient introduction of DNA into eukaryotic cells is usually referred to as transfection, a term originally coined by bacterial geneticists for introduction of DNA via phage.

Transfected cells express the introduced genes for only a relatively short period of time (typically on the order of a few days, depending on the generation time of the cell in question). This transience is due to the fact that the DNA being introduced does not constitute a stable replicon, or unit capable of autonomous replication. This is most often due to the lack of an origin of replication, but it can also be due to the lack of integration of the DNA into the genome of the cell, resulting in instability.

Transformation requires that the transfecting DNA either integrate into the genome or replicate autonomously. Such events can be rare, and the key to establishing stable transfection is a powerful selection. If integration has occurred, the selection pressure need not be maintained once those organisms that are not expressing the introduced genes have been eliminated. If the transfecting DNA is being maintained episomally, the selection must typically be maintained indefinitely to be sure that the episome is not lost owing to the randomness of segregation at mitosis. (In well characterized systems such as *Saccharomyces cerevisiae*, addition of a centromere can help stabilize an episomal element, but this option is not typically available and, even still, is not 100% effective.) Because of the added complication of needing a suitable selection for stable transformation, attempts to introduce DNA into a new cell type almost invariably start with plasmid constructs that ask only that some transient expression be seen.

Introduction of genes into eukaryotic cells is done in any of a variety of ways. Among the most common are calcium phosphate precipitation, liposome fusion, and electroporation. Of these, the one that is least dependent on the particulars of the cell being transfected is electroporation, which is even used in bacteria. Although the precise mechanism operating in electroporation is not known, the brief pulse of current is thought to result in the generation of short-lived pores in the cell membrane that allow DNA to pass through.

When attempting to establish transfection in a new system, even transient, there are several variables that must be considered. Among these, most important are the reporter gene (i.e., the one whose product will be assayed to measure success or failure of the transfection protocol), the signals necessary for expression of the gene [promoter, poly (A) addition site, etc.], the medium and buffers to be used, the electrical parameters of the pulse to be delivered, and the state or form of the cell to be transfected. For an obligate intracellular parasite, the last of these parameters is key. Does one use the extracellular parasite which presents fewer physical barriers (i.e., the transfecting DNA does not have to traverse the host cell before it can even reach the plasma membrane of the parasite) or the intracellular parasite which is in its natural habitat and might recover most successfully from the electroporation procedure?

In the case of toxoplasma the choice is not obvious, but the relative robustness of the organism outside the host cell, where it retains full metabolic activity including the ability to invade a new host cell for many hours, clearly favors the use of extracellular parasites. The choice is further influenced by the fact that toxoplasma has an unusual pellicle consisting of three lipid bilayers in place of the usual one (the inner two of these can be thought of as extremely large vesicles that are compressed up against the outermost plasma membrane but in such a way that there are no gaps between them). Collectively, the three membranes seem likely to be a significant barrier to transfection; transfection of the intracellular parasite has the added problem of getting material across the host plasma membrane and the parasitophorous vacuole in which the parasite resides.

Optimal Method for Transient Transfection

The following protocol is based on a large number of experiments in which the major variables were altered and represents our recommendation to an investigator interested in the transient expression of a given gene in toxoplasma. Following this, we discuss each of the variables so that optimization for special circumstances can be reached relatively quickly.

1. Infect a monolayer culture of human foreskin fibroblasts with sufficient parasites (RH strain tachyzoites) to give infection of most of the host cells in a single cycle. Incubate the culture at $37°C$ until the parasites have lysed the host cells spontaneously. A single T-175 flask of confluent cells should yield approximately 5×10^8 parasites, enough for 10–30 individual transfections.
2. Concentrate the freshly released parasites (which should be used within 5 hr of host cell lysis) by centrifugation at 400 g for 15 min. This centrifugation is at room temperature, as are all remaining manipulations unless otherwise stated.
3. Wash the parasites by suspension in a large volume (50 ml) of cytomix buffer (19) (120 mM KCl, 0.15 mM CaCl$_2$, 10 mM K$_2$HPO$_4$/KH$_2$PO$_4$, pH, 7.6, 25 mM HEPES, pH, 7.6, 2 mM EGTA, pH, 7.6, 5 mM MgCl$_2$, supplemented with 2 mM freshly prepared ATP and 5 mM freshly prepared glutathione) and spinning at 400 g for 15 min. Resuspend in this same buffer at a final concentration of about 10^7/ml (5-fold higher or lower concentrations also work).
4. Precipitate ~ 10–150 μg per electroporation of the highly purified DNA (CsCl banding or Quiagen purified works well) with ethanol and resuspend in cytomix at a concentration of 2 mg/ml.
5. For each electroporation, gently mix 0.8 ml of parasites and 0.1 ml of DNA and transfer the entire 0.8 ml into an electroporation cuvette (4 mm gap). Provide the

electrical shock using a BTX Electro Cell Manipulator 600 (BTX Inc., San Diego, CA) and the following settings: 2.5 kV/Resistance High Voltage, resistance 48 Ω, charging voltage 2.0 kV (delivered voltage ~ 1.4 kV; time constant 0.4 msec). A small pop is usually heard.

6. After electroporation, leave the parasites in the cuvette for 15 min at room temperature.

7. Transfer the entire 0.8 ml of electroporation mix to a fresh culture of confluent human foreskin fibroblasts. Expression of the introduced gene can be observed within a few hours (depending on the gene), but higher levels are usually seen 1–2 days later. The assay for expression will obviously depend on the nature of the reporter gene being used (i.e., what is encoded by the plasmid being introduced).

Importance of Component Parameters in Transient Transfection

Promoter

In kinetoplastida, the only other parasite systems where transfection has been achieved, identification of transcription start sites has proven almost impossible; indeed, they may not be precisely defined even in nature. In toxoplasma, however, trans-splicing has not been observed, and likely transcription start sites are readily identifiable through mapping the 5′ end of the mRNA. Three putative "promoter" regions have been analyzed via transfection for promoter activity (20): those for *SAG1* [encoding the major surface antigen, P30 (21)], *ROP1* [encoding a rhoptry protein (22)], and *TUB1* (encoding α-tubulin [23]). All three, when placed upstream of *cat* (a bacterial gene encoding chloramphenicol acetyltransferase; see below), give substantial activity compared to controls lacking such regions. *TUB1* gives about 4-fold greater activity than *ROP1* which, in turn, is about 2-fold more active than *SAG1*.

It must be emphasized that the region being used from each of the genes extends from well upstream of the mapped transcription start site to the G of the ATG start codon. It is not possible, therefore, to ascribe differences in the activity of the three constructs to promoter activity versus an effect of the 5′-untranslated region, including the context in which the ATG lies. Indeed, recent results using a different reporter (*E. coli lacZ*, see below), consistently give a different hierarchy of expression efficiency for the three promoters just mentioned. The basis for this difference is not yet clear.

Poly(A) Addition Site

Within this category we are necessarily lumping several variables: 3′-untranslated region, poly(A) addition site, and transcription terminator. As yet, there are no data on what constitutes a transcription terminator and what properties of a 3′-untranslated region might be important in gene expression. On the basis of cDNA analysis, however, several poly(A) addition sites are known. Hence, for a variety of genes, segments extending from the stop codon down through the poly(A) addition site and several hundred base pairs beyond can be tested and compared.

We have, as yet, no data comparing different downstream regions as all constructs have had either the *SAG1* downstream sequence or have simply run into the plasmid vector sequence. Comparing the latter to the former, however, shows that the *SAG1* downstream sequence is important, as without it less than 10% of the activity is seen compared to with it (D. Soldati and J. C. Boothroyd, unpublished results). The reason for this (transcriptional, posttranscriptional, translational, etc.) is not yet known.

Translation Start Site

Both *ROP1* and *SAG1* have two reasonable candidates for the ATG used to initiate translation. In both cases, there is one ATG followed by about 20 codons encoding hydrophobic amino acids, then another ATG followed by a similarly hydrophobic region. As these two genes encode proteins that must pass through the secretory pathway, a hydrophobic signal is predicted. Based on the context of the two ATGs and algorithms that predict the cleavage point for removal of the small peptide, it was predicted that the second ATG was used in both cases.

Constructs were generated where *cat* was fused to either the first or second ATG for both genes. As predicted, fusion to the first ATG gave insignificant activity compared with controls lacking a promoter (20). Fusion to the second ATG, however, gave constructs that show substantial chloramphenicol acetyltransferase (CAT) activity. To check that the difference was not due to transcriptional effects, the mRNA produced from both types of constructs was analyzed by primer extension for the *SAG1* gene: no differences were found in either abundance or transcription start point (data not shown). Hence, in toxoplasma, as for other systems, simply being the first in-frame ATG is not enough: the local sequence context is also critical. However, as yet the only clear rule is that the $^-3$ position relative to the ATG must be an adenosine.

Reporter Gene

A reporter gene encodes a product that can be easily assayed and thereby serve as a ready measure of the presence, abundance, and relative activity of the plasmid construct being introduced. There are a variety of such genes in common usage in eukaryotic transfection studies. The most common are β-galactosidase (β-Gal or *lacZ* for the *Escherichia coli* gene from which it is derived), β-glucuronidase (*gus*), luciferase, and chloramphenicol acetyltransferase (*cat*). Each has its own advantages.

We have tested *gus* and *cat* and found that the first, for unknown reasons, is not expressed in toxoplasma. The problem was not a trivial one with the construct (e.g., a PCR cloning artifact) as the identical construct is fully active when transfected into trypanosomes (the constructs used a trypanosome promoter from the *PARP* gene that shows fortuitous activity in toxoplasma). The *cat* marker, however, gives excellent activity under the control of a variety of promoters (see above).

The CAT reporter enzyme can be assayed in one of two ways. The first, which is very sensitive but also more tedious, involves the conversion of radiolabeled chloramphenicol to acetylated forms followed by thin-layer chromatography (TLC) to separate the substrate from the products. If detection is via a phosphoimager, this procedure can be quantitative. It is, however, slow, not amenable to analysis of large samples, and gives only one time point per assay (i.e., the assay must be stopped before analysis). A protocol for this assay has been reported by Gorman (24).

The second assay for CAT is less sensitive but far simpler. It uses radiolabeled acetyl-coenzyme A and measures transfer of the ^{14}C-acetyl group to unlabeled chloramphenicol. Acetyl-CoA is soluble in water and does not partition into an organic phase; chloramphenicol is soluble in both phases but preferentially partitions into the organic phase when acetylated. Hence, when the labeled acetyl group is transferred to chloramphenicol, the resulting product enters the organic phase where the scintillant allows detection. (The β particle emitted by ^{14}C travels only a short distance in water, and thus even if the aqueous medium is overlaid with scintillant, there is no appreciable signal detected.) This can be exploited in that the reaction can be set up in aqueous medium, overlaid with scintillant, and then assayed at time intervals thereafter for ^{14}C (i.e., acetylated chloramphenicol), in the scintillant.

The second partitioning procedure is less sensitive than TLC, but it is easy, fast, and more quantitative as it can give a time course of conversion versus time so that values in the linear range can be used for more accurate comparisons. Even after short incubations, samples with high CAT activity can give complete conversion in the TLC assay, precluding accurate quantitation. A description of the partion assay can be found elsewhere (25).

The luciferase gene has been used as a reporter but only low levels of activity could be detected. This may be due to the abundant NTPase in toxoplasma which may be consuming the ATP necessary for the luciferase assay *in vitro*.

This disappointing result has been off set by the very efficient expression we have recently obtained with *E. coli lacZ* encoding β-galactosidase. We find that this gene gives extremely sensitive detection using a variety of substrates including X-gal, methyl-umbelliferyl β-D-galactoside (MUG) and chlorophenol red β-D-galactoside. Activity can be readily detected in parasite lysates but more importantly, perhaps, it can be detected *in situ* in both *in vitro* culture and in tissue sections from infected animals (the parasites give an intense blue staining with X-gal in both; F. Seeber and J. C. Boothroyd, manuscript in preparation). The gene can also be used as a secretory marker (D. Soldati and J. C. Boothroyd, unpublished results; L. D. Sibley, unpublished results) which may prove important in dissecting trafficking signals and in analyzing constitutive and regulated secretion.

Enhancer

Promoters and enhancers are DNA sequences that represent the extremes in a continuum of transcription control elements. Generally, a promoter is where the RNA polymerase binds and is the element necessary for basal levels of unregulated transcription. It is, of necessity, near to or even overlapping the transcription start site. Regulation is typically the job of the enhancer. Factors that bind to the enhancer generally interact with the RNA polymerase bound to the promoter. This need not be through binding DNA adjacent to the promoter: binding can occur some distance upstream or downstream with the intervening DNA looped out. Because of the inevitable twisting of the loop of intervening DNA, enhancers are not typically orientation dependent: they can be turned around and still work.

Detailed analysis of the transcription control region of the *SAG1* gene has shown the following (37). There are apparently at least two discrete transcription start sites spanning a region of about 35 base pairs (bp). The start codon for translation (see above) lies approximately 95 bp downstream of the second initiation site. Beginning about 35 bp upstream of the first initiation site are six nearly identical, tandem, 27-bp repeats. These are obvious candidates for binding of a transcription factor.

Constructs have been generated where the repeats have been moved downstream, upstream, inverted, and varied in number (37). They have also been placed adjacent to other toxoplasma "promoters" or put upstream of a reporter with no promoter. Briefly, the results show that these are best described as promoter elements because they are absolutely necessary for transcription of the *SAG1* locus and their position determines the start site for transcription. However, they can function in either orientation, which is highly unusual for a promoter and, when placed upstream of a heterologous promoter (a truncated promoter of *TUB1*), they can result in substantially more activity (37). We do not yet know if this is due to providing a double promoter or enhancing the activity of the natural promoter (*TUB1*) on the construct.

DNA Concentration

There is a near linear relationship between the amount of CAT detectable and the concentration of the transfecting DNA up to about 220 μg, after which the improvement from more DNA tends to taper off and arcing of the electric pulse can occur.

DNA Form (Linear or Circular)

For transient transfection, circular, supercoiled DNA works no better than linear. The impact of this parameter on stable transfection is discussed in the next section.

Multiple Genes

We have not had much experience trying to express more than one gene from a single plasmid. However, in one set of experiments, we found that *cat* expression from the *SAG1* promoter was substantially reduced (i.e., by at least 90%) when the *B1* gene was placed downstream with transcription proceeding in the same direction (i.e., the *B1* gene was also pointing downstream; data not shown). When *B1* was reversed so that it was pointing upstream to give convergent transcription, there was no appreciable difference with the construct lacking *B1*. The difference when the two transcription units were in line was seen even when the plasmids linearized with a single cut between the *SAG1* and *B1* genes. When the two genes were completely separated using two cuts (i.e., one at either end of the SAG1-CAT portion), the difference was eliminated. The basis for this difference is not known, but it is not without precedent in eukaryotic transfection and is a caution to anyone contemplating double-expression constructs. Cotransfection of two or more plasmids is instead the preferred method (see below).

Strain of Parasite

Of the three strains tested, RH, CEP, and PLK, the fast-growing, common laboratory strain RH gives about 5-fold greater signal by Western blot analysis when the *ROP1* gene is transfected in comparison with the other two (D. Soldati and J. C. Boothroyd, unpublished results). One of the reasons for this difference is the low survival rate from the electroporation procedure and rate of growth thereafter. Assessing *cat* expression shows little if any difference between the three strains.

Intracellular versus Extracellular Parasites

Electroporation of extracellular parasites gives about 100-fold greater *cat* activity relative to experiments conducted with parasites still within human foreskin fibroblast cells (D. Soldati and J. C. Boothroyd, unpublished results). We cannot exclude the possibility that even the 1% signal with "intracellular parasites" came from a small number of extracellular parasites contaminating the preparation. Efforts to remove such contaminants focused on (1) use of cells infected only 24 hr previously to reduce the chance of natural lysis of the infected cell (which normally occurs ~ 48 hr postinvasion), corresponding to 6–8 divisions of the intracellular parasite; and (2) repeat washing of the infected host cells under conditions where extracellular parasites would be washed away. Despite these precautions, however, a few infected cells may have lysed between final wash and electroporation.

Preparation of Parasites for Electroporation

Extracellular parasites released from the host cells no more than 3 hr previously give substantially higher signals than those which have been extracellular for longer periods. Parasites that have lysed out of the cells naturally (i.e., after about 6–8 generations) give better results than those which are artificially released by passage through a 27-gauge needle. Cytomix is not a good maintenance buffer for the parasites, and they should not be left in it for any longer than necessary and, in any event, no longer than 1 hr.

Handling of Parasites after Electroporation

Following the electroporation, it is not absolutely necessary to allow the parasites to infect a fresh culture of host cells. Surprisingly, the levels of CAT activity obtained in the presence or absence of the host cells are comparable. Clearly, the extracellular parasites retain metabolic activity for much, possibly all, of the 12- to 24-hr period outside of the host cell.

For example, where different promoters are to be compared, transfected parasites can be added directly to 10 ml of Dulbecco's modified Eagle's medium (DMEM; GIBCO, Grand Island, NY) enriched with 20% Nu Serum (Collaborative Research, Bedford, MA) and incubated at 37°C for up to 12 hr in the absence of host cells. During this time they will remain metabolically active (and, incidentally, infectious), and thus comparisons can be made between different constructs without the need to expand the parasite population in a culture of host cells. Where greater sensitivity is desired, however, it is best to allow the transfected parasites to infect a host cell culture as described in the optimal protocol given above.

Time of Incubation after Electroporation

Monitoring of expression shortly after electroporation and for an 11-day period thereafter should reveal the following. CAT is detectable by 4 hr after electroporation. The levels increase approximately linearly over a 24-hr period, increase more slowly over the next 24 hr, and then drop off slowly for the ensuing 8 days until virtually undetectable. The kinetics must be interpreted in light of the fact that CAT is known to be highly stable.

As the plasmid being used has no known origin of replication, one would expect it to be diluted out as the parasite population grows (i.e., only one daughter cell should inherit the plasmid in each division and so the proportion of cells harboring the plasmid should drop exponentially). Monitoring of an equal number of parasites over a 9-day period (which necessarily includes passing the culture by diluting about 1:10 about every 2 days) shows the expected exponential drop in CAT activity per parasite with virtually none detectable by day 7.

Electroporation Conditions

The cytomix buffer was developed, in part, to replace the sodium ions found in most buffers with potassium. Use of ordinary media (optimem (Gibco/BRL, Grand Island, NY) or PBS) which are sodium-based results in greater than 10-fold reduction in the signal compared with cytomix.

We have tested voltages in the range 500–2500 V. At 2500 V, the parasites die; 1500–2250 V gives essentially similar results; less than 1500 V gives very poor results. Resistance setting (R1 through R4) has little effect except in the probability of electrical arcing when the pulse is administered. We routinely use R3 (48 Ω). With the BTX electroporator, capacitance is not variable (at these voltages, it is set at 50 MF). For the Bio-Rad Gene Pulser (Biorad, Hercules, CA), use 2 kV and 25 μF to achieve a time constant of 0.4 msec (K. Joiner, personal communication).

Stable Transformation of Toxoplasma Tachyzoites

Transient transfection has many experimental limitations compared with stable transformation. Situations where these limitations are greatest are as follows. (1) The biological properties of a transfected cell are of interest: the time course of doing a biological experiment (e.g., virulence, transmissibility) typically is far longer than the duration of expression of transiently transfected genes. (2) The experiments require all cells to be alike: even with the most efficient transient transfection, most cells do not take up the DNA; also, each transfected cell may take up different amounts of the DNA, which can influence the results if some component of the cellular machinery involved in expression of that gene is limiting. (3) Complex regulatory or other interactions are being studied: in addition to its temporary nature, transient transfection can result in many thousands of copies of a gene in a given transfected cell, thereby throwing off any subtle (or not so subtle) regulation. (4) Selection is being used to identify a particular gene among a mixture in the transfecting DNA as is the case when marker rescue or complementation is being used to clone a gene. This is the case in our attempts to identify genes encoding drug targets.

Because of its obligate intracellular life cycle, effective selection strategies for transformation of toxoplasma must also consider effects of the strategy on host cells. Neomycin and hygromycin, drugs used successfully in a wide variety of eukaryotic cells, including the kinetoplastida (27), were as toxic to host cells as the parasite. We thus chose to investigate novel selection strategies for stable transformation of toxoplasma.

We have taken several tacks to achieving stable transformation in toxoplasma: (1) complementation of a mutant of toxoplasma, (2) introduction of heterologus drug resistance genes, and (3) complementation of a naturally occurring auxotrophy. Each of the strategies has distinct advantages as highlighted below.

Transformation by Mutant Complementation

In the first approach (38), we have complemented the *sag1* mutant of Kasper (29), which is unable to express SAG1 owing to the presence of a nonsense mutation near the middle of the gene (30). Using conditions described for transient transfection, the wild-type *SAG1* gene was introduced into the mutant population. The 1.6 kb of genomic insert that was used contained the entire coding region, the transcription initiation site, and polyadenylation site. Analysis of live extracellular parasites labeled with a monoclonal antibody specific for SAG1 and using a fluorescence activated cell sorter (FACS) revealed that electroporation with 100 μg of circularized plasmid resulted in SAG1 expression in 15–20% of the transfected population for up to 4 days after transfection. Transfected parasites were able to divide, and SAG1 was detectable in individual daughter cells by immunofluorescence. By 11 days, however, FACS analysis revealed that less than .05% of the transfected culture continued to express the antigen. The miniscule population was fluorescently sorted and expanded by growth in cultures of human foreskin fibroblasts twice to produce a population that, by the end, was nearly 100% positive for SAG1 expression. The three clones derived from the population were siblings. Southern blot analysis revealed the presence of both the endogenous mutant gene and the transfected wild-type gene which had integrated nonhomologously into the genome.

Transformation with cat

The second selection strategy utilizes drug selection. We have exploited the previously unrecognized susceptibility of toxoplasma tachyzoites to chloramphenicol (31). Like clindamycin (32), a mechanistically similar antibiotic which inhibits prokaryotic translation, chloramphenicol has a delayed but highly potent parasiticidal effect. The effect is seen at doses of drug which have no appreciable effect on the confluent host cell monolayers used to propagate the parasite. As detailed above, CAT, a bacterial protein which confers resistance to chloramphenicol, is efficiently expressed in toxoplasma when transcription is driven by the upstream regions of three toxoplasma genes. As all three "promoters" are from single-copy genes, we were concerned that stable transformation, if via homologous recombination with replacement, would be a deleterious event in this haploid organism. We therefore created a construct (SBCAT1) with portions of the *B1* gene flanking a SAG1/CAT expression cassette. *B1* is a gene of unknown function which is tandemly repeated 35 times within the toxoplasma genome (33). Transfection of the construct (as well as the constructs originally developed for transient transfection) followed by chloramphenicol selection results in parasites stably expressing CAT.

Owing to the prolonged period required for selection and the necessity to pass parasites during selection, we have not been able to determine the frequency of stable transformation using *cat* selection. The minimum amount of DNA necessary to yield stable transformants has varied somewhat with the construct, but we have found that between 1 and 50 μg of linearized plasmid usually yields stable transformants. Although this has not been tested in a systematic fashion, it appears that *cat* driven by stronger "promoters" (e.g., *ROP1* or *TUB1*) requires less DNA to generate stable transformants and is more likely to result in incorporation of a single copy of *cat*.

All clones analyzed to date have *cat* integrated within the genome rather than maintained as an extrachromosomal element. (Because we are looking for integrative events, transformants obtained after transfection with circular plasmid have not yet been analyzed.) In our hands, linearized DNA has been more effective than circular DNA for generation of stable transformants using *cat* as a selectable marker. In contrast, results with other selectable markers (see below) may be more effective with circular rather than linear DNA. Recombinants transformed with the selectable markers appear to be more likely to carry the marker as an episomal element. So far, our experience has not been extensive enough to predict how individual constructs or selectable markers will behave. The frequency of homologous recombination appears to vary with the gene and/or the construct. Parasites stably transformed with linearized SBCAT1 integrated multiple copies of *cat* within the *B1* locus, whereas a SAG1/CAT construct has not yielded SAG1 knockouts.

In contrast, targeted insertion of *cat* into the *ROP1* locus has been successful although the majority of transformants have not been knockouts. (As the initial screening of the transformants was by Western blot analysis looking for loss of *ROP1* protein, we do not know if the non-knockouts were targeted insertions which did not delete the endogenous gene or random insertions elsewhere in the genome.) The difference in frequency of homologous recombination may be in the relative importance of the targeted genes and/or factors such as the length and nature of the sequences flanking the *cat* gene in the relevant plasmid constructs.

The following is the suggested protocol for generation of stable transformants using *cat* as a selectable marker.

1. Titrate effective chloramphenicol concentrations on untransfected parasites. To do this, add drug at various concentrations to parallel infected cultures and passage the parasites as they are released by lysis, maintaining constant drug pressure throughout. Toxic effects will not be evident for 2–3 cycles of parasite lysis of host cells. Too low concentrations of drug may allow the growth of CAT-negative parasites, whereas higher concentrations (e.g., 100 mM) may not yield any stable transformants. We have found 20 mM chloramphenicol to be optimal. For selection, choose the lowest drug concentration that kills all wild-type parasites.
2. Transfect $1–2 \times 10^7$ tachyzoites with 1–50 μg plasmid DNA using conditions established for transient transfection (see previous section). For targeted insertion, DNA should be linearized as close to, preferably within, toxoplasma genomic sequences, and DNA homologous to the targeted gene should flank the *cat* cassette. (For both *B1* and *ROP1,* slightly less than 1 kb of homologous DNA flanking *cat* is sufficient for successful homologous recombination. If toxoplasma is similar to mammalian systems, incorporating 10–15 kb of homologous DNA into the construct may increase the efficiency of targeting.) Inoculate parasites onto a T-25 flask of confluent human foreskin fibroblast cells. A control using no plasmid and/or an irrelevant plasmid should be set up in parallel.
3. Selection with drug is initiated immediately after transfection.
4. As they lyse the host cells, pass the parasites ($\sim$ 5–10% of released organisms or $2–5 \times 10^6$) into flasks containing fresh drug and medium. Each round of growth will take about 2–4 days depending on the strain of parasite. Toxic effects should be readily apparent after approximately 2–3 rounds of lysis ($\sim$ 7–12 days). Continue selection for a further 3–5 days after the toxic effects are seen.
5. Parasites can be cloned by limiting dilution (see above) in the presence of drug as soon as a chloramphenicol-resistant population emerges ($\sim$ 10 days) or even before seeing any effect ($\sim$ 4 days after electroporation). Early cloning may prevent the overgrowth and dominance of more robust clones within the transformed population. If parasites are not cloned, cultures treated with the drug which, owing to the presence of dying chloramphenicol-sensitive parasites, do not lyse host monolayers within 5–7 days should be scraped, syringed, and passaged using 20–25% of the lysate. Negative control cultures should have no survivors after approximately 10 days.
6. The CAT activity of parasites can be monitored using either the phase-partition assay or thin-layer chromatography (see above). Stable transformants have high

levels of CAT activity, and so the less sensitive but more convenient phase-partition assay is usually suitable. Depending on the amount of DNA used, the peak of transient CAT activity typically seen 1–6 days after transfection may not be evident using the phase-partition assay.

7. Once selected, the stable transformants we have generated have not required maintenance in the presence of drug to maintain CAT activity (probably because all have chromosomally integrated the transfecting plasmid). However, given that episomal DNA is generally rapidly lost in the absence of selection, clones should be tested for stability of CAT activity after several weeks of growth with and without drug before being maintained for any period of time without drug. Ideally, Southern analysis of recombinant DNA should also be performed. The mechanism of stable transformation may need to be confirmed with pulsed-field electrophoresis followed by Southern blotting.

Recently, we have found that restriction-enzyme mediated integration (REMI; 36) works extremely well in increasing the efficiency (up to 200-fold) of recovery of stable transformants (M. Black, F. Seeber, D. Soldati, K. Kim, and J. Boothroyd, manuscript in preparation). This procedure simply requires that a restriction enzyme (NotI works well) be included with the transfecting plasmid DNA in the electroporation mix. Although the technique, as originally described for yeast and Dictyostelium, is reported to facilitate integration of a plasmid cut with enzyme X into restriction sites for enzyme X in the genome, we find that integration is apparently random in toxoplasma with substantial enhancement even using plasmids that lack a site for enzyme X. Current results are consistent with enzyme X serving simply to induce a recombinogenic repair machinery.

To obtain transformants where more than one gene is stably integrated and expressed, we have already noted above that placing both genes on one plasmid can cause an interference that prevents expression of one or other gene. Instead, we now routinely use the REMI procedure just described together with cotransfection of multiple plasmids, only one of which carries a selectable marker. For example, if a *cat* expression plasmid is cotransfected at a 1 : 20 ratio with a plasmid bearing the *E. coli lacZ* gene, following chloramphenicol selection, 80–100% of the resulting clones are also expressing the *lacZ* gene. In one recent experiment, we have used *cat* with three other nonselectable plasmids (at a ratio of 1 : 15 : 15 : 15, respectively). The results showed that of ten chloramphenicol-resistant clones obtained, eight also expressed one or more of the nonselectable, cotransfecting plasmids with two of the ten expressing all three! This approach, then, greatly simplifies the steps needed to multiply engineer a strain and obviates the need for a very large number of selectable markers.

Transformation by Auxotrophy Complementation

An alternative strategy for stable transformation is based on rescuing cells from an auxotrophic condition by introducing a metabolic gene that allows growth. The production of auxotrophs and transformation by conversion to prototrophy is widely used in bacteria and yeast and provides a wide range of independent selectable markers. In the case of an obligate intracellular parasite like toxoplasma, it is not possible to culture cells in strictly defined media. It has also not been possible to generate auxotrophic mutants because of the difficulty in controlling the host cell pool of nutrients. To circumvent this difficulty, we took advantage of the fact that toxoplasma is a naturally occurring auxotroph for tryptophan. The growth of toxoplasma in human fibroblasts can be inhibited in tryptophan-depleted culture media, and the effect is increased by treatment with the cytokine (IFN-γ) (34). The basis of tryptophan starvation is enhanced degradation of tryptophan following IFN-γ treatment, and the effect can be reversed by adding excess exogenous tryptophan to the culture.

Complementation of tryptophan auxotrophy in toxoplasma was accomplished by expressing the bacterial *trpB* gene encoding tryptophan synthase, which catalyzes the formation of tryptophan from indole plus serine (40). Expression of the *trpB* gene was accomplished by replacing the coding region of *SAG1* with the *trpB* gene, thereby utilizing the flanking regions endogenous to toxoplasma to regulate expression. Transformants were obtained by electroporation followed by selection on tryptophan-depleted monolayers supplemented with indole.

One of the principal advantages of Trp selection is that it maintains the ability of toxoplasma to form plaques on host cell monolayers. Consequently, the system provides direct access to transformants without further cloning and lends itself to quantitative analysis of transformation efficiencies. It is possible to perform plaque lifts from plates containing transformants and to detect the presence of the *trpB* gene by filter hybridization (M. Messina and L. D. Sibley, unpublished). When extracellular tachyzoites are electroporated with circular forms of *SAG1/trpB*, transformation efficiencies are approximately 1 in 10^4 surviving cells. Linearized plasmid shows a slightly lower efficiency in the range of 1 in 10^5. The reason for this difference is not known but may relate to a greater resistance to exonuclease cleavage of circular molecules following electroporation.

Because plaques grow rapidly, it has been possible to analyze individual Trp clones within 10–14 days after transfection. In a majority of clones where circular DNA was used in transfection, the *SAG1/trpB* plasmid is observed as a circular episome with a low copy number. Many of the plasmid molecules are multimeric forms containing tandemly reiterated copies of the introduced plasmid. Methylation sensitivity was used to verify that the plasmids had replicated in toxoplasma and were not simply a carryover from electroporation (M. Messina and L. D. Sibley, unpublished). These plasmids were readily recovered from toxoplasma and used to transform *E. coli,* thus forming a rudimentary shuttle vector. When clones were further expanded in the presence of continual selection, they were all found to have integrated copies of the *SAG1/trpB* plasmid in the nuclear chromosomes. In over 25 separate clones examined, we have not observed stable episomal forms of *SAG/trpB* beyond 3 weeks of propagation. Consequently, it appears that the *SAG/trpB* plasmid is only inefficiently replicated in toxoplasma.

In transformants obtained from transfection with circular plasmid, integration events were always nonhomologous. We have also not detected a homologous recombination at the *SAG1* locus using plasmid that was linearized at a variety of sites upstream of the transcription unit or using purified fragments containing the *SAG1/trpB* transcription unit flanked by toxoplasma sequences. The reasons for this are not clear but may reflect a low degree of homologous recombination at this locus or an insufficient length of sequence similarity in the upstream and downstream flanking regions. The high degree of nonhomologous integration combined with efficient transformation may allow use of the system for random insertional mutagenesis, which is a powerful means of identify genes using reverse genetics.

The expression of *trpB* in toxoplasma transformants was confirmed by Western blotting with an antibody to the *E. coli* b subunit of tryptophan synthase (TS). Individual transformants expressed variable amounts of TS that were proportional to the *trpB* gene copy number. The majority of Trp transformants contain 10 or more tandemly duplicated copies of the plasmid, suggesting that a minimum level of TS expression is required for growth on indole. The result is likely a combination of promoter strength, message and protein stability, and efficiency of TS in converting indole to tryptophan. Because the vector sequences are also duplicated in Trp transformants, it may be possible to use the system to overexpress other genes in toxoplasma by including them in the same construct with *SAG1/trpB*.

The following protocol can be used for obtaining stable transformants of toxoplasma using Trp selection.

1. Purified plasmid DNAs are used to electroporate extracellular tachyzoites using a protocol similar to that described above. Typically 10^7 cells are electroporated with $10-50$ μg of plasmid DNA in cytomix buffer.
2. Following electroporation, cells are inoculated to a single T-25 tissue culture flask containing human foreskin fibroblast cells grown in complete medium containing 10% fetal bovine serum (FBS). The culture is allowed to grow for $2-3$ days during which time the parasites completely lyse the monolayer.
3. Parasites are isolated, counted, washed by centrifugation in Hanks' balanced salt solution, and plated on tryptophan-depleted monolayers of human fibroblasts. To deplete tryptophan levels in host cells, monolayers are pretreated for 24 hr with 20 U/ml IFN-γ in tryptophan-deficient DMEM supplemented with 3% dialyzed FBS (Trp$^{v/v}$ medium). Transformants are plated in Trp$^{v/v}$ medium supplemented with $50-100$ mM indole.
4. Approximately 2.5×10^4 toxoplasma cells are plated per cm^2 on a monolayer of fibroblast cells growing in flasks or culture plates. Transformants form plaques which can be picked approximately $6-8$ days later. Plaques are identified by examination at a magnification of $10-20 \times$ using an inverted tissue culture microscope and aspirated using hand held micropipettes. Plaques are transferred to single wells in 96-well plates containing monolayers of fibroblast cells and grown in Trp$-$ medium supplemented with indole.

5. Transformants typically do not require further cloning, provided the initial separation of plaques was adequate. In cases where mixed plaques are suspected, it is recommended that parasites be recloned by limiting dilution to obtain single-plaque clones in microwell plates. Transformants can be further expanded in Trp − selective medium for DNA, protein, or phenotypic analyses. Trp transformants have integrated copies of the *SAG1/trpB* gene and therefore can be maintained without selection for several weeks. Given the tandem repeated nature of the majority of transformants, long-term growth without selection may result in the loss of copies and a reduced phenotype, although this has not been fully tested.

Alternative Systems

Other selectable markers for stable transformation of toxoplasma have also been developed. Genes encoding bleomycin–phlemomycin resistance from bacteria [e.g., the *ble* gene *Streptoalloteichus hindustanus* (K. Kim, J. Kampmeier, and J. C. Boothroyd, unpublished) and the *ble* gene from Tn5 (S. Messina, I. Niesman, and L. D. Sibley, unpublished)] have also been successful. Indeed, this marker may be better in terms of the efficiency with which one can obtain transformants (with efficiencies as high as $10 - 4$ using only $1 - 2 \mu g$ of plasmid). The availability of an efficient second marker can be critical when trying to engineer a strain in which the first marker was used to knock-out a gene. For example, starting with the strain in which the *ROP1* gene has been replaced with *cat,* we have reintroduced engineered forms of the *ROP1* gene using cotransfection of the engineered plasmid and a *ble* bearing plasmid. The efficiency of this approach is high enough that it has been possible to even use cosmid DNA in the cotransfection with efficient integration (D. Soldati, K. Kim and J. C. Boothroyd, manuscript in preparation).

A number of additional strategies have been developed by Roos, Donald and colleagues. The first and best developed of these is a particularly elegant approach that uses information from plasmodium to engineer the dihydrofolate reductase (DHFR) gene of toxoplasma to confer pyrimethamine resistance (36). These workers have developed a veritable tool kit of reagents and approaches based on this (41, 42) including insertional mutagenesis, marker rescue, complementation and overexpression of heterologous genes. The most recent contribution of this group to the armamentarium has been development of mycophenolic acid selection for hypoxanthine-xanthine-guanine phosphoribosyl transferase. A very thorough and detailed collection of protocols for the methods used by Roos and colleagues can be found in reference 42.

More recently, an approach has been developed that uses information from plasmodium to engineer the dihydrofolate reductase (*DHFR*) gene of toxoplasma to confer pyrimethamine resistance (36). As with the markers described above, this has been successfully used for insertional mutagenesis, marker rescue, complementation, and overexpression of heterologous genes.

Conclusion

The inherent potential for using genetics to understand the biology of toxoplasma is now beginning to be realized. There is, however, much to be done. Among the major challenges are the development of stable plasmid replicons, physical maps of chromosomes, more detailed genetic maps, sequence tag sites or similar efficient, high density markers, catalogued cosmid or yeast artificial chromosome (YAC) banks that allow one to move from a given genetic location direction to the genomic region of interest, and methods for *in vitro* crosses (i.e., outside the cat).

Given the rapid pace of advance in this still young field, none of these developments seems very far off (with the possible exception of *in vitro* crosses). Even without further developments, however, the techniques in hand provide the tools for many exciting experiments in a system that not only represents a superb model for intracellular parasitism but also is a pathogen of major human and veterinary concern.

Acknowledgments

We are grateful to all those in and out of our laboratories who contributed to the development of the protocols and approaches described herein. In particular, we thank J. Ajioka, J. Blackwell, K. Joiner, and D. Roos for communication of results prior to publication. This work was supported by grants from the National Institutes of Health. Some of the critical, early work in developing this system was supported by the MacArthur Foundation, the Burroughs Wellcome Fund, and a postdoctoral fellowship from Merck Sharp & Dohme.

References

1. L. H. Kasper and J. C. Boothroyd, *in* "Immunology and Molecular Biology of Parasitic Infections" (K. Warren, ed.), p. 269. Blackwell, Oxford, 1993.
2. J. P. Dubey, *in* "Parasitic Protozoa" (J. P. Kreier, ed.), Vol. 6, p. 1. Academic Press, New York, 1993.
3. E. R. Pfefferkorn and L. C. Pfefferkorn, *Exp. Parasitol.* **39**, 365 (1976).
4. B. J. Luft and J. S. Remington, *Clin. Infect. Dis.* **15**, 211 (1992).
5. L. D. Sibley, A. LeBlanc, E. R. Pfefferkorn, and J. C. Boothroyd, *Genetics* **132**, 1003 (1992).
6. A. B. Sabin, *J. Am. Med. Assoc.* **116**, 801 (1941).
7. L. D. Sibley and J. C. Boothroyd, *Nature (London)* **359**, 82 (1992).
8. E. R. Pfefferkorn, L. C. Pfefferkorn, and E. D. Colby, *J. Parasitol.* **63**, 158 (1977).
9. L. H. Kasper and P. L. Ware, *J. Clin. Invest.* **75**, 1570 (1985).
10. Y. Suzuki, F. K. Conley, and J. S. Remington, *J. Infect. Dis.* **159**, 790 (1989).
11. L. C. Pfefferkorn and E. R. Pfefferkorn, *Exp. Parasitol.* **50**, 305 (1980).
12. E. R. Pfefferkorn and L. H. Kasper, *Exp. Parasitol.* **55**, 207 (1983).
13. J. Huskinson, P. Thulliez, and J. S. Remington, *J. Clin. Microsc.* **28**, 2632 (1990).

14. L. D. Sibley and J. C. Boothroyd, *Mol. Biochem. Parasitol.* **51**, 291 (1992).
15. K. Gardiner, W. Laas, and D. Patterson, *Somatic Cell Mol. Genet.* **12**, 185 (1986).
16. G. Chu, D. Vollrath, and R. W. Davis, *Science* **234**, 1582 (1986).
17. J. T. Joseph, S. M. Aldritt, T. Unnasch, O. Puijalon, and D. F. Wirth, *Mol. Cell. Biol.* **9**, 3621 (1989).
18. P. Borst, J. P. Overdulve, P. J. Weijers, F. Fase-Fowler, and M. Van den Berg, *Biochem. Biophys. Acta* **781**, 100 (1984).
19. M. J. B. Van den Hoff, A. F. M. Moorman, and W. H. Lamers, *Nucleic Acids Res.* **20**, 2902 (1992).
20. D. Soldati and J. C. Boothroyd, *Science* **260**, 349 (1993).
21. J. L. Burg, D. Perelman, L. H. Kasper, P. L. Ware, and J. C. Boothroyd, *J. Immunol.* **141**, 3584 (1988).
22. P. N. Ossorio, J. D. Schwartzman, and J. C. Boothroyd, *Mol. Biochem. Parasitol.* **50**, 1 (1991).
23. S. D. Nagel and J. C. Boothroyd, *Mol. Biochem. Parasitol.* **29**, 261 (1988).
24. C. M. Gorman, *Mol. Cell. Biol.* **2**, 1044 (1982).
25. J. R. Neumann, C. A. Morency, and K. O. Russian, *Biotechniques* **5**, 444 (1987).
26. D. Soldati and J. C. Boothroyd, manuscript in preparation.
27. A. L. ten Asbroek, C. A. Mol, R. Kieft, and P. Borst, *Mol. Biochem. Parasitol.* **59**, 133 (1993).
28. K. Kim and J. C. Boothroyd, manuscript in preparation.
29. L. H. Kasper, *Parasitol. Immunol.* **9**, 433 (1987).
30. L. H. Kasper, I. A. Khan, K. H. Ely, R. Buelow, and J. C. Boothroyd, *J. Immunol.* **148**, 9805 (1992).
31. K. Kim, D. Soldati, and J. C. Boothroyd, *Science* **262**, 911 (1993).
32. E. R. Pfefferkorn, R. F. Nothnagel, and S. E. Borotz, *Antimicrob. Agents Chemother.* **36**, 1091 (1992).
33. J. L. Burg, C. M. Grover, P. Pouletty, and J. C. Boothroyd, *J. Clin. Microsc.* **27**, 1787 (1989).
34. E. R. Pfefferkorn, *Proc. Natl. Acad. Sci. U.S.A.* **81**, 908 (1984).
35. M. Messina, I. Niesman, and L. D. Sibley, submitted.
36. R. G. K. Donald and D. Roos, *Proc. Natl. Acad. Sci. U.S.A.* **90**, 11703 (1993).
37. D. Soldati and J. C. Boothroyd, *Mol. Cell Biol.* **15**, 87 (1995).
38. K. Kim and J. C. Boothroyd, *Expt. Parasitol.* **80**, 46. (1995).
39. R. H. Schiestl and T. D. Petes, *Proc. Natl. Acad. Sci. U.S.A.* **88**, 7585 (1991).
40. L. D. Sibley, M. Messina, and I. R. Neisman, *Proc. Natl. Acad. Sci. U.S.A.* **91**, 5508, (1994).
41. R. G. Donald and D. S. Roos, *Mol. Biochem. Parasit.* **63**, 243, (1994).
42. D. S. Roos, R. G. K. Donald, N. S. Morrissette, and A. L. C. Moulton, *in* "Methods in Cell Biology" (D. G. Russell, ed.) (1995).

[2] Techniques Associated with Protozoan Parasite Genome Analysis

James W. Ajioka, Sara E. Melville, Richard M. R. Coulson,
Jose R. Espinoza, Shahid M. Khan, and Kiew-Lian Wan

Introduction

Genome analysis is a global approach centered around a large-scale synthesis of genetic and physical mapping information and has emerged as one of the most effective tools for molecular genetics. The positional cloning of genes, identification of new loci via genetic/physical linkage, and procurement of chromosome structural data are all greatly facilitated by methods developed for this kind of research. The majority of molecular techniques generally applicable to genome analysis have been described in detail elsewhere, but each species has unique genomic characteristics which can be exploited by the modification of existing protocols and by the development of new methods.

Protozoan parasites have several common genomic features: (1) asexual or clonal growth is a normal part of the life cycle; (2) the haploid genome sizes are about the size of the smallest human chromosome (50 Mb); (3) the chromosomes can generally be resolved by pulsed-field gel electrophoresis (PFGE); (4) chromosome size polymorphisms between natural isolates are relatively frequent; and (5) the genomes appear to have relatively little dispersed-repetitive DNA. The trypanosomatids, which include *Trypanosoma* and *Leishmania* species, share some unusual properties: (1) all mRNAs are trans-spliced with a 5' "spliced leader" or "miniexon" 39-nucleotide sequence, and (2) many genes are polycistronically transcribed and may exist as tandem arrays. Some of these characteristics can be used to tailor methods specifically for the analysis of protozoan parasite genomes. This chapter describes how the ability to separate whole chromosomes by gel electrophoresis simplifies the assignation of genes to chromosomes, how the large number of telomere ends may be used to generate chromosome tags, and how the spliced leader sequence aids in the construction of a directionally cloned, full-length cDNA library.

The repetitive nature of the tasks involved with genome research require that the basic materials (e.g., genomic and cDNA libraries) be of high quality and the methods of analysis reliable and robust. This chapter focuses on techniques which fulfill these criteria. Each protocol begins with a solutions, media, and materials section including specific information regarding the source/manufacturer and catalog numbers of critical reagents and materials. All enzymatic reactions use the buffer supplied by the manufacturer unless otherwise noted.

Karyotype Analysis

The protozoan parasites studied to date have linear chromosomes which do not condense in mitosis. Although the lack of condensation excludes conventional light
microscopic analysis, the chromosomes are generally small enough to resolve by
PFGE (1). Analysis of PFGE karyotypes for these organisms functions as a framework for genome organization in the same way that metaphase or polytene chromosome preparations do for higher eukaryotes. Chromosomal polymorphisms occur to
varying degrees, depending on the species (see Refs. 2–4 for review; A. Tait, unpublished observations). These polymorphisms are useful for strain identification and as
markers in genetic crosses (5, 6; A. Tait, unpublished observations). More directly,
PFGE Southern analysis is an effective method for assigning cloned genes to a chromosomal location (see Refs. 7–9 and 4 for review).

Chromosome sizes range from about 50 kb for a *Trypanosoma brucei* minichro-
mosome to about 40 Mb for the largest *Toxoplasma gondii* chromosome, but most of
the chromosomes for a given species fall between 200 kb and 5–10 Mb (9, 10). The
variance in chromosome size both within and among taxa is large enough such that
several different PFGE conditions must be employed to resolve complete karyotypes;
thus, the localization of genes on chromosomes may require several different
Southern blots. As the running conditions for resolving the larger chromosomes often
take several days, it is a tedious endeavor to analyze even a small number of samples
by Southern analysis. Dot-blots based on chromosome-specific polymerase chain reaction (PCR) pools help to circumvent this problem.

Preparation of Chromosome-Sized DNA in Agarose Blocks

Several methods for preparing chromosome-sized DNA in agarose blocks have been
published previously. The protocol outlined below is an efficient method which
works well for a variety of protozoan parasites. There are three important parameters
which will affect the quality of the preparation. (1) Always harvest cells from mid-
log to early stationary phase cultures. Dead or sick cells release nucleases which will
degrade the DNA, so it is important that the cells are kept alive until they are set in
agarose. (2) Wash the cells thoroughly because the culture medium often contains
contaminants which enhance degradation of the DNA or inhibit subsequent enzymatic reactions (e.g., restriction endonuclease digestions). (3) The relative concentration of cells in the agarose block will affect the resolution of the chromosomes on
PFGE. Taking into account the genome size of a given parasite, a final cell density in
the agarose block of $1-5 \times 10^8$ cells/ml generally produces a good karyotype with
ethidium bromide staining. However, the preparation of chromosome-sized DNA for
subsequent YAC or P1 library construction requires a higher DNA concentration
(i.e., $0.5-1 \times 10^9$ cells/ml), which results in the loss of good chromosome resolu-

tion. Accurately counting motile cells is difficult. Two or three serial 3-fold dilutions from the high end of the cell density range is usually sufficient for achieving the proper density for karyotyping or library construction on PFGE.

Solutions, Media and Materials
> $1\times$ Phosphate-buffered saline (PBS): 8 g NaCl, 0.2 g KCl, 1.44 g Na_2HPO_4, 0.24 g KH_2PO_4 per liter, adjust to pH 7.4
> LMP agarose/PBS: 1.2% low melting point agarose (SeaPlaque GTG, FMC, Rockland, ME, Cat. No. 50111) in $1\times$ PBS
> LIDS: 1% lithium dodecyl sulfate, 100 mM EDTA, 10 mM Tris-HCl, pH 8
> NDS: 0.5 M EDTA, 1% N-lauroyl sarcosine, pH 8
> Counting chamber: Thomba chamber, depth 0.02 mm (Weber Scientific International)

Preparation of Chromosomal DNA Plugs
1. Count the parasites with a Thomba counting chamber.
2. Centrifuge the parasites out of the culture medium (3000 g for 10 min at $4°$C).
3. Wash the parasites by resuspending the pellet in 25–50 ml cold PBS.
4. Repeat step 2.
5. Resuspend the parasite pellet to 1×10^8 cells/ml in PBS, noting the volume.
6. Add an equal volume of LMP agarose/PBS (kept at $37°$C), mix, and distribute into 100-μl plug molds.
7. After the plugs have solidified, incubate them in LIDS (10 plugs per 50 ml LIDS) at $50°$C overnight.
8. Transfer plugs to NDS (10 plugs per 50 ml) and incubate at $55°$C overnight.
9. The plugs can be stored in NDS at $4°$C for at least a year.

Pulsed-Field Gel Electrophoresis

Solutions, Media, and Materials
> $0.5\times$ TBE: 45 mM Tris–borate, 1 mM EDTA, pH 8
> $1\times$ TB(0.1)E: 90 mM Tris–borate, 0.2 mM EDTA, pH 8
> $0.25\times$ TAE: 10 mM Tris–acetate, 0.25 mM EDTA, pH 8
> Agarose NA (Pharmacia, Piscataway, NJ, Cat. No. 17-0554-03)
> Fastlane agarose (FMC, Cat. No. 50141)

Electrophoresis Conditions
The chromosomal size variation in protozoan parasites is sufficiently great that no single PFGE system and set of conditions will satisfactorily resolve an entire karyotype. As a result of unique design geometries, each of the currently available PFGE systems differ in their ability to resolve different size ranges. Two commercially available systems, the Bio-Rad (Richmond, CA) CHEF DRII (11) and the Beckman

TABLE I Conditions for Pulsed-Field Gel Electrophoresis Using the CHEF DRII and TAFE II Systems[a]

Parasite	Size range resolved (kb)	Agarose (%)	Buffer	Pulse/switch times (sec)	Voltage/current	Run time (hr)	Temperature (°C)
CHEF DRII system (Bio-Rad)							
Leishmania spp.	50–600	1	0.5 × TBE	30–50	200 V (6V/cm)	30	14
	200–800	1	0.5 × TBE	40–70	200 V	24	14
	700–2000	1	0.5 × TBE	80–120	200 V	40	14
	200–3000	1	1 × TB(0.1)E	1200–700	82 V (2.5 V/cm)	96	14
Trypanosoma brucei	50–100	1	0.5 × TBE	2.8–10	200 V	24	14
	300–550	1	0.5 × TBE	30–40	200 V	18	14
	1000–~3500	1.20	1 × TB(0.1)E	1400–700	85 V	96–168	16
	3500–5700	1	0.5 × TBE	5500	82 V	288	12
Toxoplasma gondii	1500–2500	1	0.5 × TBE	2500	85 V	100	12
TAFE II system (Beckman)							
Toxoplasma gondii and Trypanosoma brucei	1000–>6000	0.7 (Fastlane)	0.25 × TAE	30	50 V	36	12
				25	60 V	36	12
				22	90 V	36	12
				15	110 V	36	12
Leishmania spp.	50–2000	1	0.25 × TBE	1	330 mA	18	14
				2	350 mA	18	14

[a]The conditions listed represent adaptations of previously published procedures and personal communications (see, e.g., Refs. 9 and 10).

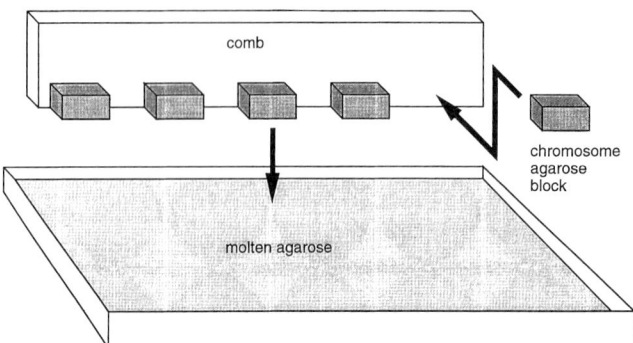

FIG. 1   Loading a gel for PFGE. Line the chromosome blocks along a solid comb (autoclave tape over a regular comb is effective). Secure them in place with a small amount of molten agarose. Allow the agarose to solidify on the comb. Place the loaded comb into the gel mold while the agarose is still molten but below 60° C (if the agarose is too hot, the blocks detach from the comb). After the gel has solidified, remove the comb and backfill the space formed by the comb with molten agarose.

(Fullerton, CA) TAFE II (12), which are based on electrophoresis in a contour-clamped homogeneous electric field (CHEF) and transverse-alternating field (TAFE), can be programmed to resolve the different chromosomal size ranges for karyotyping *T. gondii, T. brucei,* and *Leishmania* spp. (see Table I). The CHEF DRII has the advantage of linear ramping of pulse times to increase the range of chromosomes separated in a single run. The TAFE II allows the separation of larger chromosomes in a shorter total run time and produces a less marked compression of the chromosomes.

Karyotype resolution is enhanced by some additional precautions: (1) Remove the NDS or EDTA from the plugs by dialysis against the running buffer (two washes of two plugs per 50 ml, 1–2 hr per wash, room temperature). (2) Preequilibrate the gel in the electrophoresis chamber with running buffer at the correct temperature for 1– 2 hr before starting the run. (3) Load the chromosome blocks into the gel while the agarose is still molten as this prevents uneven loading and formation of air bubbles or other irregularities at the chromosome block/gel interface (see Figure 1). (4) Ensure that the electrophoresis chamber is level and the buffer circulation is adequate and at the correct temperature.

Chromosome-Specific Polymerase Chain Reaction Pools

Amplification of specific regions of chromosomal DNA has proved to be a useful tool for genome mapping. The PCR amplification of single bands dissected from *Drosophilia melanogaster* polytene chromosomes has been described by Saunders *et al.* (13). In this procedure, oligonucleotide adapters, which provide priming sites for PCR, are ligated to digested target chromosomal DNA. This technique has been modified for the amplification of individual chromosomal bands of *T. gondii* separated by PFGE in the TAFE II system.

Solutions, Media, and Materials

　　T4 DNA ligase (New England Biolabs, Beverly, MA, Cat. No. 202S)
　　*Sau*3A (New England Biolabs, Cat. No. 169S)
　　Taq DNA polymerase (AmpliTaq, Perkin Elmer, Norwalk, CT, Cat. No. N801-0060)
　　10× PCR buffer: 400 mM Tris-HCl, pH 8.9, 100 mM ammonium sulfate, 15 mM MgCl$_2$, 2.5 mM of each deoxynucleoside triphosphate (dNTP)
　　Denaturation solution: 400 μM NaOH, 20 μM EDTA
　　Geneclean II (BIO 101, La Jolla, CA, Cat. No. 3106)
　　GeneScreen Plus (NEN/DuPont, Boston, MA, Cat. No. NES-976)

Digestion of Chromosomal Band

1. Excise a single chromosomal band from a TAFE II gel (see figure 2).
2. Extract the DNA from the agarose using Geneclean II (BIO 101).

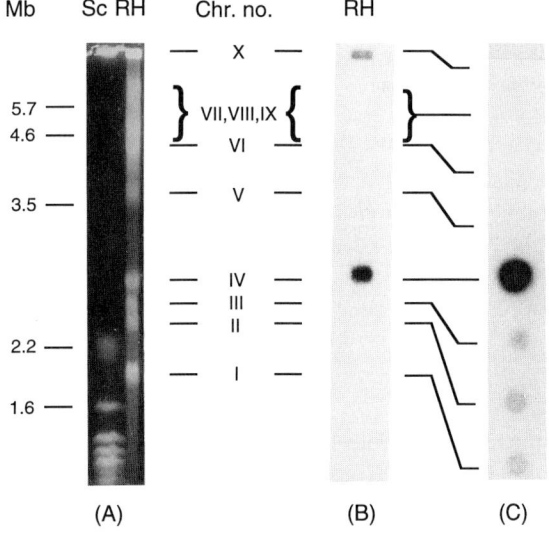

FIG. 2   Chromosome-specific PCR pool dot-blot. (A) TAFE karyotype of *T. gondii:* Sc, *S. cerevisiae;* RH, RH strain of *T. gondii.* (B) TAFE Southern blot of *T. gondii* strain RH chromosomes probed with chromosome IV-specific cosmid cB8 (10). (C) Dot-blot of chromosome-specific PCR pools probed with cosmid cB8.

3. Digest the DNA with 4 U of *Sau*3A at 37°C for 3 hr.
4. Incubate reaction at 65°C for 15 min to inactivate the enzyme.
5. Extract the DNA with phenol–chloroform and ethanol precipitate.
6. Resuspend the DNA pellet in 5 μl of distilled water.

Preparation of Adapter

The adapter consists of two oligonucleotides: a 24-mer (5′ GATCAGAAGCTTGA-ATTCGAGCAG 3′) and a 20-mer (5′ CTGCTCGAATTCAAGCTTCT 3′). Mix equimolar amounts of both oligonucleotides and anneal at 57°C for 1 hr.

Polymerase Chain Reaction Amplification of Chromosomal Band

1. Add 2 μg adapters to 1 μl of 10× ligation buffer, 1 U of T4 ligase, and 5 μl of digested chromosomal DNA from above, and bring the final volume to 10 μl.
2. Incubate the reaction at 16°C for 2 days.
3. Extract with phenol–chloroform and ethanol precipitate.
4. Resuspend the DNA in 10 μl of distilled water.
5. Add 100 ng of the 20-mer oligonucleotide to 5 μl of the purified DNA, 4 μl of 10× PCR buffer, and 1 U of *Taq* polymerase, and bring the final volume to 40 μl. Overlay with 40 μl of mineral oil.
6. Conduct the PCR in a Hybaid Omnigene thermal cycler. First program: 5 min at 37°C, 5 min at 55°C. Second program: 30 cycles of 1 min at 94°C, 1 min at 55°C, and 3 min at 72°C. Third program: 1 min at 55°C, 10 min at 72°C.
7. Prepare probe DNA by reamplifying 5 μl of the above PCR products under the same conditions, without the initial cycle.

Chromosome-Specific Dot-Blots

1. Denature 100 ng of each chromosome-specific PCR pool in 25 μl of denaturation solution for 10 min at 99°C.
2. Spot 2 μl of each denatured chromosome-specific PCR pool onto a GeneScreen Plus (NEN/Du Pont) nylon membrane. Air-dry the membrane for 15 min at room temperature and bake at 75°C, 2 hr.
3. Probe the dot-blot with a radiolabeled cosmid as per manufacturer's specifications.

Comments

The quality of the dot-blot (Fig. 2) is determined by the isolation of each chromosomal band from the adjacent bands. Also, repetitive DNA sequences will cross-hybridize between PCR pools. However, the signal strength from one pool is normally so much greater that the chromosomal assignment is unambiguous. The cross-hybridization due to repetitive DNA can be reduced by prehybridizing the probe with whole genomic DNA. Caution should be taken when mapping small DNA fragments (<1 kb) by this method, as the preparation of the chromosomal DNA for PCR amplification requires a restriction endonuclease digestion which will bias against certain chromosomal regions which have few sites (i.e., large DNA fragments do not undergo PCR amplification efficiently).

Library Construction

Genomic and cDNA libraries are the material basis for physical mapping. Data from genome analysis of *Plasmodium, Trypanosoma cruzi,* and *T. brucei brucei* suggest that no single cloning vector will faithfully replicate the entire genome (14, 15). *Escherichia coli*-based systems such as P1 (16) and cosmids (17) are currently being used for *T. brucei brucei* (S. Melville, unpublished results) and *Leishmania major* (A. Cruz, unpublished results). These vectors have the advantage that purified recombinant DNA can easily be prepared, but they have problems maintaining some sequences including large tandem arrays and AT-rich sequences (14, 15). Yeast-based systems, such as yeast artificial chromosomes (YACs; Ref. 18), suffer from the difficulty of producing quantitative amounts of recombinant DNA but will handle large tandem arrays (14, 19) and can clone fragments of DNA as large as 1 Mb (20). However, all of the cloning methods can be adapted such that they will maximize the information return when applied to a particular species.

Cloning with YACs can also be modified to clone telomeres selectively (15, 21). This adaptation of YAC cloning is an important asset for physical mapping as the ends of chromosomal contigs can be accurately defined. Because the number of chromosomal ends (40–50) for trypanosomatids is high relative to genome size (5 × 10^7 bp), telomere-associated single-copy sequences (TASCS) are an efficient and relatively easily obtainable set of anchor sequence-tagged sites (STSs; Ref. 22) for starting chromosomal contigs. For *T. brucei,* variable surface glycoprotein (VSG) expression sites are subtelomeric, so YAC–telomere cloning may prove to be a useful method for analyzing these regions.

In assembling a genome map, the genetic component includes identifying and physically mapping expressed genes with cDNA clones. These expressed site tags (ESTs) have provided a wealth of new information and represent the most rapidly expanding part of the DNA sequence databases (23). For species like *Leishmania* where transmission genetics remain enigmatic and other molecular methods are not yet routine, ESTs are the most efficient source of genetic information.

YAC–Telomere Library Construction and Identification of Telomere-Associated Single-Copy Sequences

The telomere base composition for *Leishmania* (and other trypanosomatids) consists of tandem repeats of a basic motif, TTAGGG, which is repeated from tens to hundreds of times from the chromosome end (24, 25). It is sufficiently different from other motifs including the *Saccharomyces cerevisiae* motif, that they will not cross-hybridize. The subtelomeric region is likely to be composed of more complex repetitive sequences which may lack restriction enzyme sites, making it difficult to clone the regions in conventional *E. coli*-based vectors. For this reason, the use of YAC vectors for the cloning of telomeres and telomere-associated sequences is a suitable approach. The rationale behind the method is based on the observation that telomeres from most organisms will serve as telomere addition sites in yeast (26). The YAC vector pJS97 (Ref. 27; GIBCO/BRL, Gaithersburg, MD) contains all the components for stable replication in yeast, except as a linear DNA fragment, where it lacks a telomere at one end. If, for example, a piece of DNA such as a *Leishmania* telomere is ligated onto the nontelomere end of the linear fragment, pJS97 will become a stable YAC (21). The method described below is assembled from several different sources where details are given for protocols with substantive alterations.

Solutions, Media, and Materials

> TE: 10 mM Tris-HCl, 1 mM EDTA, pH 8
> OHB (oligonucleotide hybridization buffer): 0.5 g bovine serum albumin (BSA), 0.5 g polyvinylpyrrolidone (PVP), 0.5 g Ficoll, 1 g sodium dodecyl sulfate (SDS), 1.0 g sodium pyrophosphate, 5× SSC in 1 liter
> 20× SSC: 3 M NaCl, 0.3 M sodium citrate, pH 7
> *Cla*I (New England Biolabs, Cat. No. 197S)
> *Eco*RI (New England Biolabs, Cat. No. 101L)
> T4 DNA ligase (New England Biolabs, Cat. No. 202CS)
> Alkaline phosphatase (Boehringer Mannheim, Indianapolis, IN, Cat. No. 713 023)
> T4 polynucleotide kinase (Boehringer Mannheim, Cat. No. 174 645)
> Phenol–chloroform: phenol–chloroform–isoamyl alcohol (25:24:1, v/v)
> [γ-^{32}P]ATP (Amersham, Arlington Heights, IL, Cat. No. PB 101218)
> NICK G-50 column (Pharmacia, 17-0855-02)
> pJS97 (GIBCO/BRL, Cat. No. 5387 SA)

Preparation of Telomere-Enriched Leishmania DNA

1. Wash a DNA plug (prepared as described above) in TE, equilibrate the plug in 10 ml of 1× *Eco*RI buffer (without Mg^{2+}) for 30 min at room temperature.
2. Incubate the plug in 100 μl of 1× *Eco*RI buffer (without Mg^{2+}) with 100 U of *Eco*RI for 3 hr at 4°C.
3. Start the reaction by adding Mg^{2+} to 10 mM final concentration and incubate at 37°C overnight.
4. Add 0.5 ml TE, melt the plug at 68°C, extract with phenol, extract with phenol–chloroform, and ethanol precipitate.
5. Resuspend the DNA to less than 0.5 μg/ml concentration with 1× T4 DNA ligase buffer and 5 U/ml T4 DNA ligase. Incubate at 4°C overnight.

6. Extract with phenol–chloroform, ethanol precipitate, and resuspend the DNA to 50–100 μg/ml in TE.

Preparation of pJS97

1. Digest 5–10 μg of pJS97 in 50 μl of 1× *Cla*I buffer with 20–40 U of *Cla*I at 37°C for 2 hr.
2. Extract with phenol–chloroform, ethanol precipitate.
3. Resuspend the DNA in 50 μl of 1× *Eco*RI buffer with 20–40 U of *Eco*RI. Incubate at 37°C for 2 hr.
4. Extract with phenol–chloroform, ethanol precipitate.
5. Resuspend the DNA in 50 μl of 50 mM Tris-HCl, pH 8, and 1 μl calf intestinal phosphatase. Incubate at 37°C for 30 min.
6. Extract with phenol, extract with phenol–chloroform, and ethanol precipitate.
7. Resuspend the DNA in 50–100 μl TE.

Ligation of pJS97 and Telomere-Enriched Leishmania DNA

Mix 50 ng pJS97, 50 ng *Leishmania* DNA, 2 μl of 10× ligase buffer, 2 U of T4 DNA ligase, and distilled water to a final volume of 20 μl. Incubate at 12°C for 16 hr.

Transformation and Colony Filter Preparation

The transformation of strain YPH252 is as according to Burgers and Percival (28). Colony filters are prepared as described in Brownstein *et al.* (29).

Screening for Telomere-Containing Clones

1. Mix 10 ng (TTAGGG)$_4$, 50 μCi [γ-^{32}P]ATP, 1 μl of 10× polynucleotide kinase buffer, 1 U polynucleotide kinase, and distilled water to 10 μl final volume. Incubate at 37°C for 20 min.
2. Remove unincorporated label by separation on a NICK column.
3. Prehybridize filters for at least 15 min at 45°C in 15 ml OHB.
4. Discard prehybridization OHB, add probe to 15 ml OHB, and hybridize at 45°C overnight.
5. Wash filters twice in 4× SSC, 0.1% SDS, 0.1% pyrophosphate at 45°C for 15 min.
6. Expose to X-ray film and identify positively hybridizing clones.

Preparation of YAC–Telomere DNA

DNA is prepared as described by Burke *et al.* (18).

Subcloning Telomere-Associated Single-Copy Sequences from YAC–Telomere Clones

1. Digest 10 μg of YAC–telomere DNA with 100 U of *Bam*HI in 10 μl of 10× *Bam*HI buffer diluted to 100 μl final volume.

2. Extract with phenol, extract with phenol–chloroform, and then ethanol precipitate.
3. Resuspend the DNA in 50 μl TE and dilute a portion of this to a concentration of 0.5 μg/ml in 1× T4 DNA ligase buffer.
4. Ethanol precipitate and then resuspend the DNA in 5 μl TE.
5. Transform *E. coli* DH5α by electroporation in 0.2-cm cuvettes at 25 μF, 200 Ω, and 25 kV in a Gene Pulser transfection apparatus (Bio-Rad).
6. Plate the electroporated cells on selective LB-ampicillin agar plates.

Comments

There are variables among the procedure steps which can be altered to suit a particular species. The cloning vector pJS97 allows for several restriction enzymes at the cloning site. Both *Eco*RI and *Bgl*II have been successfully used for *Leishmania* (Fig. 3). Because these enzymes have different GC contents in their recognition sequence, they will on average cut at different frequencies given the GC content of the genome. This property should allow for the production of telomere clones in which

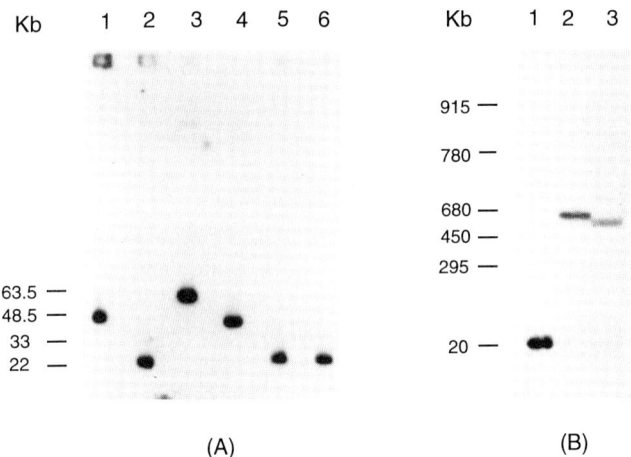

(A)   (B)

FIG. 3   (A) CHEF Southern blots of YAC–telomere clones probed with (TTAGGG)$_4$. Lane 1, *Leishmania peruviana* YTP1; lane 2, *L. peruviana* YTP2; lane 3, *L. major* YTM1; lane 4, *L. major* YTM2; lane 5, *L. major* YTM3; lane 6, *L. major* YTM4. CHEF conditions: 1–15 sec for 20 hr, 200 V, 0.5× TBE. YTP1 and YTP2 were constructed from a complete *Bgl*II digest of genomic *L. peruviana* DNA. YTM1–4 were constructed from a complete *Eco*RI digest of genomic *L. major* DNA. Accounting for 7.8 kb of pJS97 DNA, the TASCS defined by the two enzymes range from 6 to 50 kb from the end of the chromosome. (B) TAFE Southern blots of *L. peruviana* and *L. major* chromosomes probed with TASCS-P2. Lane 1, *L. peruviana;* lane 2, *L. major.* TAFE conditions: 330 mA, 1 min, 17 hr; 350 mA, 2 min, 17 hr. TASCS-P2, derived from YTP2, cross-hybridizes with an *L. major* chromosome of nearly the same size, suggesting that TASCS may be useful among species for contig anchor STSs.

the TASCS are varying distances from the end of the chromosome. Partial restriction enzyme digestion of genomic DNA is another alternative, albeit somewhat more difficult to control. Regardless of the specific method, for protozoans which have substantial subtelomeric repeat structures (which may contain restriction sites), varying the clone size may be crucial for finding the TASCS nearest the end of the chromosome. The circularization/ligation step for enriching for telomere fragments (which should remain linear) may be less important for species like the trypanosomatids, as the genome sizes are small relative to the number of chromosomes.

Perhaps the most crucial step in identifying bonafide YAC–telomere clones is *a priori* knowledge of the species-specific telomere motif. Many sequences will provide the appropriate secondary structure to behave as telomere addition sites in *S. cerevisiae*. In a given telomere–YAC ligation/transformation, only 2–10% of the clones selected for telomere function are positive for the species-specific telomere repeat motif; thus, screening by motif–oligonucleotide hybridization is essential for identifying bonafide YAC–telomere clones. The utility of these clones for physical mapping requires the subcloning of TASCS. A telomere–YAC represents such a small proportion of the total YPH252 genomic DNA that circularization/ligation and subsequent transformation must be very efficient. Electroporation of *E. coli* strain DH5α cells has proved to be the most reliable method of transformation. As with conventional YAC cloning, cocloning artifacts occur with sufficient frequency such that telomere–YACs and the subcloned TASCS must be checked for colinearity with the source genomic DNA by restriction enzyme analysis. Finally, the YAC–telomere clones can be established to be bonafide telomere ends by *Bal*31 exonuclease digestion of genomic DNA and subsequent probing of a Southern blot with the corresponding subcloned TASCS (26).

Constructing cDNA Libraries from Trypanosomatid mRNAs

The construction of cDNA libraries from trypanosomatids such as *Leishmania* spp. takes advantage of the observation that most, if not all, cytoplasmic mRNAs contain a common sequence of 39 nucleotides at the 5' ends called the spliced leader (SL) or miniexon (30). This unusual property in conjunction with polyadenylation at the 3' end provides specific priming sites for both ends of the mRNAs, selecting for the synthesis of full-length cDNAs. If the primers for each end are designed with a different restriction enzyme recognition sequence, the cDNAs can be cloned directionally into any suitable vector, in this case, λZAPII (Stratagene, La Jolla, CA; Fig. 4).

Solutions, Media, and Materials

20 mM Methyl-dCTP stock: dissolve 10 μM 5-methyl-2'-deoxycytidine 5'-triphosphate (Boheringer Mannheim, Cat. No. 757 047) in 500 μl of 10 mM Tris-HCl, pH 7.5

(A) 5′-GAGAGAGAGAGAGAGAGAGAGAACTAGT<u>CTCGAG</u>TTTTTTTTTTTTTTTTTTTT-3′

(B) 5′-GGAATTCCAT<u>GCGGCCGC</u>**AACTAACGCTATATAAGTATCAGTTTCTGTACTTTATTG**-3′

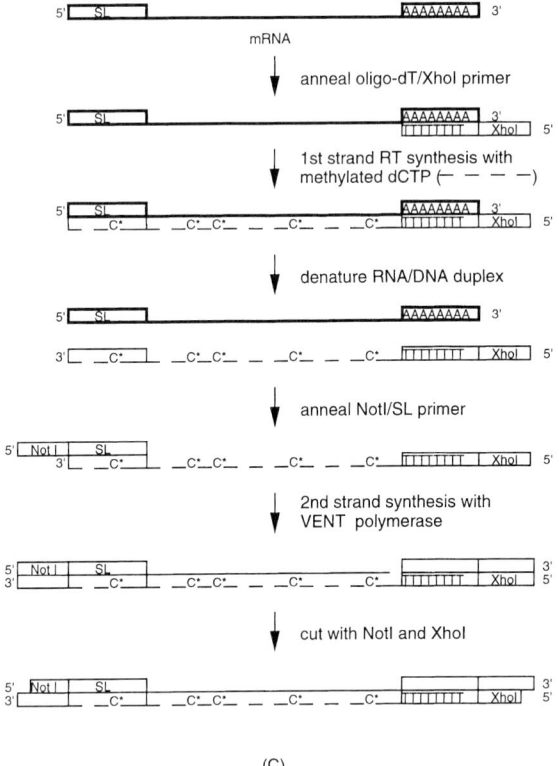

(C)

FIG. 4   Construction of cDNA libraries from trypanosomatids. (A) Oligo(dT)–*Xho*I primer. The *Xho*I site (underlined) is adjacent to the oligo(dT) [poly(A)$^+$] tail. (B) The spliced-leader primer RI/NOTI/SL. The region in bold type indicates the *Leishmania major* spliced-leader sequence and the underlined region a *Not*I site. (C) Schematic of library construction. The mRNA is reverse transcribed in the presence of 5-methyl-dCTP using the oligo(dT)–*Xho*I primer. This reaction produces single-stranded DNA molecules which contain a complementary spliced-leader sequence. The second-strand synthesis employs a thermostable DNA polymerase (Vent polymerase, New England Biolabs) with high processivity and is directed by the spliced-leader primer whose sequence is complementary to all the 5′ ends of the single-stranded cDNAs. The double-stranded cDNAs now contain a *Not*I site at the 5′ end and a *Xho*I site at the 3′ end. Notably, all the newly synthesized duplex cDNAs, except what was originally primer templated, is hemimethylated. Restriction digestion with *Not*I and *Xho*I will only cleave the sites present in the primers and not those within the remainder of the cDNA. Finally, after size-selection, cDNAs of at least 400 bp can be ligated into the *Not*I and *Xho*I sites of the bacteriophage insertion vector λZAPII. This method has two major advantages over conventional procedures: (1) use of primers specific to the terminal 3′ and 5′ ends of the mRNAs should select for the synthesis of full-length cDNAs, and (2) the cDNAs can be directionally cloned because of the different restriction sites at the 3′ and 5′ ends.

STE: 150 mM NaCl/1 mM EDTA/10 mM Tris-HCl, pH 7.5
*Not*I (New England Biolabs, Cat. No. 189S)
*Xho*I (New England Biolabs, Cat. No. 146S)
RNase H$^-$ reverse transcriptase (BRL, Cat. No. 8053SA)
Vent DNA polymerase (New England Biolabs, Cat. No. 254S)
T4 DNA ligase (New England Biolabs, Cat. No. 202S)
Oligo(dT)-cellulose spun columns (Pharmacia, Cat. No. 27-9258-01/02)
NICK G-50 spun column (Pharmacia, Cat. No. 17-0862-01/02)
CL-4B spun column (Pharmacia, Cat. No. 27-5105-01)
GeneScreen Plus (NEN/DuPont, Cat. No. NEF-976)
λZAPII DNA (Stratagene, Cat. No. 236201)
Gigapack II Gold packaging extracts (Stratagene, Cat. No. 200215)
XL1-Blue MRF' (Stratagene, Cat. No. 200301)

mRNA Purification

The quality and quantity of mRNA obtained from the parasite stage of interest is the most crucial factor in creating a large, representative library. The most important factor in obtaining high quality RNA is to immediately extract the RNA from cells that have been prepared from a healthy culture (see step 1 below). However, if it is not practical to proceed with the RNA extraction, the prepared cell pellet can be snap-frozen in N$_2$ and stored at $-70°$C.

1. Prepare cells through step 5 following the procedure for preparation of chromosomal DNA plugs (see above).
2. Extract total RNA using the acid guanidinium thiocyanate–phenol–chloroform method as in Chomczynski and Sacchi (31).
3. Starting with at least 0.25 mg of total RNA, isolate poly(A)$^+$ by two cycles on oligo(dT)-cellulose spun columns following manufacturer's specifications. This will yield approximately 10 μg poly(A)$^+$ RNA.
4. Check the quality and quantity of both the total and poly(A)$^+$ RNA by Northern blotting, as in Fourney *et al.* (32), onto GeneScreen Plus and probing with a gene that hybridizes to an abundant mRNA (e.g., a clone that encodes a cytoskeletal protein).

First-Strand Synthesis

1. Mix 5 μg poly(A)$^+$ RNA, 2.2 μl of 1318 μg/ml oligo(dT)–*Xho*I primer, and water to a final volume of 27 μl.
2. Incubate at 65°C for 5 min; cool on ice for 1 min.
3. At room temperature, add 10 μl of 5× reverse transcriptase buffer (BRL), 5 μl of 0.1 M dithiothreitol (DTT), 3 μl of 10 mM dATP/dTTP/dGTP/methyl-dCTP, 1 μl of 10 U/μl RNasin, 4 μl of 200 U/μl RNase H$^-$ reverse transcriptase.
4. Incubate at 37°C for 1 hr.

5. Add 100 μl TE and extract with an equal volume of phenol–chloroform.
6. Apply the aqueous phase to a NICK G-50 column equilibrated with TE according to the manufacturer's instructions (Pharmacia).
7. Centrifuge at 500 g for 4 min at 20°C.
8. Ethanol precipitate the eluate. (*Note:* Use ammonium acetate, as the removal of the methyl-dCTP is critical.)
9. Wash the DNA with 70% (v/v) ethanol, air-dry the RNA–DNA hybrid, and resuspend in 80.4 μl water.

Second-Strand Synthesis

1. Transfer the RNA–DNA hybrid to a PCR tube and add the following: 10 μl 10× Vent buffer, 5 μl of 50 mM MgSO$_4$, 2 μl of 25 mM dNTPs, and 1.6 μl of 8000 μg/ml RI/NOTI/SL primer (Fig. 4B).
2. Overlay with 20 μl mineral oil and place in a PCR machine. Heat to 95°C for at least 5 min.
3. Add 1 μl of 2 U/μl Vent DNA polymerase and continue to heat at 95°C for 1 min.
4. Incubate at 72°C for 15 min.
5. Extract with 120 μl chloroform.
6. Apply the aqueous phase to a NICK G-50 spun column (see above) and centrifuge at 500 g for 4 min at 20°C.
7. Ethanol precipitate the eluate.
8. Wash the DNA pellet with 70% (v/v) ethanol and air-dry the double-stranded cDNA.
9. Resuspend in 35 μl TE.

Preparation of cDNA for Ligation

1. To the double-stranded cDNA add 5 μl of 10× *Not*I buffer, 5 μl of 1 mg/ml BSA, and 5μl of 10 U/μl *Not*I.
2. Incubate at 37°C overnight.
3. Extract with phenol–chloroform and back-extract with 50 μl TE.
4. Pool the aqueous phases and desalt on a NICK G-50 spun column (see above). Centrifuge at 500 g for 4 min at 20°C.
5. Make the eluate volume up to 70 μl with TE.
6. Add 10 μl of 1 mg/ml BSA, 10 μl of 10× *Xho*I buffer, and 10 μl of 10 U/μl *Xho*I.
7. Incubate at 37°C for 6 hr.
8. Extract with 100 μl phenol–chloroform.
9. To desalt and size-fractionate the double-stranded cDNA, apply the aqueous phase to a CL-4B spun column equilibrated with STE and then according to the manufacturer's instructions.
10. Determine the yield of *Not*I/*Xho*I-cleaved size-selected cDNA in the eluate.

Preparation of Vector for Ligation

1. Digest 5 μl of 1 mg/ml λZAPII DNA by adding 30 μl TE, 5 μl of 10× *Not*I buffer, 5 μl of 1 mg/ml BSA, and 5 μl of 10 U/μl *Not*I.
2. Incubate at 37°C for 5 hr.
3. Extract with 50 μl phenol–chloroform and back-extract with 50 μl TE.
4. Pool the aqueous phases and extract with an equal volume of chloroform.
5. Ethanol precipitate.
6. Wash the DNA pellet with 70% (v/v) ethanol, air-dry, and resuspend in 35 μl TE.
7. Add 5 μl of 1 mg/ml BSA, 5 μl of 10× *Xho*I buffer, and 5 μl of 10 U/μl *Xho*I.
8. Incubate at 37°C for 5 hr.
9. Extract the digest with 50 μl phenol–chloroform and back-extract with 50 μl TE.
10. Pool the aqueous phases and apply to a CL-4B column (see above).
11. The eluate volume is approximately 100 μl (i.e., *Not*I/*Xho*I-cut λZAPII DNA is at 50 ng/μl).

Ligation of cDNA and Vector

It is important to emphasize that control ligations of vector alone and vector plus control insert are essential for assessing the quality of the subsequent cDNA library.

1. Check the integrity and concentration of the vector and insert DNAs by running 50–100 ng *Not*I/*Xho*I-cut λZAPII DNA and 0.5–1.0 μg *Not*I/*Xho*I-cleaved DNA, size-selected.
2. If the DNAs are intact, mix 200 ng cDNA and 1 μg of λZAPII DNA and coprecipitate with ethanol.
3. Wash the DNA pellet with 70% (v/v) ethanol, air-dry, and resuspend in 5 μl of ligation reaction mix: 3.5 μl water, 0.5 μl of 10× ligation buffer, 0.5 μl of 10 m*M* ATP, 0.5 μl of 2000 U/μl T4 DNA ligase.
4. Incubate at 14°C for 2 days.
5. Check the ligation reaction on a 0.4% agarose gel.

Packaging Ligations and Plating the cDNA Library

1. Package the ligation mix with Gigapack II Gold packaging extracts as per the manufacturer's instructions (Stratagene).
2. Plate the recombinant phage on XL1-Blue MRF′ host bacteria on 5-bromo-4-chloro-3-indolyl-β-D-galactoside (X-Gal)/isopropylthiogalactoside (IPTG) NZY agar plates as per the manufacturer's instructions (Stratagene).

Comments

The cloning efficiency of the library is usually about 1 × 10^6 plaque-forming units (pfu)/μg cDNA. About 80–90% of the plaques should be white on X-gal/IPTG NZY agar plates, presumably as a consequence of a cDNA insert. The size of the inserts can be determined by PCR. This method of cDNA library construction should select for full-length, directionally cloned cDNAs, eliminating many of the problems normally associated with cDNA analysis. This is particularly important for making ESTs for genome mapping, as knowledge of the polarity and terminal 5' end of the DNA sequence from a cDNA greatly facilitates database searching.

Acknowledgments

We thank Jennie Blackwell, John Boothroyd, David Sibley, Andy Tait, James Alexander, Angela Cruz and Jorge Arevalo for providing biological materials, unpublished information, and encouragement. We acknowledge Glaxo, the Wellcome Trust, the MRC, and the AFRC (BBSRC) for supporting this work.

References

1. L. H. T. Van der Ploeg, D. C. Schwartz, C. R. Cantor, and P. Borst, *Cell (Cambridge, Mass.)* **37,** 77 (1984).
2. L. H. T. Van der Ploeg, M. Smits, T. Ponnudurai, A. Vermuelen, J. H. E. T. Meuwissen, and G. Langsley, *Science* **229,** 658 (1985).
3. L. M. Corcoran, K. P. Forsyth, A. E. Bianco, G. V. Brown, and D. Kemp, *Cell (Cambridge, Mass.)* **53,** 807 (1988).
4. P. Bastien, C. Blaineau, and M. Pages, *in* "Subcellular Biochemistry, Volume 18: Intracellular Parasites" (J. L. Avila and J. R. Harris, eds.), Plenum, New York, 1992.
5. W. C. Gibson, *Parasitology* **99,** 391 (1989).
6. S. H. Giannini, M. Schittini, J. S. Keithly, P. W. Warburton, C. R. Cantor, and L. H. T. Van der Ploeg, *Science* **232,** 762 (1986).
7. T. W. Spithill and N. Samaras, *Nucleic Acids Res.* **13,** 4155 (1985).
8. W. C. Gibson and L. H. Garside, *Mol. Biochem. Parasitol.* **42,** 45 (1990).
9. K. Gottesdiener, J. Garcia-Anovernos, M. Gwo-Shu Lee, and L. H. T. Van der Ploeg, *Mol. Cell. Biol.* **10,** 6079 (1990).
10. L. D. Sibley and J. C. Boothroyd, *Mol. Biochem. Parasitol.* **51,** 291 (1992).
11. G. Chu, D. Vollrath, and R. W. Davis, *Science* **234,** 1582 (1986).
12. K. Gardiner, W. Laas, and D. Patterson, *Somatic Cell Mol. Genet.* **12,** 185 (1986).
13. R. Saunders, D. M. Glover, M. Ashburner, I. Siden-Kiamos, C. Louis, N. Monastirioti, C. Savakis, and F. Kafatos, *Nucleic Acids Res.* **17,** 9027 (1989).
14. J. W. Ajioka, J. Espinoza, and J. T. Swindle, *in* "Genome Analysis of Protozoan Parasites" (S. P. Morzaria, ed.), p. 169. ILRAD, Nairobi, Kenya, 1993.
15. D. De Bruin, M. Lanzer, and J. V. Ravetch, *Genomics* **14,** 332 (1992).
16. N. Sternberg, *Proc. Natl. Acad. Sci. U.S.A.* **89,** 103 (1990).
17. K. A. Ryan, S. Desgupta, and S. M. Beverley, *Gene* **131,** 145 (1993).
18. D. T. Burke, G. F. Carle, and M. V. Olson, *Science* **236,** 806 (1987).

19. D. Garza, J. W. Ajioka, D. T. Burke, and D. L. Hartl, *Science* **246,** 641 (1989).
20. Z. Larin, A. P. Monaco, and H. Lehrach, *Proc. Natl. Acad. Sci. U.S.A.* **88,** 4123 (1991).
21. J.-F. Chang, C. L. Smith, and C. R. Cantor, *Nucleic Acids Res.* **17,** 6109 (1989).
22. M. V. Olson, L. Hood, C. Cantor, and D. Botstein, *Science* **245,** 1434 (1989).
23. J. C. Venter, M. D. Adams, A. Martin-Gallardo, W. R. McCombie, and C. Fields, *TIB-TECH* **10,** 8 (1992).
24. L. H. T. Van der Ploeg, A. Y. Liu, and P. Borst, *Cell (Cambridge, Mass.)* **36,** 459 (1984).
25. J. Ellis and J. Crampton, *Mol. Biochem. Parasitol.* **29,** 9 (1988).
26. E. H. Blackburn and J. W. Szostak, *Annu. Rev. Biochem.* **53,** 163 (1984).
27. N. McCormick, J. S. Shero, C. J. Connelly, S. E. Antonarakis, and P. A. Hieter, *Technique* **2,** 65 (1990).
28. P. N. Burgers and K. J. Percival, *Anal. Biochem.* **163,** 391 (1987).
29. B. H. Brownstein, G. A. Silverman, R. D. Little, D. T. Burke, S. J. Korsmeyer, D. Schlessinger, and M. V. Olson, *Science* **244,** 1348 (1989).
30. K. Perry, K. P. Watkins, and N. Agabian, *Proc. Natl. Acad. Sci. U.S.A.* **84,** 8190 (1987).
31. Chomczynski and Sacchi, *Anal. Biochem.* **162,** 156 (1987).
32. R. N. Fourney, J. Mikayoshi, R. S. Day, and M. C. Patterson, *Bethesda Res. Lab. Focus* **10,** 5 (1988).

[3] *Aspergillus nidulans* as a Model System for Cell and Molecular Biology Studies

Gustavo H. Goldman and N. Ronald Morris

Introduction

Fungi are eukaryotic, heterotrophic organisms with an absorptive mode of nutrition. Most fungi are multinucleate, with a rigid chitinous cell wall, and exhibit a mycelial, yeastlike, or dimorphic (yeast and mycelial phases) type of growth. *Aspergillus nidulans* is a filamentous fungus of considerable biological importance that provides an excellent model system for genetic and biochemical analysis, as well as for studies of metabolic (1, 2) and developmental regulation (3). *Aspergillus nidulans* has a membrane-enclosed nucleus, chromatin that condenses, histones, actin, and tubulins (α, β, and γ) (4). It has cytoplasmic microtubules during interphase, an intranuclear microtubular spindle at mitosis, and spindle pole bodies. Additionally, kinesin-like and dynein motor proteins have been identified (5).

The genetic system of *A. nidulans* is sophisticated. It has eight well-marked chromosomes with many useful color, auxotrophic, and drug resistance markers. *Aspergillus nidulans* possesses a small genome ($2.6-3.1 \times 10^7$ bp), chromosomes that can be separated by pulsed-field gel electrophoresis (PFGE), and a low amount of repetitive DNA. Because of these characteristics, it is possible to order *A. nidulans* existing genomic DNA libraries according to chromosome by using PFGE-isolated chromosomes in colony filter hybridizations (6, 7). Thus, chromosomal subcollections of *A. nidulans* have been isolated, which make the cloning and characterization of previously genetically mapped genes easier. *Aspergillus nidulans* is homothallic, that is, there are no mating types, so that almost any two strains can be mated. The organism is normally haploid, but heterokaryotes and stable diploids can be produced. Genes can be mapped to chromosomes by parasexual genetics (cosegregation of genes with chromosome markers during haploidization of diploids) and to loci by sexual crosses. *Aspergillus nidulans* grows on inexpensive, defined media, either submerged or on a solid surface, and produces asexual spores (conidia) and sexual spores (ascospores) which can be stored for long periods. *Aspergillus* grows at reasonable rates between 15° and 44°C, providing an exceptionally wide range of temperatures in which to look for temperature-sensitive (ts) and cold-sensitive (cs) mutations.

This chapter describes some of the possibilities offered by *A. nidulans* as a model system for studying cell and molecular biology. It is our aim to review the approaches being adopted, when possible to use examples that illustrate the techniques, and to provide some useful protocols for working with *A. nidulans*.

Methods in Molecular Genetics, Volume 6

Transformation System for *Aspergillus nidulans*

DNA-mediated transformation using polyethylene glycol (PEG) has been the method most often exploited for *A. nidulans* and is considered the standard gene transfer system. It involves careful handling of cells because the protoplasts used are osmotically sensitive. Protoplasts are produced by removing the cell wall of either germinating spores or hyphae with cell wall-degrading enzymes. The most commonly used product for this purpose is Novozyme-234 (Novo Industries), a hydrolytic enzyme mixture secreted by the filamentous fungus *Trichoderma harzianum,* which is used alone or in combination with other lytic enzymes, such as β-glucuronidase and chitosanase. All protoplast preparations have to be protected by the presence of an osmotic stabilizer. Different osmotic stabilizers, such as sorbitol, NaCl, or $MgSO_4$, at concentrations between 0.8 and 1.2 *M,* have been used (8).

Mycelia or germlings can be used as the starting material for the production of protoplasts. When cell walls are partially or totally removed, cells are treated with a mixture of calcium chloride, PEG, and transforming DNA. The exogenous DNA molecules are internalized while a PEG-induced protoplast fusion takes place. Following fusion, the protoplasts are plated on appropriate osmotically buffered regeneration medium that will allow selection for the expression of the phenotype encoded by the transforming DNA. Because PEG-mediated transformation is time consuming, an alternative transformation method using electroporation has been developed (9, 10), but it yields lower transformation frequencies than those obtained by standard transformation methods using PEG.

Selection of Transformants

Selection of transformants from the background of nontransformed cells depends on expression of genes conferring a selectable phenotype (8). The first report of transformation of *A. nidulans* was based on the complementation of a *pyrG*⁻ mutant strain (deficient in orotidine-5′-phosphate decarboxylase, an enzyme involved in pyrimidine biosynthesis) with the orotidine-5′-phosphate decarboxylase gene of *Neurospora crassa* (11). Subsequently, many other genes suitable as prototrophic markers in recipient auxotrophic mutants have been used as selectable markers (for review, see Ref. 8). These include *trpC,* encoding a trifunctional enzyme involved in tryptophan biosynthesis from chorismate, and *argB,* encoding ornithine carbamoyltransferase. Transformation can also be achieved using dominant genes such as resistance to hygromycin B, phleomycin, oligomycin, or the fungicide benomyl as dominant markers. The *A. nidulans amdS* gene which enables the organism to grow on acetamide as the sole carbon and nitrogen source is also a useful selective marker (for reviews, see Refs. 5 and 12).

In cases where a transforming gene cannot be directly selected for, one option is to look for cotransformation with a readily selectable marker. There is a high probability that a cell that takes up one kind of DNA will also take up another, particularly if the ratio of cotransforming DNA to transforming DNA is kept high (8). The frequencies of reported cotransformation are variable and dependent on the strain and the transformation conditions.

Fate of Transforming DNA

Integration of DNA into Chromosomes

Plasmid shuttle vectors transform *Aspergillus* species by chromosomal integration (5, 13). Plasmids readily integrate either into homologous or heterologous sites. Genomic integration of circular plasmids occurs in several ways. Three types of integration events can be defined (Fig. 1). Type I involves integration of the plasmid at a region of homology (sequence similarity) within the genome and is called homologous recombination. In type II transformation, the plasmid integrates into sites within the genome where no known homology exists. This is called heterologous or ectopic integration. The mechanism by which the DNA recombines in type II events (heterologous recombination) has not been determined. The relative frequency of homologous integration versus heterologous integration is variable and depends on the specific gene being transformed. Type III integration involves gene conversion events. No plasmid sequence or selectable marker is detected within the genome but the gene being transformed is replaced by the introduced sequence.

Multiple copies integrated in tandem are a common feature of transformants. There are two mechanisms by which this might come about: either extrachromosomal plasmids first undergo homologous recombination with each other to form circular oligomers which then integrate by homology with the single chromosomal copy, or a monomeric plasmid recombines with its chromosomal homolog and then tandem repeats arise through secondary integrations (8). Although integrated plasmids are generally mitotically stable, introduced DNA sequences are often meiotically unstable. In *A. nidulans,* tandemly repeated sequences are lost at variable but readily detectable frequencies after self-fertilization or outcrossing (13).

Autonomously Replicating Vectors

Transformation frequencies in *A. nidulans* are relatively low compared to yeast, in which most transformation utilizes plasmids that replicate autonomously. Autonomously replicating plasmids offer the advantage that they are easy to recover. Most *Aspergillus* plasmids do not replicate autonomously, but an autonomous replicating plasmid was described in *Aspergillus* that can transform *A. nidulans* at high fre-

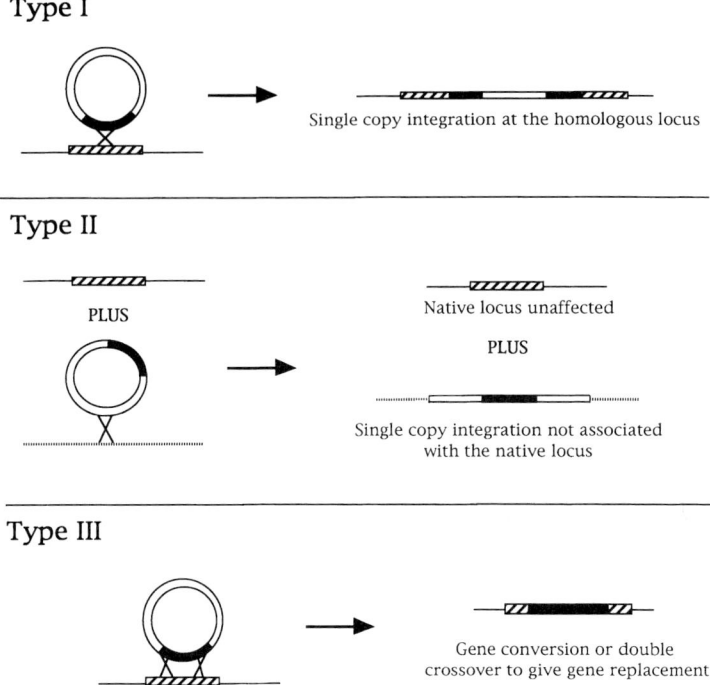

Type I

Single copy integration at the homologous locus

Type II

PLUS

Native locus unaffected

PLUS

Single copy integration not associated
with the native locus

Type III

Gene conversion or double
crossover to give gene replacement

Fig. 1   Patterns of plasmid integration in *A. nidulans* transformants. Type I (homologous integration). The plasmid integrates at a region of homology within the genome. Type II (heterologous or ectopic recombination). The plasmid integrates into sites within the genome where no known homology exists. Type III (gene conversion). No plasmid sequence is detected within the genome, but the gene is replaced by the introduced copy.

quency (14). *Aspergillus nidulans* was transformed with an *A. nidulans* gene bank, and from an unstable colony a plasmid was reisolated which transformed *A. nidulans* at a frequency of 20,000 transformants per 10^6 protoplasts at near saturation levels of transforming DNA. The plasmid, designated ARp1, is 11.5 kb in size and consists of sequences derived from the 5.4-kb gene bank vector, which carries the *A. nidulans* *argB* gene, and a 6.1-kb insert, designated AMA1. Southern analysis of transformant DNA showed ARp1 to be maintained in free form and not integrated into the chromosome. It has a mean copy number of 10–30 per haploid genome and is mitotically unstable, being lost from 65% of asexual progeny of transformants. Although the plasmid has shown similar transformational properties in *A. niger* and *A. oryzae,* the function of the chromosomal copies of the AMA1 sequence in *A. nidulans* remains to be determined.

The plasmid ARp1 has been used to construct markerless helper plasmids (15).

When the helper plasmids were added to transformation of *A. nidulans* using plasmids which normally transform by chromosomal integration, as much as a 200-fold increase in transformation efficiency resulted. The recovery in *Escherichia coli* of autonomously replicating plasmid cointegrates indicated that cotransformation involves recombination between integrating and helper plasmids at a high frequency. In effect, the addition of helper plasmids converts an integrating plasmid into an autonomously replicating gene library *in vivo* (15). Thus, the gene of interest can be recovered by transformation of total DNA of a specific fungal transformant into *E. coli*.

The helper plasmid provides a new method of gene cloning by complementation of mutant alleles, obviating the need for construction of a gene library *in vitro* and amplification in *E. coli* (16). The method involves simultaneous transformation of mutant strains of *A. nidulans* with fragmented chromosomal DNA from a donor species and DNA of a helper plasmid. Transformant colonies appear as the result of the joining of chromosomal DNA fragments carrying the wild-type copies of the mutant allele with the helper autonomously replicating plasmid. Joining may occur by either ligation (if the helper plasmid is in linear form) or recombination (if it is covalently closed circular DNA). The event occurs with high efficiency *in vivo* and generates an autonomously replicating plasmid cointegrate. More studies are necessary to assess the reliability of these systems.

Gene Cloning

There are many available strategies for cloning genes from filamenthous fungi (for review, see Ref. 12). Genes can be cloned by conventional methods such as (i) cloning directly from an *Aspergillus* library by DNA–DNA hybridization; (ii) using an antibody to the protein to clone from an expression library; (iii) polymerase chain reaction (PCR); (iv) plus–minus hybridization screening of clone banks or screening with probes prepared by subtractive or cascade hybridization; (v) physical mapping and chromosome walking techniques; and (vi) complementation in *E. coli* and *Saccharomyces cerevisiae,* using appropriate plasmids and cDNA inserts of *A. nidulans* under the control of species-related promoters.

Cloning by Complementation Using Transformation

Cloning by transformation requires two features: (i) transformation frequencies must be high enough so that a gene bank can be screened using realistic quantities of DNA and protoplasts, and (ii) it should be possible to reisolate or rescue the complementing plasmid. One additional advantage of *A. nidulans* is that chromosome-specific recombinant DNA libraries are available (7). Thus, a mutation mapped to a specific linkage group or chromosome can be complemented by transforming with the appropriate chromosome-specific library.

A transforming sequence can be recovered in three ways. First of all, the transformant DNA can be cleaved with a restriction enzyme that cuts once, but no more than once, in the transforming plasmid within the sequence duplicated through type I integration (see section on integration of DNA into chromosomes and Fig. 1). The fragments generated in this way are circularized with ligase, and the reconstituted plasmid is selected by transformation of *E. coli* (17). Second, the sequence can be recovered by constructing a cosmid library [a cosmid is a plasmid with bacteriophage λ *cos* (packaging) sequences] including a fungal marker for selection in *A. nidulans*, an ampicillin resistance marker and origin of replication for selection in *E. coli*, and a *Bam*HI cloning site which would accept fragments cleaved with the so-called four-cutter endonuclease *Mbo*I and *Sau*3AI. After fungal transformation, the transforming sequence can be recovered by incubation of the genomic DNA of the transformant with an appropriate packaging mixture and further phage infection of an *E. coli* strain. Finally, sib selection can be used. This refers to the construction of an ordered cosmid gene bank in which *E. coli* clones, each containing a plasmid with a different *A. nidulans* DNA fragment, are maintained separately. Transformation is originally by DNA extracted from pools of large numbers of the clones and subsequently by smaller and smaller positive subpools until an individual clone capable of complementing the *A. nidulans* mutation of interest is identified.

An important point about cloning by complementation is the need to confirm that the cloned gene obtained is the target gene and not some other sequence which acts as a suppressor of the target gene mutation. This can be tested by demonstrating that the transforming plasmid with the target gene integrates site-specifically at the mutated locus. The recombinatory event leading to site-specific plasmid integration generates a tandem duplication of wild-type and mutant sequences. The reverse event that triggers plasmid elimination causes the plasmid and one or the other of the mutant sequences to be lost and is therefore expected to generate both wild-type and mutant segregants. The loss of the integrated plasmid marker can be done by selection in the presence of fluorootic acid (see Section on disruption, deletion, and gene replacement). Thus, to show that the integration was at the mutated locus, a segregant showing the complementary phenotype is crossed with a wild-type strain. If integration had been in a locus other than the mutated locus, the cross would have generated both mutated and wild-type phenotypes. If, however, the DNA sequence integrated into the mutated gene locus, all segregants will show the wild-type phenotype (18).

Use of Transformation for Analysis of Gene Cloning

Disruption, Deletion, and Gene Replacement

Homologous integration of transforming plasmids allows one to disrupt, delete, or replace genes in *A. nidulans*. It often happens that a cloned gene sequence looks like a functional gene in that it is transcribed, contains an open reading frame, and perhaps has some interesting similarities to known genes in other organisms, but cannot be assigned a function because no mutations have been identified in it. In such cases, the first step in understanding the gene is to use the clone to disrupt or delete the equivalent genomic sequence, thereby creating a null mutant. In gene replacement, as opposed to gene disruption or deletion, the purpose is to retain gene activity but to modify its transcription or product.

The most common methodology for gene disruption in *A. nidulans* is similar to the one-step disruption, originally described by Rothstein (19) for *S. cerevisiae*. The procedure consists of inserting a copy of a selectable marker (*hygB*, encoding resistance to the antibiotic hygromycin, for example) into the cloned gene under investigation. The construct is then used to transform and convert a hygromycin-sensitive strain to a resistant phenotype. There are many examples of successful gene disruption in *A. nidulans* using this technique (for review, see Ref. 12).

In *S. cerevisiae*, it is possible to make a directed mutation in an essential gene in a diploid and then by tetrad analysis to prove the gene is essential. In *A. nidulans*, which can form stable diploids, tetrad analysis is not practical and the approach has limited applicability. Osmani *et al.* (20) have developed a method called hetero-karyon disruption (Fig. 2) by which it is possible to disrupt or delete essential genes in the haploid state of *A. nidulans* and subsequently recover cells containing the disrupted or deleted allele for phenotypic analysis. This is accomplished by transforming a haploid strain containing the *pyrG*$^-$ mutation with a plasmid containing an internal portion of the target gene (for gene disruption) or the sequence bordering the target gene (for gene deletion) and the *pyrG* gene of *A. nidulans*. Site-specific integration of the plasmid into the target gene locus will effectively inactivate the gene. This will generate balanced heterokaryons containing the inactivated gene, *pyrG*$^+$, *pyrG*$^-$ nuclei, and noninactivated target gene *pyrG*$^-$ nuclei. Because the nuclei complement the genetic defects of one another, both types of nuclei are maintained within the same cytoplasm and the heterokaryon is auxotrophic for uridine. On conidiation (formation of uninucleate spores), the two types of nuclei become segregated into the uninucleate spores. Thus, a mixed population of two classes of conidia is obtained. One class of conidia contains untransformed, noninactivated target gene, *pyrG*$^-$ nuclei, and the other class contains transformed, gene-disrupted (or deleted), *pyrG*$^+$ nuclei. The gene-disrupted (or deleted) conidia can then be recovered (together with the nondisrupted conidia), identified, and characterized morphologically under the microscope.

Gene replacement is a useful technique for cloning a gene of interest, altering it *in vitro*, and then replacing the existing wild-type allele of the gene in the organism with the experimentally altered version. The most used strategy for gene replacement in *A. nidulans* is indirect or two-step replacement gene replacement. In this technique, circular DNA is used for transformation, and homologous integration produces tandem duplications. Then, integration by homologous recombination can be reversed to result in loss of the integrated plasmid. Plasmid loss takes place at a

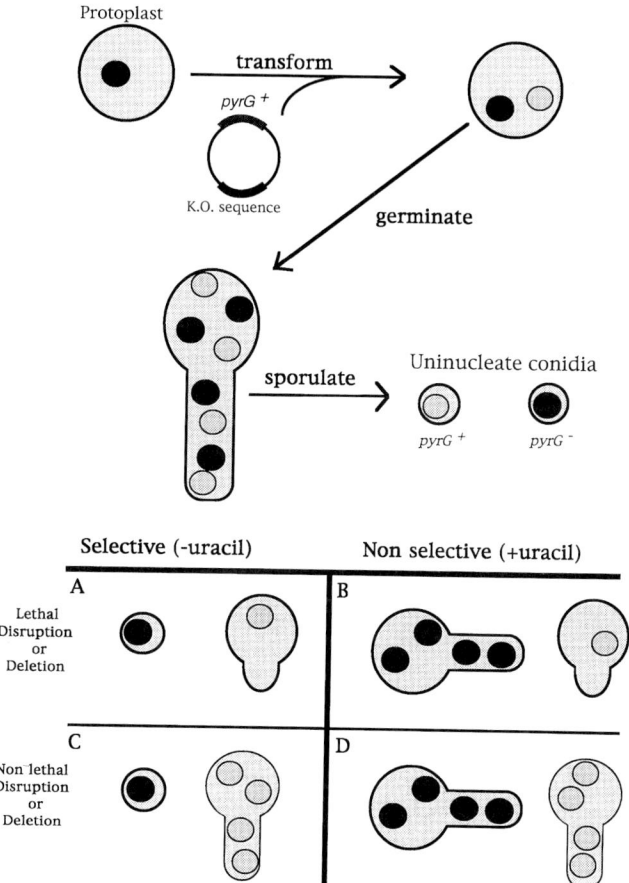

Fig. 2   Gene disruption or deletion in a heterokaryon of *A. nidulans.* If the disruption or deletion is lethal, under selective conditions (without uracil) there is no germination of the spores (A), but under nonselective conditions (plus uracil) nontransformed spores will germinate (B). If the disruption or deletion is nonlethal, in selective conditions there is only germination of the transformed spore (C), whereas in nonselective conditions both spores will germinate (D).

moderate frequency in *A. nidulans* when it is permitted to self-fertilize under nonselective conditions. Two-step gene replacements can also be achieved without passing through the sexual cycle by utilizing one of the selective markers for which there are both forward and reverse selections. Several such genes are available (21): (i) integration of the *oliC31* gene can be selected for using oligomycin and selected against, for subsequent removal, using trithyltin (19); (ii) in a *pyrG⁻* mutant strain the *A.*

nidulans pyrG gene can be selected for by growth in medium lacking uridine plus uracil and subsequently selected against by using fluoroorotic acid (FOA), to which *pyrG*⁻ mutant strains are resistant because they do not incorporate the toxic analog; and (iii) the *niaD* (nitrate reductase) gene can be selected for by complementation or against by utilizing the chlorate resistance of *niaD*⁻ mutant strains.

Overexpression and Fusion of Promoters with Reporter Genes

Overexpression of a gene may provide useful information about its function. A useful attribute of *A. nidulans* is that it has a cloned, well-characterized, inducible promoter, which can be used to regulate the expression of other genes to which it is fused. This is the *alcA* (alcohol dehydrogenase) promoter (22). The *alcA* promoter is induced when cells are grown on a variety of alcohols and is almost completely repressed on glucose. Thus, genes linked to *alcA* can not only be turned on and turned off, but can also have their expression regulated.

The expression of reporter genes under the control of *A. nidulans* promoters is a useful technique for assessing the ability of selected sequences to control gene activity. The β-galactosidase (*lacZ*) gene has been used as a reporter gene in *A. nidulans* for promoter analysis and cell specificity of gene expression (23–26). Caution must be exercised in interpreting the results of experiments with reporter genes because little is known about the effects of growth and development conditions on the stability of β-galactosidase or its mRNA. Thus, variations in enzyme levels under different conditions may reflect parameters other than transcription.

The Cell Cycle

The morphological events of mitosis in *A. nidulans* are well known and have been thoroughly described by light and electron microscopy (27, 28). The cell cycle of *A. nidulans* is similar to that of most other eukaryotes except that it is relatively short (90 min) and mitosis is closed, that is, the nuclear membrane does not break down at mitosis. *Aspergillus nidulans* forms a multicellular colony consisting of branched networks of tubular cells that constitute the mycelium. In *Aspergillus* the mycelial colony can be established either from a uninucleate haploid asexual spore (conidium) or a binucleate sexual spore (ascospore). The spores are dormant cells arrested in G_1 with an unreplicated complement of DNA. The first hyphal nuclear division takes about 400 min, depending on growth conditions (29, 30). At 32°C on rich medium, *A. nidulans* has a G_1 phase of 15 min, a 40-min S phase, a 40-min G_2 phase, and a 5-min M phase (29). Morris (31) has used direct microscope observation of temperature-sensitive, conditionally lethal mutants of *A. nidulans* to identify a large set of mitotic mutations. These have been classified as *bim* (blocked in mitosis), *nim* (never in mitosis), *nud* (nuclear distribution), and *sep* (septation) mutations. Several genes from each of the first three classes have been cloned, sequenced, and studied experimentally and show novel and conserved characteristics for the cell cycle in eukaryotes (for review, see Ref. 4).

There are some cytological markers that can be used for monitoring the cell cycle in *A. nidulans* (adapted from Ref. 30).

1. *The DNA-specific dye, DAPI:* In fixed cells stained with 4,6-diamidino-2-phenylindole (DAPI), the interphase nucleus has three discernable regions: the chromatin stains as a bright blue crescent or sphere surrounding a single dark nucleolus, and a single, small, densely stained, very bright DAPI-philic spot is observed, usually at the edge of the chromatin. In mitosis, these separate regions are no longer discrete; the chromatin region condenses and the nucleolar region disappears.

2. *Antitubulin antibodies:* Microtubule distribution cycles between an interphase organization, where longitudinally orientated cytoplasmic microtubules run down the length of the hyphae, and a mitotic organization where cytoplasmic microtubules disassemble to form the internuclear spindle. Few cytoplasmic microtubules are present at metaphase, but as the cell enters anaphase, spindle pole-associated (or astral) microtubules increase in length and abundance.

3. *MPM-2 antibody:* A monoclonal antibody raised against mammalian mitotic cells recognizes mitotic-associated phosphoproteins in a wide range of eukaryotes (32). Antibody MPM-2 reacts with the spindle pole body from middle or late G_2 phase until the end of mitosis, implying that this structure is phosphorylated during this phase of the cycle (33). Other antibodies are now available which recognize specific cellular components, for example, anti-γ-tubulin antibodies (34).

Conclusions and Perspectives

Aspergillus nidulans provides an important and useful model for studying molecular and cellular biology. One of the main advantages of *A. nidulans* as a model system is the fact that classical genetics, molecular and cellular biology, and morphology can be used in combination as a single powerful tool to gain information about a particular biological problem. *Aspergillus nidulans* is well suited for studying mechanisms controlling development and cell differentiation in multicellular eukaryotes. A great deal of information is already known about the genes that regulate entry into the condidial pathway and control the steps leading to formation of the complex, multicellular reproductive apparatus, the conidiophore, and the mature conidia (for review, see Refs. 3 and 35). *Aspergillus nidulans* is also a useful model system for analyzing the biochemical events of the cell cycle. Analysis of the genetics of mitosis in this organism has not only shown that *A. nidulans* has known homologs in other organisms but also has provided information about new mitotic proteins that have so far been found only in *A. nidulans* (for review, see Ref. 4).

Aspergillus nidulans is a saprophytic fungus, that is, it is able to utilize a wide range of materials as sources of nutrition. In the laboratory situation, a simple defined medium will support *Aspergillus* growth, and a large number of different nitrogen and carbon sources can be utilized by the fungus. Thus, many different biochemical pathways must exist for the assimilation of the different catabolites. The diversity of catabolic pathways and the associated requirement for genetic regulation makes *A. nidulans* an interesting model for gene regulation studies (for review, see Ref. 1).

Protocols

Transformation of Aspergillus nidulans

The following procedure is from Ref. 36.

1. Inoculate 2×10^8 to 1×10^9 conidia into 50 ml of YG medium (5 g yeast extract, 20 g glucose per liter). Incubate approximately 5 hr at $37°C/200$ rpm. The conidia should form clumps and begin to germinate.
2. Collect germinating spores by spinning at 2500 rpm for 3 min. Discard supernatant and resuspend germinating spores in the following freshly made protoplast solution (about 40 ml):

   0.1 ml β-glucuronidase
   20 ml solution 1 (0.8 M ammonium sulfate, 100 mM citric acid, pH 6.0)
   20 ml solution 2 (5 g yeast extract, 10 g sucrose per 500 ml)
   400 mg BSA (bovine serum albumin)
   0.5 ml of 1 M MgSO$_4$
   80 mg Novozyme-234 (Novo Industries)

3. Shake gently for 2.5 hr at $37°C$. Many protoplasts will exhibit a large vacuole after this time.
4. Collect by spinning at 2000 rpm for 3 min and wash pellet twice in cold solution 3 (0.4 M ammonium sulfate, 1% sucrose, 50 mM citric acid, pH 6.0). Resuspend pellet in 1 ml of solution 5 (0.6 M KCl, 50 mM CaCl$_2$, 10 mM 2 N-morpholinoethanesulfonic acid, pH 6.0). Allow protoplasts to rest on ice for 10 min (or overnight).
5. Add 100 μl protoplasts to 4–20 μg DNA. Mix gently and add 50 μl of solution 4

(25% PEG 6000, 100 mM CaCl$_2$·2H$_2$O, 0.6 M KCl, 10 mM Tris-HCl, pH 7.5). Leave on ice for 20 min.

6. Add 1 ml of solution 4 and leave at room temperature for 20 min. Add all the protoplast suspension to 30 ml of YGA top agar [YG plus 1% Difco (Detroit, MI) agar, 0.6 M KCl] and plate out onto six small plates containing YAG plus 0.6 M KCl.

Genomic DNA Extraction

The following protocol is from Ref. 37.

1. Inoculate 1.0×10^7 spores into 500 ml of YG medium and incubate at 37°C for 24 hr. Harvest mycelia by filtration through Whatman No. 4 paper and wash twice with TES (50 mM Tris, 50 mM NaCl, 5 mM EDTA, pH 8.0). Freeze immediately with liquid nitrogen and grind into a coarse powder using a cold mortar and pestle. Lyophilize overnight. Freeze at $-70°$C. Break the mycelia using mortar and pestle.
2. Transfer sample to a Corex tube and add 7–10 ml of proteinase K buffer (0.2 M Tris-HCl, pH 8.0, 0.1 M EDTA, 1% sarkosyl, 100 μg/ml proteinase K). Add the proteinase K only before use. Incubate at 45°–50° C for 1 hr.
3. Spin for 10 min in an HB-4 rotor at 3000 rpm. Transfer supernatant to a fresh tube. Adjust volume to 10 ml with buffer without proteinase. Divide in two and add 1 volume of 2-propanol to each. Invert to mix.
4. Spin at 10,000 rpm for 20 min in an HB-4 rotor. Discard supernatants. Resuspend pellet in 5 ml of 50 mM Tris, 50 mM NaCl, pH 8.0. Add 10 ml of 100% ethanol. Invert to mix.
5. Spin at 10,000 rpm for 20 min in an HB-4 rotor. Discard supernatant. Repeat ethanol precipitation. Air-dry pellet briefly and resuspend in 3 ml TE (50 mM Tris, 5 mM EDTA, pH 8.0).
6. Add 4.9 g ultrapure CsCl and then 66 μl bisbenzimide from a 10 mg/ml stock solution. Centrifuge at 50,000 rpm for 16–24 hr. Stop without brake. For recentrifugation, each DNA band is added to a solution of 0.94 g CsCl/ml TE. No additional bisbenzimide is necessary.
7. Remove the DNA band and add 3 volumes TE. Bisbenzimide and CsCl are removed by extracting several times (five to seven) with isobutanol equilibrated with TE. Extract twice with ether. Allow ether to evaporate.
8. Precipitate DNA by addition of $\frac{1}{20}$ volume of 8 M LiCl and add 2 volumes of 100% ethanol. After at least 2 hr at $-20°$C, the precipitated DNA is pelleted by centrifugation at 10,000 rpm in an HB-4 rotor for 30 min at 4°C. The DNA pellet is washed once with 70% ethanol. Finally, the DNA is resuspended in 200 μl TE.

DNA Minipreparations

The procedure for DNA minipreparations is from Ref. 38.

1. Inoculate a 125-ml flask, containing 25 ml YG medium, with 1.0×10^7 spores. Incubate at 37° C overnight.
2. Harvest mycelia by filtration of the culture through a Whatman filter. Lyophilize overnight. Freeze at $-70°$ C and break the mycelia using a mortar and pestle. Transfer sample to Eppendorf tubes and add extraction buffer [200 mM Tris-HCl, pH 8.5, 250 mM NaCl, 25 mM EDTA, 0.5% sodium dodecyl sulfate (SDS)]. Use 480 μl/40 mg of mycelium homogenate. Mix gently using a pipette.
3. Add 40 μl RNase stock solution (20 mg/ml) and incubate 30 min at 37° C. Add an equal volume of phenol–chloroform mixture (1:1). Extract for 10 min by mixing the Eppendorf tubes. Centrifuge for 15 min at 12,000 rpm.
4. Take the upper aqueous phase and add the same volume (500 μl) of chloroform, mix, and centrifuge for 5 min at 12,000 rpm.
5. Take the aqueous phase and add 0.54 volume of 2-propanol. Centrifuge at 12,000 rpm for 1 min. Remove the supernatant and wash the pellet with 70% ethanol for 1 min at 12,000 rpm. Discard the supernatant, dry, and resuspend in 20 μl of TE.

RNA Extraction

The procedure for RNA extraction is from Ref. 40 (adapted from Ref. 39). All the material must be sterilized and treated with 2% DEPC (diethyl pyrocarbonate), excepting the NTES solution (0.1 M NaCl, 10 mM Tris-HCl, pH 7.5, 1 mM EDTA, 1% SDS) to which DEPC cannot be added.

1. Start with approximately 2 g frozen mycelia. Grind to a fine powder in liquid nitrogen. Transfer sample to tubes and add 4.5 ml NTES and 3 ml phenol–chloroform–isoamyl alcohol (49:49:2, v/v). Vortex or shake well for at least 10 min.
2. Transfer to a tube and spin for 10 min at 8000 rpm at 4° C. Take aqueous phase to a new tube and add $\frac{1}{10}$ volume of 2 M sodium acetate and 2 volumes of 100% ethanol. Leave at $-20°$ C for 20 min.
3. Pellet precipitate at 8000 rpm for 10 min. Remove supernatant and eventually, when the pellet looks green or brownish, rinse in 70% (v/v) ethanol. Dissolve pellet in 2.5 ml distilled water. Spin at 5000 rpm for 5 min (the polysaccharides will appear as a pellet).
4. Transfer to a new tube and add 2.5 ml of 4 M lithium acetate. Leave on ice at 4° C for at least 3 hr or overnight. Centrifuge at 8000 rpm for 10 min. Dissolve pellet in 0.9 ml distilled water. Add 0.1 ml of 2 M sodium acetate and 2 volumes of ethanol and leave at $-20°$ C for 2 hr.

5. Centrifuge at 8000 rpm for 10 min. Dissolve pellet in 500 μl distilled water and check concentration in spectrophotometer. Store at $-70°$C.

DNA Staining

The method for DNA staining is from Ref. 41.

1. Put sterile coverslip in bottom of small petri dish and add 5–10 ml YG medium.
2. Inoculate with about 1.0×10^6 spores of *A. nidulans* and incubate at 37°C for 8–12 hr.
3. Remove the coverslip and fix for 15–20 min in DAPI fixation solution (50 mM H_3PO_4, pH 6.5–6.8, 0.2% Triton X-100, 5% glutaraldehyde, 0.25 μg/ml DAPI).
4. Wash twice with Tris-buffered saline (TBS) (10$\times$ TBS stock is 72.4 g Tris base, 118.6 g NaCl, per liter distilled water, adjusted to pH 7.4).
5. Wash twice with water, dry, and mount on a slide using Citifluor.

Immunofluorescence Assay for Aspergillus

The following is from Ref. 41.

1. Inoculate about 10^6 conidia into petri dish containing sterile coverslips and 15 ml YG medium. Incubate at desired temperature for 8–16 hr.
2. Blot coverslips (edge and back) and incubate on Parafilm with 200 μl fix buffer for 45 min at room temperature [fix buffer is 8% formaldehyde, 100 mM PIPES, 25 mM EGTA, 5 mM $MgSO_4$, 5% dimethyl sulfoxide (DMSO), pH 7.0].
3. Put coverslips in holders and wash four times in microtubule stabilizing buffer (MTSB; 100 mM PIPES, 25 mM EGTA, 5 mM $MgSO_4$, pH 6.9).
4. Blot coverslips and incubate on Parafilm with 100 μl Novozyme-234 for various times (15–60 min) at 28°C. Usually use 1.5% Novozyme-234 plus 10 μM aprotinin in Pembals buffer (100 mM PIPES, 1 mM $MgSO_4$, 1 mM EGTA, 1 M sorbitol, 1% BSA, pH 6.9; sterilize, and store at room temperature).
5. Wash four times in MTSB. Blot coverslip holder and dunk in 200 ml extraction buffer [100 mM PIPES, 25 mM EGTA, 5 mM $MgSO_4$, 10% DMSO, 0.2% Nonidet P-40 (NP-40), pH 6.9] for 45–60 sec. Rinse four times respectively in MTSB, 0.33$\times$ MTSB, and twice in TBS (7.24 g Tris base, 11.86 g NaCl, per liter distilled water, pH 7.4).
6. Blot coverslips as above. Incubate on Parafilm with 50–100 μl primary antibody diluted in TBS plus 1% BSA. Incubate at 28°C for 1 hr or at 4°C overnight. For overnight incubation, use a container with a lid and put a moist paper towel inside to prevent drying.

7. Wash three times in TBS plus 0.1% BSA. Blot coverslips as above and incubate on Parafilm with $50-100$ μl secondary antibody diluted in TBS plus 1% BSA.
8. Wash three times in TBS plus 0.1% BSA. Stain with DAPI (0.7 μg/ml in TBS plus 0.1% BSA and 0.02% azide) for 5 min (coverslip can be dipped or stained on Parafilm). Wash twice in TBS plus 0.1% BSA and twice with distilled water. Blot coverslips and allow to air-dry on a paper towel. Mount in Citifluor antifade compound.

For Novozyme-234 digestions, about $15-30$ min is required for microtubule staining, $20-50$ min for MPM-2 staining, $15-30$ min for anti-γ-tubulin staining, and $50-80$ min for BIMA [*bimA* gene encodes a tetratrico peptide repeat (TPR) protein that localizes to the spindle pole body in *A. nidulans;* Ref. 41].

Pulsed-Field Techniques

The application of pulsed-field gel electrophoresis (PFGE) to separate chromosomes in agarose matrices is an exciting and powerful technique for the analysis of fungal genomes. The development of molecular karyotypes, physical maps of entire genomes, and fine structure physical and genetical maps of chromosomes are possible when PFGE is used in conjunction with other techniques. Often DNA sequences from *Aspergillus* are isolated from genomic libraries by selecting for complementation of auxotrophic mutations in either homologous or heterologous systems. Hybridization analysis using cloned DNA sequences to probe the electrophoretic karyotype can be used in conjunction with genetic analysis to determine the chromosomal origin and genetic location of these sequences. In addition, the use of PFGE fractionation of *Aspergillus* chromosomes can provide a source of DNA for the construction of chromosome-specific libraries (6, 7).

Contour-Clamped Electric Field Electrophoresis

The following protocol is from Ref. 6.

1. Inoculate 1 liter of YG medium (5 g yeast extract, 20 g glucose, 1 liter distilled water) with $1-3 \times 10^9$ conidia and incubate at $32°C$ for 18 hr with vigorous shaking.
2. Mycelial cells are harvested by filtration through Miracloth (Calbiochem, San Diego, CA), rinsed briefly with glass-distilled water, and gently squeezed to remove excess liquid.
3. The cells ($10-30$ g) are suspended in 30 ml of OM buffer (1.2 M MgSO$_4$/10 mM sodium phosphate, pH 5.8), and a solution of 300 mg Novozyme-234 (Novo In-

dustries) and 150 mg BSA in 40 ml OM buffer is added. The cells are converted into protoplasts by incubation for 1–2 hr at 32°C with gentle shaking.

4. The suspension is filtered through Miracloth, and 12 ml filtrate containing the protoplasts is transferred to centrifuge tubes and overlayed carefully with 10 ml ST buffer (0.6 *M* sorbitol/10 m*M* Tris-HCl, pH 7.0). The tubes are centrifuged at 5000 rpm in an HB-4 swinging-bucket rotor (Sorvall) for 15 min at 4°C. The banded protoplasts are removed using a bent Pasteur pipette and mixed with an equal volume of STC buffer (1.2 *M* sorbitol/10 m*M* Tris-HCl, pH 7.5/10 m*M* CaCl$_2$). The protoplasts are then pelleted at 7000 rpm in an HB-4 swinging-bucket rotor (Sorvall) for 10 min at 4°C and washed twice with 10 ml STC buffer. The pellet is resuspended in GMB buffer (0.125 *M* EDTA, pH 7.5/0.9 *M* sorbitol) such that the concentration of protoplasts is 2–2.5 × 10^8 cells/ml.

5. The suspension is then placed at 37°C, an equal volume of molten 1.4% InCert agarose (FMC, Rockland, ME) in GMB buffer precooled to 42°C is added, and the agarose–protoplast mixture is poured into a plug mold (2 × 2 × 25 mm) and solidified on ice for 10 min. The agarose plugs are immersed in NDS buffer (0.5 *M* EDTA, pH 8.0/10 m*M* Tris-HCl, pH 9.5/1% sodium *N*-lauroylsarcosinate) containing proteinase K (2 mg/ml) at 50°C for 24 hr. Finally, the plugs are washed two or three times in 50 m*M* EDTA (pH 8.0) at 50°C and stored at 4°C in EDTA. The chromosomal DNA in the plugs remains intact for at least 4 months.

6. Contour-clamped electric field (CHEF) electrophoresis is performed using an apparatus described by Chu *et al.* (2), except that vertical rather than horizontal electrodes are employed. A 40-ml gel containing 0.8% agarose is poured directly into a mold (8 × 8 × 1 cm) in the apparatus. The DNA–agarose plugs are inserted into the gel wells and sealed with 0.8% SeaPlaque agarose (FMC). Gels are electrophoresed at 12°C in 0.5× TAE buffer at 45 V with three pulse intervals of 50, 45, and 37 min at durations of 72, 12, and 72 hr, respectively. The gels are stained in ethidium bromide (0.5 µg/ml) for 45 min and then destained in water for 1 hr.

7. The gel is successively soaked in 0.25 *M* HCl for two periods of 15 min, in 0.5 *M* NaOH/1.0 *M* NaCl twice for 20 min, in 0.5 *M* Tris-HCl, pH 7.5/1.5 *M* NaCl for two 20-min periods, and finally equilibrated with 10× SSC (1× SSC is 0.15 *M* NaCl, 15 m*M* sodium citrate, pH 7.0). The DNA is then transferred by capillary blotting onto Hybond-N$^+$ membrane (Amersham, Arlington Heights, IL) using 10× SSC for 12–16 hr. The DNA is fixed on the membrane with 0.4 *M* NaOH for 20 min and washed with 5× SSC.

Acknowledgments

We thank the members past and present of the Morris laboratory for developing many of the methods described; Drs. Debbie Willins and Xin Xiang for critical reading of the manuscript;

and Waqar Ali Shah for the computer work in the figures. This work was supported by grants from the Institute of General Medical Science from the National Institutes of Health.

References

1. M. A. Davis and M. J. Hynes, *Trends Genet.* **5,** 14 (1989).
2. H. N. Arst and C. Scazzocchio, *in* "Gene Manipulations in Fungi" (J. W. Bennet and L. Lasure, eds.), p. 309. Academic Press, San Diego, 1985.
3. W. E. Timberlake and M. A. Marshall, *Trends Genet.* **4,** 162 (1988).
4. N. R. Morris and A. P. Enos, *Trends Genet.* **8,** 32 (1992).
5. N. R. Morris, X. Xiang, and S. Beckwith, *Trends Cell Biol.* in press (1994).
6. H. Brody and J. Charbon, *Proc. Natl. Acad. Sci. U.S.A.* **86,** 6260 (1989).
7. H. Brody, J. Griffith, A. J. Cutichia, J. Arnold, and W. E. Timberlake, *Nucleic Acids Res.* **19,** 3105 (1991).
8. J. R. S. Fincham, *Microbiol. Rev.* **53,** 148 (1989).
9. M. G. Richey, E. T. Marek, C. L. Schadl, and D. A. Smith, *Phytopathology* **79,** 844 (1989).
10. B. N. Chakraborty, N. A. Patterson, and M. Kapoor, *Can. J. Microbiol.* **37,** 858 (1991).
11. J. Ballance, F. P. Buxton, and G. Turner, *Biochem. Biophys. Res. Commun.* **112,** 284 (1983).
12. W. E. Timberlake, *in* "More Gene Manipulations in Fungi" (J. W. Bennet and L. Lasure, eds.), p. 51. Academic Press, San Diego, 1991.
13. J. Tilburn, C. Scazzocchio, G. G. Taylor, J. H. Zabicky-Zissman, R. A. Lockington, and R. W. Davies, *Gene* **26,** 205 (1983).
14. D. Gems, I. L. Johnstone, and A. J. Clutterbuck, *Gene* **98,** 61 (1991).
15. D. H. Gems and A. J. Clutterbuck, *Curr. Genet.* **24,** 520 (1993).
16. D. Gems, A. Aleksenki, L. Belenki, S. Robertson, M. Ramsden, Y. Vinetski, and A. J. Clutterbuck, *Mol. Gen. Genet.* in press (1994).
17. M. M. Yelton, J. E. Hamer, and W. E. Timberlake, *Proc. Natl. Acad. Sci. U.S.A.* **81,** 1470 (1984).
18. M. Ward, *in* "Modern Microbial Genetics" (U. N. Streips and R. E. Yasbin, eds.), p. 455. Wiley-Liss, New York, 1991.
19. R. J. Rothstein, *in* "Methods in Enzymology" (R. Wu, L. Grossman, and K. Moldave, eds.), Vol. 101, p. 202. Academic Press, San Diego, 1983.
20. S. A. Osmani, D. B. Engle, J. H. Doonan, and N. R. Morris, *Cell (Cambridge, Mass.)* **52,** 241 (1988).
21. M. Ward, B. Wilkinson, and G. Turner, *Mol. Gen. Genet.* **202,** 265 (1986).
22. R. B. Waring, G. S. May, and N. R. Morris, *Gene* **79,** 119 (1989).
23. R. F. M. Van Gorcom, P. J. Punt, P. H. Pouwels, and C. A. M. J. J. Van den Hondel, *Gene* **48,** 211 (1986).
24. J. E. Hamer and W. E. Timberlake, *Mol. Cell Biol.* **7,** 2352 (1987).
25. T. H. Adams and W. E. Timberlake, *Mol. Cell Biol.* **10,** 4912 (1990).
26. J. Aguirre, T. H. Adams, and W. E. Timberlake, *Exp. Mycol.* **14,** 290 (1990).
27. C. F. Robinow and C. E. Caten, *J. Cell Sci.* **5,** 403 (1969).
28. B. R. Oakley and N. R. Morris, *Cell (Cambridge, Mass.)* **24,** 837 (1981).

29. L. G. Bergen and N. R. Morris, *J. Bacteriol.* **156,** 155 (1983).
30. J. H. Doonan, *J. Cell Sci.* **103,** 599 (1992).
31. N. R. Morris, *Genet. Res.* **26,** 237 (1976).
32. D. D. Vandre, F. M. Davis, P. N. Rao, and G. G. Borisy, *Eur. J. Cell Biol.* **41,** 72 (1986).
33. D. B. Engle, J. H. Doonan, and N. R. Morris, *Cell Motil. Cytoskeleton* **10,** 432 (1988).
34. B. R. Oakley, C. E. Oakley, Y. S. Yoon, and M. K. Jung, *Cell (Cambridge, Mass.)* **61,** 1289 (1990).
35. W. E. Timberlake, *Annu. Rev. Genet.* **24,** 5 (1990).
36. S. A. Osmani, G. S. May, and N. R. Morris, *J. Cell Biol.* **104,** 1495 (1987).
37. R. C. Garber and O. C. Yoder, *Anal. Biochem.* **135,** 416 (1983).
38. U. Raeder and P. Broda, *Lett. Appl. Microbiol.* **1,** 17 (1985).
39. J. D. G. Jones, P. Dunsmuir, and J. Bedbrook, *EMBO J,* **4,** 2411 (1985).
40. G. H. Goldman, R. A. Geremia, A. B. Caplan, S. Vila, R. Villarroel, M. Van Montagu, and A. Herrera-Estrella, *Mol. Microbiol.* **6,** 1231 (1992).
41. P. M. Mirabito and N. R. Morris, *J. Cell Biol.* **120,** 959 (1993).

[4]   Special Molecular Techniques for Study of Filamentous Fungi

George A. Marzluf

Introduction

Neurospora crassa and *Aspergillus nidulans* have been used extensively as model experimental organisms and have been central in the development of many important new concepts in genetics and cell biology (1). Much attention is now directed to the many varied filamentous fungi, including plant and animal pathogens, and the wealth of biochemical diversity which they represent. This chapter focuses on the molecular genetic techniques that are special, and in some cases completely unique, for research with filamentous fungi, and also briefly indicates some of the characteristics of the fungi which make them the ideal choice for studying certain questions of major importance.

Special Resources

An extremely valuable resource is the Fungal Genetics Stock Center (Department of Microbiology, University of Kansas Medical Center, Kansas City, KS 66160-7420; telephone 913-588-7295). The Stock Center maintains and distributes thousands of wild-type and mutant fungal strains, including *Aspergillus, Neurospora,* and *Fusarium* species. The stock center also has available a selection of various cloned genes from *Neurospora* and *Aspergillus,* as well as genomic and cDNA libraries and vectors for special uses. The Stock Center publishes a complete catalog, and an informative *Fungal Genetics Newsletter* is also published annually. Perkins *et al.* (2) have published an amazingly complete description of the chromosomal loci of *Neurospora crassa.* An extensive compilation of sequence-specific DNA-binding proteins of fungi and their DNA recognition elements has been prepared by Dhawale and Lane (3).

Preparation of Cellular Components and Macromolecules

Care and Feeding of Fungi

Various synthetic media and the conditions for growth of fungi in liquid or solid media as well as genetic techniques for *Neurospora* are described in an authoritative

Methods in Molecular Genetics, Volume 6

chapter which is as valuable today as when it was written over two decades ago by
Davis and de Serres (4).

Isolation of Nuclei and Nuclear Proteins

To address many new questions in the area of molecular genetics, it is necessary to
prepare nuclei, chromatin, nuclear proteins, or other nuclear components. Because of
their small size, fungal nuclei are more difficult to isolate than nuclei from mammalian cells, for example, from rat liver. Hautala *et al.* (5) described one of the earliest
techniques for isolation of crude nuclear extracts and of highly purified nuclei from
Neurospora, and a slightly modified version of the protocol (6) is outlined in Fig. 1.
A crude nuclear preparation is satisfactory for some purposes, for example, to prepare, via sonication, a nuclear protein extract for mobility shift experiments. A detailed protocol for preparation of nuclear proteins, which includes heparin-agarose
chromatography, is available (6). Purified nuclei can be obtained by Ludox (or Per-
coll) step-gradient centrifugation and appear as an opaque layer that forms between
layers of 25 and 50% Ludox (5). The quality of a purified nuclear preparation can be
evaluated by fluorescence microscopy using acridine orange (nuclei stain a brilliant
green and other cellular components, orange) or mithramycin. An alternative procedure for preparation of nuclei was described by Baum and Giles (7). The use of
isolated *Neurospora* nuclei to detect hypersensitive nuclease sites that develop
adjacent to active genes has been described (8, 9). In some cases it may be advisable

Mycelial pads from 1 liter culture are washed with distilled water, cut into thin strips, and
frozen at $-70°C$ until used.
$$\downarrow$$
Mycelial strips are placed into a bead beater chamber (Biospec Products, Bartlesville, OK)
with 50 g acid-washed 0.5-mm glass beads plus 100 ml buffer A (1 M sorbitol, 7% Ficoll 400,
20% glycerol, 5 mM $MgCl_2$, 10 mM $CaCl_2$, and 5% Triton X-100).
$$\downarrow$$
Homogenize mycelial strips with four 30-sec pulses, with 30 sec cooling intervals, transfer
homogenate to beaker, allow glass beads to settle, and decant into a 400-ml Omni-Mixer chamber (Sorvall) or alternative stirring apparatus.
$$\downarrow$$
Submerge chamber in an ice–water bath. Homogenize sample at moderately high speed (setting 6.5) for two 15-min periods, with a 10-min cooling interval. This step frees nuclei trapped
in large membrane complexes.
$$\downarrow$$
Centrifuge at low speed (e.g., in Sorvall GSA rotor at 2050 rpm) for 10 min at 4°C to pellet
cell debris. Transfer supernatant to clean tube and centrifuge at 7500 rpm (GSA rotor) for
45 min to sediment nuclei, yielding a crude nuclear preparation.

FIG. 1   Isolation of nuclei from filamentous fungi.

to isolate nuclei from spheroplasts, for example, to obtain very high molecular weight DNA.

Isolation of Other Organelles

Mitochondria are easily isolated by differential centrifugation (10). After homogenization of fungal mycelia in a osmotic stabilizing buffer, for example, containing 0.44 M sucrose plus EDTA and phenylmethylsulfonyl fluoride, the extract is centrifuged at low speed (3000 rpm in a Sorvall SS-34 rotor for 10 min) to remove cell debris, nuclei, and large membrane aggregates. The supernatant is then centrifuged for 20 min at 12,000 rpm in the same rotor, which pellets the mitochondria (10).

Davis, Weiss, and colleagues (11–13) have elegantly demonstrated that filamentous fungi are ideally suited to study compartmentation of biosynthetic and catabolic reactions and in the storage of amino acids and other metabolites. Amino acids such as arginine are sequestered in a specialized organelle, the vesicle, which prevents the degradation which would otherwise occur by catabolic enzymes present in the cytoplasm even in uninduced states. Vesicles, which are very delicate organelles, can be prepared with extreme care (13, 14).

Isolation of DNA

A number of different methods have been described for isolation of DNA from *Neurospora,* all of which are useful (15–18). An early procedure developed by Metzenberg and Baisch (15) provides large amounts of high molecular weight DNA. Schechtman (16) devised a simple protocol which yields 200–300 μg of reasonably pure DNA of greater than 50 kb in length from a single 40-ml culture. The DNA is sufficiently pure to be efficiently cut by various restriction enzymes. At least a dozen samples can be processed in 1 day; moreover, the procedure can readily be scaled up. The major steps in the protocol are outlined in Fig. 2, but one should consult the original description (16) for important details. Similarly, various protocols also work for the isolation of DNA from *Aspergillus;* large amounts of highly purified *Aspergillus* DNA can be prepared from 5 g of mycelia as described by Oakley *et al.* (17). Rapid minipreparations of *Aspergillus* DNA can be used for Southern blotting and related techniques (18).

RNA Isolation

The isolation of high quality RNA from any organism requires vigilance to prevent partial degradation or complete loss caused by ever-present ribonucleases. In addi-

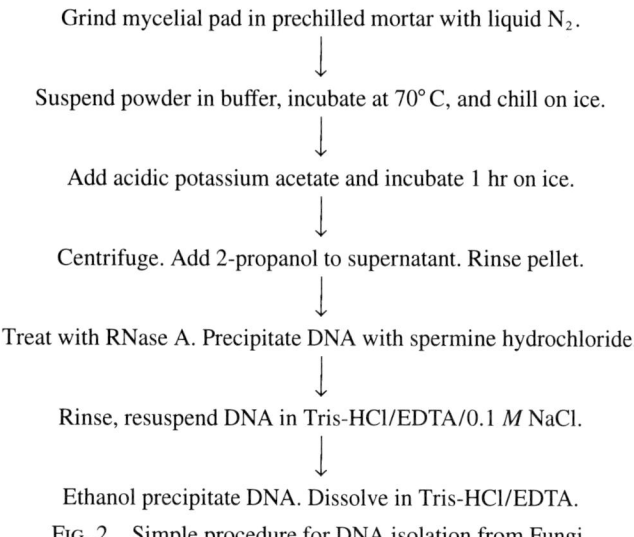

Grind mycelial pad in prechilled mortar with liquid N_2.

$\downarrow$

Suspend powder in buffer, incubate at $70°$ C, and chill on ice.

$\downarrow$

Add acidic potassium acetate and incubate 1 hr on ice.

$\downarrow$

Centrifuge. Add 2-propanol to supernatant. Rinse pellet.

$\downarrow$

Treat with RNase A. Precipitate DNA with spermine hydrochloride.

$\downarrow$

Rinse, resuspend DNA in Tris-HCl/EDTA/0.1 M NaCl.

$\downarrow$

Ethanol precipitate DNA. Dissolve in Tris-HCl/EDTA.

FIG. 2   Simple procedure for DNA isolation from Fungi.

tion to concern for various constitutive and inducible nucleases in any filamentous fungus, one should assume ribonuclease contamination of ordinary chemical reagents and glassware. Thus, all reagents and glassware for RNA work should be kept separate from general laboratory supplies; glassware should be baked overnight at $250°$ C prior to each use, and buffers should be treated with 0.1% dimethyl pyrocarbonate overnight and then autoclaved. Phenol must be distilled prior to use and treated with extreme care.

Reinert *et al.* (19) developed a procedure which has found widespread and effective use for large-scale isolation of RNA from *Neurospora* mycelia. The first step involves blending frozen mycelial chunks in a mixture consisting of equal volumes of a buffer containing EDTA and 4% sodium dodecyl sulfate and of a mixture of phenol, chloroform, and isoamyl alcohol (50:48:2, v/v). Another method that employs guanidinium isothiocyanate/cesium chloride gradients, originally devised for isolation of RNA from mammalian tissues rich in ribonuclease (20), also works effectively for RNA isolation from filamentous fungi. A single-step method (21) that employs acidic guanidinium thiocyanate–phenol–chloroform extraction is effective for the isolation of total RNA from *N. crassa*. Another simple and rapid method for preparation of RNA from multiple mycelial samples devised by Lindgren *et al.* (22) employs aurintricarboxylic acid, which binds irreversibly to RNA and is an extremely potent enzyme inhibitor. RNA prepared by this rapid method is satisfactory for DNA–RNA and RNA–RNA hybridizations but cannot be used for *in vitro* translation or reverse transcription. Osmani *et al.* (23) have described an efficient procedure for isolation of RNA from *Aspergillus*.

Transformation

Transformation Procedures

The ability to readily transform filamentous fungi with exogenous DNA has proved to be revolutionary in providing the pivotal step required for cloning, precise identification, characterization, and sophisticated manipulation of individual genes. The basic procedure was developed for *Neurospora* in the pioneering work of Case *et al.* (24, 25). The standard protocol has been improved by modifications (26, 27), and very similar techniques have proved efficient for *Aspergillus* (17).

Germinated conidiospores are collected, suspended in 1 *M* sorbitol, and used for the preparation of spheroplasts by treatment with Novozyme (InterSpex, Inc., Foster City, CA) in 1 *M* sorbitol for 30–60 min at 30°C with gentle shaking. Different batches of Novozyme will differ in efficiency so that each new lot must be standardized. The degree of spheroplasting can be checked by placing a 5-μl sample on a slide, adding a coverslip, and then adding water. When a sufficient amount of the cell wall has been digested away, the resulting spheroplasts will swell and burst. It is important not to prolong the Novozyme digestion step because the viability of the spheroplasts will decrease. An extremely convenient feature is that after the spheroplasts are resuspended in Tris buffer containing sorbitol, calcium choride, polyethylene glycol, and dimethyl sulfoxide, they may be dispensed in aliquots to microtubes and frozen at −80°C and will remain competent for transformation for at least several months (26, 27). The transformation step itself is straightforward. Spheroplasts are thawed, incubated with DNA in the presence of heparin, and, after brief incubations at 4° and 25°C, are mixed with top agar and plated onto suitable selective medium (24–27). Transformed colonies appear in approximately 2 days.

An alternative simple protocol can be used when it is necessary to transform many different strains and a low transformation efficiency is acceptable (28). Germinated conidia are treated with lithium acetate, incubated with DNA, followed by the addition of polyethylene glycol and a brief heat shock, and plated on selective medium. This protocol can be employed with relatively impure DNA preparations, such as that obtained with rapid minipreparation procedures (28).

Selectable Markers

The *qa-2*⁺ gene was the first selective marker used to transform *Neurospora,* and plasmids carrying the gene were made widely available by Giles, Case and colleagues (24, 25, 29). However, because the recipient strain used for selection must be a *qa-2 arom-9* double mutant, the system is difficult to employ in many cases. The *am*⁺ gene, which encodes glutamate dehydrogenase, has also been used as a selectable marker (30). The gene which encodes nitrate reductase (*niaD*⁺ in *Aspergillus,*

nit-3⁺ in *Neurospora*) represents a selective marker which is potentially useful for
many different filamentous fungi, as mutants lacking the enzyme are readily selected
via chlorate resistance (31–34). The *N. crassa pyr4⁺* gene, which encodes orotidine-
5'-phosphate decarboxylase, has been used as a selectable marker for *A. nidulans*
using the *pyrG* mutant as the host strain for transformation (17).

A major breakthrough was the development of a dominant selectable marker,
namely, a mutant *N. crassa* β-tubulin gene, for benomyl resistance by Orbach *et al.*
(27), which makes it possible to transform virtually any wild-type or mutant *Neurospora* strain. Similarly, Austin *et al.* (35) have engineered the bleomycin resistance
gene, *ble,* from the bacterial transposon Tn5 for expression in *N. crassa* by fusing it
to the *am* gene promoter and terminator regions. The *ble* gene fusion works in both
N. crassa and in *A. nidulans,* and thus it may be a useful dominant transformation
marker for many other filamentous fungi (35). Another useful selective marker that
works very well with *A. nidulans* and *N. crassa* is the bacterial hygromycin-resistance
gene, *hph* (36, 37).

It is often desired to introduce a manipulated gene or even a wild-type gene for
which no selective condition is available, and yet the lack of suitable restriction sites
may prevent its introduction into a vector with a useful selectable marker. In these
cases, cotransformation with two vectors works effectively, because a high percentage of the transformants will contain both vectors. It is also possible to transform
with linear DNAs.

Nature and Characterization of Transformants

Most transformants of *N. crassa* or *A. nidulans* have one or more copies of the transforming DNA integrated into ectopic genomic sites, and only infrequently will the
DNA be integrated at the homologous site via homologous recombination (25,
38, 39). Moreover, *N. crassa* conidia and the spheroplasts derived from them are
multinucleate and possess an average of two to three nuclei. Usually, only one of the
nuclei is transformed by integration of one or more copies of the exogenous DNA.
Thus, the transformed colonies are heterokaryotic, which seriously complicates their
use for any sophisticated studies of the function of the inserted genes. One cannot
use genetic crosses to purify *Neurospora* transformants because of repeat-induced
point mutations as described below. Homokaryons can be generated by filtration to
enrich for microconidia, which are then plated onto selective medium (40). Southern
blotting is necessary to identify homokaryons with a single copy of the transformed
gene for subsequent analysis of its function. Another complication is the finding that
the integrated transforming DNA may be methylated or otherwise modified and its
function impaired (39, 41).

When examining various manipulated forms of a cloned gene for function via
transformation, a concern is the variability introduced by the integration of the trans-

forming DNA at various different *N. crassa* genomic sites, because the immediate environment may affect gene expression. To examine a set of transformants which all possess DNA integrated at the same location, one can use a truncated selectable marker which can function only after homologous recombination with the endogenous mutant gene. In this technique, for example, a truncated *his-3* gene is targeted to the *his3* locus, and the vector carrying it integrates and thus also inserts any passenger gene as well (42). Although transformation with *A. nidulans* may occur more often by homologous recombination than that with *N. crassa,* similar concerns about the genomic location and copy number of the transforming DNA species must be addressed. It is obvious that careful attention must be given to the nature of the transformants obtained in the case of any filamentous fungus. Reporter genes can be employed to examine gene expression or its control in fungi (42).

Gene disruption, a strategy in which a specific gene is partially or totally deleted, represents an extremely valuable use of transformation. Gene disruption allows determination of the phenotype conferred by the complete loss of function of any cloned gene, and also provides an ideal host strain for analysis of various manipulated forms of the gene in question. The most direct technique for gene disruption involves replacement of a desired central region of the cloned gene with a dominant selectable marker, such as, the benomyl resistance gene, leaving sufficient 5' and 3' flanking DNA for homologous integration. The construct is then transformed into the wild-type strain with dual selection for the dominant marker and against the gene being disrupted, if possible. Transformants with the desired characteristics must be carefully analyzed via Southern blotting to ensure that the designed deletion was obtained. An early demonstration of gene disruption in *Neurospora* by targeted integration was accomplished with the *am*[+] gene (55).

Repeat-Induced Point Mutation

A remarkable phenomenon which occurs in *Neurospora,* and possibly other closely related fungi, known as repeat-induced point mutation (RIP) was discovered by Selker *et al.* (43, 44). When a second copy of a gene is introduced via transformation into *Neurospora,* both copies are "ripped," that is, suffer multiple GC to AT mutations throughout the length of the duplicated segments during a genetic cross as a premeiotic event (44). This occurs whether the duplicated or even triplicated copies of the same sequence are closely linked or are even located on different chromosomes (45). Moreover, two copies of a completely foreign DNA introduced by transformation will both be ripped, although a single copy would not be affected. Previously ripped duplicated genes will be subject to additional ripping during subsequent crosses, until they differ sufficiently not to be identified as repeated copies (46). The multiple point mutations introduced by RIP can be so extensive that the DNA fragment cannot be recognized via Southern blot hybridization when probed with the original wild-type sequence (47).

The molecular basis for repeat-induced point mutation of duplicated sequences is not entirely understood despite intensive study (46), but it appears to involve, as a first step, an extensive methylation of cytosine residues in duplicated DNA segments, whose recognition may depend on their pairing during meiosis. A mutation, *dim-2,* results in loss of all detectable DNA methylation and leads to defective chromosome behavior (e.g., aneuploidy), but its effect on RIP, if any, has yet to be defined (48). One interesting exception to the RIP of duplicated sequences is the feature that the multiple copies, approximately 180 repeats, of the rDNA genes are not affected, nor are the multiple copies of 5 S rRNA or tRNA genes (49–51). This lack of RIP of naturally occurring duplicated sequences may be related to the fact that recombination is highly suppressed in the rDNA region (52), which may indicate that premeiotic sequence pairing represents the mechanism for recognition of duplicate sequences.

Consequences of Repeat-Induced Point Mutation

Repeat-induced point mutation of duplicated DNA segments of *Neurospora* is a serious hindrance for genetic crosses involving transformants, but it also provides a special molecular tool that can be a powerful ally. When it appears that a specific structural gene has been cloned, RIP can be used to verify its identity. The cloned gene is introduced by transformation and mutated by RIP. The identity of the cloned gene is confirmed if the newly created RIP mutant yields only mutant progeny when crossed with a traditional mutant at that locus. Moreover, if conventional mutants of the cloned gene have never been obtained, its identity may be clear when a RIP mutant gives an obvious phenotype (e.g., an expected auxotrophy or the lack of the appropriate enzyme activity). The first example of the use of RIP to obtain a mutant phenotype for an unknown gene was to identify a gene required for acetate utilization (53). In some cases, it is not possible to predict with any confidence the phenotype that a mutant of a cloned gene would cause; in such cases, RIP can be used to determine the phenotype that results from a complete loss-of-function allele for that gene. When only a cloned gene from a different organism is available, it can be used to transform and RIP the endogenous *Neurospora* gene, provided sufficient homology exists between the two genes (46).

In many cases, the loss of an essential gene via mutation will result in inviability. A new sophisticated technique, called sheltered RIP, makes use of suitably marked heterokaryons so that the total loss or altered copy number of an essential gene can be followed (10). In a study by Harkness *et al.* (10), sheltered RIP was used to examine mutants of *mom-19,* which encodes an import receptor for nuclear-encoded mitochondrial precursor proteins. Homokaryons carrying the RIPed gene exhibited a complex and extremely slow growth phenotype, implying that the *mom-19* gene product, although not completely essential, is quite important for normal growth (10).

When transforming a *Neurospora* mutant with a cloned gene that has been manipulated *in vitro* (e.g., by site-directed mutagenesis), a concern is that a product of the mutant gene may interfere with the function of the introduced gene, or that a functional allele might be obtained via homologous recombination. Another use of RIP is to mutate a gene throughout its entire length, thereby creating an ideal host strain for transformation with manipulated forms of the cloned gene (47, 54).

Repeat-induced point mutation does not occur in *Aspergillus* and probably will not be found with most filamentous fungi. However, integration of cloned DNA by homologous recombination appears to occur with higher frequency and fidelity in *Aspergillus,* so that gene disruption and gene replacement can be achieved by targeted integration.

Identification and Cloning of Fungal Genes

Isolation of Fungal Genes

Hundreds of fungal genes, including structural, regulatory, tRNA, and rRNA genes, have been isolated since the 1980s and have led to a revolution in the understanding of molecular and cellular processes. Some fungal genes have been isolated by approaches in widespread use, for example, with the use of heterologous or synthetic probes (56), differential screening of cDNA libraries (57, 58), or chromosome walking. The *am* gene of *Neurospora* was obtained by employing a synthetic DNA probe based on the amino acid sequence of the encoded protein, glutamate dehydrogenase (56); the *ars* gene, which encodes arylsulfatase, was isolated via chromosome walking by Paietta (59).

The wealth of mutants of *A. nidulans* and *N. crassa* makes possible the isolation of many genes by complementation. However, the recovery of a cloned gene following complementation via a library, by simply recovering a recombinant plasmid, is not possible because of the lack of any suitable autonomously replicating vectors. Rather, transforming DNA integrates into chromosomal locations, making recovery of the complementing DNA problematic for the filamentous fungi. One approach is marker rescue, in which the genomic DNA is cut with an appropriate restriction enzyme and the fragments ligated to form circular molecules; the complex mixture can then be transformed via the selective marker into *Escherichia coli,* thus recovering the vector, which carries the desired gene. This technique was used to isolate the *pyrG* gene of *Aspergillus nidulans* (17).

A more generally applicable technique, known as sib selection, was devised by Akins and Lambowitz (26). In this procedure, a gene is isolated from a genomic library by successive rounds of transformation of a mutant *N. crassa* strain with pools of plasmid DNAs which are progressively reduced in complexity until a single candidate clone is obtained. Volmer and Yanofsky (60) developed a cosmid possessing the dominant selectable marker for benomyl resistance to generate libraries which contain large *Neurospora* DNA inserts. Individual cosmid clones are ordered in the wells of microtiter dishes, to allow rapid and efficient screening of the libraries via sib selection; this approach has led to the isolation of many different genes which were identified via complementation or molecular probes. The libraries are available from the Fungal Genetics Stock Center (see above). Similar approaches should be applicable with any filamentous fungi for which mutants and a transformation system are available.

Identification of Cloned Genes

Once a candidate gene has been isolated, for example, with a heterologous probe or by complementation of a conventional mutation, it is important to confirm that the isolated DNA segment actually contains the gene of interest. Obviously, if a recombinant plasmid carrying a fungal DNA sequence complements a defined mutation, this strongly suggests that it contains the desired gene, although additional evidence is required to be absolutely certain of its identity. If a truncated form of the cloned gene can be used to disrupt the genomic copy in a wild-type strain by homologous recombination or by RIP its identity is confirmed. Alternatively, if the vector carrying the cloned DNA targets a selectable marker to the locus of the desired gene, as demonstrated by a subsequent genetic cross, the cloned gene can be considered positively identified. The nucleotide sequence of the cloned gene may immediately confirm its identity by comparison with other sequences in databases.

Restriction fragment length polymorphism (RFLP) mapping is an efficient technique that can readily demonstrate whether the isolated DNA segment maps to the genetic locus of the desired gene. Metzenberg *et al.* (61, 62) have developed and made available a set of strains which makes it possible for investigators studying *Neurospora* to apply RFLP mapping rapidly to any isolated gene. Analysis via RFLP mapping should be possible for positive identification of a cloned gene in any fungal species, provided simple genetic crosses can be carried out.

Special Attributes of Filamentous Fungi

It deserves at least brief mention that filamentous fungi are without question ideal experimental organisms for a sophisticated molecular analysis of a number of extremely important and interesting biological phenomena. In these cases, quite ordinary molecular techniques plus the special techniques described above are employed to dissect significant questions, which are arguably best examined with fungi. Global regulatory circuits which allow selective expression of entire sets of unlinked genes which specify catabolic enzymes for carbon, nitrogen, sulfur, or phosphorus metabolism have been extensively studied in *Aspergillus* and in *Neurospora* (63–65). Many major and pathway-specific regulatory genes have been cloned and the function of their protein products in sequence-specific DNA binding studied (65–69).

Dunlap, Loros, and colleagues have made remarkable progress in a molecular genetic analysis of the circadian biological clock of *Neurospora* (70–73). This includes the isolation and characterization of genes that affect the periodicity of the clock (71) plus the identification and cloning of several clock-controlled genes (72, 73).

Timberlake, Adams, and colleagues (74–76) have used a molecular genetic approach to define a complex regulatory pathway that controls the development of the conidiophore of *Aspergillus.* Their elegant work has revealed that the sequential expression of three regulatory genes, *brlA, abaA,* and *wetA,* regulate conidiophore development, a complex differentiation and morphological process. Additional similar areas of considerable interest for which filamentous fungi are ideally suited include light regulation (77), mating type and incompatibility factors (78), aging and senescence (39), and complete molecular dissection of mitochondrial gene structure and function (79) as well as the nature and origin of introns (80, 81).

Postscript

Spectacular progress has been made in the understanding of many different phenomena in various filamentous fungi. The application of newly emerged molecular techniques, particularly the isolation of individual genes and detailed analysis, by gene manipulation *in vitro* and transformation back into the original host organism, has led to this revolution. It has provided precise answers to long-standing problems and has also identified entirely new and unexpected areas for investigation. This molecular revolution promises to continue to expand our knowledge and horizons, and to provide new insights with the many divergent filamentous fungi.

Acknowledgment

Research in the author's laboratory is supported by Grant GM23367 from the National Institutes of Health.

References

1. D. D. Perkins, *Genetics* **130,** 687 (1992).
2. D. D. Perkins, A. Radford, D. Newmeyer, and M. Bjorkman, *Microbiol. Rev.* **46,** 426 (1982).
3. S. S. Dhawale and A. C. Lane, *Nucleic Acids Res.* **21,** 5537 (1993).
4. R. H. Davis and F. de Serres, *in* "Methods in Enzymology" (H. Tabor and C. W. Tabor, eds.), Vol. 17A, p. 79. Academic Press, New York, 1970.
5. J. A. Hautala, B. H. Connder, J. W. Jacobson, G. L. Patel, and N. H. Giles, *J. Bacteriol.* **130,** 704 (1977).
6. M. N. Kanaan and G. A. Marzluf, *Mol. Gen. Genet.* **239,** 334 (1993).
7. J. A. Baum and N. H. Giles, *J. Mol. Biol.* **182,** 79 (1985).

8. J. A. Baum and N. H. Giles, *Proc. Natl. Acad. Sci. U.S.A.* **83,** 6533 (1986).
9. N. Brito, C. Gonzalez, and G. A. Marzluf, *J. Bacteriol.* **175,** 6755 (1993).
10. T. A. Harkness, R. L. Metzenberg, H. Schneider, R. Lill, W. Neupert, and F. E. Nargang, *Genetics* **136,** 107 (1994).
11. K. N. Subramanian, R. L. Weiss, and R. H. Davis, *J. Bacteriol.* **115,** 284 (1973).
12. R. L. Weiss, *J. Biol. Chem.* **248,** 5409 (1973).
13. R. H. Davis, *Microbiol. Rev.* **50,** 280 (1986).
14. R. L. Weiss and R. H. Davis, *J. Biol. Chem.* **248,** 5403 (1973).
15. R. L. Metzenberg and T. J. Baisch, *Neurospora Newslett.* **28,** 20 (1981).
16. M. Schechtman, *Fungal Genet. Newslett.* **33,** 45 (1986).
17. B. R. Oakley, J. E. Rinehart, B. L. Mitchell, C. E. Oakley, C. Carmona, G. L. May, and G. S. May, *Gene* **61,** 385 (1987).
18. C. E. Oakley, C. F. Weil, P. L. Kretz, and B. R. Oakley, *Gene* **53,** 293 (1987).
19. W. R. Reinert, V. B. Patel, and N. H. Giles, *Mol. Cell. Biol.* **1,** 29 (1981).
20. J. M. Chirgwin, A. E. Przybyla, R. J. MacDonald, and W. J. Rutter, *Biochemistry* **18,** 5294 (1979).
21. P. Chomczynski and N. Sacchi, *Anal. Biochem.* **162,** 156 (1987).
22. K. M. Lindgren, A. Lichens-Park, J. L. Loros, and J. C. Dunlap, *Fungal Genet. Newslett.* **37,** 21 (1990).
23. S. Osmani, G. W. May, and N. R. Morris, *J. Cell Biol.* **104,** 1495 (1987).
24. M. E. Case, M. Schweizer, S. R. Kushner, and N. H. Giles, *Proc. Natl. Acad. Sci. U.S.A.* **76,** 5259 (1979).
25. M. E. Case, *in* "Genetic Engineering of Microorganisms for Chemicals" (A. Hollaender, ed.), p. 87. Plenum, New York, 1982.
26. R. A. Akins and A. M. Lambowitz, *Mol. Cell. Biol.* **5,** 2272 (1985).
27. M. J. Orbach, E. B. Porro, and C. Yanofsky, *Mol. Cell. Biol.* **6,** 2452 (1986).
28. S. S. Dhawale, J. V. Paietta, and G. A. Marzluf, *Curr. Genet.* **8,** 77 (1984).
29. M. Schweizer, M. E. Case, C. C. Dykstra, N. H. Giles, and S. R. Kushner, *Proc. Natl. Acad. Sci. U.S.A.* **78,** 5086 (1981).
30. J. A. Kinsey and J. A. Rambosek, *Mol. Cell. Biol.* **4,** 117 (1984).
31. I. L. Johnstone, P. C. MacCabe, P. Greaes, S. J. Gurr, G. E. Cole, M. A. Brow, S. E. Unkles, A. J. Clutterbuck, J. R. Kinghorn, and M. A. Innis, *Gene* **90,** 181 (1990).
32. S. E. Unkles, E. I. Campbell, P. J. Punt, K. L. Hawker, R. Contreras, A. R. Hawkins, C. A. M. Van den Hondel, and J. R. Kinghorn, *Gene* **111,** 149 (1992).
33. Y. H. Fu and G. A. Marzluf, *Proc. Natl. Acad. Sci. U.S.A.* **84,** 8243 (1987).
34. P. M. Okamoto, Y. H. Fu, and G. A. Marzluf, *Mol. Gen. Genet.* **227,** 213 (1991).
35. B. Austin, R. M. Hall, and B. M. Tyler, *Gene* **93,** 157 (1990).
36. D. Cullen, S. A. Leong, L. J. Wilson, and D. J. Henner, *Gene* **57,** 21 (1987).
37. C. Staben, B. Jensen, M. Singer, J. Pollock, M. Schechtman, J. Kinsey, and E. Selker, *Fungal Genet. Newslett.* **36,** 79 (1989).
38. S. S. Dhawale and G. A. Marzluf, *Curr. Genet.* **10,** 205 (1985).
39. N. C. Mishra, *Adv. Genet.* **29,** 1 (1991).
40. D. Ebbole and M. S. Sachs, *Fungal Genet. Newslett.* **37,** 17 (1990).
41. M. J. Orbach, W. P. Schneider, and C. Yanofsky, *Mol. Cell. Biol.* **8,** 2211 (1988).
42. D. Ebbole, *Fungal Genet. Newslett.* **37,** 15 (1990).
43. E. U. Selker, E. B. Cambareri, B. C. Jensen, and K. A. Haack, *Cell (Cambridge, Mass.)* **51,** 741 (1987).

44. E. U. Selker and P. W. Garrett, *Proc. Natl. Acad. Sci. U.S.A.* **85,** 6870 (1988).

45. J. R. S. Fincham, I. F. Connerton, E. Notarianni, and K. Harrington, *Curr. Genet.* **15,** 327 (1989).

46. E. U. Selker, *Annu. Rev. Genet.* **24,** 579 (1990).

47. P. M. Okamoto, R. H. Garrett, and G. A. Marzluf, *Mol. Gen. Genet.* **238,** 81 (1993).

48. H. M. Foss, C. J. Roberts, K. M. Claeys, and E. U. Selker, *Science* **262,** 1737 (1993).

49. S. J. Free, P. W. Rice, and R. L. Metzenberg, *J. Bacteriol.* **137,** 1219 (1979).

50. R. Krumlauf and G. A. Marzluf, *J. Biol. Chem.* **255,** 1138 (1980).

51. R. L. Metzenberg, J. N. Stevens, E. U. Selker, and E. Morzycka-Wroblewska, *Proc. Natl. Acad. Sci. U.S.A.* **82,** 2067 (1985).

52. P. J. Russell, R. C. Petersen, and S. Wagner, *Mol. Gen. Genet.* **211,** 541 (1988).

53. S. Marathe, I. F. Connerton, and J. R. S. Fincham, *Mol. Cell. Biol.* **10,** 2638 (1990).

54. G. Jarai and G. A. Marzluf, *Curr. Genet.* **20,** 283 (1991).

55. J. V. Paietta and G. A. Marzluf, *Mol. Cell. Biol.* **5,** 1554 (1985).

56. J. H. Kinnaid, M. A. Keighren, J. A. Kinsey, M. Eaton, and J. R. S. Fincham, *Gene* **20,** 387 (1982).

57. V. Berlin and C. Yanofsky, *Mol. Cell. Biol.* **5,** 849 (1985).

58. G. E. Exley, J. D. Colandene, and R. H. Garrett, *J. Bacteriol.* **175,** 2379 (1993).

59. J. V. Paietta, *Mol. Cell. Biol.* **9,** 3630 (1989).

60. S. J. Vollmer and C. Yanofsky, *Proc. Natl. Acad. Sci. U.S.A.* **83,** 4869 (1986).

61. R. L. Metzenberg, J. N. Stevens, E. U. Selker, and E. Morzycka-Mroblewska, *Neurospora Newslett.* **31,** 35 (1984).

62. R. L. Metzenberg and J. S. Grotelueschen, *Fungal Genet. Newslett.* **39,** 50 (1992).

63. P. Kulmburg, D. Sequeval, F. Lenouvel, M. Mathieu, and B. Felenbok, *Mol. Cell. Biol.* **12,** 1932 (1992).

64. G. A. Marzluf, *Annu. Rev. Microbiol.* **47,** 31 (1993).

65. S. Kang and R. L. Metzenberg, *Mol. Cell. Biol.* **10,** 5839 (1990).

66. G. Burger, J. Strauss, C. Scazzocchio, and B. Lang, *Mol. Cell. Biol.* **11,** 5746 (1991).

67. Y. H. Fu and G. A. Marzluf, *J. Biol. Chem.* **265,** 11942 (1990).

68. Y. H. Fu and G. A. Marzluf, *Proc. Natl. Acad. Sci. U.S.A.* **87,** 5331 (1990).

69. P. Kulmburg, N. Judewicz, M. Mathieu, F. Lenouvel, D. Sequeval, and B. Felenbok, *J. Biol. Chem.* **1267,** 21146 (1992).

70. J. C. Dunlap, *Trends Genet.* **6,** 159 (1990).

71. C. R. McClung, B. A. Fox, and J. C. Dunlap, *Nature (London)* **339,** 558 (1989).

72. J. J. Loros, S. A. Denome, and J. C. Dunlap, *Science* **243,** 385 (1989).

73. J. J. Loros and J. C. Dunlap, *Mol. Cell. Biol.* **11,** 558 (1991).

74. T. C. Sewall, C. W. Mims, and W. E. Timberlake, *Plant Cell* **2,** 731 (1990).

75. S. Han, J. Navarro, R. A. Greve, and T. H. Adams, *EMBO J.* **12,** 2449 (1993).

76. A. Andrianopoulos and W. E. Timberlake, *Mol. Cell. Biol.* **14,** 2503 (1994).

77. V. Y. Sokolovskyh, F. R. Lauter, B. Mueller-Rober, M. Ricci, T. J. Schmidhauser, and V. E. Russo, *J. Gen. Microbiol.* **138,** 2045 (1992).

78. N. L. Glass and G. A. Kuldau, *Annu. Rev. Phytopathol.* **30,** 201 (1992).

79. U. Kamper, U. Kuck, A. D. Cherniack, and A. M. Lambowitz, *Mol. Cell. Biol.* **12,** 499 (1992).

80. R. Saldanha, G. Mohr, M. Belfort, and A. M. Lambowitz, *FASEB J.* **7,** 15 (1993).

81. A. M. Lambowitz and M. Belfort, *Annu. Rev. Biochem.* **62,** 587 (1993).

Section II

Yeast Chromosomes: Transcription, Recombination, Replication

[5] *In Vitro* Fragmentation of Yeast Chromosomes and Yeast Artificial Chromosomes at Artificially Inserted Sites and Applications to Genome Mapping

Hervé Tettelin, Agnès Thierry, Cécile Fairhead, Arnaud Perrin, and Bernard Dujon

Introduction

Techniques for large-scale genome mapping, cloning, and sequencing have been developed in the baker's yeast, *Saccharomyces cerevisiae,* a particularly suitable model and tool for such experiments for several reasons. First, yeast chromosomes can easily be prepared intact in agarose plugs and separated and purified by pulsed-field gel electrophoresis (PFGE). Second, integrative transformation mediated by homologous recombination is so simple and efficient that yeast chromosomes or genes can be modified as desired. Third, techniques applied on the yeast genome are of general interest because of the use of yeast artificial chromosomes (YACs) as vectors for cloning large exogenous genomic fragments.

The discovery of endonucleases cutting very rarely (meganucleases) and encoded by mobile group I introns has permitted us to develop new tools for genome mapping that are particularly well suited to the case of yeast chromosomes and YACs. We report here the use of intron-encoded endonucleases for site-directed chromosomal fragmentation and its applications to physical mapping of intact yeast chromosomes, YACs, or fragments thereof. The methods described include integration of I-*Sce*I recognition sites in yeast chromosomes and in YACs, preparation of yeast DNA in agarose plugs for PFGE after I-*Sce*I and/or I-*Ppo*I digestion, purification of chromosome fragments and labeling of chromosomal DNA probes in agarose plugs, and hybridization of cosmid, plasmid, or λ clones with chromosome fragments. Examples of applications of such methods are summarized.

Basics of Yeast Integrative Transformation

Because yeast achieves very efficient homologous recombination, transforming DNA can be easily inserted within yeast chromosomes or YAC inserts, provided a segment homologous to the target sequence is carried on the transforming vector. The vector is a yeast integrative plasmid (YIp), that is, a bacterial plasmid containing a yeast

selectable marker but no yeast replication origin (called ARS for autonomously replicating sequence) (1). The homologous segment may be relatively short, but the
efficiency of recombination tends to be reduced if it is shorter than 200 nucleotides.

The presence of a double strand break in the transforming DNA within the region
of homology significantly increases the frequency of recombination (2). Thus, it is
recommended whenever possible to linearize the vector *in vitro* within the region of
homology prior to transforming yeast. This can easily be achieved if there are appropriate restriction sites in the target sequence homolog (TSH). Two topologies can be
used: "ends-in" and "ends-out" (Fig. 1).

The ends-in topology is the easiest one for construction of transforming DNA. A
TSH is inserted into a YIp. The restriction site used to cleave within the TSH prior
to transformation should be chosen so that it is absent from the vector sequence. After
transformation and selection of yeast clones (see below), the entire transforming

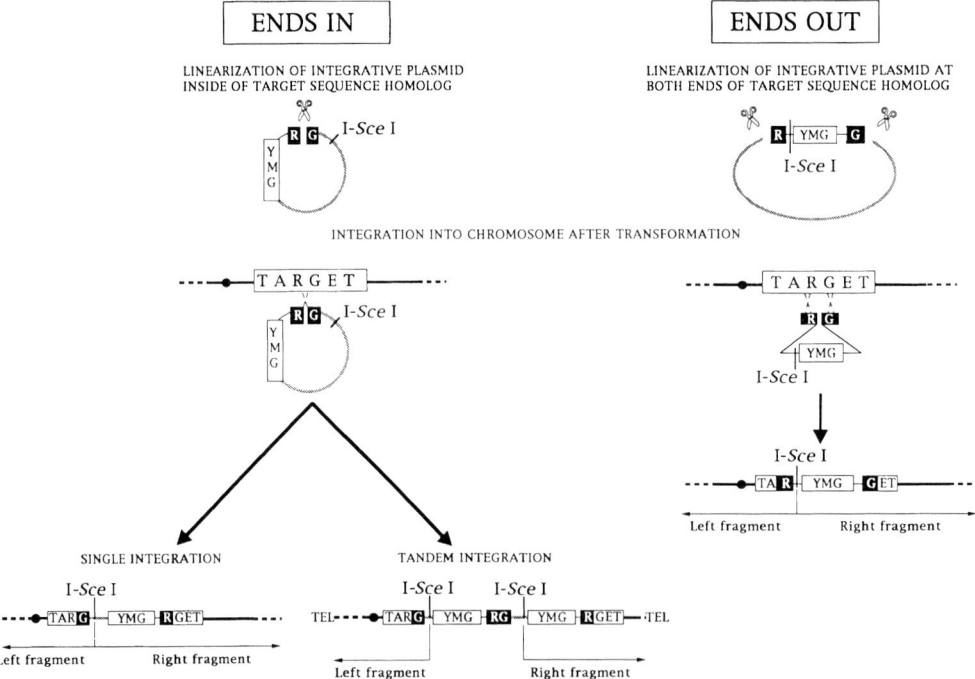

FIG. 1   Topologies used for yeast integrative transformation using an I-*Sce*I cassette. Yeast
chromosomes are shown as black lines with a black circle to represent the centromere. The
targeted sequence is boxed. Transforming DNA is shown as YMG (yeast marker gene) and
RG (target sequence homology) sequences.

DNA becomes integrated at a given site of the yeast chromosome or YAC insert, and the homologous region forms a direct repeat with the resident yeast sequence. Such transformants are essentially stable. However, reciprocal recombination between the two direct repeats can result in looping out of the transforming DNA at a frequency depending on the size and nature of the repeats (3). In such a case, reciprocal exchange occurs between the chromosome and the plasmid which may result in modification of the chromosomal sequence if the TSH is not completely identical to the chromosomal target. Conversely, the chromosomal sequence can be retrieved on the excised plasmid (plasmid rescue), which is also a useful trick.

The ends-out topology requires one more step to construct transforming DNA, but it results in completely stable transformants in which a segment of a yeast chromosome or YAC is replaced by the transforming DNA. Construction involves a first step in which the TSH is inserted into a bacterial plasmid and a second step in which the cassette containing the yeast selectable marker is inserted into the TSH. The plasmid must be linearized at both ends of the TSH before transformation of yeast.

A major condition has to be fulfilled for successful integrative transformation: the transforming DNA should not contain any active ARS, as the presence of a single ARS is sufficient to warrant autonomous replication to the transforming DNA on recircularization *in vivo*. An ARS occurs every 40 kb in the yeast nuclear genome on average. In addition, such sequences have been found in most exogenous DNAs (1). If the region where integration is sought is not characterized, ARS can only be diagnosed *a posteriori* (see later). If an ARS happens to be present in the chosen TSH, a neighboring segment should be used as TSH instead. If the site of integration cannot be displaced, the ARS should be mapped and inactivated.

Properties of Intron-Encoded Meganucleases

Meganucleases are a novel class of double-strand cutting endonucleases that recognize very long target sites (typically 15–18 bp), as opposed to bacterial restriction endonucleases which have recognition sites of 8 bp at the most. Such enzymes were originally discovered in mobile group I introns (4). At present, several meganucleases have been characterized. They belong to at least three different protein families, defined from their internal amino acid motifs and from the precise nature of the cuts they generate (5). The major family is composed of proteins with two very well conserved dodecapeptide motifs. They generate 4-nucleotide staggered cuts with 3'-OH overhangs within their nonpalindromic recognition sequences.

The biological role of such endonucleases is to propagate the group I intron that codes for them by a mechanism called intron homing (4). Some intron encoded endonucleases, including I-*Sce*I [the first one to be discovered in this family, (6)] and I-*Ppo*I (7), are extremely stringent and can recognize a single site in the entire yeast genome. Others, such as I-*Sce*II (8), are less specific as their tolerance for degeneracy

is higher. Neither the biological nor the biochemical reasons for such a high specificity are understood at present, but the existence of the nucleases offers a new type
of tool for genome mapping.

Proteins belonging to the same dodecapeptide-containing family have also been
found independently of introns. Some are expressed out of regular genes [HO (9),
Endo-*Sce*I (10)], and others [PI-VDE I (11), PI-*Tli*I (12)] are protein inserts from a
variety of genes. Some are also highly specific endonucleases that can be used *in
vitro.*

As I-*Sce*I was available before the others, we have developed a set of mapping
tools with this nuclease. I-*Sce*I cleaves DNA *in vitro* after a two-step recognition of
its site (5). Assay conditions to use the enzyme on large DNA molecules have been
optimized (13). Under such conditions, there is no natural I-*Sce*I site in the entire
yeast nuclear genome. When an artificial site is introduced in a chromosome, complete cleavage at that site can be obtained without digestion of other chromosomes
(14). On the basis of this principle, we have developed a strategy which allows fragmentation of yeast chromosomes at any desired location. This strategy has also been
extended to YAC inserts.

I-*Sce*I and other such endonucleases, like I-*Ppo*I and I-*Ceu*I (15), are encoded by
group I introns of rDNA genes and have been selected by evolution to cleave rDNA
of their host organism. However, I-*Sce*I cleaves at a site which is sufficiently different
between yeast mitochondrial rDNA and nuclear rDNA of most species; hence, it does
not cleave yeast nuclear rDNA. By contrast, I-*Ppo*I cleaves at a site in *Physarum
polycephalum* nuclear rDNA which happens to be present in yeast nuclear rDNA as
well. This offered us a tool to cleave rDNA repeats of yeast. This is useful because
free circular copies of rDNA (3 μm) exist in yeast cells [resulting from looping out
of chromosome XII (16)] and tend to contaminate preparations of other chromosome
fragments or YACs.

Strains and Growth of Cells

Escherichia coli Hosts

All integrative vectors used for yeast transformation are propagated in *Escherichia
coli* strain TG1 [*supE, hsd*Δ5, *thi,* Δ(*lac-proAB*), *F'* (*traD36, proAB$^+$, lacI^Q,
lacZ*ΔM15] (17).

Saccharomyces cerevisiae Strains

A series of isogenic strains derived from S288C and carrying different combinations
of auxotrophic requirements useful for transformation has been constructed (Table I).

TABLE I List of *Saccharomyces cerevisiae* strains[a]

Strain	Genotype	Origin of strain	Strain construction
S288C	MATα mal gal2	YGSC Berkeley	
FY23	MATa ura3-52 trp1Δ63 leu2Δ1 GAL2+	Fred Winston	
FY73	MATα ura3-52 his3Δ200 GAL2+	Fred Winston	
FY1679	MATa/MATα ura3-52/ura3-52 trp1Δ63/+ leu2Δ1/+ his3Δ200/+ GAL2+/GAL2+	This work	Diploid from FY23 × FY73 cross
FY1679-18B	MATα ura3-52 trp1Δ63 leu2Δ1 his3Δ200 GAL2+	This work	Meiosis of FY1679
FY1679-28C	MATa ura3-52 trp1Δ63 leu2Δ1 his3Δ200 GAL2+	This work	Meiosis of FY1679
FY1679-28A	MATα ura3-52 trp1Δ63 leu2Δ1 GAL2+	This work	Meiosis of FY1679
FY1679-1D	MATα ura3-52 trp1Δ63 his3Δ200 GAL2+	This work	Meiosis of FY1679
FY1679-5D	MATα ura3-52 trp1Δ63 his3Δ200 GAL2+	This work	Meiosis of FY1679
FY1679-1C	MATa ura3-52 leu2Δ1 his3Δ200 GAL2+	This work	Meiosis of FY1679
FY1679-5C	MATa ura3-52 leu2Δ1 his3Δ200 GAL2+	This work	Meiosis of FY1679
FY1679-18D	MATα ura3-52 his3Δ200 GAL2+	This work	Meiosis of FY1679
FY1679-5A	MATα ura3-52 trp1Δ63 GAL2+	This work	Meiosis of FY1679
FY1679-1B	MATa ura3-52 trp1Δ63 GAL2+	This work	Meiosis of FY1679
FY1679-6A	MATα ura3-52 leu2Δ1 GAL2+	This work	Meiosis of FY1679
FY1679-1A	MATa ura3-52 leu2Δ1 GAL2+	This work	Meiosis of FY1679
FY1679-18C	MATα ura3-52 GAL2+	This work	Meiosis of FY1679
FY1679-28B	MATa ura3-52 GAL2+	This work	Meiosis of FY1679

Note: YGSC, yeast genome stock center.

[a] All strains are isogenic to S288C except for the markers indicated. *ura3-52* is a Ty insertion (reversion rate $<10^{-8}$) [M. Rose and F. Winston, *Mol. Gen. Genet.* **193**, 557 (1984)].

TABLE II   Set of Transgenic Strains with *I-Sce*I Sites Used for Genome Mapping

Chromosome[a]	Strain[b,c]	Integrative vector[d]	Disrupted gene	Ref.
IV	FY23/RD112			
	FY23/RD248			
XV	FY23/RO181(*)			
	FY1679/O321	pAF412		
	FY1679/O272	pAF403		
	FY1679/O471	pAF421		
	FY1679/O306	pAF411		
	FY1679/O477	pAF426		
	FY1679/O491	pAF430		
	FY1679/O323	pAF416		
	FY1679/O497	pAF435		
VII	FY23/RG007			
	FY23/RG162			
	FY1679/G05	pAF453	*cdc43*::*URA3*–I-*Sce*I	
	FY1679/G06	pAF454	*cdc20*::*URA3*–I-*Sce*I	
	FY1679/G11	pAF460	*dst1*::*URA3*–I-*Sce*I	
	FY1679/G09	pAF457	*pdr1*::*URA3*–I-*Sce*I	
	FY1679/G12	pAF461	*gcd2*::*URA3*–I-*Sce*I	
	FY1679/G13	pAF462	*spt6*::*URA3*–I-*Sce*I	
	FY1679/G15	pAF465	*hip1*::*URA3*–I-*Sce*I	
	FY1679/G19	pAF471	*mes1*::*URA3*–I-*Sce*I	
XVI	FY23/RP142			
	FY23/RP270			
XIII	FY23/R255			
II	FY23/RB231(*)			
XIV	FY23/RN208(*)			
X	FY1679/JC41	pC41		*e*
	FY1679/JC82	pC82		*e*
	FY1679/JT21	pAF304	*tif2*::*URA3*–I-*Sce*I	
XI	FY1679/KA302	pAF302	*fas1*::*URA3*–I-*Sce*I	*f*
	FY1679/KD304	pAF304	*tif1*::*URA3*–I-*Sce*I	*f*
	FY1679/KE40	pAF305		*f*
	FY1679/KG41	pAF306	*mak11* or *cdc16*::*URA3*–I-*Sce*I	*f*
	FY1679/KH81	pAF021		*f*
	FY1679/KM57	pAF307		*f*
	FY1679/KT62	pAF308		*f*
V	FY23/R216(*)			
VIII	FY23/R165			
IX	FY23/RI117			
	FY1679/I20	pAF473	*suc2*::*URA3*–I-*Sce*I	
III	FY23/CC6	pAF201	*ycr034W*::*URA3*–I-*Sce*I	*g*
	FY73/CC11	pAF201	*ycr034W*::*URA3*–I-*Sce*I	*g*
	FY1679/CC14	pAF201	*ycr034W*::*URA3*–I-*Sce*I	*g*
	FY1679/CC23	pAF202	*ycr035C*::*URA3*–I-*Sce*I	*g*
	FY1679/CA34	pAF228	*ycr037W*::*URA3*–I-*Sce*I	*g*

(continued)

TABLE II   *(continued)*

Chromosome[a]	Strain[b,c]	Integrative vector[d]	Disrupted gene	Ref.
VI	FY23/RF222(*)			
I	FY23/RA170			

[a] Yeast chromosomes carying I-*Sce*I cassettes are listed in order of decreasing size (chromosome XII is omitted).

[b] Nomenclature for transgenic strains indicates transformed host strain (refer to Table I) followed, after the slash, by a combination of letters and numbers to specify the integrated site. Cassettes inserted at random are indicated with R as the first letter.

[c] Asterisk (*) indicates reduced mobility on PFGE.

[d] Integrative vectors are derived from pAF101 or pAF109 (see Fig. 2).

[e] M.-E. Huang, J.-C. Chuat, A. Thierry, B. Dujon, and F. Galibert, DNA Sequence, *The Journal of Sequencing and Mapping,* **4**, 293 (1994).

[f] A. Thierry, A. Perrin, J. Boyer, C. Fairhead, and B. Dujon, *Nucleic Acids Res.* **19**, 189 (1991).

[g] A. Thierry, C. Fairhead, and B. Dujon, *Yeast* **6**, 521 (1990).

The strains can be easily transformed to integrate cassettes into their chromosomes. Strain FY1679 is used for sequencing chromosomes XI, X, VII, XV, XIV, and IV in the European yeast sequencing program (18). Transgenic strains containing I-*Sce*I sites integrated at various locations into their chromosomes and used for mapping and purification of chromosome fragments are listed in Table II. The YACs are usually propagated in strain AB1380 (*MATα, ade2-1, can1-100, lys2-1, trp1, ura3, his5*) (19).

Growth Media and Conditions

The composition and preparation of bacterial and yeast media are given in Tables III and IV, respectively. Bacteria are grown at 37° C on plates or in broth with aeration.

TABLE III   Composition and Preparation
of Standard Bacterial Medium (LB)[a]

Component	Amount (%, w/v)
Bacto-tryptone (Difco)	1
Bacto-yeast extract (Difco)	0.5
NaCl	1
Bacto-agar (Difco)	2 (for solid medium)

[a] Components (available from Difco, Detroit, MI) are dissolved in distilled water and autoclaved 20 min at 120° C. For solid medium, cool to around 60° C and pour plates. For LB medium containing ampicillin, 50 μg/ml ampicillin is added after cooling to around 60° C.

TABLE IV   Composition and Preparation of Yeast Media[a]

Medium	Component	Amount (%, w/v)
YPGlu	Bacto-peptone	1
	Bacto-yeast extract	1
	D-Glucose	2
	Bacto-agar	2 (for solid medium)
WO	Yeast nitrogen base without amino acids (Difco)	0.67
	D-Glucose	2
	Bacto-agar	2 (for solid medium)
SC	Same as WO plus 0.2% (w/v) amino acid mix[b]	
SC − X, Y ...	Same as SC but use amino acid mix[b]	
(selective medium)	lacking indicated ingredients X, Y ...	
5-FOA	Yeast nitrogen base without amino acids (Difco)	0.67
	D-glucose	2
	5-Fluoroorotic acid	0.1
	Amino acid mix[b]	0.2
	Bacto-agar	2 (for solid medium)

[a] For preparation of media, components are dissolved in water and autoclaved 20 min at 110°C. For solid medium, cool to around 60°C and pour plates.

[b] To prepare amino acid mix, weigh appropriate amounts of each of the 20 L-amino acids plus adenine and uracil in the following proportions, mix thoroughly, and keep in a dry cool place. For selective media, one component (or several) is omitted.

Amino acid	Weight proportion
L-Leucine	2
All other L-amino acid	1
Adenine	1
Uracil	1

Yeast cells are grown at 30°C on plates or in broth with aeration. Yeast colonies take 2 days to form on complete medium and 3 days on selective medium. Yeast transformants take 3 to 5 days to form colonies on selective medium.

Basic Vectors Containing I-SceI–URA3 Cassette

Vectors are shown in Fig. 2. A standard vector, pAF101, containing the I-SceI site and the URA3 selectable marker was constructed from pUC19 (20). This vector con-

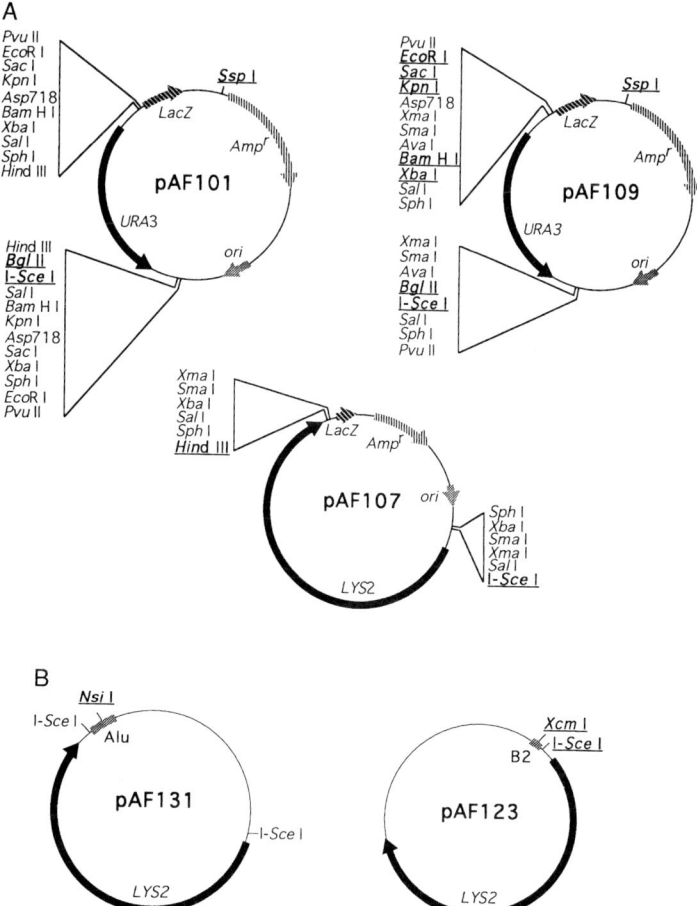

FIG. 2   Maps of basic integrative vectors. (A) Vectors for integration of the I-*SceI–URA3* or I-*SceI–LYS2* cassettes. All vectors are pUC19 derivatives. Unique sites are underlined and appear in bold type. Duplicated sites available to prepare the cassettes are also indicated. Absent sites from pAF101 and pAF 109 are those for *BclI*, *ClaI*, *EagI*, *HpaI*, *NheI*, *NruI*, *SacII*, *SnaBI*, and *XhoI*. Absentsites from PAF107 are those for *ApaI*, *ClaI*, *EagI*, *MluI*, *NotI*, *NsiI*, *PacI*, and *XcmI*. (B) Vectors for multiple integration of the I-*SceI–LYS2* cassette into YACs. All vectors are pBluescriptSK+ derivatives (Stratagene, La Jolla, CA). Unique sites are underlined. pAF123 contains the *B2* repetitive element from mouse DNA; pAF131 contains the *Alu* repetitive element from human DNA in which an *NsiI* site has been inserted by *in vitro* mutagenesis.

tains the whole pUC19 multiple cloning site, duplicated on either side of the *URA3* gene. It can be used for both ends-in and ends-out constructions. For the ends-out topology, the I-*SceI*–*URA3* cassettes can be prepared with a variety of sticky ends. For the ends-in topology, two unique cloning sites (*Bgl*II and *Ssp*I) are available. Therefore, another vector, pAF109, was constructed by cloning the pAF101 *Sal*I cassette at the *Sal*I site of pUC19. pAF109 contains seven different cloning sites for ends-in constructions. Absent sites in pAF101 and pAF109 can be used to linearize TSH. [*Note: Cla*I (an absent site in both vectors) is sensitive to methylation which occurs in *dam*+ *E. coli* strains if its recognition site ATCGAT is preceded by a G or followed by a C. *Nru*I (also an absent site in both vectors) is sensitive to methylation which occurs if its recognition site TCGCGA is preceded by GA or followed by TC.]

Because most YAC clones already available are propagated into the yeast strain AB1380 and use the two selectable markers *URA3* and *TRP1,* cassettes carrying the *LYS2* marker were constructed. pAF107 is similar to pAF101 except that the *LYS2* gene replaces the *URA3* gene and the sites flanking the cassette are partly different. Absent sites in pAF107 are *Apa*I, *Cla*I, *Eag*I, *Mlu*I, *Not*I, *Nsi*I, *Pac*I, and *Xcm*I. Two vectors have been constructed for random integration into YACs containing human or murine inserts, respectively. pAF123 contains a mouse *B2* repeat element whose linearization site is *Xcm*I. pAF131 contains a human *Alu* repeat element whose linearization site is an artificially introduced *Nsi*I site.

Construction of Transforming DNA

Integration of I-SceI Cassette into Yeast Chromosomes

Depending on desired applications, integration of the cassette in the yeast genome can be made into a predetermined site of interest, into random locations along a given chromosome, or at random in the entire yeast genome.

Directed Integration of I-SceI Cassette in Given Yeast Chromosome Using Known Genes or Sequences, or Random Fragments

To target the cassette into a predetermined site in a yeast chromosome, a segment homologous to the targeted site must be cloned in the ends-in or ends-out topology (see Fig. 1). The segment can be either a polymerase chain reaction (PCR) amplification product, or a cloned chromosome fragment or gene. Alternatively, the cassette can be integrated at an *a priori* unknown position along a given chromosome by using unmapped chromosome fragments or genes. In all cases, the TSH must be chosen so that after construction of the transforming plasmid from pAF101 or pAF109, unique sites remain for linearization inside the TSH (ends-in) or at both ends (ends-out).

Target Sequence Homolog Already Cloned into Bacterial Vector

For "ends-in" topology, approximately 50 ng of the fragment used as TSH is digested by *Bgl*II or isocaudamers for cloning into pAF101, or by *Eco*RI, *Bgl*II, etc., or isocaudamers for cloning into pAF109. DNA is purified on an agarose gel and ligated to 50 ng of dephosphorylated plasmid DNA digested with *Bgl*II (or other enzyme). For ends-out topology, the TSH must first be cloned into a small bacterial vector. Then, 50 ng of the cassette is prepared from pAF101 or pAF109 by digestion and gel purification and ligated to 50 ng of vector which has been linearized inside the cloned TSH. The resulting inserted cassette must have regions of homology to the chromosomal target of at least 100 bp on each side. Ligated DNA is transformed into TG1 cells and transformants are selected on LB–ampicillin plates. Transformant clones containing the appropriate recombinant plasmid are verified by restriction digests.

Target Sequence Homolog Obtained by Polymerase Chain Reaction Amplification

Fragments of, typically, 400 to 1000 bp containing a restriction site which will be used for linearization in the TSH prior to yeast transformation (ends-in topology), or which allows insertion of the yeast selectable marker within the TSH (ends-out), can be prepared by PCR amplification from total yeast DNA (see later section on preparation of total yeast DNA). Primers must contain 5' extensions including appropriate restriction sites for subsequent cloning. For example, if using pAF101 in the ends-in topology, a *Bam*HI site is most appropriate (isocaudamer of *Bgl*II). (Remember that the site should not be at the very end of the fragment for efficient digestion. For example, 5' CGC<u>GGATCC</u>. . . 3' is recommended by New England Biolabs, Beverly, MA). Fifty nanograms of the PCR product is digested by the enzyme(s) whose site(s) is present in the primers, then ligated to 50 ng of linearized, dephosphorylated bacterial vector. In the ends-out topology, an additional step is required: the cassette is extracted from pAF101 or pAF109 and ligated as above. Bacterial transformation and selection are performed as described above.

Random Integration of I-SceI Cassette in Total Yeast Genome

I-*Sce*I cassettes can be integrated at random into the entire yeast genome in a single transformation experiment, using a shotgun library of total yeast DNA cloned directly into pAF101, pAF109, or pAF107 in the ends-in topology. The method used in our laboratory is as follows: 15 μg of genomic DNA from the FY1679 strain is digested by *Bam*HI and ligated with 300 ng of *Bgl*II-digested, dephosphorylated pAF101 DNA. *Bam*HI generates fragments of an average size of 5 kb in yeast. Ligated DNA is transformed into TG1 cells and transformants are selected on LB–ampicillin plates. A total of 2000 independent transformant colonies are carefully mixed, and plasmid DNA is extracted. Twenty micrograms of this DNA is digested by *Hpa*I and used to transform yeast strain FY23. *Hpa*I sites occur every 3 kb on the average in yeast. It is assumed that the linearized TSH should be integrated into the yeast genome more efficiently than other inserts lacking the restriction site. Yeast transformants are selected on SC − uracil plates. Results of this experiment are described below (see Applications). A selected set of strains containing I-*Sce*I sites in all chromosomes is listed in Table 2. Such strains are useful to separate chromosomes or prepare fragments (see Fig. 4 and later).

Integration of I-SceI Cassette into Yeast Artificial Chromosome Inserts

Homologous recombination can also be used to integrate I-*Sce*I cassettes into YAC inserts, provided a segment of the YAC insert is cloned into a yeast integrative vector. The segment can be either a unique sequence of the YAC insert (a regular TSH) or a repetitive element. In the first case, the pAF107 vector can be used to prepare the I-*Sce*I cassette as described above. In the second case, vectors pAF123 and pAF131 have been constructed to transform any yeast clone containing a mouse or a human YAC insert, respectively. No TSH cloning is needed, as the *B2* or *Alu* sequence will serve as TSH on transformation. After a single yeast transformation, the I-*Sce*I cassette is inserted at the various *B2* or *Alu* sequences along the YAC insert, one per clone.

Yeast Transformation Procedure

Prior to yeast transformation, the transforming DNA must be linearized. For the ends-in topology, the site of linearization is unique and located within the TSH. For the ends-out topology, the transforming DNA is cut out from the plasmid using two restriction sites located at both ends of the TSH.

Although any yeast strain with appropriate markers can be used, we recommend yeast strains FY1679, FY23, and FY73, or isogenic derivatives, for which integrative vectors of the pAF101 and pAF109 type were constructed and which are being used in the European program for sequencing the yeast genome. In the case of YACs, the host yeast strain is generally AB1380.

The quantity of transforming DNA is very important. For practical reasons, many routine experiments are carried out with relatively large amounts of transforming DNA (∼1 to 3 μg). Systematic analysis of the effects of DNA concentration on the structure of the transformants has been done for the ends-in topology (21). An excess of transforming DNA results in tandem-arrayed integrations at the target site. In diploids, the tandem sequences are generally all clustered into one of the two chromosome homologs only. Similarly, in cases of duplicated targets in the yeast genome (paralogous genes), or in cases of multiple targets in YACs (repeated sequences), integration occurs at one site per nucleus. We have observed that the integration of I-*Sce*I sites sometimes results in the retardation in PFGE of the chromosome carrying the site. The retardation corresponds to an apparent size increase varying from a few dozen kilobases to around 100 kb, consistent with a high number of tandem repeats of the insert. For most applications, tandem arrays do not affect subsequent mapping if the I-*Sce*I digestion is complete (see Fig. 1). To avoid multiple integration in tandem arrays, the concentration of transforming DNA should be kept minimal.

Using the electroporation method, the frequency of transformants obtained with a given transforming DNA and a given yeast strain is a very reproducible figure. The appropriate concentration of transforming DNA can easily be determined from a saturation curve where the number of transformants per microgram of transforming DNA is plotted as the function of the DNA concentration. At concentrations corresponding to the linear portion of the curve, the risk of multiple tandem integration is very limited. However, the frequency of transformants for a given DNA concentration also varies with the recipient yeast strain and the nature of the transforming DNA (both the length and the nature of the homologous region play a role).

The following electroporation protocol is adapted from Meilhoc *et al.* (22) and gives satisfactory and reproducible results.

1. Grow yeast cells in 50 ml of YPGlu (Table IV) at $30°C$ with aeration until the culture reaches about 10^7 cells/ml or an optical density (OD) of $0.8–1$ (cells are counted with optical microscope or the OD of the culture is measured).
2. Centrifuge cells 5 min at 3000 *g,* resuspend in 25 m*M* dithiothreitol (DTT), and incubate 10 min at $30°C$.
3. Centrifuge cells 5 min at 3000 *g* and resuspend at a concentration of 10^9 cells/ml in a solution of 0.27 *M* sucrose, 10 m*M* Tris-HCl, pH 7.5, 1 m*M* $MgCl_2$.
4. Place 100 μl of cell suspension into 0.2-cm electroporation cuvettes (Bio-Rad, Richmond, CA, Cat. No. 165-2086).
5. Add linearized vector DNA (usually $0.1–3$ μg) and perform electroporation as shown in Table V.
6. Immediately after electroporation, pipette out cells and plate on SC − uracil me-

TABLE V   Electric Parameters for Electroporation of Yeast Cells Using Small Plasmids[a]

Electroporator	Electric wave	Parameters	
Bio-Rad	Exponential decrease	Electric field between electrodes:	2250 V/cm
		Total capacitance discharge:	250 μF
		External derivation:	200 Ω
		Time constant:	30 to 35 msec
Jouan	Square	Electric field between electrodes:	2500 V/cm
		Wave time:	30 to 35 msec

[a] Other electroporators have not been tested by the authors.

dium (if using a *URA3* cassette) or SC − lysine medium (*LYS2* cassette) (see Table IV for medium composition).

7. Incubate plates at 30°C for 4 to 5 days until colonies appear.
8. Pick colonies individually using sterile toothpicks or a sterile platinum loop and streak on SC − uracil or SC − lysine agarose plates for subclone isolation.
9. Incubate plates at 30°C for 3 to 4 days. Pick single clones for further experiments.

Note: For integrative transformation using a single TSH, it should be sufficient to pick up only one colony in step 8. In practice, we recommend picking four colonies for most routine experiments. (This should be enough even if genes are duplicated.) If the target is repeated (Ty or δ elements for yeast, mammalian repeats for YACs) and the goal is to cover a whole chromosome or YAC with cassettes, many colonies are obviously needed. If the TSH is uncharacterized and may contain an ARS, a dozen colonies should be picked and screened genetically as described below.

Genetic Screening of Yeast Transformants

Homologous recombination is so efficient that genetic screening is not necessary for integrative transformants and molecular screening should be applied directly. However, in cases of an unknown TSH that may contain an ARS, it may be useful to distinguish between stable (integrative) and unstable (replicative) transformation. If an ARS is present on the transforming DNA, the DNA will often recircularize instead of integrating. (Integration is possible, however, but at a lower efficiency.) Transformants obtained will easily lose the plasmid, and this can be detected by loss of the selected phenotype. Streak transformed clones on complete medium and grow for about 20 generations (a colony size). Plate each clone on complete medium. When colonies have appeared, replica-plate on SC − uracil or SC − lysine.

In the case of integration into YACs, a genetic screening is recommended prior to molecular verification. Because clones transformed with pAF131 or pAF123 are selected on SC − lysine medium, it is necessary to reisolate them first on SC − lysine and uracil medium to select for the maintenance of the YAC itself. In the case of random integration into repeated sequences, a sufficient number of independent transformants must be picked (typically, 30 to 50 transformants for around 500 kb YAC inserts, if *B2* or *Alu* sequences are used). The fraction of correct integrants can easily be screened by testing the simultaneous loss of the markers *URA3* and *LYS2*. This is achieved by plating each transformant in parallel on SC − lysine (to store the clones) and on plates containing 5-fluoroorotic acid (5-FOA), which select for Ura⁻ cells. Plates are incubated at 30°C for 2–3 days. The 5-FOA^R colonies are picked and subcloned on YPGlu plates. Subclones are then tested on SC − uracil, SC − lysine, and SC − tryptophan media. All transformants that become simultaneously

Ura$^-$, Lys$^-$, and Trp$^-$ must have integrated the cassette into the YAC. They are picked from the SC–LYS plates, stored at 4°C, and used for further experiments.

Molecular Verification of Integrative Transformants by Southern Blot Analysis

In all cases, integration of cassettes must be verified at the molecular level. Southern blot analysis is the most convenient method. Various techniques needed at this point are listed below. Each of the techniques is obviously of broader application.

Small-Scale Preparation of Total Yeast DNA

1. Grow strains in YPGlu at 30°C to stationary phase.
2. Centrifuge a 1.5-ml culture aliquot ($3-5 \times 10^8$ cells) at 3000 g for 5 min at 4°C.
3. Resuspend cells in 200 μl of solution I [50 mM sodium phosphate buffer (0.5 M sodium phosphate buffer contains 0.5 M Na$_2$HPO$_4$ and 4 ml orthophosphoric acid per liter), pH 7.4, 25 mM EDTA, pH 7.4, 1% (v/v) 2-mercaptoethanol], add 15 units of lyticase (Boehringer Mannheim, Indianapolis, IN, Cat. No. 1 372 467), and incubate 1 hr at 37°C.
4. Add 200 μl of solution II [0.2 M Tris-HCl, pH 9, 80 mM EDTA, pH 9, and 1% (w/v) sodium dodecyl sulfate (SDS)], place 30 min at 65°C, and cool 10 min at 4°C.
5. Add 100 μl of 5 M potassium acetate, incubate 45 min at 4°C, and centrifuge at 13,000 g for 2 min at 4°C.
6. Add 200 μl of 7.5 M ammonium acetate and 1.2 ml of ethanol to the supernatant. Place 5 min at -70°C and centrifuge 15 min at 13,000 g.
7. Rinse the pellet with 70% (v/v) ethanol, dry, and resuspend DNA in 50 μl of TE. [10 mM Tris-HCl, pH 7.5 1 mM EDTA]. Store at -20°C.

Restriction Digest of Yeast DNA

Digest around 0.5 to 1 μg of DNA from each transformant to be tested. For classic restriction digests, follow the manufacturer's recommendations. For I-*Sce*I digests, use the following protocols.

1. Add components at 0°C to the DNA in the following order: 10× incubation buffer [1 M diethanolamine hydrochloride, pH 9.5, 10 mM DTT, 2 mg/ml bovine serum albumin (BSA)], water to complete reaction volume, enzyme, then MgCl$_2$ (to start the reaction). Routine small-scale digestions (for plasmid restriction, veri-

fication of constructs, etc.) can be performed with the following amounts: about 1 μg of DNA, 2 μl of 10× incubation buffer, 1.6 μl of 0.1 M MgCl$_2$ (8 mM final concentration), 1 unit of I-SceI (Boerhinger Mannheim, Cat. No. 1 362 399), and water to a final volume of 20 μl.

2. Incubate 1 hr at 37°C and stop the reaction by adding 6 μl of 0.5 M EDTA, pH 9.

Southern Blot Transfer of Yeast DNA

1. Load digested DNA on 0.9% agarose gel and run in 1× TBE buffer (10× TBE buffer, pH 8.3, is 1 M Tris base, 1 M boric acid, 25 mM EDTA), at 1–3 V/cm.
2. Transfer gel onto a positively charged nylon membrane (HybondN+, Amersham, Arlington Heights, IL). In our laboratory, transfer is performed under vacuum at 55 Pa using a Vacugene apparatus (Pharmacia, Piscataway, NJ), but other transfer techniques work as well.
3. Hybridize the membrane with a random primed probe, labeled as described below. Hybridization and washing conditions are described later.

Labeling of Probes by Random Priming

The following procedure for labeling probes by random priming is adapted from Hodgson and Fisk (23).

1. To 25–50 ng of DNA in solution (not more than 11 μl), add 2 μl of 10× random priming buffer (0.5 M Tris-HCl, pH 7, 0.1 M MgSO$_4$, 1 mM DTT), 1.25 μl of pd(N)$_6$ primers (Pharmacia) at 0.1 units/μl, and water to a final volume of 14 μl.
2. Heat 5 min at 95°C and incubate 10 min at 37°C to allow annealing.
3. Add 1 μl of a mix of dCTP, dTTP, and dGTP (10 mM each), 3 μl of [α-^{32}P]dATP at 370 MBq/ml (110 TBq/mmol), and 1–3 units of Klenow polymerase.
4. Incubate 2 hr at 37°C. Count probe activity as described in a later section.

Interpretation of Molecular Analysis of Transformants

Southern blot analysis allows distinction between integration and replication of ARS-containing plasmids after *in vivo* recircularization. The presence of a recircularized transforming DNA containing an ARS is detected by a band much more intense than the endogenous locus band. The band has the size expected for the recircularized plasmid. The higher signal results from the fact that a plasmid containing an ARS

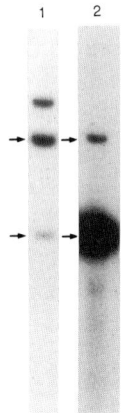

FIG. 3   Hybridization of a Southern blot of transgenic yeast DNA digested by *Ssp*I with the *URA3* cassette. The endogenous *ura3-52* locus is indicated by arrows. Lane 1, Proper integration into a yeast chromosome. Lane 2, Example of a transformant by an ARS-containing plasmid. The plasmidic band comigrates with one of the *ura3-52* bands.

but no centromere is multicopy. Low intensity signals comparable to the endogenous signal indicate integration.

Restriction enzymes used for digestion should be chosen so that they allow distinction between proper integration at the desired locus (when known) and possible abnormal integration (usually a rare event). Double digests by I-*Sce*I and another enzyme can be performed to check for the presence of the I-*Sce*I site. For example, Fig. 3 shows results of integrations of the *URA3*–I-*Sce*I cassette in natural yeast chromosomes. The enzyme used was *Ssp*I, which is often chosen because it cuts yeast DNA into 1- to 3-kb pieces. The blot was probed with the *URA3*–I-*Sce*I cassette. The endogenous *ura3-52* locus contains *Ssp*I sites, and two bands are revealed by the probe. The integration of the *URA3*–I-*Sce*I cassette into the homologous site is vizualized by one extra, low intensity band on the autoradiogram (*Ssp*I does not cut into the insert).

To distinguish between unique insertions and tandem repeats (see Fig. 1) which can happen in the ends-in topology if using large amounts of transforming DNA, a Southern blot should be performed with the DNA of the transformed strain digested with an enzyme which cuts only once in the integrated fragment, and probed with the same fragment. If the insertion is single, the digestion will yield two fragments (junction fragments) whose sizes depend on the distance of the next site in the resident DNA. If the insertion is multiple, there will be an additional fragment of the same size as the transforming DNA. The intensity of this additional fragment will allow estimation of the number of repeats. A case involving a very high number of tandem integrations might be confused with a replicative transformation because of the high intensity of the band, but the presence of junction bands, of intensity corresponding to one copy, indicates integration.

Molecular Analysis of Integration of I-*Sce*I Cassettes by Pulsed-Field Gel Electrophoresis

The PFGE method allows mapping of the integration site of the cassette after digestion of intact chromosomes in agarose plugs. In the case of YACs, it also allows verification of YAC integrity.

Preparation of Intact Yeast Chromosomes or Yeast Artificial Chromosomes in Agarose Plugs

1. Grow transgenic yeast clones in 50 ml SC − uracil medium to a concentration of about 5×10^7 cells/ml. Estimate the effective number of cells by counting with an optical microscope.
2. Centrifuge cells in a 50-ml tube at 3000 g for 10 min at 4°C, resuspend cells, and wash successively in 10 and 5 ml of 50 mM EDTA, pH 9.
3. Resuspend cells at 2.5×10^9 cells/ml in 50 mM EDTA, pH 9. To 1 ml of cell suspension held at 37°C, add 330 μl of solution I [1 M sorbitol, 20 mM EDTA, pH 8, 0.1 M sodium citrate, pH 5.8, 5% (v/v) 2-mercaptoethanol] containing 150 units of lyticase (Boehringer Mannheim Cat. No. 1 372 467) and add 1.67 ml of molten agarose [1% (w/v) InCert GTG agarose (FMC, Rockland, ME) in 125 mM EDTA, pH 9] kept at 50°C.
4. Pour into prewarmed mold (see description below), let gel set at 4°C, and cut into approximately 80-μl plugs (5 × 8 × 2 mm) (~1 μg of DNA/plug for a haploid strain).
5. Place plugs in a small volume of solution II [0.45 M EDTA, pH 9, 10 mM Tris-HCl, pH 9, 7.5% (v/v) 2-mercaptoethanol], just enough to cover plugs, at 37°C overnight.
6. Replace solution II with an excess volume of solution III [0.45 M EDTA, pH 9, 10 mM Tris-HCl, pH 9, 1% (v/v) sarkosyl, 1 mg/ml proteinase K (Boehringer Mannheim, Cat. No. 1 092 766)], so that diffusion occurs, and incubate overnight at 50°C or 7 hr at 65°C.
7. Cool plugs to 0°C, rinse in 0.5 M EDTA, pH 9, and store at 4°C in the same solution.

The mold is composed of a Teflon or Delrin comb between glass plates held together by clamps. The comb is 2 mm thick and 10 cm long, and the spaces between teeth are 8 mm wide. The mold is held upright and the gel poured between the teeth. After the agarose is set the mold is taken apart, and the agarose strips are cut into 5-mm long plugs.

In all experiments, never manipulate agarose plugs with metallic objects. EDTA in the presence of metal ions acts as a nonspecific nuclease.

Digestion of Yeast Chromosomes or Yeast Artificial Chromosome Inserts Containing I-SceI Site in Agarose Plugs

1. Equilibrate plugs in an excess volume of 0.1 M diethanolamine hydrochloride, pH 9.5, overnight at 4°C.
2. Rinse in 1× incubation buffer (0.1 M diethanolamine hydrochloride, pH 9.5, 1 mM DTT, 0.2 mg/ml BSA) for 1 hr at 4°C.
3. Place one plug per microcentrifuge tube containing 150 μl of 1× incubation buffer at 4°C. Add 20 units of I-SceI together with 20 ng of its enhancer and incubate 1.5 hr at 4°C to allow diffusion of the enzyme.
4. Start the reaction by adding 1.6 μl of 1 M MgCl$_2$ solution (8 mM final concentration) and incubate 30 min to 1 hr at 37°C (do not exceed 1 hr as nonspecific degradation may occur).
5. Cool plugs at 0°C and stop the reaction by adding 500 μl of 0.5 M EDTA, pH 9.

Note: Diffusion of the enzyme is needed prior to digestion because I-SceI is unstable in the presence of Mg^{2+} ions (13, 5). Reducing the Mg^{2+} concentration to 2 mM, or using 1 mM Mn^{2+} ions instead, results in the appearance of additional fragments corresponding to secondary sites, not cleaved under optimal conditions (14).

Separation and Purification of Intact Yeast Chromosomes, Yeast Artificial Chromosomes, or Fragments Thereof

Separation of Yeast Chromosomes and Yeast Artificial Chromosomes by Electrophoresis

All PFGE procedures are performed in our laboratory on a Rotaphor (Biometra, Göttingen, Germany) apparatus. Other equipment has not been tested thoroughly by the authors, but this has no implication for their performance and quality. Agarose plugs containing 1 μg of yeast DNA are introduced into wells and immobilized with molten agarose. Gels are 1% agarose [SeaKem GTG (FMC) for analytical gels and SeaPlaque GTG (FMC) for preparative gels] dissolved in 0.25× TBE buffer and run in the same buffer.

TABLE VI   Pulsed-field Gel Electrophoresis Conditions
on a Rotaphor (Biometra) Apparatus

To separate small chromosomes of *S. cerevisiae*

Voltage applied:	140 V
Pulse duration:	Linear slope from 160 to 80 sec
Angle:	120° C constant
Buffer temperature:	12° C
Migration duration:	65 hr

To separate preferentially large chromosomes

Voltage applied:	140 V
Pulse duration:	Linear slope from 200 to 100 sec
Angle:	120° C constant
Buffer temperature:	12° C
Migration duration:	86 hr

Variation of the pulses (which is the duration of voltage applied in each orientation) during a run allows better separation of chromosomes of different sizes. Long pulses separate large chromosomes, and short ones separate small chromosomes. The conditions we use are listed in Table VI. After electrophoresis, staining of the gel is performed by immersion for 1 hr in 0.2 μg/ml of ethidium bromide in water (for a 20 × 20 cm gel, use 1 liter). The gel is photographed under short UV illumination (254 nm) if analytical or long UV light (312 nm) if preparative (to minimize DNA breakage). Figure 4 shows examples of yeast chromosomes separated by PFGE. For purification, the band corresponding to the DNA of interest is cut out of a preparative gel. (Cut agarose slices as small as possible and store them at −20° C if subsequent manipulations are not undertaken immediately.)

Selective Digestion of rDNA (on Chromosome XII and 3 μm) with I-PpoI

I-*Ppo*I cleaves the approximately 1100-kb rDNA stretch from chromosome XII of *S. cerevisiae* and the 3 μm circular extrachromosomal rDNA into 9-kb pieces. This allows elimination of the 3 μm DNA circles which contaminate chromosomes purified after PFGE. An I-*Ppo*I digestion of the yeast DNA prior to PFGE generates a 9-kb band at the bottom of the gel and two remaining parts of chromosome XII of approximately 450 and 600 kb, respectively (Fig. 4C). All steps and buffers are the same as for I-*Sce*I digest with the following exceptions. In step 3, diffusion of 240 units of I-*Ppo*I (Promega, Madison, WI, Cat. No. R7031) is carried out for only 30 min. In step 4, the reaction is carried out for at least 2 hr.

Note: Diffusion can proceed during digestion because the enzyme is stable in the

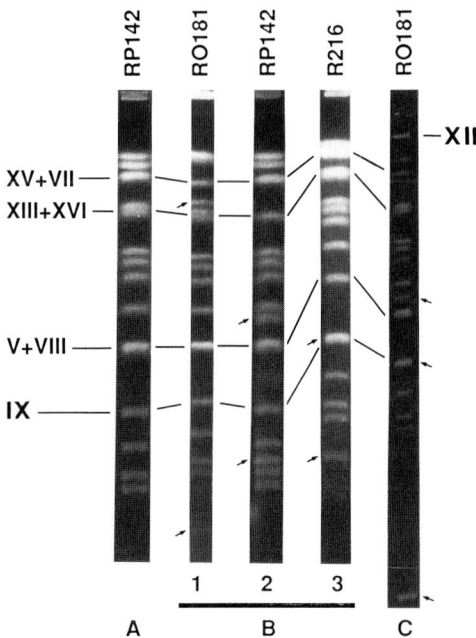

FIG. 4   Pulsed-field gel electrophoresis (PFGE) of yeast chromosomes. (A) Transgenic strain FY23/RP142 before cutting with I-*Sce*I shows the wild-type pattern of yeast chromosomes. (B) Example of separation of comigrating chromosomes after digestion with I-*Sce*I. Lane 1, Separation of chromosomes VII and XV (digestion of chromosome XV) using the transgenic strain FY23/RO181. Lane 2, Separation of chromosomes XIII and XVI (digestion of chromosome XVI) using the transgenic strain FY23/RP142. Lane 3, Separation of chromosomes V and VIII (digestion of chromosome V) using the transgenic strain FY23/R216. Note that the larger fragment of chromosome V comigrates with chromosome IX. (C) Example of elimination of rDNA from PFGE by digestion with I-*Ppo*I which generates a 9-kb rDNA band and two chromosome XII residual fragments of around 600 and 450 kb, respectively. Note that the 450-kb fragment of chromosome XII comigrates with chromosome IX. Arrows indicate chromosomal fragments resulting from the I-*Sce*I or I-*Ppo*I cuts.

presence of MgCl$_2$, and the duration of digestion can exceed 2 hr because no chromosomal degradation occurs.

Double Digestion with I-SceI and I-PpoI

All steps and buffers are the same as for I-*Sce*I digestion except in step 3, diffusion of 20 units of I-*Sce*I together with 20 ng of its enhancer and 240 units of I-*Ppo*I is carried out for 1.5 hr.

Southern Blotting and Hybridization of Pulsed-field Gels

Alkali blotting of DNA on nylon HybondN$^+$ (Amersham) membranes is performed under vacuum (~55 Pa) with a Vacugene apparatus (Pharmacia). Other equipment has not been tested by the authors, but this has no implication for quality. The gel is first treated for 30 to 40 min with 0.25 M HCl poured on the gel; then the acid is replaced by 0.4 M NaOH, and transfer is performed for 2 hr. Membranes may be used directly for hybridization or may be dried and stored.

Labeling of Entire Chromosomes, Yeast Artificial Chromosomes, or Fragments Thereof in Agarose Slices

Purified chromosomes, YACs, or fragments thereof can be used as probes for identification and mapping of cosmid, plasmid, or λ clones (see Applications).

1. Rinse agarose slices twice in 40 volumes of water for 1 hr at 4°C to eliminate TBE buffer and ethidium bromide.
2. Melt 50-μl agarose slices containing DNA (use 15 ng for 200-kb fragments, 160 ng for 2200-kb fragments, or proportional) at 65°C for 15 min, then add 14 μl of water, 6 μl of pd(N)$_6$ primers (Pharmacia) at 0.1 units/μl, and 10 μl of 10× random priming buffer (0.5 M Tris-HCl, pH 7, 0.1 M MgSO$_4$, 1 mM DTT). Heat 5 min at 95°C and incubate 10 min at 37°C to allow annealing of primers to DNA.
3. Add 5 μl of a mix of dCTP, dTTP, and dGTP (10 mM each), 10 μl of [α-^{32}P]dATP at 10 mCi/ml, and 20 units of Klenow polymerase. Incubate 2 hr at 37°C.
4. Probe activity is estimated as follows: an aliquot of the probe is adsorbed on a HA type membrane (Millipore, Bedford, MA) and precipitated by three immersions of 1 min each in 5% trichloroacetic acid followed by a 1-min immersion in 80% (v/v) ethanol. The membrane is then placed in a scintillation counter.
5. Denature labeled DNA at 95°C for 5 min and use directly as probe.

Hybridization Using Labeled Yeast Chromosomes, Yeast Artificial Chromosomes, or Fragments Thereof as Probes

The following hybridization protocol is adapted from Church and Gilbert (24). Volumes are given for large or small bottles in an Appligene (Illkirch, France) hybridization oven. Other equipment has not been tested thoroughly by the authors, but this has no implication for performance and quality.

1. Perform prehybridization of membranes for at least 1 hr at 65°C in 15 ml of hybridization buffer [0.25 M sodium phosphate buffer, pH 7.4, 1 mM EDTA, pH 8, 7% (w/v) SDS, 1% (w/v) BSA] in a large bottle, or 10 ml in a small bottle.
2. Add the probe, previously denatured at 95°C for 5 min, in the same buffer and incubate at least 12 hr at 65°C to allow hybridization.
3. Subject the membranes to two or three washes in an excess volume of washing buffer [20 mM sodium phosphate buffer, pH 7.4, 1 mM EDTA, pH 8, 1% (w/v) SDS] for 15 min each at 65°C. Seal wet membranes in Saran Wrap and process for autoradiography.

Note: Nylon HybondN+ membranes can be dehybridized as follows. Add an excess volume of boiling 0.5% (w/v) SDS solution and allow to cool at room temperature. If removal of the probe from membranes is to be performed, never allow hybridized membranes to dry.

Applications

Previous protocols and strategies offer a variety of applications that range from mapping single genes into yeast chromosomes or YAC inserts to general mapping procedures. They are also useful for preparing any fragment of yeast chromosomes or YACs needed for cloning experiments or probe preparation. In our laboratory, we have used these methods for mapping yeast chromosomes as a part of the European sequencing program. A random integration experiment of I-*Sce*I sites and mapping of chromosome VII are summarized here as possible examples of applications. The same strategy has also been applied to mapping a YAC containing a mouse insert.

Random Integration of Cassettes into Yeast Genome

Application of random integration has been made in our laboratory in order to integrate I-*Sce*I sites at a variety of locations in the entire yeast genome. This strategy requires subsequent analysis of many transformant clones. In the experiment, performed as described earlier, of 204 transformant clones analyzed by Southern blotting, 40% proved to be an integration of the pAF101 vector into the *ura3-52* locus on chromosome V (this could be avoided by using a yeast strain in which the *URA3* gene is entirely deleted). The remaining clones were analyzed by PFGE after I-*Sce*I digestion. About 30% of them showed no chromosomal insert (no I-*Sce*I site) and were probably transformed by ARS-containing plasmids, 60% showed one insert of the pAF101 recombinant plasmid into one of the chromosomes, and 10% showed insertions into several chromosomes. A total of 39 transgenic yeast strains were obtained (see Table II).

Nested Chromosomal Fragmentation for Separation and Physical Mapping of Yeast Chromosome VII

The physical mapping of chromosome VII first implies its separation from the comigrating chromosome XV and, second, implies its nested fragmentation. The first was achieved by insertion of an I-*Sce*I cassette in chromosome XV and purification of chromosome VII by PFGE after digestion with I-*Sce*I (see Fig. 4B). The next step was the screening of a total yeast genomic cosmid bank using the purified chromosome as probe. Figure 5A shows colony hybridization with a chromosome VII probe.

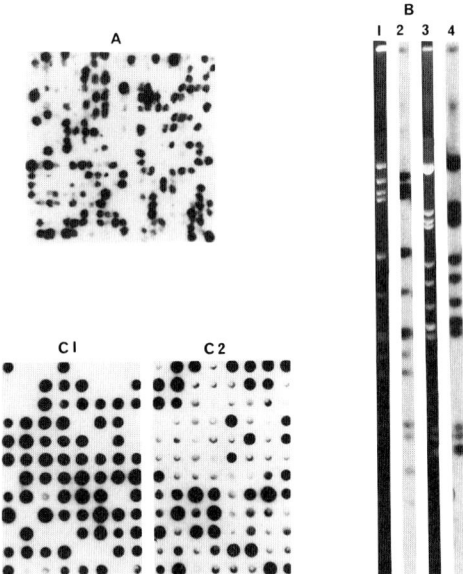

FIG. 5   Hybridization experiments using chromosomes or fragments labeled in agarose plugs. (A) Hybridization of a cosmid library with a mixture of chromosomes VII and XV. Cosmid colonies are deposited on HybondN+ membranes and treated as described previously (17), prior to hybridization. (B) Lanes 1 and 2, Hybridization of a Southern blot of an *Eco*RI-digested cosmid clone with purified chromosome VII. Lanes 3 and 4, Cosmid from chromosome XV hybridized with purified chromosome XV. (C) Hybridization of chromosome VII-specific cosmid clones with purified chromosome VII nested fragments. Part C1, Probe was the left fragment (~600 kb) of chromosome VII, purified from transgenic yeast strain FY23/RG162. Part C2, Probe was the right fragment (~450 kb) of the same transgenic yeast.

Positive cosmid clones were digested by *Eco*RI, and the corresponding blot was hybridized with a chromosome VII probe (Fig. 5B). The nested fragmentation of chromosome VII was achieved by integration of a set of 20 I-*Sce*I cassettes along that chromosome. After I-*Sce*I digestion and PFGE purification of nested chromosomal fragments, the chromosome VII-specific cosmid clones were ordered in groups along the chromosome using successive hybridizations (Fig. 5C).

Nested Chromosomal Fragmentation for Physical Mapping of Yeast Artificial Chromosome Inserts

Mapping of a 450-kb YAC insert from chromosome X of mouse was achieved by integration of pAF123 into the various *B2* repeats of the insert (25). This experiment involved a single transformation of the YAC-containing strain, followed by genetic screening and molecular analysis of the transformants. Each transformant contained a single insertion of the cassette. In total, I-*Sce*I sites were found inserted at 11 different locations along the YAC insert, allowing us to map a set of λ clones using nested YAC fragments as probes.

Conclusion

The integration of I-*Sce*I sites at several places along a chromosome or a YAC provides real molecular milestones that allow the rapid construction of physical maps (25, 26). Advantages of the approach are several. First, as the map is an artificial one, its resolution can be chosen as needed. At any moment in a mapping program, a higher resolution can be easily obtained by additional integrations of the I-*Sce*I cassette within the region of interest. Second, contrary to bottom-up methods (fingerprinting, sequence-tagged site mapping, and end-specific probes), it is the genomic DNA itself which is mapped (yeast chromosomes or YACs) rather than the clones, thus reducing cloning artifacts. Third, contigs can be placed on the global chromosome or YAC map at any stage of map construction, facilitating contig assembly and chromosome walking.

Apart from applications in yeast, the top down chromosomal fragmentation strategy using I-*Sce*I or similar enzymes can also be used for other organisms such as prokaryotes (27) or for eukaryotic genomes of low complexity.

Acknowledgments

We thank members of the Unité de Génétique Moléculaire des Levures for stimulating discussions and particularly L. Colleaux and A. Plessis whose experiments were relevant to the present chapter. We also thank C. Rougeulle and P. Avner for collaboration on YAC mapping experiments, M. Olson and F. Winston for the generous gifts of strains AB1380, FY23, and FY73, and A. Goffeau for collaboration on yeast chromosome VII mapping and for interest and support of this work. This work was supported in part by grants from the European Commission [BRIDGE, BIOTECH and Human Genome Analysis programs] and from the Ministère de la Recherche et de l'Espace (92C0038). B.D. is Professor of Molecular Genetics at the University Pierre et Marie Curie, Paris.

References

1. C. S. Newlon, *Microbiol. Rev.* **52,** 568 (1988).
2. T. L. Orr-Weaver, J. W. Szostak, and R. J. Rothstein, *Proc. Natl. Acad. Sci. U.S.A.* **78,** 6354 (1981).
3. R. J. Rothstein, *in* "Methods in Enzymology" (C. Guthrie and G. R. Fink, eds.), Vol. 194, p. 281. Academic Press, San Diego, 1991.
4. B. Dujon, *Gene* **82,** 91 (1989).
5. A. Perrin, M. Buckle, and B. Dujon, *EMBO J.* **12,** 2939 (1993).
6. L. Colleaux, L. d'Auriol, F. Galibert, and B. Dujon, *Proc. Natl. Acad. Sci. U.S.A.* **85,** 6022 (1988).
7. D. E. Muscarella, E. L. Ellison, B. M. Ruoff, and V. M. Vogt, *Mol. Cell. Biol.* **10,** 3386 (1990).
8. B. Sargueil, B. Hatat, A. Delahodde, and C. Jacq, *Nucleic Acids Res.* **18,** 5659 (1990).
9. R. Kostriken, J. N. Strathern, A. J. S. Klar, J. B. Hicks, and F. Heffron, *Cell (Cambridge, Mass.)* **35,** 167 (1983).
10. H. Watabe, T. Shibata, T. Iino, and T. Ando, *J. Biochem. (Tokyo)* **95,** 1677 (1984).
11. C. K. Shih, R. Wagner, S. Feinstein, C. Kanik-Ennulat, and N. Neff, *Mol. Cell. Biol.* **8,** 3094 (1988).
12. F. B. Perler, D. G. Comb, W. E. Jack, L. S. Moran, B. Quiang, R. B. Kucera, J. Benner, B. E. Slatko, D. O. Nwankwo, S. K. Hempstead, C. K. S. Carlow, and H. Jannasch, *Proc. Natl. Acad. Sci. U.S.A.* **89,** 5577 (1992).
13. C. Monteilhet, A. Perrin, A. Thierry, L. Colleaux, and B. Dujon, *Nucleic Acids Res.* **18,** 1407 (1990).
14. A. Thierry, A. Perrin, J. Boyer, C. Fairhead, and B. Dujon, *Nucleic Acids Res.* **19,** 189 (1991).
15. A. Gauthier, M. Turmel, and C. Lemieux, *Curr. Genet.* **19,** 43 (1991).
16. M. V. Olson, *in* "The Molecular Biology of the Yeast *Saccharomyces,* Volume I" (J. R. Broach, J. R. Pringle, and E. W. Jones, eds.), p. 1. Cold Spring Harbor Laboratory, Cold Spring Harbor, New York, 1991.
17. J. Sambrook, E. F. Fritsch, and T. Maniatis, "Molecular Cloning: A Laboratory Manual," 2nd Ed. Cold Spring Harbor Laboratory, Cold Spring Harbor, New York, 1989.
18. A. Goffeau and A. Vassarotti, *Biofutur* **128,** 33 (1993).
19. D. T. Burke, G. F. Carle, and M. V. Olson, *Science* **236,** 806 (1987).
20. A. Thierry, C. Fairhead, and B. Dujon, *Yeast* **6,** 521 (1990).
21. A. Plessis and B. Dujon, *Gene* **134,** 41 (1993).

22. E. Meilhoc, J.-M. Masson, and J. Teissié, *Bio/Technology* **8,** 223 (1990).
23. C. P. Hodgson and R. Z. Fisk, *Nucleic Acids Res.* **15,** 6295 (1987).
24. G. M. Church and W. Gilbert, *Proc. Natl. Acad. Sci. U.S.A.* **81,** 1991 (1984).
25. L. Colleaux, C. Rougeulle, P. Avner, and B. Dujon, *Hum. Mol. Genet.* **2,** 265 (1993).
26. A. Thierry and B. Dujon, *Nucleic Acids Res.* **21,** 5625 (1992).
27. M. Itaya, J. J. Laffan, and N. Sueoka, *J. Bacteriol.* **16,** 5466 (1992).

[6] Basic Analysis of Transcription Factor Binding to Nucleosomes

Jacques Côté, Rhea T. Utley, and Jerry L. Workman

Introduction

Both genetic and biochemical studies illustrate a pivotal role of nucleosomes in the regulation of transcription by RNA polymerase II. *In vitro* studies as well as studies in yeast have illustrated that nucleosomes suppress basal transcription from promoters leading to higher levels of transcriptional regulation by upstream regulatory factors (reviewed in Ref. 1). Thus, it is in this repressive background of chromatin structures where maximum function of transcriptional regulators occurs. This implies that overcoming nucleosome repression is a natural component of transcriptional regulatory pathways which contributes to the selectivity of gene transcription in eukaryotic cells. Nucleosomes repress both the initial binding of upstream activators to their cognate sequence elements and the binding of the basal factors to form preinitiation complexes at the TATA box and transcription initiation site (2). The latter step has thus far been primarily revealed in functional transcription studies. These methods have been described previously (3, 4). In contrast, nucleosome repression of upstream activator binding has been most extensively studied in assays of direct binding to nucleosomal DNA.

Direct analysis of transcription factor binding to nucleosomes has revealed many general properties of factor–nucleosome interactions. Different factors are inhibited in binding to different degrees by occupancy of their binding sites in nucleosomes. For example, the glucocorticoid receptor exhibits very efficient nucleosome binding abilities (5), whereas similar analyses indicate that the binding of NF1 and human heat-shock factor are severely inhibited by nucleosomes (6, 7). Other factors demonstrate intermediate levels of affinity for nucleosomal DNA. GAL4 derivatives, Sp1, USF, and Myc/Max and Max/Max dimers have also been shown to bind nucleosomes (7–10). Differences in affinity for nucleosomal DNA suggest discrete functions of different regulatory factors during gene activation in chromatin. Transcription factor binding to nucleosomes is affected by nucleosome position (e.g., the location of the binding site on the nucleosome) (5, 8, 11). Binding can also be enhanced by histone acetylation (12), cooperative binding (7, 11), and by the function of accessory protein complexes including the histone binding protein nucleoplasmin (9) and the SWI/SNF protein complex (13). The multiplicity of these mechanisms challenge old premises that nucleosomes are insurmountable barriers to the transcription apparatus.

Methods in Molecular Genetics, Volume 6

In this chapter we focus on basic preparations of reagents and their application to the analysis of transcription factor binding to nucleosomes. These are the methods for which we most often receive inquiries. We present protocols for the preparation of nucleosomes and histones, nucleosome reconstitution, and the analysis of transcription factor binding. Our goal is to provide protocols which are applicable to any nucleosome-length DNA fragment that can be easily performed by any biochemist or molecular biologist with minimal protein experience. The only limitations to performing these experiments is the availability of sufficient quantities of the transcription factors of interest.

Reagents for Nucleosome Reconstitution

In this section we describe the preparation of histones and nucleosome cores for the reconstitution of nucleosomes. Initially we discuss criteria for selection of DNA fragments for nucleosome reconstitution. Protocols for radiolabeling the DNAs can be found in many molecular biology manuals. We do not discuss the purification of transcription factors, as those used in our laboratory (i.e., GAL4 derivatives, USF, Sp1, NFkB) may not apply for specific experiments of the reader. We do note, however, that it is helpful to procure large quantities of purified factors as nucleosome binding generally occurs with a reduced affinity relative to DNA.

DNA Fragments for Nucleosome Reconstitution

Reconstitution of individual nucleosomes should be performed with DNA fragments near the length of the nucleosome, 150–200 base pairs (bp). Longer fragments may allow additional histones or nucleosomes to bind the DNA, resulting in a heterogeneous population. For nucleosome cores, which contain an octamer of the four core histones (two H2A/H2B dimers and one H3/H4 tetramer) wrapped with 146 bp of DNA, it is best to use fragments of approximately 150 bp. This helps to reduce the heterogeneity of translational positions of the nucleosome core on the DNA (14). For chromatosomes, a nucleosome core also bound by the linker histone H1, DNA fragments greater than 167 bp are needed as H1 binding requires extra DNA (15, 16).

Further criteria for choice of DNA fragment depends on the goal of the experiment. One approach is to mimic *in vitro* the nucleosome positioning observed on a promoter *in vivo* in order to test the binding of regulatory factors in this configuration. This has been accomplished with the mouse mammary tumor virus (MMTV) promoter and allowed an analysis of glucocorticoid receptor and NF1 binding (6). However, little is known about the mechanism of nucleosome positioning *in vivo* or its relationship to nucleosome positioning *in vitro*. Indeed, some sequences like the

5 S repeats of *Xenopus* demonstrate strong nucleosome positioning *in vitro* (17) which is not readily apparent *in vivo* (18). If nucleosome positions have been mapped *in vivo*, a practical approach to this dilemma is to test for similar positioning *in vitro* on fragments of 200–250 bp (19); if not observed *in vitro*, similar positions can be achieved *in vitro* by restricting the end of the nucleosome length fragments (∼150 bp) to the end of the desired nucleosome location.

The use of artificial DNA fragments has allowed experimental manipulation of binding sites and direct tests of various parameters. For example, these approaches have demonstrated nucleosome position effects on glucocorticoid receptor, GAL4-AH, and Sp1 binding by movement of the binding sites relative to the nucleosome core (5, 8, 11). The use of fragments with multiple binding sites has demonstrated cooperative binding of GAL4 derivatives to nucleosomes (7, 11, 15). Similarly, analysis of natural enhancer and promoter sequences promises an examination of the cooperativity of binding of the multiple factors to those elements in nucleosomes.

Preparation of H1-Depleted Cellular Nucleosome Cores

The simplest protocol to reconstitute nucleosome cores is by transfer of histone octamers from cellular nucleosome cores onto radiolabeled DNA probes (see below). Cellular nucleosome cores are readily prepared from nuclei by cleaving the cellular chromatin into oligonucleosome-size fragments and removing the nonhistone proteins and histone H1 by gel filtration at 0.6 M salt. At that salt concentration only the four core histones (H2A, H2B, H3, and H4) remain bound to the DNA as a histone octamer and elute with the DNA from the gel-filtration columns.

Note: Every step is performed on ice or in a cold room.

Preparation and Low-Salt Extracted Nuclear Pellets

1. HeLa cells (3 liters) are grown to exponential phase in Joklik's minimum essential medium (MEM) plus 10% (v/v) calf serum in a spinner flask ($0.5 - 1 \times 10^5$ cells/ml). Cells are collected, concentrated, and washed with phosphate-buffered saline (137 mM NaCl, 2.7 mM KCl, 4.3 mM Na$_2$HPO$_4$, 1.4 mM KH$_2$PO$_4$, pH 7.3).
2. Resuspend cells in 20 pellet volumes (∼40 ml) of lysis buffer [20 mM HEPES (pH 7.5), 0.25 M sucrose, 3 mM MgCl$_2$, 0.2% (v/v) Nonidet P-40 (NP-40), 3 mM 2-mercaptoethanol, 0.4 mM phenylmethylsulfonyl fluoride (PMSF), 1 μM pepstatin, A, 1 μM leupeptin] and Dounce homogenize 10–20 strokes with a type B pestle. Cell lysis can be monitored under a light microscope.
3. Pellet nuclei for 15 min at 3000 g (4°C) and wash twice with 50 ml of the same buffer and once with buffer B [20 mM HEPES (pH 7.5), 3 mM MgCl$_2$, 0.2 mM EGTA, 3 mM 2-mercaptoethanol, 0.4 mM PMSF, 1 μM pepstatin A, 1 μM leupeptin]. Homogenize gently if necessary to resuspend the pellet (2–3 strokes).

4. Resuspend nuclei in 1–3 volumes of buffer B. While gently stirring, add dropwise 1 total volume of buffer B containing 0.6 M KCl and 10% (v/v) glycerol. Continue gentle stirring for 10 min at 4°C. If necessary, gentle homogenization can be repeated.
5. Pellet nuclei for 30 min at 17,500 g (15,000 rpm, in a Sorvall SS-34 rotor). Nuclear pellets can be frozen in dry ice at this step and kept at -80°C.

Solubilization and H1 Depletion of Oligonucleosomes

1. Nuclear pellets are washed in a large volume (40 ml) of MSB [20 mM HEPES (pH 7.5), 0.4 M NaCl, 1 mM EDTA, 1 mM 2-mercaptoethanol, 0.5 mM PMSF, 5% glycerol]. Gentle Dounce homogenization can be used (5 strokes, B pestle) to resuspend the nuclei. Pellet nuclei for 20 min at 10,000 g (11,500 rpm, SS-34 rotor).
2. Resuspend nuclei (typically 2 ml) in 4 volumes of HSB [20 mM HEPES (pH 7.5), 0.65 M NaCl, 1 mM EDTA, 1 mM 2-mercaptoethanol, 0.5 mM PMSF, 0.34 M sucrose] and homogenize with several strokes of B pestle (~50). This produces high molecular weight chromatin fragments which can be separated from nuclear debris and subsequently shortened by nuclease digestion.
3. Pellet nuclei for 20 min at 10,000 g (11,500 rpm, SS-34 rotor) and dialyze the supernatant overnight at 4°C against 4 liters of LSB [20 mM HEPES (pH 7.5), 0.1 M NaCl, 1 mM EDTA, 1 mM-2 mercaptoethanol, 0.5 mM PMSF] with 6–8 kDa cutoff dialysis membrane, Spectra/Por.
4. As some precipitation of the samples will occur on dialysis to 0.1 M salt, the sample must be mixed well before removal from the dialysis bag.
5. Add 0.03 volumes of 100 mM CaCl$_2$ and warm the sample for 2 min at 37°C.
6. Add 10 units/ml micrococcal nuclease (Sigma, St. Louis, MO) and incubate for 5 min at 37°C.
7. Quench the digestion with 0.1 volume of 0.5 M EGTA (pH 8.0) and put on ice. Insoluble materials should have disappeared at this point.
8. Add 2 M NaCl dropwise while gently vortexing to obtain 0.6 M final concentration.
9. Spin the sample for 30 min at 150,000 g (40,000 rpm, in a Beckman, Fullerton, CA, SW55 rotor) and apply the supernatant onto a Sepharose CL-6B column (1.6 × 58 cm), Pharmacia, Piscataway, NJ) equilibrated in HSB without sucrose. The amount of the sample loaded should not exceed 10% of the column volume. Any extra supernatant is frozen and stored and can be used in subsequent column runs. Run the column at 12 ml/hr while collecting 2-ml fractions.
10. Monitor the absorbance at 260 nm of every third fraction by diluting an aliquot 100-fold in 2 M NaCl. Aliquots (10–20 μl) are also analyzed by sodium dodecyl sulfate–polyacrylamide gel electrophoresis (SDS-PAGE) (15% acrylamide gel) and by 1.5% native agarose gel electrophoresis (after proteinase K treatment for the DNA gel) (see Fig. 1).

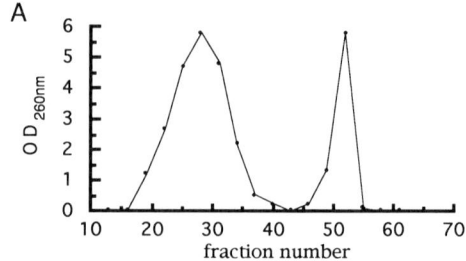

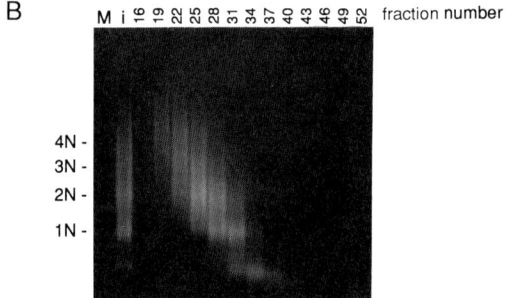

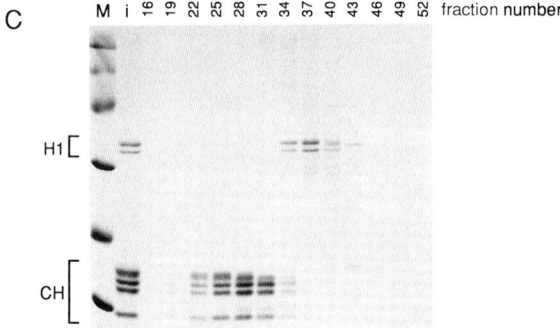

FIG. 1   Purification of H1-depleted oligonucleosome cores. (A) Elution profile of gel filtration on Sepharose CL-6B as measured by absorbance at 260 nm. (B) Visualization of oligonucleosome cores eluted from the gel-filtration column. Aliquots (10 μl, of every third fraction) were deproteinized and run on a 1.5% native agarose gel. Samples were treated with proteinase K (0.5 mg/ml in 0.5% SDS) for 1 hr at 50° C prior to loading on the gel. Mobility of mono-, di-, tri-, and tetranucleosome core fragments is indicated (note that the digestion products are from H1-depleted chromatin which reduces the regularity of nucleosome spacing). The initial sample loaded on the column is also shown (i). Markers (M) are from the 123-bp DNA ladder (GIBCO/BRL Gaithersburg, MD). (C) Monitoring of histone elution by protein gel electrophoresis. Aliquots (12.5 μl) of the same fractions as in (B) were loaded on 15% acrylamide gels and subjected to SDS-PAGE. After 6 hr of migration at 150 V, the gel was stained with coomassie brilliant blue. Markers (M) are low-range SDS-PAGE standards (Bio-Rad). Protein bands corresponding to core histones (CH) and histone H1 (H1) are indicated.

11. The fractions of interest are those which contain short oligonucleosomes (apparent on the agarose gel; Fig. 1B) and the four core histones but lack histone H1 (apparent on the SDS-PAGE gel; Fig. 1C). Pool the fractions and dialyze against 20 mM HEPES (pH 7.5), 1 mM EDTA, 1 mM 2-mercaptoethanol, 0.5 mM PMSF.
12. Concentrate samples by dialysis against solid sucrose, constrict the volume of the dialysis bag, and then dialyze against buffer without sucrose. If necessary the samples can be further concentrated using Centriprep-10 concentrators (Amicon Danvers, MA), to achieve a DNA concentration near 1 mg/ml. Divide into aliquots, freeze on dry ice or in liquid nitrogen, and store at $-80°$ C.

Figure 1 illustrates the chromatographic separation of oligonucleosome cores from histone H1. The fractions pooled as H1-depleted oligonucleosome cores are contained in fractions 19–28. These fractions contain short DNA fragments bound by histone octamers (e.g., all four core histones) but are depleted of histone H1.

Preparation of Core Histones and Histone H1

In addition to octamer transfer from H1-depleted oligonucleosome cores, nucleosome reconstitution can also be achieved with pure histone proteins when mixed at the appropriate stoichiometries with DNA probes in the presence or absence of carrier DNA (see below). Below we describe a rapid procedure for purification of core histones based on hydroxylapatite chromatography (20). In addition, H1 can be purified from the flow through of the column using the following protocols and adapted from Stein and Mitchell (21).

Purification of Core Histones by Hydroxylapatite Chromatography

1. Nuclear pellets are prepared as described above. Nuclear pellets containing approximately 12 mg DNA (as evaluated by OD_{260} of a 100-fold dilution in 2 M NaCl) are resuspended in 50 ml HAP buffer [50 mM sodium phosphate (pH 6.8), 0.6 M NaCl, 1 mM 2-mercaptoethanol, 0.5 mM PMSF] and stirred gently for 10 min at 4° C.
2. While gently stirring, add 20 g dry BioGel HTP powder (Bio-Rad, Richmond, CA, adsorption capacity of 0.6 mg DNA per g dry powder).
3. Add just enough HAP buffer to allow pouring the resin into a column (~30 ml). Pour the column (2.5 × 20 cm) and collect the eluant, which contains partially pure histone H1.
4. Wash the resin with 10 column volumes of HAP buffer (~600 ml at 60 ms/hr).
5. Elute the core histones with a step of the same buffer containing 2.5 M NaCl while collecting 8-ml fractions. Monitor the protein concentration by absorbance at 230 or 280 nm and pool the peak fractions.

6. Concentrate the core histones to 2–10 mg/ml using Centriprep-10 concentrators (Amicon). Divide into aliquots, freeze on dry ice, and store at −80°C.

An example of the purification of core histones by hydrolxylapatite chromatography is shown in Fig. 2A. The material loaded on the column contains many proteins in addition to the core histones and histone H1. The core histones are purified on the column, and the partially pure H1 fraction is further purified by gel filtration.

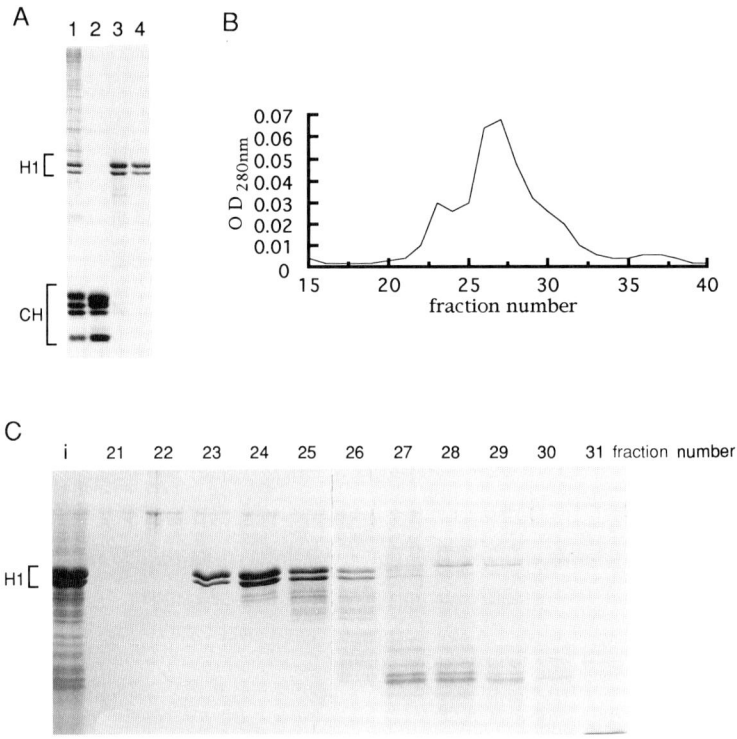

FIG. 2   Purification of core histones and histone H1. (A) Purity of core histones and histone H1 pools after fractionation. Lane 1 shows the initial sample (10 μg) used in batch-binding with hydroxylapatite resin; lane 2, pool of the 2.5 M NaCl step elution (10 μg); lane 3, flow through of the hydroxylapatite column (3.5 μg) and thus the initial sample loaded on the Superose 12 gel-filtration column; lane 4, pool of the pure H1-containing fractions (1 μg) from Superose 12 gel filtration (see B and C). (B) Elution of the Superose 12 gel-filtration column monitored by absorbance at 280 nm. (C) Protein content of the Superose 12 fractions. Aliquots (12 μl) were subjected to 12% SDS-PAGE. Fraction 23, corresponding to the first peak in (B), contains pure histone H1.

Note: An alternative approach uses the H1-depleted oligonucleosome cores from above as a source of core histones. These are dialyzed against HAP buffer and then loaded on a prepacked BioGel HTP column and eluted as above.

Purification by Histone H1 by Gel Filtration

1. The flow-through fraction from the hydroxylapatite column or the histone H1-containing fractions from the gel-filtration columns used to deplete oligonucleosome cores of H1 can be used as a source of H1. The samples are concentrated to 2 ml using centriprep-10 concentrators (Amicon).
2. Histone H1 is separated from contaminants by loading the sample on a Sephacryl S-200 HR column (1 × 50 cm, Pharmacia) equilibrated in 20 mM HEPES (pH 7.5), 0.6 M NaCl, 1 mM EDTA, 1 mM 2-mercaptoethanol, 0.5 mM PMSF. Then 2-ml fractions are collected at a flow rate of 6 ml/hr. If a fast protein liquid chromatography (FPLC) or high-performance liquid chromatography (HPLC) system is available, 0.4-ml aliquots can also be loaded on a superose 12 HR 10/30 column (Pharmacia) and 0.5-ml fractions collected at a flow rate of 12 ml/hr.
3. Fractions are monitored by OD_{280} and analyzed by SDS-PAGE on 12% acrylamide gels.
4. Fractions containing more than 90% pure histone H1 are pooled, concentrated if necessary to about 0.5 μg/μl, divided into aliquots, frozen on dry ice, and stored at $-80°$C.

The purified H1 is illustrated in Fig. 1A. Figure 1B illustrates a chromatogram from H1 purification on a Superose 12 column (Pharmacia). The proteins present in the resultant fractions are illustrated in the SDS-PAGE gel shown in Fig. 1C. The purest fractions of H1 are in the initial peak (fraction 23). These are used for reconstitution of chromatosomes.

Nucleosome Reconstitution and Analysis of Factor Binding

There are several methods to reconstitute nucleosome cores onto nucleosome length DNA fragments. These can be divided into two groups: those done at physiological salt concentrations utilizing histone chaperones and those done at high salt utilizing only pure histones and DNA. Protocols describing the use of histone chaperones in nucleosome reconstitution have been reported elsewhere (3, 22, 23). Here we focus on procedures utilizing high salt because these are commonly used in transcription factor binding assays and require only the reagents described above. The three protocols for nucleosome reconstitution described below include octamer transfer, reconstitution by dialysis from high salt utilizing pure histones and carrier DNA,

TABLE I   Comparison of Nucleosome Reconstitution Methods[a]

	Octamer transfer	Pure histones carrier DNA	Pure histones homogeneous DNA
Time (hr)	3	18	18
Efficiency (%)	80–95	~50	~50
Gradient purification	Optional	Required	Required
Nonspecific DNA	Yes	Variable	No
Concentration of probe	$<1\,nM$	$<1\,nM$	$\leqslant 100\,nM$
Protein analysis	No	No	Yes

[a] The advantages and disadvantages of three different nucleosome core reconstitution methods are compared. The simplest approach is octamer transfer, which is accomplished in a short period of time and does not require gradient purification of the nucleosome cores. The method has the disadvantage, however, that H1-depleted oligonucleosome cores remain in subsequent binding reactions which can compete for transcription factor binding. Reconstitution of DNA probes with pure histones in the presence of carrier DNA can alleviate the problem if the carrier DNA is sufficiently different in size from the probe DNA such that it is largely removed on sucrose gradients. Reconstitution of nucleosome cores with pure histones and large amounts of homogeneous probe DNA is free of any carrier DNA. However, the high concentration of nucleosome cores bearing factor binding sites can drive factor binding, making an assessment of relative binding affinities difficult. Homogeneous sequence nucleosome cores are required for analysis of the protein composition of factor/nucleosome ternary complexes (Vettese-Dadey et al., Chapter [7], this volume). See text for details of reconstitution methods.

and reconstitution with homogenous sequence DNA. Each has particular advantages dependent on the type of analysis performed and the transcription factors tested for binding (Table I).

Nucleosome Reconstitution by Histone Octamer Transfer and Analysis of Factor Binding

By far the quickest and simplest procedure for nucleosome reconstitution is transfer of histone octamers from purified H1-depleted nucleosome cores onto radiolabeled DNA fragments (24). The procedure results in efficient reconstitution of nucleosomes onto probe DNAs which can be used directly in binding assays. Thus, gradient purification of the reconstitutes is not necessary (Table I). This approach is recommended for most initial studies of transcription factor binding and has worked effectively in our hands in analyzing nucleosome binding of GAL4 derivatives, USF, and NFkB (9, 11, 25). The primary disadvantage to the approach is that the nonspecific DNA in the HeLa oligonucleosome cores remain in the binding reactions and may compete for binding of factors with low sequence specificity. For example, the HeLa oligonucleosome cores interfered with the binding of Sp1 to nucleosome cores bearing Sp1 sites. This prompted us instead to utilize reconstitution with pure histones onto the probes in the presence of carrier DNA (8).

Nucleosome Reconstitution by Octamer Transfer

1. Approximately 2–50 ng of nucleosome-length end-labeled DNA probe [0.04–1 $\times$ 10^6 counts/min (cpm)] is mixed with 5–10 μg of HeLa H1-depleted oligonucleosome cores (as determined by DNA concentration). The NaCl concentration of the mix is brought to 1 M in 20 μl total volume.
2. Incubate for 20 min at 37°C.
3. Serially dilute the transfer reaction to 0.85, 0.65, 0.5, and 0.3 M NaCl by adding 3.6, 7, 9.4, and 26 μl of 50 mM HEPES (pH 7.5), 1 mM EDTA, 5 mM dithiothreitol (DTT), 0.5 mM PMSF with incubations at 30°C for 30 min at each dilution step.
4. Bring the reaction to 0.1 M NaCl with final dilution buffer [10 mM Tris-HCl (pH 7.5), 1 mM EDTA, 0.1% NP-40, 5 mM DTT, 0.5 mM PMSF, 20% glycerol, 100 μg/ml bovine serum albumin (BSA)] and incubate at 30°C for 30 min.
5. The final diluted mixture can be used directly in binding reactions. Alternatively, it can be divided into aliquots, frozen on dry ice, and stored at −80°C.

Note: Naked DNA control samples are prepared by mixing all of the components except the probe DNA (preferably in a mock transfer including all incubations). The probe DNA is then added after the sample has reached 0.1M NaCl so that all of the components of the mixture are the same; however, octamer transfer does not occur. The control DNA samples cannot be freeze–thawed without at least some transfer of histones onto the probe DNA.

The efficiency of nucleosome core reconstitution is dependent on the amount of H1-depleted oligonucleosome cores included in the binding reactions (i.e., the ratio of oligonucleosomes to probe DNA). This is illustrated in Fig. 3A, which shows results of a mobility shift assay (see below) of nucleosome core reconstitution. The fraction of the probe reconstituted into nucleosome cores increases with the amount of oligonucleosomes included in the transfer reactions until almost all of the probe is reconstituted.

Nucleosome reconstitution can also be revealed by rotational phasing of the DNA helix on the surface of the histone octamer, which gives rise to a 10 to 11-bp repeating pattern of DNase I cutting (see below) which is not apparent on the naked DNA controls. However, a given DNA fragment only becomes rotationally phased if it has a preference to bend in one direction around the nucleosome core. DNA fragments do not demonstrate rotational phasing also efficiently reconstitute into nucleosome cores. This is illustrated in Fig. 3B, C. Figure 3B shows nucleosome core reconstitution of three different DNA fragments by the mobility shift assay. However, DNA

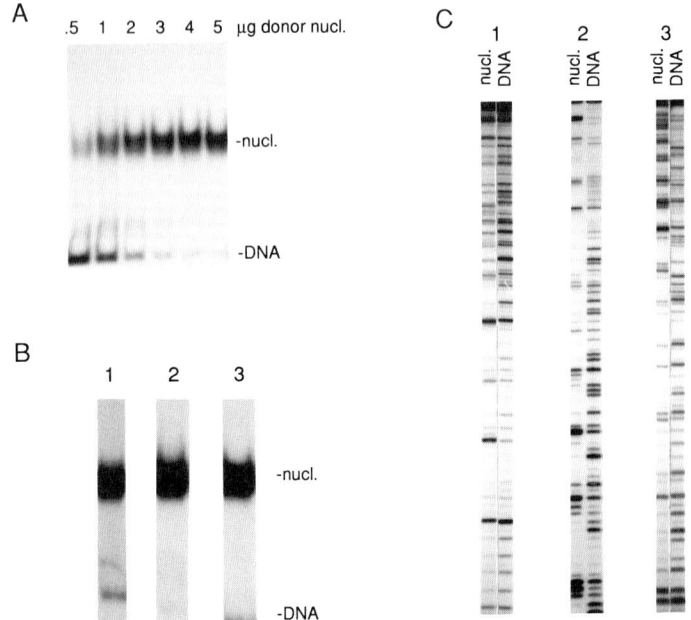

FIG. 3   Nucleosome reconstitution by octamer transfer. (A) Titration of the amount of H1-depleted nucleosome cores used in the transfer reaction. Six different transfer reactions were done using a 154-bp end-labeled DNA fragment (2.5 ng, 40,000 cpm) and the indicated amount of donor nucleosome cores (measured by the DNA content). Aliquots (10 μl) of the transfer reactions were then analyzed on a native 4% acrylamide, 0.5× TBE gel. Bands corresponding to reconstituted nucleosome cores and free DNA are indicated. (B) Three slightly different 154-bp DNA probes (50 ng, 8 × 10^5 cpm) were reconstituted using 5 μg of donor nucleosome cores. Then, 1-μl aliquots were diluted and analyzed as in (A). Lane 1 shows reconstitution of a DNA probe containing a centered synthetic 30-bp bending sequence; lane 2, a DNA probe containing the same bending sequence but at the end of the fragment; lane 3, a DNA probe without artificial bending sequences. (C) The same reconstituted nucleosome cores numbered as in (B) were analyzed by DNase I digestion as described for footprinting assays. Mock reconstitution DNA lanes for each were also included to visualize the 10-bp periodicity of DNase I sensitivity in rotationally phased nucleosomes (lanes 2 and 3) compared to naked DNA.

fragment 1 does not demonstrate rotational phasing of the DNA helix on the nucleosome core by DNase I digestion (Fig. 3C). Little cleavage periodicity was observed beyond that on the naked DNA. By contrast, a 10- to 11-bp cleavage periodicity was observed for reconstituted nucleosomes 2 and 3. The rotational phasing on nucleosome 2 was established by the inclusion of an artificial piece of bent DNA on one side of the fragment. Nucleosome 1 also contained this bent DNA sequence, but it was located across the dyad axis (center of the nucleosome) where it was unable to establish rotational phasing. Nucleosome 3 does not contain bent DNA sequences and fortuitously took up a specific rotational phase on nucleosome core reconstitution.

Analysis of Factor Binding to Transferred Nucleosome Cores

Once nucleosome reconstitution is achieved, the analysis of transcription factor binding can proceed by approaches also used on naked DNA substrates (i.e., mobility shifts and DNase I footprinting).

Mobility Shift Assays

1. Typically each binding reaction is performed with $2-5 \times 10^3$ cpm of reconstituted probe containing 25–250 ng of donor HeLa nucleosome cores (a fraction of the initial reconstitution mixture) in volumes of 10–20 μl for 30 min at 30°C [in most cases no additional carrier DNA like poly(dI-dC) is added]. Binding conditions vary with different transcription factors, but normally 50–100 mM NaCl, 5% glycerol, and 250 μg/ml BSA are included. Concentrations of transcription factors have to be determined and can range from 1 nM to 10 μM.
2. Binding reactions are directly loaded on a native 4% acrylamide (acrylamide to bisacrylamide ratio of 29 : 1), 0.5 × TBE gel (1.5 mm thickness) and run at 150 V for 2.5–3 hr at room temperature.
3. The gel is then dried for 1 hr at 80°C and exposed overnight with an intensifying screen at −80°C.

An example of factor binding to nucleosome cores as assayed by mobility shifts is shown in Fig. 4A. A nucleosome core containing a single GAL4 site 32 bp from one end was incubated with increasing concentrations of GAL4-AH. As can be seen in the mobility shift gel, the residual naked DNA in the binding reaction was readily bound by GAL4-AH at the lower GAL4-AH concentrations. Nucleosome binding, however, required higher concentrations of GAL4-AH and is revealed by the formation of a GAL4-AH/nucleosome complex with reduced mobility relative to GAL4-AH/DNA.

DNase I Footprinting Assays

1. Binding reactions are performed as above but are scaled up 2- to 3-fold to obtain enough final counts for sequencing gel exposures.
2. DNase I is added at the end of the binding reactions at a concentration of 100 units/ml for nucleosome cores and 10 units/ml for mock-reconstituted DNA controls.
3. Incubate at room temperature for 1 min.

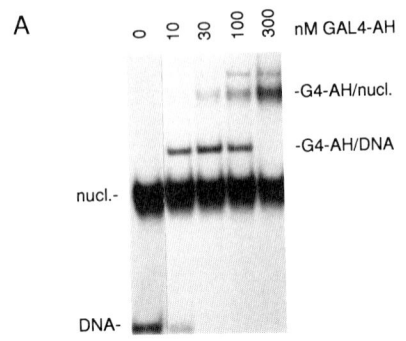

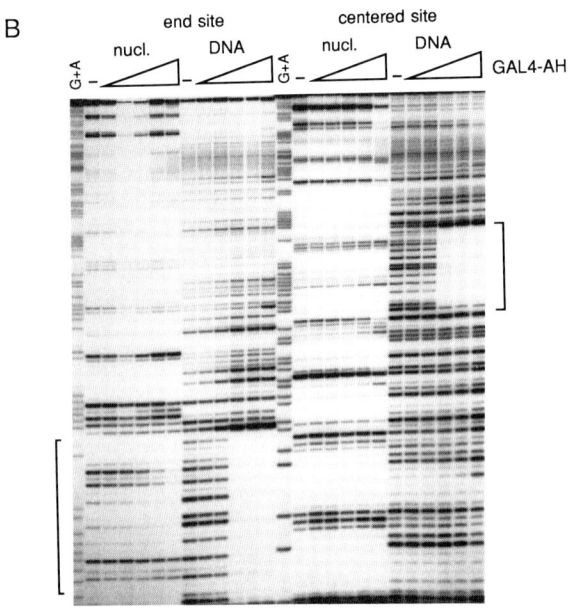

FIG. 4   GAL4-AH binding to transferred nucleosome cores. (A) A mobility shift assay was performed as described in the text using 2000 cpm of transferred nucleosome cores (0.1 ng DNA probe and 25 ng cold donor nucleosomes) and 0–300 nM of GAL4-AH. Binding reactions included 50 mM NaCl and 1 μM ZnCl$_2$ and were performed for 30 min at 30°C. The GAL4 binding site was located 32 bp from the end of the fragment. Identities of the different complexes are indicated. Because a higher GAL4-AH concentration is needed to bind nucleosomal DNA, under these conditions nonspecific binding of a second GAL4-AH dimer to naked DNA begins to appear. (B) Binding reactions with increasing amounts of GAL4-AH were analyzed by DNase I footprinting assays as described in the text. The DNA probe used in lanes 2–13 has the GAL4 site 32 bp from the end (end site), whereas the one in lanes 15–26 has it 75 bp from the end (centered site). GAL4-AH titrations were 0, 1, 3, 10, 30, and 100 nM with the mock-reconstituted DNA templates (lanes 8–13 and 21–26, respectively) and 0, 10, 30, 100, 300, and 1000 nM with the nucleosome cores (lanes 2–7 and15–20 respectively). G+A markers are included (lanes 1 and 14).

4. Stop the digestion by adding 1 volume of 20 mM Tris-HCl (pH 7.5), 50 mM EDTA, 2% SDS, 0.25 mg/ml yeast tRNA, and 0.2 mg/ml proteinase K.
5. Incubate samples for 1 hr at 50°C.
6. Precipitate samples by adding 0.5 volume of 7.5 M ammonium acetate and 3 volumes of ethanol. Mix well to avoid SDS precipitation.
7. Wash pellets with 80% (v/v) ethanol and resuspend them in 2 μl water.
8. Add 3 μl of 95% formamide, 10 mM EDTA, 0.1% xylene cyanol, 0.1% bromophenol blue, heat for 5 min at 90°C, and cool on ice.
9. Load samples on an 8% acrylamide (acrylamide–bisacrylamide ratio 19:1), 8 M urea, 1X TBE sequencing gel, which is run at 60 W constant power for 1 hr 45 min and exposed wet with an intensifying screen for 2 days at −80°C.

GAL4-AH binding to two different nucleosome cores as assayed by DNase I footprinting is illustrated in Fig. 4B. Both nucleosome reconstitutes were rotationally phased as revealed by the periodic digestion pattern of the nucleosomes versus the DNA controls. Binding of GAL4-AH to the nucleosome core with the GAL4 site 32 bp into the nucleosome occurred at between 30- and 100-fold higher concentration of GAL4-AH than required for binding the same fragment as naked DNA. When the GAL4 site was located in the center of the nucleosome core, a clear footprint was not observed at even 100-fold more GAL4-AH than required to bind naked DNA. The difference in GAL4-AH affinity for two nucleosomes is due to the nucleosome position effect (11). The analysis of position effects on factor binding is described elsewhere in this volume (26).

Reconstitution of Nucleosome Cores with Pure Histones

Reconstitution of nucleosome cores with pure core histones involves mixing the histones and DNA at 2 M salt followed by slow reduction of salt by dialysis (see below) or by dilution (24) to allow formation of nucleosome cores and reduce nonspecific interactions of histones with DNA which leads to precipitation. Carrier DNA is added to allow workable histone concentrations and a better match of histone to DNA ratios (which is difficult with trace amounts of labeled probe DNA only). A good starting histone–DNA ratio is 1:1. However, even at those ratios only about 50% of the probe DNA will be reconstituted into nucleosome cores. The remainder will either escape association with the histones or will be precipitated. Thus, purification of the reconstituted nucleosome cores on sucrose gradients is necessary prior to using them in binding studies (Table I). The amount of nonspecific DNA remaining in the collected gradient fractions depends on its molecular weight. If high molecular weight carrier DNA is used, it can be effectively separated as reconstituted oligonucleosomes from the mononucleosome cores bearing the probe DNA on the sucrose gradients. In addition, although the following protocol utilizes calf thymus DNA, it is also possible to use other carrier DNAs which have a lower capacity to compete for factor binding, for example, poly(dI·dC).

Nucleosome Core Reconstitution by Dialysis from High Salt and Gradient Purification

1. Approximately 150–300 ng of nucleosome-length end-labeled DNA fragment ($2-3 \times 10^6$ cpm) is mixed with 75 μg of calf thymus DNA (>600 bp as checked by agarose gel electrophoresis) and 75 μg of core histones (as prepared above) in 50 μl total volume containing 10 mM HEPES (pH 7.5), 2 M NaCl, 1 mM EDTA, 1 mM 2-mercaptoethanol, 1 mg/ml BSA. The amount of histones and DNA can be reduced. However, smaller volumes are difficult to handle in standard dialysis procedures, and more dilute samples do not reconstitute as efficiently.

2. The reaction is dialyzed for 2 hr at 4°C against 200 ml of the same buffer without BSA.

3. Continue dialysis overnight with the slow addition of 600 ml of the same buffer without salt to reduce the NaCl concentration to 0.5 M. Add the buffer without salt via a peristaltic pump at 36 ml/hr.

4. Dialyze against 500 ml of the same buffer containing 250 mM NaCl for 2 hr at 4°C.

5. In SW55 tubes (Beckman) pour 5-ml gradients of 5–25% (top–bottom) sucrose containing 10 mM HEPES (pH 7.5), 1 mM EGTA, 0.1% NP-40, 0.1 mM PMSF (pour at a flow rate of ~0.25 ml/min using a small two-chamber gradient maker, e.g., SG series from Hoefer, San Francisco, CA).

6. Carefully load the dialyzed sample at the top of one gradient, balance with another gradient, and centrifuge for 18 hr at 34,500 rpm (110,000 g) in a SW55 rotor at 4°C (or 45,000 rpm for 11 hr 30 min).

7. Collect 175- to 200-μl fractions from the bottom of the gradient and count 5 μl of each in scintillation liquid to identify the nucleosome core and naked DNA peaks (Fig. 5A).

8. Aliquots (10 μl) of each fraction are also analyzed by migration on a native 4% acrylamide–0.5 × TBE gel (under same conditions as described above for analysis of transcription factor binding to nucleosome cores). The fractions containing the reconstituted nucleosomes are readily identified by their change in mobility (Fig. 5B).

9. Fractions containing nucleosome cores reconstituted on the DNA probe with little contamination by free DNA are pooled, divided into aliquots, frozen on dry ice, and stored at −80°C.

The gradient-purified nucleosome cores can be used in binding reactions similar to those described above for nucleosomes formed by octamer transfer. An example of GAL4-AH binding to nucleosome cores formed with pure histones following gradient purification is shown in Fig. 5C. However, DNase I footprinting assays using these should require 10-fold less nuclease compared to transferred nucleosome cores.

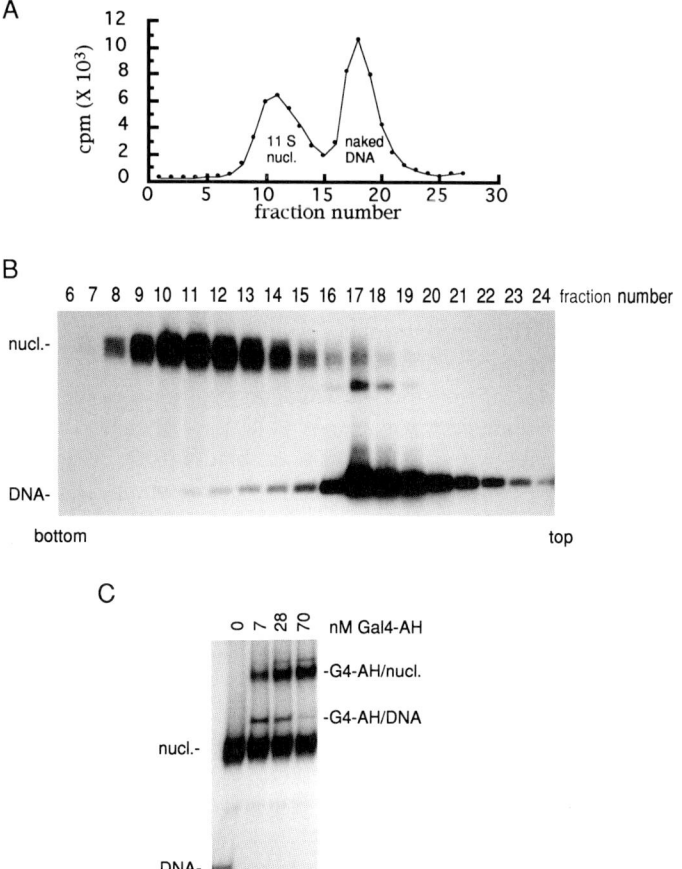

Fig. 5   Purification of nucleosome cores reconstituted with pure histones. (A) Fractionation on sucrose gradients of the same DNA probe used in Fig. 4A reconstituted with pure core histones and carrier DNA as described in the text. Gradient fractionation is monitored by liquid scintillation counting of 5-μl aliquots of each fraction (with the first fraction coming from the bottom of the gradient). (B) Aliquots (10 μl) of each fraction are also analyzed on a native 4% acrylamide gel to visualize contamination of nucleosome core fractions with naked DNA. (C) Mobility shift assays were performed using purified nucleosome cores from (B) and increasing amounts of GAL4-AH (0–70 nM). Dilution of the purified nucleosome cores is kept minimal during the binding reaction to avoid breakdown of the nucleosome cores. Here, 9 μl purified nucleosome cores was used in each 15-μl binding reaction.

Note: Simply by virtue of dilution on the gradient, some nucleosome cores will dissociate into naked DNA. Thus, it is impossible to achieve a purified nucleosome fraction completely free of naked DNA.

Reconstitution of Homogenous Sequence Nucleosome Cores

It is possible to produce large quantities of homogeneous sequence nucleosome cores by dialysis from high salt with purified histones. This requires microgram quantities of a specific nucleosome-length DNA fragment. Such fragments can be raised from multiple polymerase chain reaction (PCR) cycles with appropriate primers (26, 27). In this instance, carrier DNA is omitted and histone concentrations are matched to the specific fragment. The primary advantage of this approach is that it yields high concentrations of purified homogeneous sequence nucleosome cores (Table I). If enough purified factors are available that bind the nucleosomes, the final protein composition of the factor/nucleosome ternary complexes can be directly assessed by SDS-PAGE of the mobility shifted complexes followed by silver staining (9, 26, 27). It is important to note that the actual concentration of the nucleosomes bearing binding sites prepared this way can be as high as 100 nM (Table I). Thus, factor binding can be driven by the concentration of binding sites. Use of homogeneous sequence nucleosome cores in the analysis of protein composition of factor/nucleosome ternary complexes is described elsewhere in this volume (26).

Reconstitution of Chromatosomes and Transcription Factor Binding

In addition to analyzing transcription factor binding to nucleosome cores, it is also possible to reconstitute chromatosomes (a nucleosome core bound by histone H1) to investigate the effects of H1 on factor binding. As shown in Fig. 6, H1 binding to the nucleosome core can further suppress transcription factor binding at the edge of the nucleosome. Histone H1 binds to nucleosome cores where the DNA enters and exits the nucleosome core as well as at the dyad axis (the center of the nucleosome). Reconstitution of chromatosomes was achieved with mixed sequence cellular nucleosome cores years ago (28), and the technique has been applied to nucleosome cores bearing specific DNA fragments (15, 16). In the following protocol we describe H1 binding to nucleosome cores generated by octamer transfer. However, in principle similar approaches could be taken with nucleosome cores reconstituted with pure histones by adding H1 at similar salt concentrations.

H1 Reconstitution onto Nucleosome Cores

1. Initially, chromatosome reconstitution is the same as a nucleosome core reconstitution reaction (see above) where labeled probe is incubated in the presence of H1-depleted oligonucleosomes in 10 mM HEPES (pH 8.0), 1 mM EDTA, 1 M NaCl for 20 min at 37°C.
2. The reaction is then successfully diluted to 0.8 and 0.6 M NaCl with 10 mM HEPES (pH 8.0), 1 mM EDTA and 15-min incubations at 37°C following each dilution.
3. At this point the 0.6 M NaCl transfer reaction can be divided in various ways to add H1. For H1 titrations (Fig. 6A), the reaction is typically separated into 5 to 10

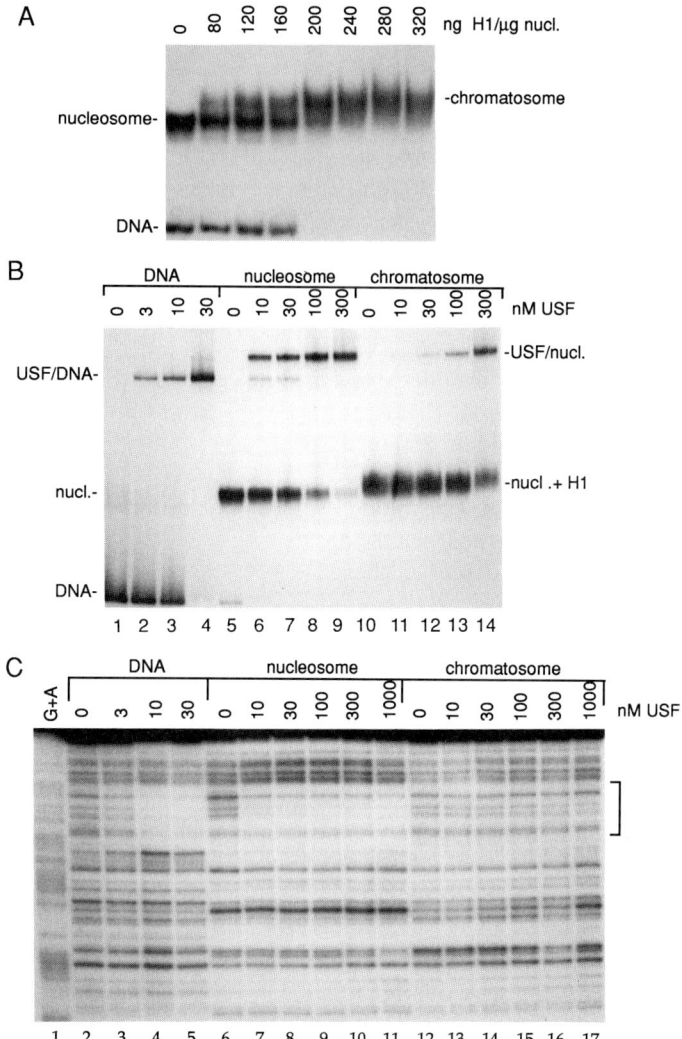

FIG. 6 (A) H1 binding to nucleosome cores to form chromatosomes. A 172-bp nucleosome core reconstituted on a sea urchin 5 S RNA gene probe by octamer transfer was reconstituted in the presence of increasing concentrations of histone H1 (presented here as nanograms H1 per microgram total nucleosomal DNA). Note that under these conditions, a molar ratio of 1 : 1 between nucleosome cores and H1 should require around 200 ng of the linker histone. H1 binding to the nucleosome core resulted in its shifting in the gel to a chromatosome complex. H1 binding aggregated the free DNA which remained in the well (not shown). (B, C) H1 binding to nucleosome cores inhibits the binding of USF to the edge of the nucleosome. A 183-bp end-labeled probe was mock reconstituted (B: lanes 1–4; C: lanes 2–5) or reconstituted into nucleosome cores (B: lanes 5–9; C: lanes 6–11) or into chromatosomes by binding of H1 (B: lanes 10–14; C: lanes 12–17). Binding reactions contained 0.3 μg nucleosomes ($\pm$ 60 ng H1) and increasing amounts of USF. Repression of USF binding was apparent by both mobility shift and DNase I footprinting analysis. For the mobility shift gel shown in (B), samples were directly loaded on a 4% native acrylamide gel. For DNase I footprinting (C), the binding reactions were digested with DNase I as described in the text. Samples were resolved on an 8% acrylamide, 8M urea sequencing gel. G+A markers are shown in lane 1.

samples, and increasing amounts of H1 [0–320 ng H1/μg total nucleosomes (including the H1-depleted oligonucleosomes)] are added in equal volumes of 0.6 M NaCl, 50 mM sodium phosphate (pH 6.0), 0.1% NP-40, 1 mM PMSF and incubated at 37°C for 30 min. To assess factor binding to chromatosomes it is often desirable to choose one amount of H1 (usually enough to give complete binding) and complete the transfer with two samples (without and with H1).

4. Subsequent dilutions to 0.4 and 0.2 M NaCl are done with 10 mM Tris (pH 7.8), 1 mM EDTA, 0.1% NP-40, 1 mM 2-mercaptoethanol and incubating at 37°C for 30 min.

5. The samples are diluted to a final concentration of 0.1 mM NaCl with 10 mM Tris (pH 7.8), 1 mM EDTA, 0.1% NP-40, 1 mM 2-mercaptoethanol, 20% glycerol, and incubated for 30 min at 30°C.

Note: For free DNA controls, mock reconstitutions are performed such that the labeled probe is added at the final 0.1 M NaCl step to prevent nucleosome transfer to probe DNA as described above.

Analysis of Factor Binding to Chromatosomes

1. Binding reactions are prepared by dividing the final 0.1 M NaCl reactions into samples containing 0.3 μg total nucleosomes. The same range of factor concentrations used for nucleosomal templates is usually sufficient for chromatosome templates, although H1 can cause some inhibition of binding.

2. Binding reactions are commonly carried out at 30°C for 30 min.

3. At this point all or part of the sample can be directly loaded on a 4% native acrylamide gel and resolved at 150 V for 3 hr (for optimal separation of H1-containing nucleosomes). Reactions can also be assayed by DNase footprinting as described above.

Figure 6A shows an H1 titration and its binding to nucleosome cores. Note that the free DNA is generally shifted to the well by aggregation with H1. The binding of H1 reduces the mobility of the nucleosome. The binding of GAL4 derivatives to nucleosome cores is unaffected by H1 binding (15). However, binding of the human factor USF to a site on the edge of a nucleosome core is repressed by H1 binding forming a chromatosome. Inhibition of USF binding by histone H1 is apparent by both mobility shift (Fig. 6B) and DNase I footprinting assays (Fig. 6C).

Conclusion

In this chapter we have described the preparation of H1-depleted oligonucleosome cores, core histones, and H1 and the use of these reagents in the initial analysis of transcription factor binding to nucleosomes. If such activity is observed, a more detailed analysis of the interactions of the transcription factor with nucleosome cores can ensue. These include analysis of nucleosome position effects on factor binding, cooperative binding with the same or different factors, stimulation of binding by accessory protein complexes, and analysis of the protein composition of factor/ nucleosome ternary complexes. These methods are described elsewhere in this volume (26).

References

1. J. Svaren and W. Horz, *Curr. Opin. Genet. Dev.* **3**, 219 (1993).
2. J. L. Workman and A. R. Buchman, *Trends Biochem.* **18**, 90 (1993).
3. J. L. Workman, I. C. A. Taylor, R. E. Kingston, and R. G. Roeder, *Methods Cell. Biol.* **35**, 419 (1991).
4. G. E. Croston, L. M. Lira, and J. T. Kadonaga, *Protein Expression Purif.* **2**, 162 (1991).
5. Q. Li and O. Wrange, *Genes Dev.* **7**, 2471 (1994).
6. T. K. Archer, M. G. Cordingley, R. G. Wolford, and G. L. Hager, *Mol. Cell. Biol.* **11**, 688 (1991).
7. I. C. A. Taylor, J. L. Workman, T. J. Schuetz, and R. E. Kingston, *Genes Dev.* **5**, 1285 (1991).
8. B. Li, C. C. Adams, and J. L. Workman, *J. Biol. Chem.* **269**, 7756 (1994).
9. H. Chen, B. Li, and J. L. Workman, *EMBO J.* **13**, 380 (1994).
10. D. S. Wechsler, O. Papoulas, C. V. Dang, and R. E. Kingston, *Mol. Cell. Biol.* **14**, 4097 (1994).
11. M. Vettese-Dadey, and P. Walter, H. Chen, L.-J. Juan, and J. L. Workman, *Mol. Cell. Biol.* **14**, 970 (1994).
12. D. Y. Lee, J. J. Hayes, D. Pruss, and A. P. Wolffe, *Cell (Cambridge, Mass.)* **72**, 73 (1993).
13. J. Côté, J. Quinn, J. L. Workman, and C. L. Peterson, *Science* **265**, 53 (1994).
14. G. Meersseman, S. Pennings, and E. M. Bradbury, *EMBO J.* **11**, 2951 (1992).
15. L.-J. Juan, P. Walter, I. C. A. Taylor, R. E. Kingston, and J. L. Workman, *Cold Spring Harbor Symp. Quant. Biol.* **58**, 213 (1993).
16. J. J. Hayes and A. P. Wolffe, *Proc. Natl. Acad. Sci. U.S.A.* **90**, 6415 (1993).
17. A. P. Wolffe and J. J. Hayes, *Methods Mol. Genet.* **2**, 314 (1993).
18. D. R. Engelke and J. M. Gottesfeld, *Nucleic Acids Res.* **18**, 6031 (1990).
19. B. Neubauer and W. Hörz, *in* "Methods in Enzymology" (P. M. Wassarman and R. D. Kornberg, eds.), Vol. 170, p. 630. Academic Press, San Diego, 1989.
20. R. H. Simon and G. Felsenfeld, *Nucleic Acids Res.* **6**, 689 (1979).
21. A. Stein and M. Mitchell, *J. Mol. Biol.* **203**, 1029 (1988).
22. A. Stein, *in* "Methods in Enzymology" (P. M. Wassarman and R. D. Kornberg, eds.), Vol. 170, p. 585. Academic Press, San Diego, 1989.
23. L. Sealy, R. R. Burgess, M. Cotton, and R. Chalkley, *in* "Methods in Enzymology" (P. M. Wasserman and R. D. Kornberg, eds.) Vol. 170, p. 612. Academic Press, San Diego, 1989.
24. D. Rhodes and R. A. Laskey, *in* "Methods in Enzymology" (P. M. Wassarman and R. D. Kornberg, eds.), Vol. 170, p. 575. Academic Press, San Diego, 1989.

25. C. C. Adams and J. L. Workman, *Mol. Cell Biol.,* **15**, 1405 (1995).

26. M. Vettese-Dadey, C. C. Adams, J. Côté, P. Walter, and J. L. Workman, Chapter [7], this volume.

27. J. L. Workman and R. E. Kingston, *Science* **258**, 1780 (1992).

28. P. P. Nelson, S. C. Albright, J. M. Wiseman, and W. T. Garrard, *J. Biol. Chem.* **254**, 11751 (1979).

[7] Experimental Analysis of Transcription Factor–Nucleosome Interactions

Michelle Vettese-Dadey, Christopher C. Adams, Jacques Côté,
Phillip Walter, and Jerry L. Workman

Introduction

It is becoming increasingly clear that the cell possesses mechanisms to disrupt pre-existing nucleosomes at enhancer and promoter elements prior to or concurrent with the activation of gene transcription. *In vivo* studies indicate that such mechanisms can function independently of DNA replication (1–6), which has often been suggested as a pathway for chromatin remodeling. Nucleosome disruption and/or displacement on inducible genes has been shown to be initiated by the binding of inducible transcription factors (reviewed in Refs. 7 and 8). It follows, then, that apparent nucleosome-free regions which are constitutively present may form as a result of the binding of constitutive factors. These observations implicate transcription factor binding in initiating nucleosome displacement from gene regulatory elements. Indeed, both inducible and constitutive transcription factors have been shown to bind to their sequence elements when reconstituted into nucleosome cores *in vitro*. These include such diverse factors as the glucocorticoid receptor (9–11), derivatives of the yeast factor GAL4 (12, 13), Sp1 (14), USF (15, 16), Myc/Max and Max/Max dimers (17), and NFkB (16). Thus, nucleosomes do not present an insurmountable barrier to transcription factor access. However, the affinities of factors for nucleosomal DNA differs (11, 12), suggesting that factors may perform a distinct function in initiating binding and in disrupting nucleosomes (8).

In addition to the inherent ability of a transcription factor to bind nucleosomal DNA, there are several mechanisms which can further enhance nucleosome binding. These include cooperative nucleosome binding with the same or different factors which can increase the affinity of individual transcription factors from 10-fold to greater than 100-fold (12, 13, 16). Inhibition of factor binding is often mediated via the core histone amino termini, and binding can be stimulated by acetylation of lysine residues within these domains (13, 18). Transcription factor binding can also be driven by the function of additional cellular proteins. The histone-binding protein nucleoplasmin stimulates transcription factor binding to nucleosomes and can remove H2A/H2B dimers from factor/nucleosome ternary complexes (15). ATP-dependent nucleosome disruption *in vitro* by the GAGA factor implicates accessory activities (19). The yeast SWI/SNF complex has been shown to utilize the energy of ATP hydrolysis to disrupt nucleosome structure and boost the binding of GAL4 derivatives (20). The observed *in vitro* functions of nucleoplasmin and the SWI/SNF complex suggest that induction of transcription factor binding is directly linked to nucleosome disruption and disassembly. Consistent with this idea, the binding of multiple GAL4-derivatives to a nucleosome core destabilizes the histone octamer, allowing histone displacement onto competitor molecules (15, 21).

In this chapter we describe advanced techniques to analyze the interactions of an individual or group of factors with nucleosome cores. We present the basic techniques to analyze transcription factor binding to nucleosomes elsewhere in this volume (22). The methods presented here allow an investigation of the parameters governing the binding of a particular factor to nucleosomes, including nucleosome position effects and the role of the core histone amino termini. In addition, we describe approaches to analyze cooperativity of factor binding and the function of accessory protein complexes (i.e., nucleoplasmin, SWI/SNF) in factor binding. Finally, we present protocols to directly assess the proteins present (factors and histones) in ternary complexes resulting from binding of transcription factors to nucleosomes. These approaches can be applied to analyze the binding of an individual transcription factor. Alternatively, the methods can be used to address the function of multiple factors in binding and displacing nucleosomes reconstituted on complex enhancer or promoter elements.

Analysis of Nucleosome Position Effects on Transcription Factor Binding

Nucleosome positioning is often suggested to play a major role in the regulation of gene transcription. However, nucleosome position effects may often be subverted by regulatory mechanisms (23). For example, the *in vitro* positioning of nucleosomes on the *Xenopus borealis* and *Xenopus laevis* 5 S RNA genes differs substantially (18), suggesting that any potential effects of the different nucleosome positions *in vivo* are readily overcome by additional mechanisms [nucleosome sliding, histone acetylation, etc. (18, 24)]. Another mechanism by which inhibition of factor binding to the very center of a nucleosome core can be overcome is by cooperative binding with additional factors (13). To determine the effect of nucleosome position on factor binding it is necessary to manipulate experimentally the position of the binding site within the nucleosome (25). This information can then be used to assess the ability of additional activities to circumvent nucleosome position effects.

There are two components to nucleosome positioning that can be readily tested for an effect on transcription factor binding. The first is the translational position of the site in the nucleosome (i.e., at the beginning versus the center of the nucleosomal DNA). Translational position will determine if a binding site is in a position of H2A/H2B dimer interactions or in a position of interaction of the H3/H4 tetramer. The second is the rotational orientation of the binding site on the surface of a nucleosome core. Rotational phasing of DNA on the surface of a nucleosome occurs if a particular

DNA fragment has a preference for bending around the histone octamer in a particular direction. This exposes one face of the helix (and potentially a binding site) to the solvent while the other faces the histone octamer.

pBend vectors were designed to facilitate the analysis of bent DNA by allowing the movement of an inserted DNA sequence relative to the ends of a larger fragment (26). These constructs have also been used to address DNA bending by the binding of transcription factors (27). These plasmids consist of tandemly repeated polylinker regions separated by unique restriction sites into which protein binding sites can be inserted (see Fig. 1A). Cutting these constructs with any one enzyme within the polylinkers will generate a DNA fragment containing the inserted binding site, and the location of the binding site with respect to the ends of the excised fragment will differ depending on its proximity to the restriction sites used. These fragments are useful for the reconstitution of nucleosomes having binding sites at different positions on their surface, provided the distance between each pair of restriction sites is greater than nucleosome-core length, 146 bp (see Fig. 1A). If the distance between pairs of restriction sites is considerably longer, fragments can be shortened by cutting again with a second enzyme near one end of the fragment (see Fig. 1B). Below we describe a general strategy to construct such a vector.

Construction of Vectors for Analysis of Translational Positioning

1. Bend vectors are transformed and maintained in recombination-deficient *Escherichia coli* host strains (i.e., Sure strain, Stratagene, La Jolla, CA) to reduce recombination resulting from the repeated polylinkers. Prepare plasmid stocks from these strains.
2. Prepare a double-stranded oligonucleotide 20–30 bp in length bearing the recognition sequence for the protein of interest with compatible ends for one of the unique restriction sites located between the repeated polylinkers.
3. Following linearization and dephosphorylation of the recipient pBend vector, ligate the phosphorylated double-stranded oligonucleotide to insert it between the repeated polylinkers. Transform and amplify the ligation products in a recombination-deficient strain of *E. coli*.
4. The ligation products should be sequenced to determine the number and orientation of the oligonucleotides inserted and tested with several restriction enzymes to confirm that the two polylinkers are intact.

Fragments from these plasmids can be used directly to measure translational position effects on transcription factor binding to nucleosomes. However, the rotational phasing of the fragments when reconstituted into nucleosome cores may vary dependent on where the fragment is cut, or multiple rotational phases may exist in a single reconstituted sample. Nucleosome binding by the factor tested may be independent

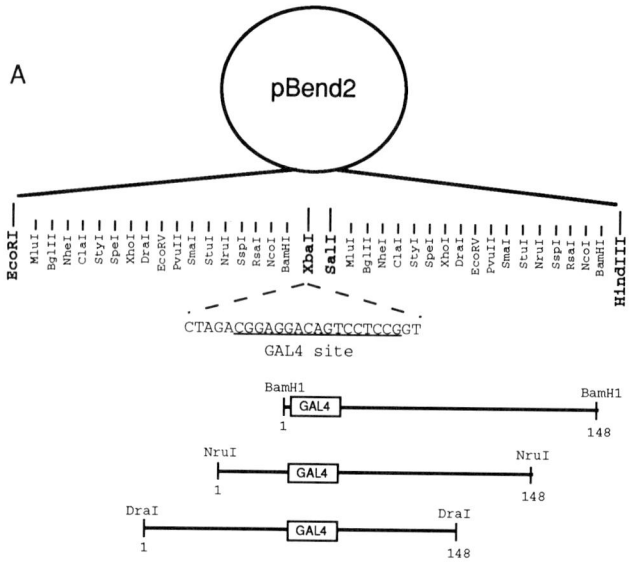

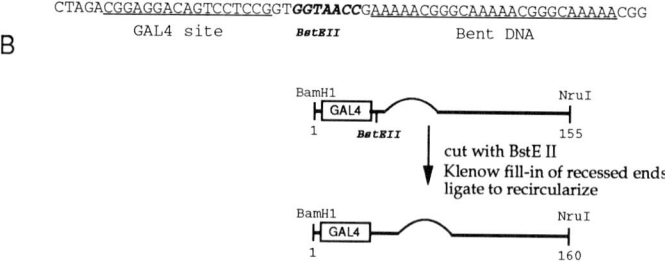

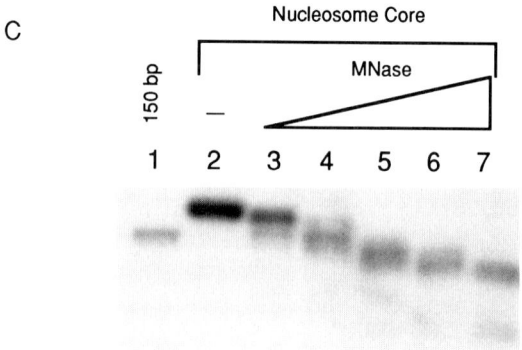

FIG. 1 Construction of pBend derivative plasmids for evaluating and nucleosome position effects on the binding of GAL4 derivatives. (A) Engineering DNA probes for investigating

of rotational orientation, as observed for GAL4 derivatives (12). This can be tested directly if the rotational orientation of the DNA helix in the reconstituted nucleosomes is predetermined. This is done by including a bent DNA sequence, in addition to the desired protein binding site, in the inserted oligonucleotide which, by virtue of their direction of bending, will determine the orientation of the DNA helix on the surface of the histone octamer (25, 28). The orientation of the DNA helix at the binding site will then be determined by the distance between the bent DNA sequences and the binding site. The binding site can be turned over by inserting or deleting 4–5 bp between it and the bent DNA (see example in Fig. 1B).

Construction of Vectors for Analysis of Rotational Positioning

1. The pBend plasmid is linearized at a unique restriction site in between the repeated polylinkers and dephosphorylated with phosphatase enzyme.

---

translational position effects. A 24-bp oligonucleotide containing a 17-bp consensus GAL4 binding site was inserted into the unique *Xba*I site separating the repeated polylinkers of plasmid pBend2. Nucleosome-length DNA fragments containing the GAL4 binding site are excised from the resulting plasmid by cleavage with a single enzyme cutting within the repeated polylinkers. All excised fragments are of equal length regardless of which restriction enzyme is selected; however, the position of the GAL4 binding site on the surface of reconstituted nucleosomes will vary. (B) Engineering DNA probes for investigating rotational position effects. A 60-bp oligonucleotide containing a 17-bp consensus GAL4 binding site and a 25-bp bent DNA sequence, separated by a unique restriction site (*Bst*EII), was inserted into the unique *XBa*I site between the repeated polylinkers of plasmid pBend2 as in (A). Owing to the length of the inserted oligonucleotide, cleavage by two different restriction enzymes within the repeated polylinkers is necessary to generate nucleosome-length DNA fragments for reconstitution into nucleosome core particles. The rotational alignment of the GAL4 site with respect to the surface of reconstituted nucleosome cores can be altered by inserting 5 bp (one-half helical turn of DNA) in between the GAL4 site and the DNA bend. This is accomplished by simply linearizing the plasmid with *Bst*EII, filling in the recessed ends, and recircularizing the plasmid by blunt-end ligation. Similarly, 5 bp can be deleted by removal of the 5-bp overhang with S1 nuclease followed by the blunt-end ligation to recircularize (not shown). (C) Micrococcal nuclease test for complete occupancy of 146-bp DNA by reconstituted nucleosome cores. Nucleosome cores were reconstituted onto a 160-bp DNA fragment containing an internal label. The reconstituted nucleosomes (lanes 2–7) were digested with 0.01 U/ml micrococcal nuclease for 0, 1, 3, 9, 15, and 30 min, respectively, followed by deproteinization of the DNA and analysis on a polyacrylamide nondenaturing gel. The nucleosome cores were initially trimmed (lane 4) to a size slightly smaller than the 150-bp marker (lane 1). This digestion intermediate is not observed with naked DNA (not shown), indicating that approximately 146 bp was protected by the histone octamer.

2. A double-stranded oligonucleotide is prepared containing bent DNA sequences (28) plus the protein binding site of interest, separated by the recognition sequence for a unique restriction enzyme that cleaves the DNA to produce a 4 to 5-bp 5'-overhang. The oligonucleotide should also contain compatible ends for one of the unique restriction sites located between the repeated polylinkers.

3. Ligate the oligonucleotide into the linearized recipient plasmid, transform cells, and check the transformants as above. (The inclusion of longer bent DNA sequences than the one shown in Fig. 1B may result in an even more dramatic rotational positioning.)

4. To alter the helical distance between the protein binding site and the DNA bend, grow and prepare the constructed plasmid from the transformation above.

5. Linearize the plasmid with the unique restriction enzyme engineered between the DNA bend and the protein binding site.

6. Take a portion of the linearized plasmid DNA and fill in the 5' overhangs with Klenow fragment. This will add 4–5 bp between the DNA elements.

7. Take another part of the digested plasmid and trim the single-stranded ends with S1 nuclease. This will remove 4–5 bp between the elements.

8. Recircularize the Klenow-treated and S1-treated plasmid by blunt-end ligation and retransform.

9. The transformants should be sequenced to determine the exact number of base pairs added or deleted between the binding site and bent DNA sequence.

Factor Binding to Different Positions on Nucleosome Cores

To analyze transcription factor binding to different translational positions on nucleosome cores, nucleosome-length fragments are cut out of the pBend vectors constructed above using different pairs of restriction enzymes. The first enzyme should cut at the required distance from the binding site, whereas the second enzyme chosen should shorten the fragment to near nucleosome-core length. It is convenient to end-label the plasmid with Klenow polymerase or kinase before cutting with the second enzyme so that DNase I footprinting can be performed on the reconstitutes. The probe DNA fragments are then purified on nondenaturing polyacrylamide gels prior to use.

To analyze rotational phasing effects, the same fragment should be cut out of the three different plasmids bearing both the binding site for the factor of interest as well as bent DNA sequences. By simultaneously analyzing both fragments with 4 bp inserted and deleted along with the fragment from the parent plasmid, rotational phasing effects on factor binding should be revealed independent of 4-bp changes in translational position.

The fragments are reconstituted into nucleosome cores by octamer transfer or by dialysis from high salt with carrier DNA as described by Côté et al. (22). The extent of reconstitution is revealed by the fraction of the probe contained in the nucleosome core complex on mobility shift gels, which should be greater than 80%. The extent of rotational phasing is determined by DNase I digestion of the nucleosome reconstitutes. Rotational phasing is revealed by a 10 to 11-bp periodicity in cleavage of the reconstituted nucleosome cores which is not apparent in control naked DNA digests (22).

In our experience, the ends of DNA fragments reconstituted into nucleosome cores have efficiently constrained the ends of the nucleosome core particle. In other words, the nucleosome cores occupied 146 bp of DNA rather than some smaller fraction of the fragment with naked DNA extending from the end of the nucleosome core. It is possible to confirm whether the ends of the fragment are constricting the location of the nucleosome core such that it occupies 146 bp of DNA by micrococcal nuclease trimming the reconstitutes, followed by the analysis of the lengths of the resulting double-stranded digestion products (13). If the nucleosome core contains 146 bp of DNA, this length digestion product will appear in the digestion time course. An example of such an analysis is shown in Fig. 1C. It is important to note that for the micrococcal nuclease digestion test, it is necessary to use probe fragments where the label is internal so that it is not rapidly removed by the nuclease. Such a probe can be made by kinasing a unique restriction site on the plasmid which is internal to the final fragment desired and then ligating the plasmid back into closed circles (described in Ref. 29). Double digestion of the labeled plasmid with the flanking enzymes defining the fragment of interest will release the probe with an internal label.

Transcription factor binding reactions to reconstituted nucleosome cores with binding sites at different positions can be performed under most conditions in which factor binding to naked DNA is observed. The reconstituted nucleosome cores can either make up a small or large fraction of the final binding reaction depending on the buffer conditions and volume of solutions of transcription factors added. An example of nucleosome position effects on GAL4-AH binding is illustrated in Fig. 2A, B.

Analysis of Cooperative Transcription Factor Binding to Nucleosomes

Cooperative transcription factor binding to nucleosomes was first observed with derivatives of GAL4 (12, 13). In more recent experiments utilizing GAL4 derivatives, USF, and NFkB, we have observed cooperative binding between any pair or all three of the factors to nucleosome cores bearing the respective binding sites (16). Thus, cooperative nucleosome binding appears to be a general mechanism which facilitates transcription factor access to nucleosome cores.

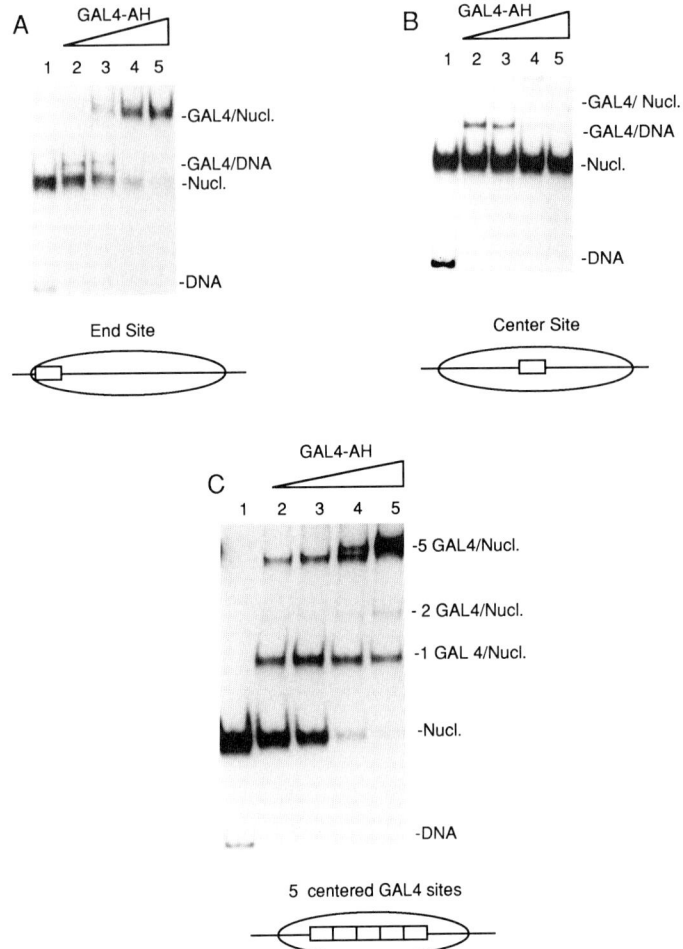

FIG. 2   Nucleosome position effects and cooperative binding of GAL4-AH to nucleosome cores. (A) Binding of GAL4-AH to nucleosome cores bearing a single GAL4 site near the edge of the nucleosome. A 160-bp probe with a single GAL4 site centered at 21 bp from an end was reconstituted into nucleosome cores (lanes 1–5). Nucleosome cores were incubated with increasing amounts of GAL4-AH, and binding was assayed by the electrophoretic mobility shift assay. The mobility of the reconstituted cores (Nucl.), a GAL4-AH dimer bound to DNA (GAL4/DNA), and a GAL4 dimer bound to the nucleosome core (GAL4/Nucl.) are indicated. The final concentrations of GAL4 used in the binding reactions were 0, 5.3, 15.9, 53, and 159 nM for lanes 1–5, respectively. Note that effective GAL4-AH binding was observed at the site at this location on the nucleosome core. (B) Binding of GAL4-AH to nucleosome cores bearing a single GAL4 site near the center of the nucleosome. Binding reactions were the same as above except that the center of the single GAL4 site was located 74 bp from an end. Note that binding is greatly decreased as the GAL4 site is moved from the end to the

Cooperative Binding of Same Factor to Nucleosome Cores

Cooperative binding of the same factor to multiple sites in nucleosome cores can be tested in comparison to single sites at different positions. This analysis will provide an estimation of the fold increase in affinity of a factor for a particular location in a nucleosome resulting from cooperative binding. An example of the increase in the affinity of GAL4-AH for nucleosome cores resulting from cooperative binding to multiple sites is illustrated in Fig. 2C.

To generate DNA fragments bearing multiple copies of a binding site, double-stranded oligonucleotides are produced which contain the binding site of interest and complementary ends for a convenient restriction site in pBend. Twenty base pairs is a convenient size for the oligonucleotides, although any size long enough to prevent steric occlusion of transcription factor binding should suffice. The double-stranded oligonucleotides are kinased and then ligated into multimers. Multimers containing different numbers of the oligonucleotide are isolated on nondenaturing acrylamide gels and ligated into phosphatased pBend vectors. Nucleosome-length fragments for nucleosome reconstitution can be excised from the resulting plasmids by proper selection of restriction enzymes.

Following reconstitution of the multisite fragments into nucleosome cores, titrations of transcription factor concentrations in binding reactions followed by mobility shift and/or DNase I footprinting analysis (22) should reveal the extent of cooperative binding that occurs. For example, in Fig. 2C, binding of five GAL4-AH dimers to nucleosome cores bearing five sites (the center 95 bp of the fragment) occurs at greater than 10-fold higher affinity than binding of a single dimer to the center of the nucleosome core (Fig. 2B). Thus, cooperative binding of multiple GAL4-AH dimers overcomes nucleosome positional inhibition of GAL4-AH binding.

Cooperative Binding of Unrelated Factors to Nucleosomes Cores

A variation on the above experiment can determine the ability of more than one different factor to bind to nucleosome cores cooperatively. DNA probes for such an

---

center of the nucleosome core. (C) Binding of GAL4-AH to nucleosome cores bearing five GAL4 sites across the center of the nucleosome. Binding reaction conditions were the same as in (A) except that a probe with five centered GAL4 sites was used. Note that following the binding of the first GAL4 dimer, the next most prominent complex is 5 GAL4 dimers bound to the nucleosome core indicating a cooperative mode of binding. Binding of all five sites occurred at a lower concentration than to a single site (refer to A and B). This fragment contains five centered sites and therefore contains less accessible central sites. Thus, cooperative binding to multiple sites alleviated the repression to binding at the center-most sites.

analysis can be generated by the cloning strategy described above for generating probes containing one factor binding site by simply inserting a longer oligonucleotide that contains binding sites for more than one factor. Alternatively, individual factor binding sites can be inserted separately into the unique restriction sites. Nucleosome-length fragments containing multiple binding sites are excised and labeled with the appropriate restriction enzymes. These fragments are then reconstituted into nucleosome cores as described (22).

An alternative approach is to generate nucleosome-length probes of natural promoter or enhancer sequences which bear binding sites for the factors of interest. These can be generated by restriction digestion of plasmids bearing the promoter or enhancer sequences if appropriate restriction sites are available to generate proper length fragments. The polymerase chain reaction (PCR) can also be used to generate appropriate length fragments by using oligonucleotides with 5′ ends approximately 150 bp apart and flanking the binding sites of interest (see below).

Cooperative nucleosome binding is tested by titrations of each factor in the presence or absence of the other and assayed by mobility shifts and DNase I footprinting. There are several variations on this experiment which should render useful data. For example, if one of the factors contains enough inherent ability to bind nucleosomal DNA on its own, the nucleosomes can be saturated with this factor and then the relative affinity of the second factor for the factor/nucleosome complex can be compared to its affinity to nucleosome cores alone. We have observed over a 100-fold increase in USF binding to a nucleosome core next to a previously bound GAL4-AH dimer (16). If neither factor efficiently binds nucleosome cores alone, one can be titrated in the presence of the other, in which case cooperative binding of both factors may be observed. It is important to note that the ability of two factors to bind a nucleosome core cooperatively may depend on the proximity of the binding sites. The most dramatic effects are observed when factor binding sites are within 20 bp of one another.

Function of Core Histone Amino Termini in Controlling Factor Access

In principle, there are several aspects of nucleosome structure which could hinder the binding of transcription factors. The most obvious possibility is steric occlusion by the histone octamer on one face of the DNA helix. However, a striking finding is that the amino-terminal domains of the core histones are responsible for a large fraction of the inhibition of binding of TFIIIA and GAL4 derivatives to nucleosome cores (13, 18). These relatively unstructured domains are not essential for nucleosome core formation or stability (reviewed in Ref. 30). They are highly charged, and modification of lysine residues within these domains by acetylation, which neutralizes the charge of these residues, reduces their interaction with DNA (31) and increases transcription factor binding to nucleosomes (18) (see below). Acetylation of these do-

mains is the histone modification which most closely correlates with transcription
activity (30).

The interactions of many transcription factors with nucleosome cores may be regulated by the core histone amino termini. Below we describe a protocol to test the role
of the amino termini in the binding of a transcription factor. The amino-terminal
domains are susceptible to proteolysis and can be easily removed by treatment with
trypsin (32). Proteolytic removal of the amino termini followed by inactivation of the
protease allows an assessment of their function in factor binding by comparison to
intact nucleosome cores.

Removal of Amino Termini

To address the function of the histone amino-terminal tails on factor binding, we
utilize a standard proteolysis protocol which selectively removes the amino termini
of the core histones. Nucleosome cores are treated with trypsin, to remove the
N-termini, followed by the addition of an excess of trypsin inhibitor. This protocol
utilizes H1-depleted oligonucleosome cores and octamer transfer nucleosome reconstitution procedures described by Côté *et al.* (22).

1. Six microliters short oligonucleosome core particles (1 mg/ml) is digested with
   4 μl of trypsin (Sigma, St. Louis, Mo, 0.1 mg/ml in 100 mM Tris-HCl, pH 8.0)
   in a final volume of 12 μl for 5 min at room temperature (final trypsin concentration of 30 μg/ml).
2. Reactions are stopped by adding 4 μl of 1 mg/ml soybean trypsin inhibitor
   (Sigma, 100-fold excess by weight).
3. Trypsinized cores are prepared for octamer transfer to radiolabeled probe by adding 5 μl of 5 M NaCl and radiolabeled probe yielding a final starting volume of
   25 μl and 1 M salt concentration [see Côté *et al.* (22)].
4. Trypsin inhibitor is included in all buffers used for dilutions and factor binding
   reactions at a concentration of 200 μg/ml.

Analysis by sodium dodecyl sulfate–polyacrylamide gel electrophoresis (SDS-
PAGE) of the histones from control and trypsinized nucleosomes is shown in Fig. 3A.
The stimulation of GAL4-AH binding resulting from the tryptic removal of the
amino-terminal tails is illustrated in Fig. 3C.

Controls for Trypsin Inactivation

All control reactions from which the amino termini are not removed include the same
amount of trypsin, but its addition is preceded by the addition of the trypsin inhibitor.
Nucleosome reconstitution and transcription factor binding reactions which contain
trypsin and are preceded by the addition of inhibitor are considered to produce intact

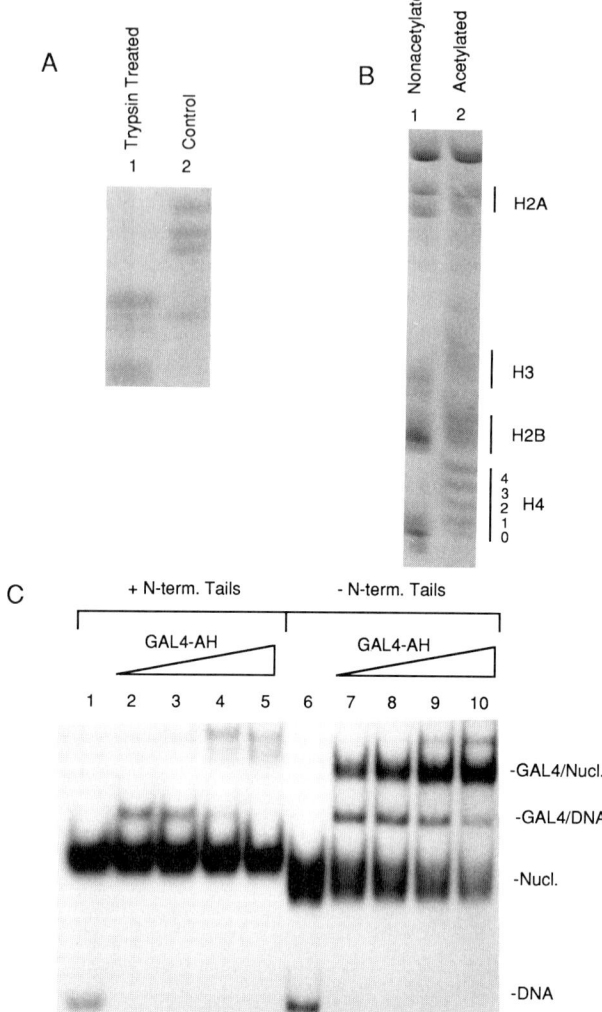

Fig. 3 Analysis of the function of the core histone amino-terminal tails in factor binding. (A) Protein gel of the histones from nucleosome core particles used for reconstitution assays (lane 2) and the trypsin-treated nucleosome cores (lane 1). The four core histones in lane 2 are H4, H2A, H2B, and H3, from bottom to top. For a description of the bands resulting from trypsin removal of the amino-terminal tails (lane 1), see Ref. 32. (B) Triton–acid–urea gel of nonacetylated core histones (lane 1) and hyperacetylated nucleosome cores (lane 2). The four acetylated forms of H4 are shown as well as the position of nonacetylated H4. (C) Stimulation of GAL4-AH binding to nucleosomal DNA by removal of the core histone amino-terminal tails. A DNA probe with a single GAL4-AH site centered at 40 bp from an end was reconstituted into nucleosome cores with (lanes 1–5) or without (lanes 6–10) the amino-terminal tails and incubated with increasing amounts of GAL4. The complexes and concentrations of GAL4-AH are the same as in Fig. 2. On removal of the amino-terminal tails, binding to the site is greatly enhanced, which suggests that binding to this site is largely dictated by the contact of the N-terminal tails with the DNA wrapped into the nucleosome core.

nucleosomes, and therefore contain the amino termini. Protein gels of these controls and the test samples should illustrate that the core histones are not degraded by trypsin during the reconstitution or binding reactions, once the inhibitor is present (13).

To test for the possible trypsin degradation of transcription factors during subsequent binding reactions, DNA binding controls are performed under identical conditions in the presence or absence of the added trypsin.

1. For the DNA control reactions, mix 6 μl short oligonucleosome core particles (1 mg/ml) with 4 μl soybean trypsin inhibitor and 2 μl distilled water. The DNA probes are omitted until after the transfer reactions.
2. Add 4 μl trypsin or the corresponding buffer and 5 μl of 5 M NaCl. These controls are then serially diluted as in the octamer transfer reactions [see Côté et al. (22)].
3. Labeled DNA probes are added into binding reactions containing the mock-reconstituted mixtures containing or omitting trypsin, followed by addition of the transcription factor of interest.
4. Comparison of the extent of transcription factor binding which occurs in reactions with or without trypsin will reveal whether the factor is degraded by residual trypsin activity.

Use of Hyperacetylated Nucleosome Cores

If removal of the core histone amino termini results in a substantial increase in transcription factor binding, binding of the particular factor may also be stimulated by histone acetylation. For example, stimulation of TFIIIA binding to nucleosome cores by removal of the amino termini is partly mimicked by hyperacetylation of the core histones (18). GAL4 derivative and USF binding is also stimulated by hyperacetylation, but to a much lesser extent than by removal of the amino termini. This is most likely due to the fact that hyperacetylated histones are a heterogeneous population of acetylated forms. Thus, whereas acetylated forms are increased, only a fraction of the nucleosomes may be acetylated on specific lysine residues (30) which participate in stimulating factor binding.

To test directly whether increased levels of histone acetylation will increase the binding of a transcription factor, nucleosome cores are reconstituted with hyperacetylated and control histones. Hyperacetylated H1-depleted oligonucleosome cores or core histones are prepared from HeLa nuclei as described (22). However, nuclei are prepared from cells (1 $\times$ 10^6 cells/ml) subjected to 20–24 hr of treatment with 10 mM sodium butyrate (an inhibitor of histone deacetylase, reviewed in Ref. 30). Sodium butyrate at 2 mM is included in all purification steps of the acetylated nucleosomes or histones as well as parallel preparations of control nucleosomes or histones from cells not treated with sodium butyrate. The level of acetylation resulting from butyrate treatment is revealed in Triton–acid–urea gels, an example of which is shown in Fig. 3B.

Stimulation of Transcription Factor Binding by Accessory Protein Complexes

In the milieu of the cell, transcription factors do not confront nucleosomes in isolation. Several other cellular components could facilitate the binding of transcription factors and the displacement of nucleosomes. *In vitro* studies have illustrated two types of protein complexes which induce transcription factor binding to nucleosomes. The first was the nucleosome assembly protein nucleoplasmin, which is a histone chaperone with *in vivo* specificity for H2A/H2B (33). By interacting with the histones in nucleosome cores, nucleoplasmin stimulates the binding of GAL4 derivatives, USF, and Sp1 to nucleosomes and also facilitates the displacement of H2A/H2B dimers from transcription factor/nucleosome complexes (15). The SWI/SNF complex contains 10 polypeptides including 5 of the SWI/SNF gene products which are required for many activators to stimulate transcription in yeast. The purified yeast SWI/SNF complex stimulates the binding of GAL4 derivatives to nucleosomes in a reaction that utilizes the energy of ATP hydrolysis to disrupt nucleosome structure (20). It is likely that additional protein components will regulate transcription by stimulating transcription factor binding to nucleosomes.

The ability of nucleoplasmin or other components to stimulate the binding of a particular transcription factor to nucleosomes can be tested directly. Thus, two different types of experiments are possible which are variations of the basic nucleosome binding assays described by Côté *et al.* (22). First, if a transcription factor(s) is not observed to bind directly to nucleosome cores, its ability to bind when assisted by a histone chaperone like nucleoplasmin can be addressed. Nucleoplasmin is readily purified from *Xenopus* eggs (34) in substantial quantities. Inclusion of nucleoplasmin at concentrations between $10-20$ μg/ml in binding reactions can have a dramatic effect on the affinity of a transcription factor for nucleosomal DNA.

The second type of experiment is to test whether a specific protein or complex induces transcription factor binding to nucleosomal DNA. Candidate proteins might include other histone-binding proteins or other DNA-dependent ATPases like the SWI/SNF complex. In this instance, a model system of factors and nucleosome substrates is established in which to assay changes in factor affinity. A good system would be the binding of GAL4-AH [which is readily expressed and purified from *E. coli* strains (35)] and nucleosome cores containing GAL4 binding sites generated by octamer transfer (22). Inclusion or omission of test protein (or complex) as well as potential cofactors (i.e., ATP) in binding reactions with increasing amounts of GAL4-AH will reveal changes in GAL4-AH affinity for the nucleosome core. It is important to note that the amount of H1-depleted nucleosome cores used in reconstitutions may have to be titrated for binding assays (22) if an accessory protein complex is capable of also interacting with the donor nucleosome cores (20). An example of the stimulation of GAL4-AH binding by nucleoplasmin and by the yeast SWI/SNF complex is shown in Fig. 4.

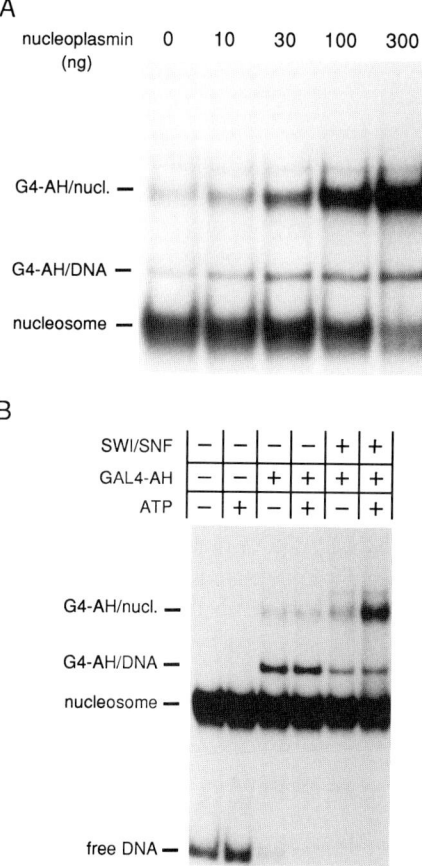

FIG. 4 Accessory protein complexes which stimulate transcription factor binding to nucleosomes. (A) Nucleoplasmin stimulates the formation of GAL4-AH/nucleosome ternary complexes. A 154-bp DNA probe harboring a single GAL4 binding site 32 bp from one end was reconstituted by the octamer transfer method and incubated with 7 nM GAL4-AH and increasing amounts of purified nucleoplasmin (0–300 ng). The 14-μl binding reactions, also containing 250 ng carrier DNA from donor nucleosomes, were incubated for 30 min at 30°C and analyzed by the mobility shift assay as described. The different protein/DNA complexes are identified. The small fraction (<10%) of DNA probe remaining as naked after the reconstitution is all bound by GAL4-AH under these conditions (G4-AH/DNA). The presence of increasing amounts of nucleoplasmin in the binding reaction drastically stimulates the formation of the GAL4-AH/nucleosome complex (G4-AH/nucl.) (B) The yeast SWI/SNF protein complex stimulates the binding of GAL4-AH to nucleosomes in an ATP-dependent fashion. Nucleosome cores were prepared as in (A), and 1 μl of the reconstitution product (containing 25 ng donor HeLa nucleosomal DNA) was used for binding reactions in the presence or absence of 1 mM ATP, 10 nM GAL4-AH, and 100 ng purified yeast SWI/SNF protein complex. The 10-μl binding reactions were analyzed as in (A). Clearly, the binding of GAL4-AH to its site within a nucleosome core is enhanced by the presence of the purified SWI/SNF protein complex, but only if ATP is also present.

Analysis of Protein Composition of Factor/Nucleosome Complexes

Additional information regarding the interactions of transcription factors and nucleosome cores can be generated by analyzing the proteins present in the ternary transcription factor/nucleosome complexes formed. For example, binding of five GAL4-AH dimers in the presence of nucleoplasmin results in a loss of H2A/H2B dimers from the complex, suggesting that stimulation of factor binding by nucleoplasmin is linked to removal of these histones (15). Similar disruption of nucleosome content may accompany stimulation by the SWI/SNF complex (20) or might occur during the binding of particular transcription factors alone. Moreover, it is possible to investigate histone modifications and/or variants which facilitate factor binding by analyzing the histones preferentially located in the factor-bound population (see below).

We describe two approaches to analyze the protein components in factor/nucleosome ternary complexes. The first involves immunoblotting mobility shift gels, and the second is SDS-PAGE of mobility-shifted complexes. Each approach utilizes homogeneous sequence nucleosome cores [see Côté et al. (22)] such that only histones associated with nucleosomes bearing binding sites for the factor(s) of interest will be analyzed.

Nucleosome Core Reconstitution on DNA Fragments Generated by Polymerase Chain Reaction

1. Polymerase chain reactions producing the nucleosome-length fragment of interest are done in numerous repeats to obtain at least $20-30$ μg total [measured by comparison with an aliquot with DNA mass ladders (GIBCO/BRL, Gaithersburg, MD) on an agarose gel]. [α-^{32}P]dCTP and [α-^{32}P]dATP (3000 Ci/mmol) are included in the reactions at a ratio of 1 : 60 with cold nucleotides to label the product.
2. Purify the correct-length product by migration on a native 8% acrylamide (acrylamide–bisacrylamide ratio of 29 : 1), 1 $\times$ TBE gel.
3. Place an estimated amount of probe in three tubes and do nucleosome core reconstitution by dialysis from high salt (22) using histone-to-DNA (w/w) ratios of 1, 1.5, and 2.
4. Check the efficiency of reconstitution by running an aliquot of each sample on a native 4% acrylamide–0.5$\times$ TBE gel (as above for mobility sift assays).
5. The sample containing the best yield of reconstituted nucleosome cores compared to free DNA and nonspecific aggregation is then purified by sucrose gradient centrifugation.

Note: If the yield of the PCR amplifications is reproducible, the best DNA-to-histone ratio found in step 4 can then be used for subsequent reconstitutions.

Western Blot Detection of Proteins in Factor/Nucleosome Complexes

To better assess the contribution played by acetylation when transcription factors bind to nucleosomes, we developed a method to incorporate Western blotting in nucleosome mobility shift assays. In this section we discuss the conditions for running and blotting an electrophoretic mobility shift assay (EMSA) and the methods used to detect the DNA and antigen on blots. The core histones used for these experiments are homogeneous and purified by sucrose gradient ultracentrifugation (as described above) for purposes of quantitation and to remove contaminating unbound histones that will increase background.

Protocol for Running Mobility Shift Gel

Once nucleosome cores have been purified on a sucrose gradient and fractions are collected and pooled, factor binding can be performed as described elsewhere in this volume (22). Typically, 20–40 ng per lane of nucleosome cores is used for a transcription factor binding reaction. This is a sufficient amount of protein for detection with a chemiluminescence system. This amount can be determined by running a sample of the radiolabeled probe prior to dialysis reconstitution on a 1.2–1.5% agarose gel with the Bio-Rad (Richmond, CA) DNA mass markers. The band to be quantitated is then excised from the gel and counted in a scintillation counter. This will give an estimate of the specific activity (cpm/ng) of the probe DNA. Because nucleosomes contain approximately equal weights of histones and DNA, an estimation of the amount of protein present can also be determined by counting samples. Reconstituted nucleosome cores are first titrated with increasing amounts of transcription factor, subjected to EMSA, dried, and exposed to film to determine the best factor concentrations and degree of separation of nucleosome and factor-bound complexes.

Mobility Shift Assays

1. The purified nucleosome cores are divided equally into 20 to 40-ng aliquots, and dilutions of the factor are added in the dilution buffer appropriate for each factor, then incubated at 30°C for 20–30 min. Addition of glycerol is not necessary because the nucleosome cores are present in 15–20% sucrose. The volume of the binding reaction can range from 20 to 50 μl depending on the concentration of the pooled gradient fractions.
2. Binding reactions are immediately loaded on a native 6% acrylamide (acrylamide–bisacrylamide 29:1), 0.5 × TBE gel (1.5 mm thickness) and run at 165–200 V for 2.5–3.5 hr at room temperature.
3. After running, the gel is cut down to encompass only the utilized lanes and prepared for Western blotting. The gel is left attached to one plate in order to prevent ripping during the equilibration in transfer buffer prior to blotting.

Western Blotting of Mobility Shift Gels

The blotting method described is based on the method of Demczuk *et al.* (36) with some changes for nucleosomal complexes. This method uses two membranes for blotting; one for detection of protein and one for detection of DNA. The earlier study found that the presence of DNA on a single membrane interfered with the ability of the antibody to detect the shifted protein-bound DNA complex. We encountered similar problems and found that the use of two membranes resulted in recovery of 60–80% of the DNA on the positively charged DEAE membrane.

Note: Before the mobility shift gel has finished running, the blotting membrane must be cut and ready for equilibration in the transfer buffer. Equilibration buffers are cooled to 4°C prior to blotting, and all blotting procedures are carried out in a cold room.

1. Nitrocellulose (Hybond ECL, Amersham, Arlington Heights, IL, or USB, Cleveland, OH) and DEAE (Schleicher & Schuell, Keene, NH) membranes are cut to the exact dimensions of the gel and then briefly prewet in distilled water according to manufacturers' instructions. Both membranes are next equilibrated in 4°C Bjerrum and Schafer-Nielson transfer buffer [48 mM Tris, 39 mM glycine, 20% (v/v) methanol, pH 8.8–9.0] for 15–20 min with gentle shaking. Additionally, cut three pieces of VWR filter paper per gel, 1 cm in dimension larger than the gel.

2. At 4°C or in a cold room, the trimmed mobility shift gel is equilibrated with two changes of Bjerrum and Schafer-Nielson buffer plus SDS (48 mM Tris, 39 mM glycine, 20% methanol, 0.0375% w/v SDS, pH 8.8–9.0) for 5 min. This is the only time the equilibration buffer with SDS is used.

3. Blotting is performed in the plate electrode tank transfer chamber from Bio-Rad, and the blot is set up as follows.

    The blotting cassette is built up from the cathode starting with wetting one of the pads in equilibration buffer and placing it down.
    Soak a piece of filter paper (VWR, South Plainfield, N.J.) in equilibration buffer and place it on top of the pad. Use a test tube to roll out any air bubbles with the addition of each successive layer.
    Place the preequilibrated gel on the VWR filter paper (a 6% gel can be handled with little fear of tearing).
    The nitrocellulose is next placed on the gel starting from the center outward.
    Place a second piece of prewet VWR filter paper on the nitrocellulose.
    Layer the DEAE membrane onto the VWR filter paper, directly matching the outline of the nitrocellulose.
    Place the final piece of VWR filter paper and roll complete stack carefully before placing the final prewet pad down and completing the sandwich.

4. The cassette is fit into the tank chamber, surrounded completely with 4°C equilibration buffer (48 mM Tris, 39 mM glycine, 20% methanol, pH 8.8–9.0), and run

using the Bio-Rad power supply for 2 hr at 100 V (0.45 A increasing to 0.75–0.9 A).

5. After 2 hr, the cassette is dismantled and the nitrocellulose membrane is placed on VWR filter paper to dry overnight before immunostaining. The DEAE membrane is sealed in plastic wrap and exposed overnight with an intensifying screen at $-80°$C. DNA is exposed to Kodak (Rochester, NY) XAR film for 16–24 hr.

Detection of Antigen on Blots

For detection of the blots, we use the enhanced chemiluminescence (ECL) Western blotting protocol of Amersham. Before performing the immunodetection, the primary and secondary antibodies should first be optimized. This can be achieved by performing SDS-PAGE or a dot-blot of the antigen and testing it with different dilutions of the antibodies to be used.

1. After drying the membrane overnight between filter paper, using a pair of forceps place the membrane in phosphate buffered saline supplemented with 0.1% (w/v) Tween 20 (PBS; 137 mM NaCl, 2.7 mM KCl, 4.3 mM Na$_2$HPO$_4$, 1.4 mM KH$_2$PO$_4$) slowly at a 45° angle and let shake gently for 20 min.
2. Block nonspecific sites by immersing the membrane in a blocking solution for 30–60 min with gentle shaking. We use 1% w/v gelatin (Sigma; swine type A), 0.1% Tween 20.
3. Wash the membrane in PBS, 0.1% Tween 20 once for 15 min and twice for 5 min with fresh changes of buffer.
4. Incubation with the primary and the secondary antibodies is for 30–60 min, which varies with the antibodies used. The membrane is washed between incubation steps once for 15 min and three times for 5 min with fresh changes of wash buffer. The final wash before the detection has an additional 5 min wash to help further reduce background.
5. Detection of the antibody on the membrane is performed as per the manufacturer's instructions, depending on which chemiluminescence kit is used.

An example of such an analysis of GAL4-AH bound nucleosome complexes is shown in Fig. 5A, B, which illustrates the distribution of DNA probe and acetyllysines on histones between the factor-bound and unbound nucleosomes.

Analysis of Protein Composition of Factor/Nucleosome Complexes by Second-Dimension Electrophoresis

A complementary approach to analyzing the protein composition of factor nucleosome complexes is by second-dimension SDS-PAGE following the mobility shift separation of factor-bound nucleosome complexes (21). This requires even more of

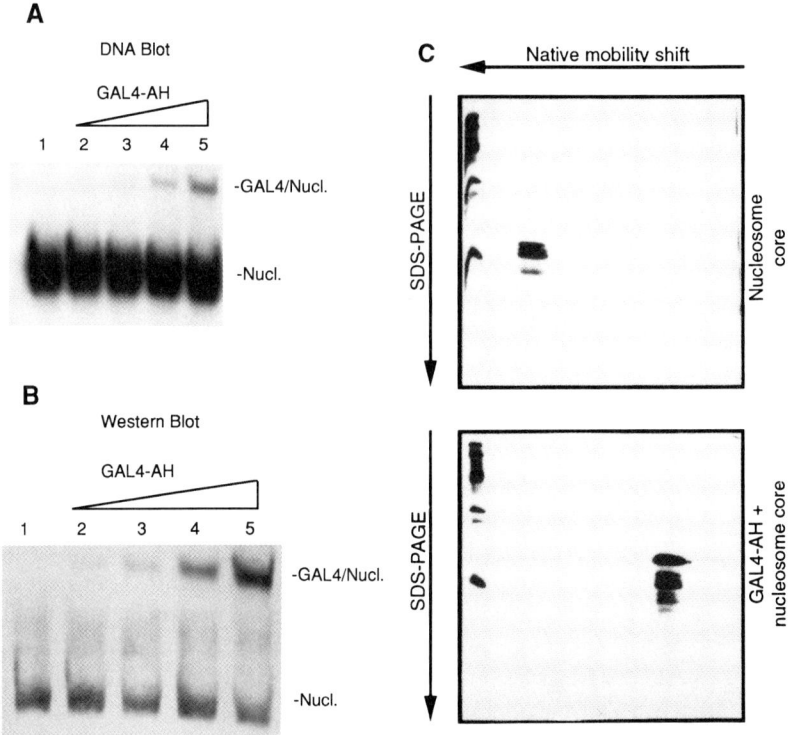

FIG. 5   Analysis of proteins in transcription factor/nucleosome ternary complexes. (A, B) Protein analysis by Western blots of mobility-shifted complexes. Gradient-purified homogeneous-sequence nucleosome cores bearing a single GAL4 site were incubated without (lane 1) or with 4.8 n*M* (lane 2), 9.5 n*M* (lane 3), 23.8 n*M* (lane 4), and 47.6 n*M* (lane 5) of GAL4-AH followed by separation of the bound and unbound nucleosomes on a 6% acrylamide nondenaturing gel. The gel was blotted to nitrocellulose and DEAE membranes as described in the text. An autoradiogram of the DEAE membrane (DNA blot) is shown in (A) and illustrates the distribution of the DNA probe. A Western blot of the nitrocellulose membrane is shown in (B), which utilized pan-acetyl lysine antibody (provided by Tim Hebbes and Colyn Crane-Robinson). Note that acetyl-lysine is enriched in the GAL4-AH bound nucleosome fraction relative to the total nucleosomes illustrated by the DNA blot in (A). (C) Two-dimensional gel analysis of proteins in factor/nucleosome complexes. Purified homogeneous nucleosome cores bearing five GAL4 sites were run on a native mobility shift gel following incubation in the absence (top) or presence (bottom) of GAL4-AH. The mobility shift lanes were separated in the second dimension by SDS-PAGE and subsequently stained with silver as described in the text. The top gel illustrates the four core histones which migrate in the first-dimension nucleosome core. The bottom gel illustrates the migration of the core histones and GAL4-AH in a single factor/nucleosome ternary complex. [Figure 5C is reproduced from J. L. Workman and R. E. Kingston, "Nucleosome displacement *in vitro* via a metastable transcription factor–nucleosome complex," *Science* **258**, 1781 (1992).]

the homogeneous sequence nucleosome cores bearing the factor binding sites because histones in general stain poorly with silver. Thus, correspondingly high concentrations of purified transcription factors are required for these analyses. An example of two-dimensional analysis of nucleosome and GAL4-AH/nucleosome complexes is illustrated in Fig. 5C.

Binding Reaction and Electrophoretic Mobility Shift Assay

1. Approximately 1–3 μg nucleosome cores is used in the binding reactions, which are usually performed with 100–200 μl of the pooled fractions from sucrose gradients.
2. The samples are incubated with enough purified transcription factor under the appropriate conditions to bind 50% or more of the nucleosomes such that they will be shifted in the EMSA gel. (*Note:* The reactions can contain up to 100 μM nucleosome cores, which may exceed the K_D for binding of many factors to nucleosomes. Thus, factor binding may be driven by the concentration of binding sites in these reactions.)
3. Following incubation, the binding reactions are loaded directly on a 1.5 mm-thick 4 or 6% acrylamide gel into wells large enough to hold the entire sample.
4. Dye is loaded into blank lanes flanking the lanes containing samples, and the gels are electrophoresed as for standard binding reactions.

Preparation of Second-Dimension Gel

While the EMSA gel is running, a 15% SDS-polyacrylamide separating gel is poured. This gel should be slightly thicker than the EMSA gel. This can be achieved by adding 0.2-mm spacers on top of 1.5 mm spacers between the plates.

1. Pour the separating gel, leaving a space approximately three times the width of the lanes in the EMSA gel (~3 cm) above the top of the separating gel.
2. Pour a stacking gel that is approximately equal to the width of the EMSA gel lanes on top of the separating gel.
3. Extensively clean and rinse with ethanol a pair of tweezers, scalpel, a small flat spatula, and a large flat spatula which is longer and wider than the lanes of the EMSA gel.

Loading and Running Second-Dimension Gel

1. When the EMSA gel is finished running, pry apart the gel plates and, using the scalpel or spatula, cut the length of the gel on either side of the lanes containing the samples (i.e., the lane between the loaded dyes).
2. Place the SDS gel at a 20° angle from horizontal and, using the solution for preparing the stacking gel (without acrylamide), wet the glass plates above the short stacking gel.
3. The slice containing the full-length lane is carefully lifted with the large spatula, then placed onto the top of the SDS gel.

4. Using the spatulas, work the EMSA slice down between the glass plates until it is snug against the top of the short stacking gel.
5. Remove excess buffer around the slice with a Pasteur pipette, being careful not to introduce bubbles between the EMSA slice and the stacking gel.
6. Pour additional stacking gel around and above the EMSA slice and allow to polymerize. A single lane for dyes and markers can be placed at one end of the EMSA slice.
7. Electrophorese until the bromphenol blue dye runs off the bottom.
8. Following electrophoresis, cut off the stacking gel and stain the gel with silver according to Wray *et al.* (37), with the following modifications. To increase the staining of proteins in the interior of the gel, stain in solution C for 30 min. Develop the gel until the entire gel becomes very dark to ensure development of stained protein in the center of the gels. Following washing in water and 50% methanol, the gel is cleared (destained) in Kodak Rapid Fix (undiluted) until the background staining is completely removed, allowing visualization of the stained proteins. During destaining, excess silver can be gently rubbed off the outside of the gel with clean gloved fingers. When adequately destained, photograph the gel immediately. *Note:* H2A and H2B stain more intensely than H3 and H4 at limiting protein concentrations.)

In variation of this type of analysis (15), the first-dimension mobility-shift gel is briefly exposed wet to film to localize the factor/nucleosome ternary complexes. These complexes are then excised from the wet gel in small slices and placed into wide lanes of an SDS–polyacrylamide gel as described above. The slices are submerged in sample buffer, and the gel is run and stained as described above.

Conclusion

The *in vivo* function of any eukaryotic transcription factor will largely depend on its interaction with nucleosomes. In this chapter we have described advanced techniques to analyze the parameters governing the interaction of transcription factors with nucleosomal DNA. In addition, we have described methods to analyze the functions of accessory complexes in inducing transcription factor binding and displacing nucleosomes. These analyses of individual factors and groups of factors which bind to complex promoter and enhancer elements will shed new light on the function of these proteins in chromatin disruption during transcription activation.

Acknowledgments

We thank Craig Peterson and Janet Quinn for providing purified yeast SWI/SNF complex and Tim Hebbes and Colyn Crane-Robinson for providing antibodies to acetyl-lysine. This work was supported by National Institutes of Health Grant GM47867 to J.L.W., a Leukemia Society Scholars Award to J.L.W., an NIH Postdoctoral Fellowship to C.C.A., and a Canadian Medical Research Council Postdoctoral Fellowship to J.C.

References

1. H. Richard-Foy and G. L. Hager, *EMBO J.* **6**, 2321 (1987).
2. K. D. Carr and H. Richard-Foy, *Proc. Natl. Acad. Sci. U.S.A.* **87**, 9300 (1990).
3. A. Schmid, K.-D. Fasher, and W. Horz, *Cell (Cambridge, Mass.)* **71**, 853 (1992).
4. J. D. Axelrod, M. S. Reagan, and J. Majors, *Genes Dev.* **7**, 857 (1993).
5. E. Verdin, P. J. Paras, and C. Van Lint, *EMBO J.* **12**, 3249 (1993).
6. H. Lee and T. K. Archer, *Mol. Cell. Biol.* **14**, 32 (1994).
7. J. Svaren and W. Horz, *Curr. Opin. Genet. Dev.* **3**, 219 (1993).
8. C. C. Adams and J. L. Workman, *Cell (Cambridge, Mass.)* **72**, 305 (1993).
9. T. Perlmann and O. Wrange, *EMBO J.* **7**, 3073 (1988).
10. B. Pina, U. Bruggemeier, and M. Beato, *Cell (Cambridge, Mass.)* **60**, 719 (1990).
11. T. K. Archer, M. G. Cordingley, R. G. Wolford, and G. L. Hager, *Mol. Cell. Biol.* **11**, 688 (1991).
12. I. C. A. Taylor, J. L. Workman, T. Schuetz, and R. E.Kingston, *Genes Dev.* **5**, 1285 (1991).
13. M. Vettese-Dadey, P. Walter, H. Chen, L.-J. Juan, and J. L. Workman, *Mol. Cell. Biol.* **14**, 981 (1994).
14. B. Li, C. C. Adams, and J. L. Workman, *J. Biol. Chem.* **269**, 7756 (1994).
15. H. Chen, B. Li, and J. L. Workman, *EMBO J.* **13**, 380 (1994).
16. C. C. Adams and J. L. Workman, *MCB* **15**, 1405 (1995).
17. D. S. Wechsler, O. Papoulas, C. V. Dang, and R. E. Kingston, *Mol. Cell. Biol.* **14**, 4097 (1994).
18. D. Y. Lee, J. J. Hayes, D. Pruss, and A. P. Wolffe, *Cell (Cambridge, Mass.)* **72**, 73 (1993).
19. T. Tsukiyama, P. B. Becker, and C. Wu, *Nature (London)* **367**, 525 (1994).
20. J. Côté, J. Quinn, J. L. Workman, and C. L. Peterson, *Science* **265**, 53 (1994).
21. J. L. Workman and R. E. Kingston, *Science* **258**, 1780 (1992).
22. J. Côté, R. T. Utley, and J. L. Workman, Chapter [6], this volume.
23. R. D. Kornberg and Y. Lorch, *Cell (Cambridge, Mass.)* **67**, 833 (1991).
24. G. S. Meersseman, S. Pennings, and E. M. Bradbury, *EMBO J.* **11**, 2951 (1992).
25. Q. Li and O. Wrange, *Genes Dev.* **7**, 2471 (1994).
26. J. Kim, C. Zweib, C. Wu, and S. Adhya, *Gene* **85**, 15 (1989).
27. T. K. Kerppola and T. Curran, *Cell (Cambridge, Mass.)* **66**, 317 (1991).
28. T. E. Shrader and D. M. Crothers, *Proc. Natl. Acad. Sci. U.S.A.* **86**, 7418 (1989).
29. F. Razvi, G. Gargiulo, and A. Worcel, *Gene* **23**, 175 (1983).
30. B. M. Turner, *J. Cell Sci.* **99**, 13 (1991).
31. L. Hong, G. P. Schroth, H. R. Matthews, P. Yau, and E. M. Bradbury, *J. Biol. Chem.* **268**, 305 (1993).
32. L. Bohm and C. Crane-Robinson, *Biosci. Rep.* **4**, 365 (1984).
33. S. M. Dilworth, S. J. Black, and R. A. Laskey, *Cell (Cambridge, Mass.)* **51**, 1009 (1987).
34. L. Sealey, R. R. Burgess, M. Cotten, and R. Chalkley, *in* "Methods in Enzymology" (P. M. Wassarman and R. D. Kornberg, eds.), Vol. 170, p. 612. Academic Press, San Diego, 1989.

35. Y. S. Lin, M. Carey, M. Ptashne, and M. R. Green, *Cell (Cambridge, Mass.)* **54**, 659 (1988).

36. S. Demczuk, M. Harbers, and B. Vennstrom, *Proc. Natl. Acad. Sci. U.S.A.* **90**, 2574 (1993).

37. W. Wray, T. Boulikas, V. P. Wray, and R. Hancock, *Anal. Biochem.* **118**, 197 (1981).

[8] *In Vivo* Analysis of Nucleosome Structure and Transcription Factor Binding in *Saccharomyces cerevisiae*

John Svaren, Ulrike Venter, and Wolfram Hörz

Introduction

Much progress has been made in the identification of trans-acting factors that control gene activation. However, in many cases, there is no direct proof that an isolated or cloned factor is actually involved in activating a certain promoter. Additionally, it is often not clear why a factor binds only to certain sites *in vivo* but not to others which may bind the factor with high affinity *in vitro*. To address these kinds of questions, methods have been developed to analyze the mechanism of action of transcription factors *in vivo*. For example, *in vivo* footprinting has become a powerful technique in determining the proteins that bind to cis-acting elements in the nucleus.

It has become clear that in many cases the ability of trans-acting factors to bind is influenced by the nucleosome structure of the locus (1–4). Additionally, an increasing amount of genetic evidence has indicated that chromatin structure is an important determinant of gene activity (5, 6). Consequently, if one is interested in determining how gene activity is established on a given promoter or enhancer, it is worthwhile to determine both the pattern of trans-activator binding as well as the nucleosome structure of DNA sequences within and surrounding the regulatory element being studied.

In this chapter, we summarize methods which we have employed to study several promoters in the yeast *Saccharomyces cerevisiae*. The *PHO5* and *PHO8* genes code for phosphatases which are regulated by the amount of phosphate in the medium (7). On phosphate starvation, each of the genes is induced, and the induction is accompanied by a chromatin structure change in the promoter region that is characterized by an increased accessibility of the promoter to nucleases (8, 9). A number of results have indicated that the chromatin structure change is required for gene activation (6). We have also applied these techniques to the *TDH3* promoter, which is constitutively active (10).

The first section contains protocols for using DNase I, restriction nucleases, and micrococcal nuclease in yeast nuclei. These methods allow one to detect nucleosome structure on specific sequences as well as to map nucleosome-free regions which might be involved in gene regulation. The second part contains a protocol for footprinting *in vivo* using dimethyl sulfate (DMS), which we have employed to determine sites of trans-activator binding *in vivo*.

Methodology

Use of Nucleases to Assay for Absence or Presence of Nucleosomes

Three different procedures employing nuclease digestion to assay for the absence or presence of nucleosomes are described below, namely, DNase I hypersensitivity analyses, digestion with restriction nucleases, and extensive digestion with micrococcal nuclease to visualize directly mononucleosome signals. All three procedures require prior isolation of nuclei. In the first section, there is a description of how we isolate nuclei from *S. cerevisiae* and use them for nuclease digestion. Our method for the isolation of nuclei is based on the original procedure of Wintersberger *et al.* (11).

Isolation of Nuclei from Saccharomyces cerevisiae
Solutions

> Preincubation solution: 0.7 M 2-mercaptoethanol, 2.8 mM EDTA
> Sorbitol solution: 1.0 M sorbitol
> Lysis solution: 1.0 M sorbitol, 5 mM 2-mercaptoethanol
> Ficoll solution: 18% (w/v) Ficoll, 20 mM KH$_2$PO$_4$, pH 6.8, 1 mM MgCl$_2$, 0.25 mM EGTA, 0.25 mM EDTA
> Zymolyase solution: 20 mg Zymolyase 100T (ICN, Costa Mesa, CA), dissolved in 1 ml water

Procedure

1. A 1-liter yeast culture is grown to early logarithmic phase ($2-4 \times 10^7$ cells/ml). This is approximately 2–4 optical density (OD) units at 600 nm, although this measurement of cell density often varies between spectrophotometers. It is best to determine the conversion factor for a particular spectrophotometer by counting cells. Collect cells by centrifugation (3000 g for 10 min). It is important to process a culture immediately after taking cells out of the incubator and not to store the culture first at low temperature; otherwise, lysis of the cells is greatly impeded.
2. Wash cells in cold water and suspend in 50 ml water.
3. Transfer into preweighed 50-ml centrifuge tubes and centrifuge (3000 g for 5 min). Determine wet weight.
4. Add 2 volumes preincubation solution (relative to wet weight of cells) and shake for 30 min at 28° C.
5. Collect by centrifugation (3000 g for 5 min) and wash in 50 ml of 1 M sorbitol.
6. Collect again (3000 g for 5 min) and resuspend in 5 ml lysis solution per 1 g of cells (wet weight).
7. Dilute 20-μl aliquots 100-fold in water and read optical density at 600 nm. This should be in the range between 1 and 2.

8. Add $\frac{1}{50}$ volume freshly prepared Zymolyase solution to the cells.
9. Incubate with slight agitation at 28°C.
10. Measure optical density at 600 nm after 15 and 30 min as in step 7. Values should drop to 5–20% of the original measurement.
11. Centrifuge (2000 g for 5 min at 5°C) and wash in 50 ml of 1 M sorbitol.
12. Centrifuge (3000 g for 10 min at 5°C) and resuspend in 7 ml Ficoll solution per 1 g cells (original wet weight). Cells lyse at this stage, but nuclei are stabilized by the Ficoll.
13. Distribute into as many aliquots as desired and centrifuge (30,000 g for 30 min at 5°C). Aliquots equivalent to 0.5 or 1 g wet weight cells are suitable for subsequent digestion experiments. It is convenient to use 10-ml polypropylene centrifuge tubes.
14. Decant supernatant, and freeze the nuclear pellet in liquid nitrogen, and store at −70°C.

Comments

The protocol for the preparation of nuclei balances speed and purity. The nuclei are sufficiently purified so that the extent of nuclease digestion is reproducible between experiments. There are procedures to obtain highly pure yeast nuclei, but they involve substantially more time (see Refs. 12, 13, and references therein). On the other hand, there are also other procedures that involve treating nuclei with nucleases in crude lysates (14–17). The advantage of such procedures is that they are faster, and loss of trans-acting factors is minimized. However, for the analysis of hypersensitivity, loss of factors may even be advantageous, as in many cases hypersensitive sites have been shown to bind trans-activators *in vivo,* which might protect against nucleases.

Lysis is strain dependent and also depends on growth conditions. Stationary cells are more difficult to lyse than dividing cells. The values as determined in step 7 are only relative measures of lysis and do not reflect the actual percentage of unlysed cells. We have successfully used nuclei from cells which gave OD_{600} values that dropped to only 60% of the starting value at the end of Zymolyase treatment. In those cases we monitor the accessibility of a constitutively accessible restriction site in chromatin as a control (see below).

The preincubation step facilitates digestion of the cell wall with Zymolyase. In the case of the acid phosphatase genes, it is fortunate that phosphate starvation is required for gene induction because all buffers used are phosphate-starvation buffers. In other cases, it may be required to maintain induction during Zymolyase treatment by adding the appropriate inducer as discussed in Ref. 18. It is also possible to treat cells with Zymolyase in medium supplemented with sorbitol and a reducing agent. We have also used Lyticase (Boehringer Mannheim) instead of Zymolyase with similar results.

As for quantity of cells, we usually start with 1 liter of a culture at 2×10^7 cells/ml. This gives approximately 2 g of cells (wet weight) and approximately 0.2 mg of DNA.

Hypersensitivity Analysis with DNase I

As mentioned above, DNase I can also be used to search for footprints of factors in nuclei. The methods involved in digesting nuclei with DNase I to that end have been summarized (15, 16). Such protocols differ from the mapping of nucleosome positions with DNase I in that nuclei are treated with DNase I as soon as possible after cell lysis, and digestions are generally performed at lower temperature and salt concentrations to minimize dissociation of DNA-binding proteins. The extent of digestion is also much lower because the single-stranded cuts of DNase I are mapped rather than the sites in which DNase I has made a double-strand break. In contrast, the following procedure is designed to detect nucleosomes and to locate hypersensitive sites as shown schematically in Fig. 1. There is no DNA denaturation step prior to gel electrophoresis, so only double-strand cuts are scored.

Solutions

    Digestion buffer: 15 mM Tris-HCl, pH 7.5, 75 mM NaCl, 3 mM MgCl$_2$, 50 μM CaCl$_2$, 1 mM 2-mercaptoethanol

    DNase I dilution buffer: 10 mM Tris-HCl, pH 7.4, 0.1 mg/ml bovine serum albumin (BSA)

    Stop solution: 1.0 M Tris-HCl, pH 8.8, 80 mM EDTA, pH 8.0

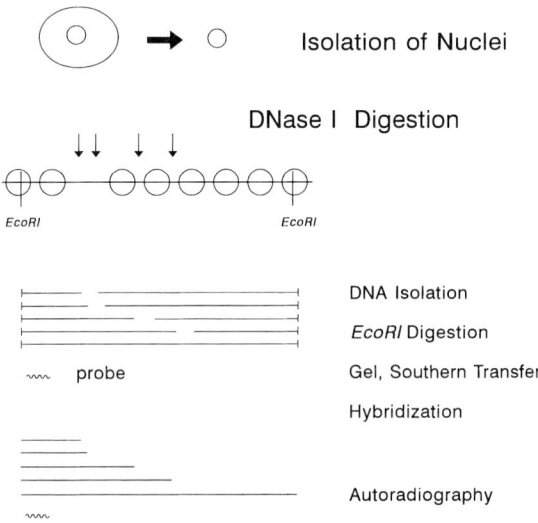

FIG. 1   Schematic for the analysis of DNase I hypersensitivity.

Proteinase K solution: 10 mg/ml proteinase K, dissolved in 10 mM Tris-HCl, pH 8.0

Chloroform solution: chloroform–isoamyl alcohol (24 : 1, v/v)

RNase solution: 5 mg/ml ribonuclease A (DNase-free) dissolved in 5 mM Tris-HCl, pH 7.5, and heated for 10 min at 100°C

Procedure

1. Suspend pelleted nuclei from approximately 500 mg cells (wet weight) for one experiment in 3 ml digestion buffer by vortexing.
2. Centrifuge (2000 g for 5 min at 5°C) and resuspend in 1.2 ml digestion buffer. Transfer 200-μl aliquots to microcentrifuge tubes.
3. Add DNase I at four different concentrations in the range from 0.5 to 20 U/ml. The extent of digestion is monitored by ethidium bromide-stained gels. Keep one sample on ice and one at 37°C without nuclease.
4. Terminate digestion by adding 10 μl stop solution, 5 μl of 20% (w/v) sodium dodecyl sulfate (SDS), and 20 μl proteinase K solution. Incubate for 30 min at 37°C.
5. Add $\frac{1}{5}$ volume of 5 M NaClO$_4$, 1 volume phenol, and mix well; then add 1 volume chloroform solution and mix well.
6. Centrifuge for 5 min in microcentrifuge.
7. Remove supernatant; reextract with 1 volume chloroform solution.
8. Remove supernatant and add 2.5 volumes ethanol to precipitate nucleic acids.
9. Collect by centrifugation (5 min) and resuspend in 125 μl TE.
10. Add 10 μl RNase solution and incubate for 1 hr at 37°C.
11. Add 5 μl of 5 M NaCl and 0.6 volume of 2-propanol and centrifuge immediately for 2 min at room temperature.
12. Wash the pellet with 70% (v/v) ethanol and dissolve in 80 μl TE.
13. Analyze 5-μl aliquots in 1% agarose gels and stain with ethidium bromide.
14. Select appropriate samples and use 20 μl for secondary digestion and indirect end-labeling.

Comments

The protocol is designed to detect histone–DNA interactions rather than the binding of transcription factors to DNA. Typical results from the authors' laboratory are found for the *PHO5* promoter in Refs. 8 and 18–21 and the *PHO8* promoter in Ref. 9. Analysis of the *GAL1/10* promoter from galactose-grown cells has yielded a hypersensitive region across the upstream activating sequence (UAS) elements with no evidence of Gal4 binding. In contrast, using the digestion protocol of Lohr and Hopper (22) with the same nuclei gave a distinct footprint within the hypersensitive region (W. Hörz, unpublished data). Therefore, at least Gal4 appears to remain associated with its cognate sites throughout the nuclear isolation procedure, but it is lost when the nuclei are digested with DNase I as described in our protocol. The trans-activator Pho4 also remains complexed to its sites during nuclear isolation (U. Venter and W. Hörz, unpublished data).

It is important to do free DNA controls in parallel, although the sequence specificity of DNase I is much less pronounced than that of micrococcal nuclease. Obviously, the sequence specificity is less of a problem whenever chromatin changes are investigated and two chromatin digests can be compared.

Measuring Accessibility of Restriction Sites in Chromatin

One of the limitations of DNase I analyses of chromatin structure resides in the fact that the susceptibility of a region is always relative to neighboring regions. There can be no absolute measurements of sensitivity. As a consequence, it is impossible to draw quantitative conclusions from DNase I experiments. We have therefore in many instances used restriction nucleases to complement DNase I. Especially in cases when a given chromatin region undergoes structural transitions, digestion with restriction nucleases has contributed valuable information as to the precise boundaries of such transitions and the question whether all or most cells in a population undergo the transition or only a small percentage. The strategy used is shown in Fig. 2 with a

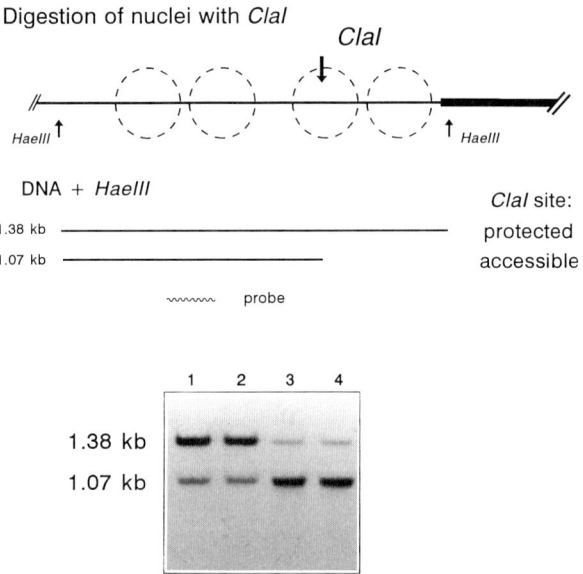

FIG. 2   Measuring the accessibility of a *Cla*I site at the *PHO5* promoter. In the repressed state, the *PHO5* promoter is organized in four positioned nucleosomes (broken circles) which are disrupted on activation of the promoter (8). Nuclei from repressed (lanes 1, 2) and induced cells (lanes 3, 4) were digested with 100 U (lanes 1, 3) or 300 U *Cla*I (lanes 2, 4). DNA was isolated, cleaved with *Hae*III, analyzed in a 1% agarose gel, blotted, and hybridized with probe D (8) as schematically shown at the top.

representative result from the *PHO5* promoter. The results from an analysis using a large number of restriction enzymes are found in Ref. 8 for the *PHO5* promoter and in Ref. 9 for the *PHO8* promoter.

Solutions

> Digestion buffer: 10 mM Tris-HCl, pH 7.4, 50 mM NaCl, 10 mM MgCl$_2$, 0.5 mM spermidine, 0.15 mM spermine, 0.2 mM EDTA, 0.2 mM EGTA, 5 mM 2-mercaptoethanol

Procedure

1. Suspend pelleted nuclei in digestion buffer by vortexing.
2. Centrifuge (2000 g for 5 min at 5°C) and resuspend in digestion buffer. Nuclei from approximately 50 mg cells (wet weight) are used for one experiment and suspended in 200 μl. Transfer to microcentrifuge tubes.
3. Add restriction nuclease at two different concentrations (ranging between 150 and 1500 U/ml). Incubate at 37°C for 30 min.
4. Terminate digestion as described for DNase I, except that the EDTA concentration is raised to 12 mM.
5. Follow steps 5 to 14 of the protocol given for DNase I analysis.

To test the accessibility of individual restriction sites, the indirect end-labeling protocol as schematically shown in Fig. 2 is applied.

1. Use 10 μl of the DNA from restriction nuclease-digested nuclei for secondary digestion with the appropriate restriction enzyme.
2. Analyze by Southern transfer and hybridization.

Comments

One of the main advantages of restriction nuclease analyses is that, for any enzyme, different sites can be assayed in the same digest. They can serve as internal controls. Accessibility can be quantitated by determining the ratio of the two bands generated after secondary digestions. This is exemplified for the *PHO5* promoter in Ref. 8 and for the *PHO8* promoter in Ref. 9. In general, most sites will be largely inaccessible in the nucleus, because most sequences are incorporated into nucleosomes. This is only conclusive, however, if it can be shown that at least one site was cut by the enzyme in the particular digest, because some enzymes seem to work better than others in chromatin digestion. We have demonstrated constitutively accessible sites for a large number of enzymes at the *PHO5* (19), *PHO3* (19), *PHO8* (9), and *TDH3* loci (10). Probes suitable to assay these sites are available from our laboratory and can serve as controls in such experiments.

Typically, intranucleosomal sites are about 5 to 10% accessible. It is not clear if the residual accessibility reflects alternative nucleosome positions or cutting within the nucleosome. However, these values seem to be true plateau values since they do not change appreciably on raising the nuclease concentrations by a factor of 2 to 4. Sites located within hypersensitive regions are usually 80–100% accessible, and again the significance of residual protection is not clear. At the *PHO5* promoter, sites located within short linker regions between positioned nucleosomes are about 50% accessible.

For certain nucleosomes, it appears that there is intermediate accessibility (40–60%) to restriction nucleases (e.g., Ref. 9). This may indicate that there exist two different populations of cells in which the nucleosome is present or absent. Such states may represent a true binding equilibrium *in vivo* or intermediate states with rather different subunit compositions of the nucleosome. For example, loss of one or both histone H2A/H2B dimers could result in intermediate accessibility.

The digestion buffer is a compromise between optimizing enzyme activity and preserving the chromatin structure. The addition of spermine and spermidine has been very useful as the polyamines stimulate activity of the vast majority of restriction nucleases. At the same time, endogenous, nonspecific nucleases are usually suppressed. A salt concentration of 50 mM NaCl works for almost all restriction enzymes in chromatin digestion.

There can be degradation of DNA during preparation of the nuclei. This is detected by the 0°C control incubation. By comparing the 37°C control with the 0°C control samples, one can also determine if endogenous nucleases are active during incubation with the restriction enzymes. It is useful to have the site used for secondary digestion (see Fig. 2) not too far away from the restriction site actually assayed. If limited degradation by endogenous nucleases does occur, comparison of the relative amounts of the restriction fragments is still meaningful because fragments of similar size should be similarly diminished.

We always use two nuclease concentrations that differ by a factor of 3–4. The results should be quite similar (from our experience they almost always are), which means that true plateau values were reached during restriction nuclease digestion of the chromatin.

Analyzing for Absence or Presence of Nucleosomes
by Micrococcal Nuclease Digestion

The principle of the assay is to convert essentially all of the chromatin into mononucleosomes and core particles. DNA isolated from this fraction is analyzed by Southern blotting and hybridization to very short DNA probes (see Fig. 3). Examples of this technique are found in Refs. 8, 9, and 23. The assay is not sensitive to nucleosome positioning since there is no indirect end-labeling step; instead, the absence or presence of a nucleosome on a given stretch of DNA is analyzed. The virtue of the procedure is that a single blot can be hybridized with many different probes and the patterns directly compared.

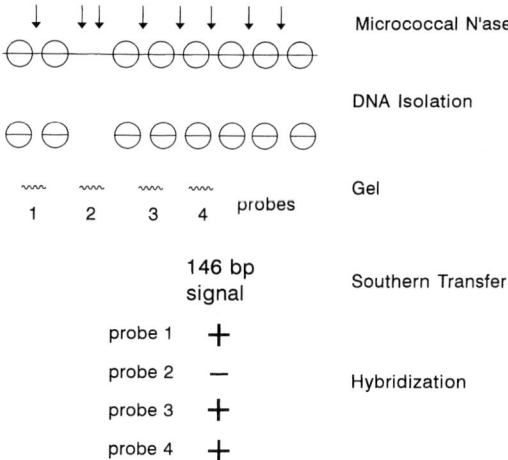

FIG. 3   Assaying for absence or presence of nucleosomes. This technique is designed to determine whether a particular sequence is organized in a nucleosome or not. It does not discriminate between randomly and specifically positioned nucleosomes.

Solutions

   Digestion buffer: 15 mM Tris-HCl, pH 8.0, 50 mM NaCl, 1.4 mM CaCl$_2$,
   0.2 mM EDTA, 0.2 mM EGTA, 5 mM 2-mercaptoethanol
   Nuclease dilution buffer: 10 mM Tris-HCl, pH 7.4, 0.1 mg/ml BSA

Procedure

1. Suspend pelleted nuclei from approximately 500 mg cells (wet weight) in 3 ml
   digestion buffer by vortexing.
2. Centrifuge (2000 g for 5 min at 5° C) and resuspend in 1.2 ml digestion buffer.
   Transfer 200-μl aliquots to microcentrifuge tubes.
3. Add micrococcal nuclease at four different concentrations in the range from 5
   to 100 U/ml, and incubate for 20 min at 37° C. The extent of digestion is
   monitored by ethidium bromide-stained gels. Select samples which have mostly
   mononucleosomes and also samples that still have di-, tri-, and tetranucleo-
   somal DNA.
4. Terminate digestion by adding 10 μl stop solution, 5 μl of 20% SDS, and 20 μl
   proteinase K solution. Incubate for 30 min at 37° C.
5. Follow steps 5 to 14 of the protocol given for DNase I digestion.
6. Analyze by agarose gel electrophoresis (2% gel) and Southern transfer without
   secondary digestion.

Comments

Micrococcal nuclease digests DNA with higher sequence specificity than DNase I (24). Therefore, it is formally possible that a nucleosome which incorporates sequences that are highly sensitive to micrococcal nuclease might be converted to subnucleosomal material more rapidly than the rest of the genome. For that reason, a series of digests with increasing nuclease concentrations should be employed, so that samples with predominantly mononucleosomal material as well as samples with DNA fragments from oligonucleosomes (3–5) can be analyzed on the same gel. The possibility of rehybridizing digests with different probes makes this approach a convenient tool for looking at different nucleosomal structures.

The units used in the protocol given above are as described for micrococcal nuclease obtained from Boehringer Mannheim (also listed as nuclease S7). The unit definition for micrococcal nuclease from Sigma (St. Louis, MO) is 85 times higher.

In Vivo Dimethyl Sulfate Footprinting

A valuable reagent that has been successfully employed in yeast for *in vivo* footprinting is dimethyl sulfate (25–31). Its principal advantage is that it is freely diffusible through the yeast cell wall and membrane and therefore can be used on intact cells. It appears to be relatively insensitive to nucleosomes (32) and is ideal for detecting the interaction of a trans-acting factor with a specific site. Its disadvantage is that protection of only guanines and, to a lesser extent, adenines can be measured. An example of *in vivo* footprinting in the *PHO8* promoter is shown in Fig. 4. Both protection and enhancement of DMS methylation is observed in the presence of a DNA-binding protein. Enhancement of methylation could be caused by a hydrophobic pocket that is created by the protein, thereby concentrating DMS next to the DNA (33). The following is a method which we have employed in our system that is based on one developed by Giniger *et al.* (25).

Dimethyl Sulfate Treatment and DNA Purification

1. Grow the desired yeast strain in a 500-ml culture to approximately $2-4 \times 10^7$ cells/ml. Centrifuge the cells (2000g for 5 min at room temperature), resuspend in 6 ml medium, and then divide into four aliquots of 1.5 ml in 50-ml tubes. Smaller amounts of cells can be treated with DMS, but the volumes should be reduced accordingly.
2. Add 2 μl DMS to each aliquot at room temperature for 5–20 min. A time course should be used to determine the optimal exposure time.
3. Stop the DMS reaction by adding 40 ml cold TEN buffer (10 mM Tris-HCl, pH 7.1, 1 mM EDTA, 40 mM NaCl) to each tube. Centrifuge the cells (2000g for 5 min at 5°C), resuspend in 1 ml of 1 M sorbitol, 0.1 M sodium citrate, pH 5.8, 10 mM EDTA, 2 mM dithiothreitol (DTT), and digest with 1 mg/ml Zymolyase 100T (ICN) for 30 min at 37°C.

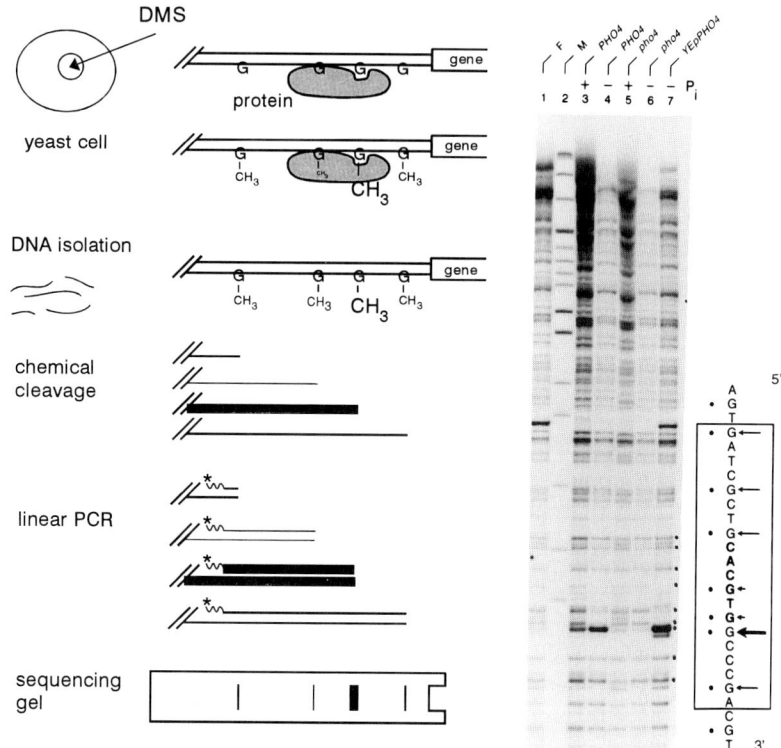

FIG. 4   *In vivo* footprinting with dimethyl sulfate. The strategy of the procedure is shown at left and an *in vivo* footprint of Pho4 on UASp2 of the *PHO8* promoter at right. Strains containing the *PHO4* gene (lanes 3, 4, and 7) or lacking it (lanes 5 and 6) were analyzed with a *PHO8*-specific primer [5'-CGTCCAGTCATGTCGTACAACGGAA, positions −598 to −574 of the *PHO8* promoter; Y. Kaneko, N. Hayashi, A. Toh-e, I. Banno, and Y. Oshima, *Gene* **58**, 137 (1987)]. The $+P_i$ and $-P_i$ refers to conditions that repress and activate the *PHO8* gene, respectively. In lane 7 the strain contains the Pho4 overexpression plasmid YEpPHO4 (20). F is free DNA and M a labeled *Hpa*II digest of pBR322 that serves as a molecular weight reference. G residues within UASp2 are marked by dots in the gel and on the sequence at right in which the Pho4-binding site, as detected by *in vitro* footprinting with DNase I (9), is boxed. The thick arrow designates the guanine residue which is hypersensitive to DMS in Pho4-containing cells relative to free DNA. The short arrows indicate guanines which are less reactive, and the medium arrows denote guanines whose reactivity with DMS is unchanged relative to free DNA.

4. Centrifuge the yeast spheroplasts (2000g for 5 min at 5°C) and resuspend each aliquot in 1 ml of 20 mM EDTA/50 mM Tris-HCl, pH 8.0. Add 50 μl of 20% SDS and incubate at 65°C for 30 min to lyse the cells. Add 0.4 ml of 5 M potassium acetate (pH 8.0) to each tube and place the tubes on ice for 1 hr.

5. Remove precipitated cell protein by centrifugation for 5 min in a microcentrifuge. Withdraw the supernatant and combine with 0.65 ml of 2-propanol in a new tube to precipitate nucleic acids. Wash each pellet with cold 70% ethanol, resuspend in 0.6 ml TE, and digest with 6.7 μg/ml RNase A (DNase-free) at 37°C for 1 hr.

6. Precipitate the DNA by adding 2 μl of 5 M NaCl and 0.4 ml of 2-propanol, then wash the pellets with 70% ethanol.

7. Resuspend each pellet in 0.4 ml restriction enzyme buffer and digest with an appropriate restriction enzyme to decrease viscosity. We generally use EcoRV, but other enzymes can be used as long as they do not cut between the priming site and the site of interest. If the distance between the primer and a specific restriction site is 300–500 bp, the resulting DNA fragment in untreated DNA samples can then be used as a control to evaluate how well the Taq polymerase extension works.

8. Precipitate the DNA in each tube, resuspend in 0.1 ml piperidine diluted 1:10 in water, and incubate 30 min at 90°C.

9. Precipitate the DNA by adding 0.1 ml of 0.3 M sodium acetate, 0.1 mM EDTA, pH 7.0, and 0.8 ml ethanol. A second ethanol precipitation or extensive lyophilization is required to remove all traces of piperidine. Dissolve the DNA in 0.4 ml TE for a 500-ml culture of 4 × 10⁷ cells/ml.

Primer Extension Using Taq Polymerase

1. The primer should be about 16–20 nucleotides long and should lie approximately 60–100 bp from the site of interest. The primer should be gel-purified to ensure that only full-length primer is present in the reaction. Label 50 ng of the primer using T4 polynucleotide kinase and [γ-³²P]ATP (6000 Ci/mmol); purify the primer from unincorporated label via a 1-ml column of Sephadex G-50.

2. Mix approximately 50 μg DNA (or 3 ng plasmid DNA that has been treated with DMS *in vitro* for a free DNA control) with 150,000 counts/min (cpm) of the radiolabeled primer in 50 μl of a buffer containing 10 mM Tris-HCl, pH 8.3, 10 mM MgCl₂, 50 mM KCl, 0.1 mg/ml gelatin, and 0.25 mM of each deoxynucleoside triphosphate (dNTP). Add 5 U Taq polymerase and incubate the samples for 30 cycles of 95°C for 1 min, 45°C for 1 min, and 70°C for 2 min.

3. Add 6.6 μl of 1% SDS, 100 mM EDTA, 1 mg/ml proteinase K and incubate the samples for 30 min at 45°C. Precipitate the DNA by adding 4 μl of 3 M NaCl and 0.15 ml ethanol. Care should be taken that the NaCl is well mixed before the addition of ethanol.

4. Wash each pellet with 70% ethanol, dissolve in a denaturing gel loading buffer, and boil for 3 min. Load the samples on a 6–10% polyacrylamide sequencing gel. After electrophoresis, the gel can either be dried or simply covered with plastic wrap and then exposed to X-ray film with an intensifying screen at −70°C. Depending on the specific activity of the primer, the methylation pattern can often be seen after an overnight exposure.

Comments

The piperidine cleavage step cleaves principally at guanines which are methylated on N-7 in the major groove by DMS. DMS also methylates adenine residues on N-3 in the minor groove, but this modification does not lead to efficient strand cleavage in the presence of piperidine. To obtain information on A residues, it is possible to use the G > A cleavage reaction for DMS-treated DNA (34). This has been used *in vitro* (35, 36) and should in principle work with DNA that has been treated with DMS *in vivo*. It has been reported that the piperidine cleavage step is not required for DMS-treated DNA that is used for *Taq* polymerase extension (37).

The quality of *Taq* polymerase from different sources appears to vary considerably. We have had good results using *Taq* polymerase from Boehringer Mannheim. Occasionally, some inhibitors of the *Taq* polymerase are present in the purified DNA samples. In such cases, a lane may be totally blank, but extension of half the amount of DNA results in some incorporation. The DNA can be further purified using commercially available DNA purification agents (Qiagen tips, Quiagen 40724 Hilden, Germany, etc.). Ten cycles or even fewer may be sufficient to observe the methylation pattern. If nonspecific products are observed, the temperature of the annealing step can be increased.

The footprints of Pho4 that we have observed are presumably fairly stable throughout the time course of the DMS treatment. However, certain DNA-binding proteins may have a faster off-rate which may inhibit detection of their binding by DMS. In such cases, it may help to use lower concentrations of DMS with shorter time points. In addition, DMS can be more efficiently quenched if 2-mercaptoethanol is included in the stop buffer.

A number of controls can be used to analyze the pattern revealed by DMS. First of all, the *in vivo* pattern should be compared with free DNA that has been treated with DMS *in vitro*. If any protection or hypersensitivity is observed, the identity of the responsible protein can be elucidated in at least two ways. If one suspects that a certain DNA-binding protein is responsible, then it is possible to measure the DMS pattern in a yeast strain lacking the protein, provided that the protein is nonessential. The absence of DMS modification in such a strain is a good indication that the DNA-binding protein actually interacts with the site *in vivo*, although it would remain formally possible that that DNA-binding protein is required to produce a second trans-acting factor that binds to the site. If purified protein is available, the DMS protection pattern of the protein *in vitro* can be compared with the *in vivo* pattern.

Although it is preferable to use these techniques on genes which are in their native chromosomal locus, we have observed that nucleosomes assume the same positions on a copy of the *PHO5* gene that is on an ARS/CEN plasmid (1–3 copies per cell) (21). Furthermore, we have also been able to obtain clear *in vivo* footprints with DMS on such a plasmid (J. Svaren, U. Venter, and W. Hörz, unpublished data).

Conclusions

The techniques described above can be used to provide a much more detailed picture of how a promoter looks *in vivo.* Such data complement genetic studies of promoter function. For example, determination of nucleosome structure allows one to design promoter mutagenesis studies so that a deletion does not destabilize a nucleosome and modify promoter activity in an unforeseen way. A combination of these and other techniques makes it possible to address the question of how the chromatin structure of a promoter is correctly configured in order to facilitate efficient initiation of transcription. To elucidate the function and mechanism of a chromatin transition, it is ideal to study an inducible gene because active and inactive states can be studied in the same cell type. However, the same techniques can be applied to constitutively active genes in order to ascertain how the active state is established and maintained.

Acknowledgments

Work from the authors' laboratory was supported by the DFG (SFB 190) and Fonds der Chemischen Industrie. J.S. was supported by a grant from the National Science Foundation Program for Medium and Long Term Research at Foreign Centers of Excellence and by a National Science Foundation NATO Postdoctoral Fellowship.

References

1. G. Felsenfeld, *Nature (London)* **355,** 219 (1992).
2. J. L. Workman and A. R. Buchman, *Trends Biochem. Sci.* **18,** 90 (1993).
3. A. P. Wolffe, *FASEB J.* **6,** 3354 (1992).
4. R. D. Kornberg and Y. Lorch, *Annu. Rev. Cell Biol.* **8,** 563 (1992).
5. F. Winston and M. Carlson, *Trends Genet.* **8,** 387 (1992).
6. J. Svaren and W. Hörz, *Curr. Opin. Genet. Dev.* **3,** 219 (1993).
7. M. Johnston and M. Carlson, *in* "The Molecular and Cellular Biology of the Yeast *Saccharomyces cerevisiae:* Volume 2, Gene Expression" (E. W. Jones, J. R. Pringle, and J. R. Broach, eds.), p. 193. Cold Spring Harbor Laboratory, Cold Spring Harbor, New York, 1992.
8. A. Almer, H. Rudolph, A. Hinnen, and W. Hörz, *EMBO J.* **5,** 2689 (1986).
9. S. Barbaric, K. D. Fascher, and W. Hörz, *Nucleic Acids Res.* **20,** 1031 (1992).
10. B. Pavlovic and W. Hörz, *Mol. Cell. Biol.* **8,** 5513 (1988).
11. U. Wintersberger, P. Smith, and K. Letnansky, *Eur. J. Biochem.* **33,** 123 (1973).
12. J. P. Aris and G. Blobel, *in* "Methods in Enzymology" (C. Guthrie and G. R. Fink, eds.), Vol. 194, p. 735. Academic Press, San Diego, 1991.
13. D. Lohr, *in* "Yeast: A Practical Approach" (I. Campbell and J. H. Duffus, eds.), p. 125. IRL Press, Oxford, 1988.

14. M. J. Fedor, N. F. Lue, and R. D. Kornberg, *J. Mol. Biol.* **204,** 109 (1988).
15. M. W. Hull, G. Thomas, J. M. Huibregtse, and D. R. Engelke, *Methods Cell Biol.* **35,** 383 (1991).
16. J. M. Huibregtse and D. R. Engelke, *in* "Methods in Enzymology" (C. Guthrie and G. R. Fink, eds.), Vol. 194, p. 550. Academic Press, San Diego, 1991.
17. N. A. Kent, L. E. Bird, and J. Mellor, *Nucleic Acids Res.* **21,** 4653 (1993).
18. A. Schmid, K. D. Fascher, and W. Hörz, *Cell (Cambridge, Mass.)* **71,** 853 (1992).
19. A. Almer and W. Hörz, *EMBO J.* **5,** 2681 (1986).
20. K. D. Fascher, J. Schmitz, and W. Hörz, *EMBO J.* **9,** 2523 (1990).
21. K. D. Fascher, J. Schmitz, and W. Hörz, *J. Mol. Biol.* **231,** 658 (1993).
22. D. Lohr and J. E. Hopper, *Nucleic Acids Res.* **13,** 8409 (1985).
23. M. J. Fedor and R. D. Kornberg, *Mol. Cell. Biol.* **9,** 1721 (1989).
24. W. Hörz and W. Altenburger, *Nucleic Acids Res.* **9,** 2643 (1981).
25. E. Giniger, S. M. Varnum, and M. Ptashne, *Cell (Cambridge, Mass.)* **40,** 767 (1985).
26. D. McDaniel, A. J. Caplan, M. S. Lee, C. C. Adams, B. R. Fishel, D. S. Gross, and W. T. Garrard, *Mol. Cell. Biol.* **9,** 4789 (1989).
27. J. D. Axelrod and J. Majors, *Nucleic Acids Res.* **17,** 171 (1989).
28. D. S. Gross, K. E. English, K. W. Collins, and S. W. Lee, *J. Mol. Biol.* **216,** 611 (1990).
29. J. Mellor, W. Jiang, M. Funk, J. Rathjen, C. A. Barnes, T. Hinz, J. H. Hegemann, and P. Philippsen, *EMBO J.* **9,** 4017 (1990).
30. L. Densmore, W. E. Payne, and M. Fitzgerald Hayes, *Mol. Cell. Biol.* **11,** 154 (1991).
31. B. Ganter, S. Tan, and T. J. Richmond, *J. Mol. Biol.* **234,** 975 (1993).
32. J. D. Axelrod, M. S. Reagan, and J. Majors, *Genes Dev.* **7,** 857 (1993).
33. R. T. Ogata and W. Gilbert, *J. Mol. Biol.* **132,** 709 (1979).
34. A. M. Maxam and W. Gilbert, *Proc. Natl. Acad. Sci. U.S.A.* **74,** 560 (1977).
35. N. L. Craig and H. A. Nash, *Cell (Cambridge, Mass.)* **39,** 707 (1984).
36. D. K. Lee, M. Horikoshi, and R. G. Roeder, *Cell (Cambridge, Mass.)* **67,** 1241 (1991).
37. A. C. Brewer, P. J. Marsh, and R. K. Patient, *Nucleic Acids Res.* **18,** 5574 (1990).

[9] Mapping of Yeast Nucleosomes *in Vivo*

Memmo Buttinelli, Giorgio Camilloni, Giovanna Costanzo,
Rodolfo Negri, Patrizia Venditti, Sabrina Venditti,
and Ernesto Di Mauro

Introduction

The localization of nucleosomes in yeast chromatin has so far been analyzed on isolated chromatin or nuclei, as intact cells and spheroplasts are impermeable to macromolecules, thus preventing penetration of enzymes normally used *in vivo* or *in vitro* as analytical tools (deoxyribonuclease I, DNase I; micrococcal nuclease, MN; exonuclease III, ExoIII; restriction endonucleases; etc.). The use of nystatin as permeabilizing agent (1) has opened the possibility of enzymatic treatment of chromatin in living cells and the development of new analytical approaches to chromatin analysis (Fig. 1).

Protocol 1: Spheroplast Preparation and Permeabilization by Nystatin

1. Grow yeast cells in 100 ml medium at $30°C$ to a cell density of $1-3 \times 10^7$ cells/ml (for *Saccharomyces cerevisiae*, $0.3-1$ OD_{260} units/ml).
2. Harvest in 50-ml tubes by spinning at 3000 g for 5 min.
3. Resuspend the pelleted cells in $5-10$ ml of buffer containing 20 mM EDTA and 600 mM 2-mercaptoethanol. Keep at room temperature for 15 min. This step improves the efficiency of spheroplast production.
4. Spin cells at 3000 g for 5 min.
5. Resuspend in 10 ml of 1 M sorbitol and spin at 3000 g for 5 min. This washing step removes EDTA and excess 2-mercaptoethanol and is important if *in vivo* enzymatic reactions are going to be performed.
6. Resuspend cells in 4 ml of 1 M sorbitol and 5 mM 2-mercaptoethanol; add $0.1-0.5$ mg/10^8 cells Zymolyase 100T dissolved in 1 ml of 1 M sorbitol. Incubate 30 min at $30°C$ under agitation. The percentage of spheroplasts produced is evaluated by counting in a Thoma chamber. Typically, 10 μl of cells is diluted 10-fold in 1 M sorbitol or water. In water, spheroplasts undergo osmotic lysis and disappear, whereas they remain round and visible in sorbitol. The efficiency of spheroplast generation varies depending on the yeast strain, type of medium, and growth phase. The Zymolyase treatment can be extended for 30 min longer, if necessary.
7. Spin spheroplasts at 4000 g for 5 min.
8. Resuspend the pellet in 10 ml of 1 M sorbitol and repeat step 7.

Methods in Molecular Genetics, Volume 6

9. Resuspend the spheroplasts in nystatin buffer containing 50 mM NaCl, 20 mM Tris-HCl, pH 8.0, 1 M sorbitol, 50 μg/ml nystatin. At this concentration, nystatin is in suspension. The polyene antibiotic permeabilizes spheroplasts by a sterol-binding mechanism which creates pores in the membrane. Depending on whether micrococcal nuclease, DNase I, or restriction enzymes are to be introduced into the spheroplasts, 1.5 mM CaCl$_2$ for the first treatment or 3 mM MgCl$_2$ for the others is added to the buffer. Spheroplasts are usually resuspended at a concentration of 10^8 per ml.

Permeabilization with nystatin allows mapping in spheroplasts of nucleosomes located on large chromosomal segments by the traditional indirect end-labeling procedure (2). The resulting picture is not substantially different from that conventionally obtained with isolated chromatin. The advantage here is the fact that the risks of protein rearrangements and/or modifications are kept to a minimum. An example of a low-resolution analysis of nystatin-permeabilized yeast chromatin is reported elsewhere (1).

Protocol 2: Low-Resolution Mapping of Large Chromosomal Region by Indirect End-Labeling Technique

In Vivo Micrococcal Nuclease Treatment of Chromatin

1. Prepare $1-3 \times 10^9$ spheroplasts from exponentially growing cells as described in Protocol 1 (steps 1–8).
2. Resuspend spheroplasts in 1.5 ml nystatin buffer plus 1.5 mM CaCl$_2$ (final concentration of 6×10^8 spheroplasts/ml).
3. Divide into several 250-μl aliquots and add increasing amounts of MN (0–200 U/ml).
4. Incubate samples for 15 min at 37°C.
5. Add $\frac{1}{10}$ of the final volume of a 10× stop mix [2% (w/v) sodium dodecyl sulfate (SDS), 20 mM EGTA, 200 mM Tris-HCl, pH 8.0]. Add proteinase K to 400 μg/ml.
6. Incubate 1 hr at 56°C.
7. Extract three times with phenol–chloroform–isoamyl alcohol (24:24:1, v/v). Each time transfer the aqueous phase to a new tube.
8. Add 3 M sodium acetate ($\frac{1}{10}$ of the final volume) and precipitate with 2 volumes ethanol.
9. Resuspend the samples in 100 μl water and add RNase A to a final concentration of 200 μg/ml. Incubate 1 hr at 37°C.
10. Extract with phenol–chloroform–isoamyl alcohol and add $\frac{1}{10}$ of the volume of 3 M sodium acetate and 0.6 volume of 2-propanol. Incubate 15 min at room

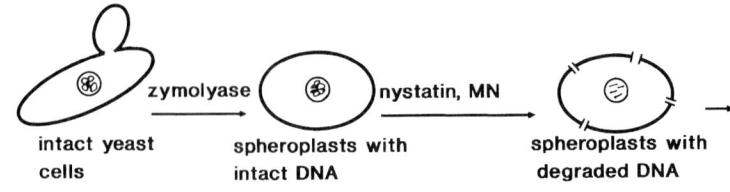

intact yeast cells → zymolyase → spheroplasts with intact DNA → nystatin, MN → spheroplasts with degraded DNA →

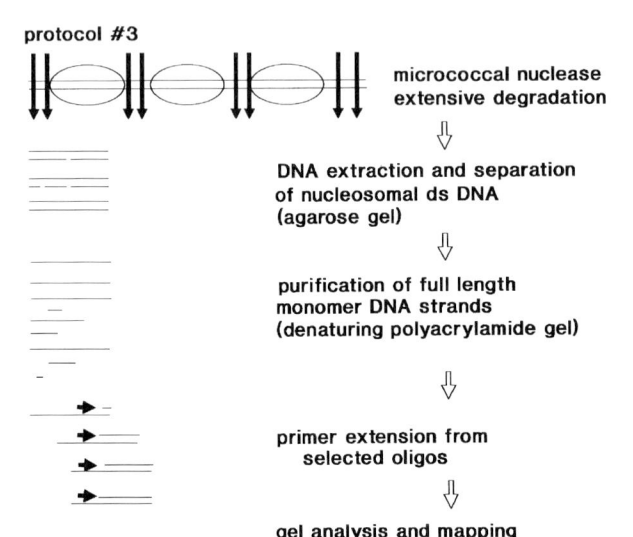

protocol #3

micrococcal nuclease extensive degradation

⇩

DNA extraction and separation of nucleosomal ds DNA (agarose gel)

⇩

purification of full length monomer DNA strands (denaturing polyacrylamide gel)

⇩

primer extension from selected oligos

⇩

gel analysis and mapping of nucleosomal borders

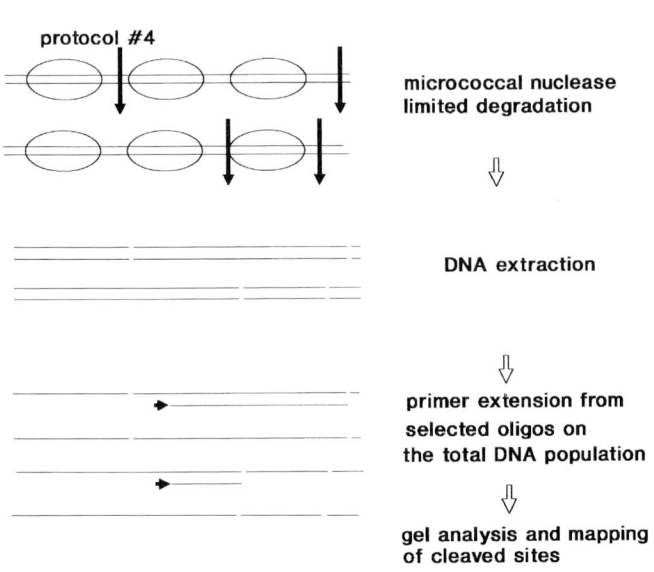

protocol #4

micrococcal nuclease limited degradation

⇩

DNA extraction

⇩

primer extension from selected oligos on the total DNA population

⇩

gel analysis and mapping of cleaved sites

temperature; spin in microcentrifuge 15 min; wash pellet with 70% (v/v) ethanol; dry in a Speed-Vac.

11. Resuspend the DNA in 100 μl water and determine the concentration by OD_{260}.

12. Digest 1–10 μg (depending on whether single-copy or multicopy genes are being analyzed) of DNA from each sample with a restriction enzyme which cuts the DNA in the vicinity of the region under study. Usually 200–300 U/ml of restriction enzyme is used to digest completely the genomic DNA in an overnight reaction.

13. Purify the samples with one phenol extraction and ethanol precipitation. Wash once with 70% ethanol and dry in a Speed-Vac.

Preparation of in Vitro Micrococcal Nuclease-Treated Purified DNA

14. Digest 1–10 μg of purified genomic DNA with up to 100 U/ml of the same restriction enzyme. This will be used as an *in vitro* control of DNA sequence-dependent hypersensitivity to MN.

15. Purify by phenol extraction and ethanol precipitation. Wash once with 70% ethanol and dry in a Speed-Vac.

FIG. 1   Schematic representation of strategies for mapping nucleosomes *in vivo* in yeast. Zymolyase-treated yeast cells are permeabilized with nystatin (Protocol 1), then treated with a degrading enzyme, namely, micrococcal nuclease (Protocol 2). The digestion is carried out under moderate to extensive conditions, so that the same chromatin sample can be analyzed in the following ways. (a) Low-resolution mapping, carried out as described in Protocol 2. A region encompassing at least 10 nucleosomes can typically be mapped. Using two overlapping, divergently oriented terminal probes, the localization of 20 nucleosomes can be defined in a single gel run. (b) Determination of nucleosomal borders in mononucleosomes by primer extension (PE). Two oppositely oriented overlapping oligonucleotides are selected on the basis of low-resolution mapping described in (a). A central position on the nucleosome to be analyzed is chosen. PE from the oligonucleotides is carried out on purified mononucleosomal DNA fragments of 146 ± 2 bp, the size of the resulting fragments providing a direct measure of the nucleosomal borders. (c) Determination of nucleosomal borders in oligonucleosome-sized DNAs by PE. The localization obtained in (a) is also used to select the oligonucleotides to be used as primers in the PE analysis of DNA purified from moderately digested chromatin. The oligonucleotides are selected according to the following two criteria. (i) Position: The sequence should be 5'-upstream relative to the nucleosome(s) to be analyzed, possibly within the central part of a 5'-located nucleosome, in order to ensure the presence of the complementary chromosomal sequences after MN degradation. (ii) Size and base composition: A GC-rich 20-mer is preferable to minimize misannealing and pausing effects in the extension by *Taq* polymerase.

16. Resuspend in 100 μl of 20 mM Tris-HCl, pH 8.0, 1.5 mM CaCl$_2$, 50 mM NaCl. Add 2–10 U/ml MN.
17. Incubate 3 min at 37°C.
18. Stop the reaction with 10× stop mix ($\frac{1}{10}$ of the volume) (see above). Add proteinase K to 400 μg/ml. Incubate 1 hr at 56°C.
19. Phenol extract and ethanol precipitate. Wash with 70% ethanol and dry in a Speed-Vac.

Indirect End-Labeling

20. Resuspend the *in vivo* and *in vitro* samples in 10–15 μl loading buffer (1.5 M sucrose, 50 mM EDTA, 0.01% (w/v) bromophenol blue, 0.01% xylene cyanol) and load on a 30-cm-long agarose gel, the concentration depending on the desired range of analysis (typically, from 1.2 to 1.8% agarose is used) (2).
21. Run the gel overnight at 2 V/cm.
22. Transfer DNA to a nitrocellulose or nylon membrane, using the standard Southern procedure.
23. Incubate 2 hr at 80°C.
24. Prepare a DNA probe by labeling 25 ng of a purified DNA fragment, mapping at one end of the region to be analyzed (previously prepared by restriction cleavage). The DNA can be labeled by nick-translation or random priming, to a specific activity of 10^9 counts/min (cpm)/μg.
25. Hybridize the filter and wash it according to standard procedures (3).
26. Analyze by autoradiography.

The low-resolution nucleosomal map obtained by Protocol 2 provides a general overview of the organization of a chromatin segment and is particularly useful for comparative analyses (e.g., determination of chromatin differences in gene induction studies and in cell-cycle-dependent metabolic variations, analysis of changes due to different genetic backgrounds). In addition to the determination of nucleosomal positions, information on the presence of nonnucleosomal protein complexes is obtained. In this respect, MN analysis is conveniently complemented by the analysis with DNase I performed in the same experimental frame (not detailed). The major limit of the approach described in Protocol 2 is the low resolution power, which prevents the determination of important facts like the uniqueness or the multiplicity of nucleosome localization on slightly different positions and their rotational orientation. To understand the relationship between DNA sequence and nucleosome position, absolute mapping precision is needed. Low-resolution mapping (Protocol 2) is a prerequisite for the programming of high-resolution studies.

High-Resolution Mapping

We provide here two different protocols for high-resolution mapping. Protocol 3 entails extensive MN degradation of spheroplast chromatin, isolation of monomer-sized DNAs, and determination of the nucleosomal borders by primer extension (PE). Provided that pausing effects in the primer extension step are minimized, this technique is a no-background assay and allows unambiguous determination of the nucleosomal borders. The prerequisite for the analysis is previous knowledge of the approximate nucleosome position, in order to program the oligonucleotides to be used for primer extension. The best position for oligonucleotides is the center of the nucleosomal particle. In principle, two elongations are performed, each with one of a pair of divergently oriented overlapping oligonucleotides. The borders so determined are matched by a simple numerical procedure (see below), matches of 146 being referred to a bona fide nucleosomal core (4).

In our experience, determination of one of the two borders is routinely sufficient, the upstream border always being consistent with and confirmed by the localization of the downstream ones (and vice versa). Therefore, only one elongation is sufficient for each nucleosome, unless higher precision is required in the delimitation of a defined linker region. The major advantages of this approach are its no-background nature and the consequent lack of ambiguity, its precision, and its independence from the sequence specificity of the degrading enzyme (owing to the extensive degradation step). An intrinsic limit of the technique is that it is based on the isolation of monomer-sized DNAs of 146 ± 2 bp, thus precluding gathering of information on surrounding proteins (nucleosomal or other).

Protocol 4 is planned to overcome that limitation. It consists of PE from specific oligonucleotides on DNA fragments derived from partial digests of spheroplast chromatin. The advantages of this approach are that the same precision in the localization of enzymatic cuts as in Protocol 3 is achieved, several borders of one or more nucleosomes can be mapped in the same electrophoretic run, and information on additional nonnucleosomal proteins can be obtained. A limit of the technique is that it is a normal background assay (the DNA fragments being produced by partial digestion) which suffers from possible ambiguities and requires comparison with the digestion profile of naked DNA. A detailed example of the results which can be obtained by this technique is given in Ref. 5, where both upstream and downstream borders of the three nucleosomes located upstream of the ABF1 binding site of the *S. cerevisiae* ARS1 region are described.

Each of the procedures (Protocols 2 to 4) suffers from the described limitations, but their combination is a powerful tool for chromatin analysis *in vivo,* combining precision of nucleosome localization with an overall view of a large region.

Protocol 3: Determination by Primer Extension of Nucleosomal Borders on DNA Isolated from Mononucleosomes

1. Prepare $1-3 \times 10^9$ spheroplasts from exponentially growing cells as described in Protocol 1 (steps 1–8). Resuspend spheroplasts in nystatin buffer plus 1.5 mM $CaCl_2$ at a final concentration of 6×10^8 per ml. Remove an aliquot [$\frac{1}{4}$ of the volume to be used as a control, namely, untreated chromatin DNA (control 1)] and add to the rest MN to 100 U/ml (extensive digestion).

2. Incubate both samples for 15 min at 37° C.

3. Treat both samples as described in steps 5 to 10 of Protocol 2.

4. Resuspend the MN-treated sample in 100 μl water; read OD_{260}; load on a 1.5% agarose gel, 1.5 mm thick and at least 30 cm long, taking care to load not more than 5 μg of DNA per lane. Run gel at 2 V/cm overnight.

5. Stain gel 15 min in ethidium bromide bath (1 μg/ml).

6. Identify the nucleosomal monomer band by its size (150 ± 5 bp) and cut it from the gel (care should be taken to avoid prolonged exposure to UV light in order to prevent DNA nicking).

7. Elute the band by freezing–thawing or by electroelution. Ethanol precipitate the recovered DNA.

8. Resuspend the monomer DNA in water and quantify recovery (by OD_{260} or by quantitative standards in agarose gel).

9. Prepare monomer-sized marker as follows: remove 100 ng monomer DNA and label it by polynucleotide kinase and [γ-^{32}P]ATP to a specific activity of 0.1–0.5 μCi/pmol.

10. Prepare a control for MN-digested naked DNA (control 2) by digesting 10 μg of the DNA fragment with 1 U/ml MN (10 min at 37° C in 100 μl of nystatin buffer plus 1.5 mM $CaCl_2$). Stop the reaction with 5 mM EGTA (final). This control is needed to check for possible strong sequence-specific cuts by MN.

11. Phenol extract and ethanol precipitate the monomer-sized DNA sample (step 8), the monomer-sized marker (step 9), and DNA control 2 (step 10). Wash the pellets with 70% ethanol, dry, resuspend the samples in loading buffer (90% formamide, 5 mM EDTA, 0.01% bromophenol blue, and 0.01% xylene cyanol); heat samples for 2 min at 90° C and load on a 0.4-mm-thick 6% polyacrylamide denaturing gel. Do not load more than 2 μg/lane.

12. Run the bromophenol blue to the bottom, expose the gel, and cut the band corresponding to the full-length monomer-sized denatured DNA (146 ± 2 nucleotides long, as in Fig. 2, lanes N) using the monomer-sized marker to locate the correct position. Cut out also the monomer-sized marker and the band corresponding to the position of monomer-sized DNA in the lane of control 2. (*Note:* In the monomer-sized control lane most of the DNA should be full-length. Shorter sized nicked DNA is often present.)

13. Elute DNA by diffusion in TE buffer overnight at 45° C.
14. Resuspend the recovered monomer DNA sample (typically 10–40 μg per 10^9 spheroplasts) in water. Measure DNA concentration.
15. Select a couple of oligonucleotides (see text for selection criteria) and label them to high specific activity (1–3 μCi/pmol) with polynucleotide kinase.
16. Extend each oligonucleotide primer by 30 cycles of denaturation, annealing, and elongation with *Taq* DNA polymerase in 50 mM KCl, 10 mM Tris-HCl, pH 9.0, 0.1% (w/v) Triton X-100, 1.5 mM MgCl$_2$, 200 μM of each deoxynucleoside triphosphate (dNTP) using as template 5–10 μg of monomer DNA for single-copy gene analysis or 0.5–1 μg for multicopy plasmids and 0.05 pmol of labeled primer.
17. Primer-extend control DNAs, using 5–10 or 0.5–1 μg of undigested purified chromatin DNA (control 1) for single-copy or multicopy analysis, respectively, and 1–3 ng for gel-purified control 2 DNA. Prepare marker lanes by subjecting the same DNA to Sanger sequencing using the same primers.
18. Resuspend the samples in loading buffer and load on a 6% polyacrylamide denaturing gel together with 1000–5000 cpm of gel-purified monomer-sized DNA marker (control of template integrity and size).

Control 1 is intended to assess *Taq* polymerase pausing; control 2 should reveal strong sequence-specific MN cuts (see text).

Figure 2 shows examples depicting three different cases of nucleosomal distribution. Figure 2a shows analysis of a nucleosome with a unique position (nucleosome -3 of the *Hansenula polymorpha* MOX promoter). In addition to the major individual particle, only a quite minor alternative position is detected. The example in Fig. 2b shows multiple alternative positions on the same rotational phase on the *S. cerevisiae* 5 S rRNA gene. Figure 2c displays analysis of a limited set of alternative positions, on the *S. cerevisiae* ARS1 B-domain nucleosome. After their localization, the upstream and the downstream borders of each nucleosome are coupled as follows. The numerical values are added two by two, selecting couples of values so that their sum is closest to 166 $\pm$ 2 (146 $\pm$ 2 + 20, i.e., the length in base pairs of the starting monomeric DNA plus the length of the overlap of the oligonucleotides). In the example shown in Fig. 2b, the coupling will be as follows (from bottom to top for oligonucleotide 1, lane 8, and from top to bottom for oligonucleotide 2, lane 9:

$$(30–34) + (141 \pm 2) = 169–177 \text{ (ncp, nucleosomal core particle, 1)}$$
$$(39–43) + (132 \pm 1) = 170–176 \text{ (ncp 2),} \qquad \text{etc.}$$

Relevant information can be inferred from experiments such as the one reported in Fig. 2b. First, genes exist on which nucleosomes occupy multiple alternative

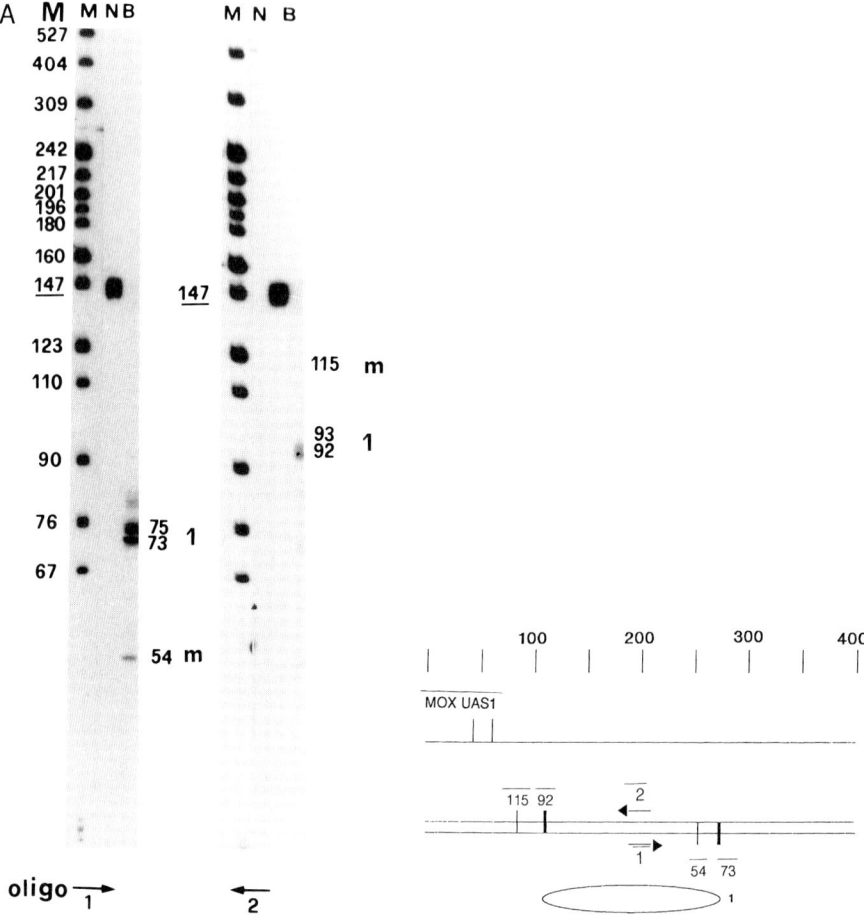

FIG. 2   Localization of the position(s) of nucleosomes in cellular chromatin by MN digestion and primer extension of the resulting products. Primer extension of mononucleosomes was performed according to Protocol 3. (a) A uniquely positioned nucleosome, namely, nucleosome −3 of the *Hansenula polymorpha* MOX promoter. The mapping procedure was as described in Protocol 3. *Left:* Experimental lanes show size markers (M), nucleosomal end-labeled monomer-sized DNA (N), and elongation products (B) from oligonucleotide 1 to the border (left-hand gel autoradiogram) and from oligonucleotide 2 (right-hand gel autoradiogram). The small-sized numbers to the right of lanes B refer to the distance (in base pairs) of the indicated band from the labeled extremity of the oligonucleotide. The large-sized numbers identify nucleosomes; m denotes a minor nucleosome. *Right:* Oligonucleotides and borders are mapped in the scheme. The coupling of the nucleosomal borders is performed as described in the text. The 5′-labeled oligonucleotides used are 5′-GGCTCTGTTTGCTGGCGTAG (oligonucleotide 1) and the complementary 5′-CTACGCCAGCAAACAGAGCC (oligonucleotide 2) [map position from −350 to −331 relative to the ATG of the *H. polymorpha* MOX

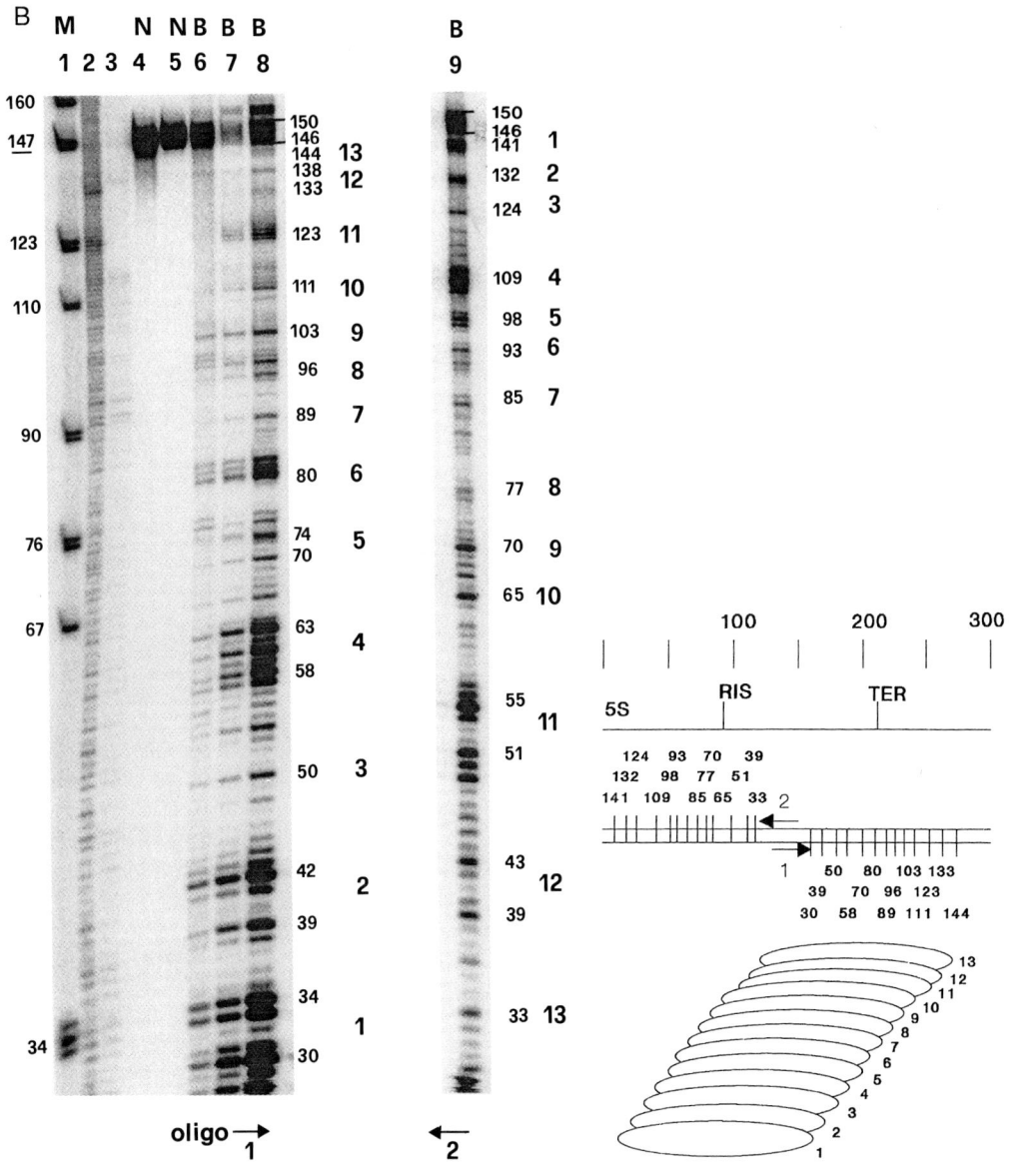

promoter, described in S. Goedecke, M. Eckart, Z. A. Janovics, and C. P. Hollenberg, *Gene* **139,** 35 (1994)]. (b) Multiple nucleosomes on alternative positions with unique rotational setting on the *S. cerevisiae* 5 S rRNA gene. *Left:* Experimental lanes show size markers (M, lane 1), primer extension from oligonucleotide 1 (see map at right) on purified chromosomal

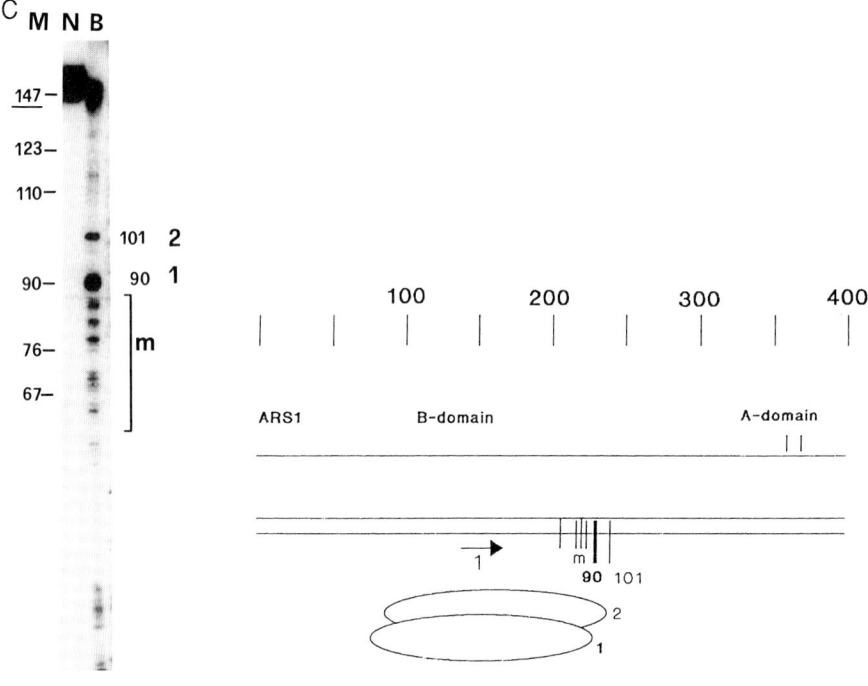

DNA (lane 2; only background pausing occurs, not corresponding to specific MN signals), same on plasmid DNA (lane 3), 5'-end-labeled nucleosomal monomer-sized DNA (100 ng) obtained by extensive MN digestion on nystatin-treated spheroplasts (N, lane 4), 5'-end-labeled monomer DNA (core size; 2 ng) obtained by extensive MN digestion of *in vitro* reconstituted nucleosomal core particles on 5 S DNA (N, lane 5), determination of borders via PE of 100 ng monomer DNA obtained from cellular chromatin (same sample shown in lane 4, unlabeled) starting from 5'-labeled oligonucleotide 1 (B, lane 6), PE of 0.5 ng monomer DNA from *in vitro* reconstituted material (same sample shown in lane 5, unlabeled), from oligonucleotide 1 (B, lane 7), same, using 2.0 ng monomer DNA (B, lane 8), and same as in lane 7, from oligonucleotide 2 (see map) (B, lane 9). *Right:* Map of the multiple nucleosomes on the 5 S gene. [For additional data on this system, see M. Buttinelli, E. Di Mauro, and R. Negri, *Proc. Natl. Acad. Sci. U.S.A.* **90,** 9315 (1993)]. RIS and TER denote transcription initiation and termination sites. (c) Example of nucleosome distribution with intermediate multiplicity: the *S. cerevisiae* ARS1 B-domain nucleosome. Only the primer extension data from one side are shown. Lane M contains size markers; lane N, as above; lane B, PE borders from the oligonucleotide 5'-GCTGGTGGACTGACGCGAAGA-3' [map position −653 to −634; G. Tschumper and J. Carbon, *Gene* **10,** 157 (1980)]. The two major and the minor (m) positions are indicated.

positions, usually spaced by an average of 10 bp, therefore being located on the same rotational phase. Second, for the *S. cerevisiae* 5 S rRNA genes, *in vivo* and *in vitro* profiles coincide (lane 6 versus lanes 7 and 8).

A relevant control is the comparison of MN digestion of purified DNA versus native or reconstituted chromatin. For this purpose, purified chromosomal or plasmid DNA is treated with MN (at a concentration $\frac{1}{200}$ relative to that of the mononucleosomal DNA to be analyzed, as MN cleaves purified DNA at a higher rate than chromatin) and a 146 ± 2 bp subpopulation is prepared (control 2 DNA in Protocol 3). The degradation profile (not shown) of this purified DNA is partially different from that obtained with chromatin.

It is known that MN has a certain degree of sequence specificity (6), although it may eventually completely degrade DNA. Therefore, with purified DNA it produces a defined cleavage pattern that changes as a function of the intensity of the enzymatic treatment (first hypersensitive sites appear, then the rest). With chromatin, MN cleaves internucleosomal regions with high preference, as shown by the typical ladder of monomers, dimers, etc., that is normally obtained. This is true also under the digestion conditions described here. Therefore, we should expect a difference between the MN patterns of naked DNA and chromatin only in the case of uniquely localized nucleosomes [as in yRp7 (see Ref. 7 and references therein) and PHO5 (8)]. In these cases, MN gives typical footprintlike patterns, similar to those produced by DNase I.

In the case of multiply positioned nucleosomes [as the ones that form on 5 S DNA (9, 10)], the situation is different. If nucleosomes occupy every available rotationally phased position (i.e., from position 1 to 146, or from 10 to 156, or from 20 to 166, etc.), every DNA site will have a chance to be located in linker or uncomplexed regions and will be exposed to MN cleavage. Therefore, in multiply positioned nucleosomes we do not expect major qualitative differences between the profiles obtained on purified DNA and the ones from *in vivo* chromatin or *in vitro* reconstituted samples. The differences will only be due to different quantitative occupancy of each position. This type of comparison is difficult to interpret and, when dealing with multiple alternative nucleosomes, could generate confusion, losing analytical value. Therefore, this control provides useful information only in the case of uniquely localized nucleosomes.

Use of Micrococcal Nuclease Instead of DNase I

Degradation with DNase I can be useful for the localization of proteins on DNA, essentially in the form of footprints. The smaller the protein, the clearer the information. DNase I can also provide information on helical periodicity of DNA–protein interactions. As for nucleosomes, given that the information is provided by the comparison between two cleavage patterns (naked DNA versus chromatin or reconsti-

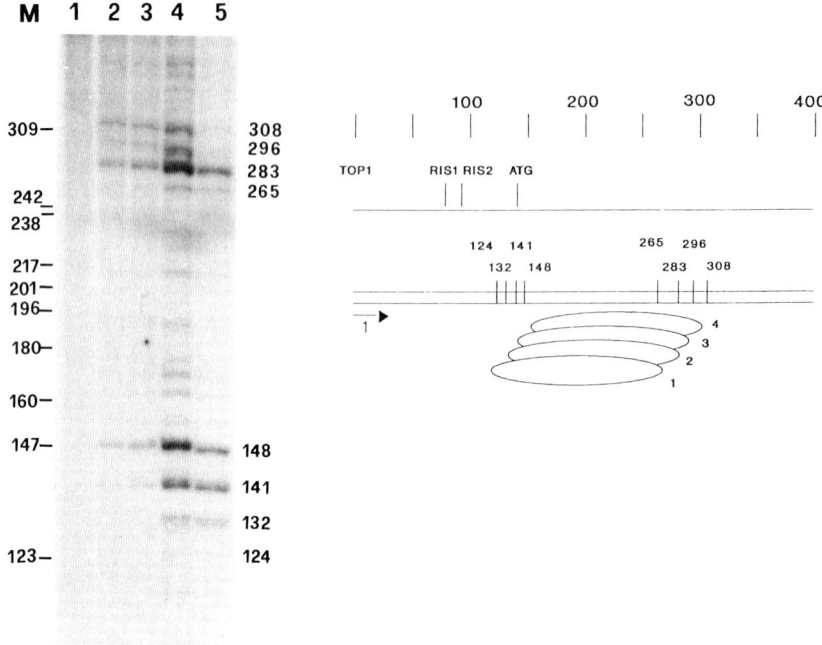

FIG. 3 Localization of the position(s) of nucleosomes in cellular chromatin by limited MN degradation and PE of the resulting products. PE of oligonucleosomal DNA was performed according to Protocol 4. The experimental lanes show the PE products on purified DNA from cells treated with 0, 2.5, 5, 10, and 20 U of MN, respectively. Lane M shows size markers. The map shows the position of the oligonucleotide used [5′-GTTTCAACACTAACACGA-GCGCAATA, from −142 to −117; numbering according to C. Thrash, A. T. Boukier, B. G. Barrel, and R. Sternglanz, *Proc. Natl. Acad. Sci. U.S.A.* **82**, 4374 (1985)] and the coupling between the two nucleosomal borders. The example shown here refers to the upstream and downstream borders of nucleosome +1 of the *S. cerevisiae TOP1* gene, located immediately downstream relative to the ATG.

tuted material), results are meaningful and interpretable for single nucleosomes or for uniquely located proteins but become almost uninterpretable in the case of partially overlapping messages, as is often the case (see above, Figs. 2b,c and 3). In these cases, difficulties arise from the use of DNase I. In particular, the periodic overlap of alternative nucleosomes or of alternative nucleosomal core particles (ncps) obscures the information relative to the nucleosomal borders obtained from DNase I digestions. One should also consider that in the case of multiple alternative positions on the same rotational setting, a modulation of the cutting pattern suggestive of a unique nucleosomal positioning will be observed, owing to overlap of phased cutting patterns on different nucleosomes. In this case, a careful analysis will show that the area encompassing the cutting patterns is shorter than 146 bp and definitely bell-shaped in its central part, proving the presence of nonuniquely positioned particles.

In summary, the mapping *in vivo* of nucleosomes by determination of the borders of mononucleosomes (Protocol 3) is a no-background assay and allows the unambiguous attribution of the occupied position(s). The limitation of the technique lies in its short analytical range (by definition equal or lower than 146–150 bp). This technique is profitably coupled with a protocol consisting of primer extension from upstream-located oligonucleotides performed on a partial MN digest of chromosomal DNA, avoiding the mononucleosome isolation step. This approach not only yields the localization of the closest nucleosomal borders, but footprints of nonnucleosomal proteins can also be observed and the borders of neighboring nucleosomes can be localized.

Protocol 4: Determination by Primer Extension of Nucleosomal Borders on DNA Purified from Oligonucleosomes

Yeast cell growth, limited digestion with MN, and DNA purification are performed as in Protocols 1 and 2. For high-resolution detection of MN cleavage sites, a single oligonucleotide is selected in proximity to the region under study. PE is performed according to Protocol 3. Note that the amount of template DNA in this analysis is preferably that used for low-resolution studies, namely, $1–10\ \mu g$/sample.

With the term oligonucleosomes we do not indicate here precisely defined oligomers. We refer to fragments of various and heterogeneous sizes produced by preferential MN cleavage in the linker regions.

Figure 3 shows an example of determination of two groups of borders, namely, the upstream and downstream borders belonging to a single family composed of the nucleosomes present on alternative helically phased positions.

Comparison of *in Vivo* and *in Vitro* Nucleosome Localization

The comparison of *in vivo* and *in vitro* nucleosome localization is relevant for the study of the mechanisms which govern the interactions of histone octamers with DNA, the topological and physiological behavior of the resulting particles, and the role of such effectors as boundary proteins, context effects, covalent modifications of DNA and of histones, and dynamic processes. The localization of nucleosomal core particles *in vitro* depends exclusively on the interaction between histones and DNA (i.e., on the properties of the proteins and on the conformational information, translational and/or rotational, used by the DNA sequence in this type of interaction). The comparison of the localization of nucleosomes *in vivo* and of nucleosomal core particles *in vitro* helps to evaluate the role of the DNA information versus the involvement of epigenetic factors.

On the 5 S rRNA gene, for instance, the multiplicity and the similarity of positions present *in vivo* can also be observed *in vitro* (Fig. 2b, lane 6 versus lane 7) using the border determination by PE of MN-produced mononucleosomes (Protocols 3 and 4). An easy and informative complementary technique for *in vitro* localization is the determination of the borders with ExoIII.

The necessity of an independent technique to determine the localization of nucleosomal particles *in vivo* is generated by the fact that each enzyme used has a certain degree of sequence specificity, thus generating ambiguities. Therefore, to achieve the highest precision, coupling mappings by an endonuclease and an exonuclease is recommended. In several instances, the distribution of nucleosomes *in vivo* has been observed to be similar to that of nucleosomal core particles *in vitro*, pointing out the predominance of the DNA information relative to epigenetic or external factors in determining their localization.

Protocol 5: Localization of *in Vitro* Reconstituted Nucleosomal Core Particles by Exonuclease III Degradation

1. Approximately 100–500 ng of 5′-labeled DNA fragment (specific activity 0.2–1.0 μCi/pmol) is reconstituted according to the salt exchange protocol, as described elsewhere (10, 11). Reconstitution is performed at a molar ratio between donor particles and acceptor DNA of approximately 2–10. At the end of the reconstitution process, DNA is in 100 μl of 20 mM Tris-HCl, pH 8.0, 0.01% Nonidet P-40, 100 mM NaCl, 2% glycerol. Prepare a mock sample by omitting nucleosome donor particles from the reconstitution mix (control 1). The control sample is treated in parallel with the experimental sample.
2. Add 2 μl of 0.1 M MgCl$_2$ (final concentration 2 mM).
3. Remove a 20-μl aliquot and add to it 2.2 μl of 10$\times$ stop mix (25 mM EDTA, 2% SDS).
4. Transfer the residual 82 μl to a 30°C water bath, let the samples equilibrate, and add 4 U of ExoIII.
5. Remove 20-μl aliquots at 5, 20, and 40 min; add to each 2.2 μl of 10$\times$ stop mix.
6. Add to every aliquot 1 μl proteinase K and incubate 30 min at 56°C.
7. Phenol extract, ethanol precipitate, wash pellets with 70% ethanol, and dry.
8. Resuspend samples in loading buffer (see Protocol 3) and load 10–50 $\times$ 10^3 cpm/sample on a 6% polyacrylamide denaturing gel with size markers.

Figure 4 shows an example of border determination of multiple alternative nucleosomal core particles on the *S. cerevisiae* 5 S rRNA gene, as obtained by Protocol 5. Only limit digests (40 min) are shown. (For a kinetic analysis, see Ref. 10).

The multiplicity of nucleosomal core particle positions on the same rotational

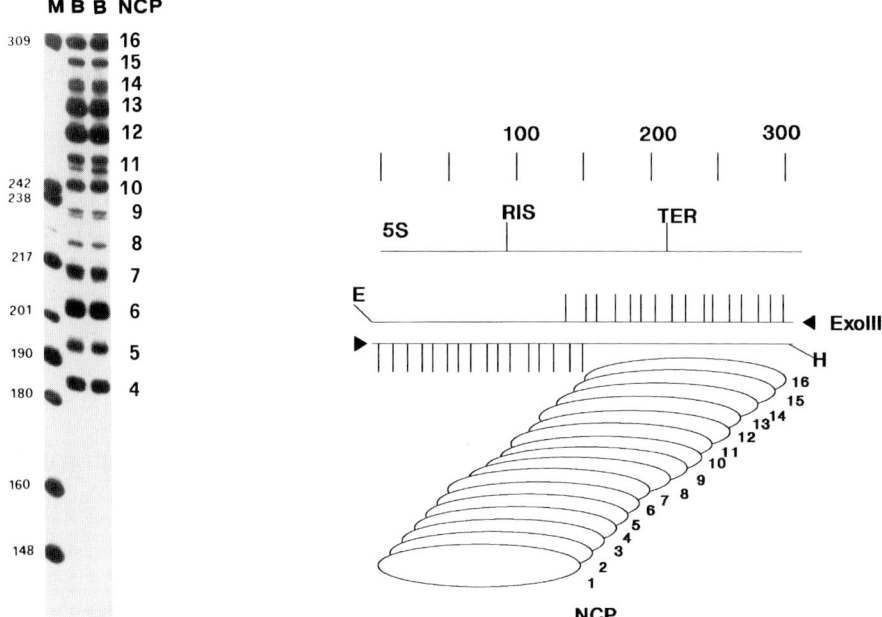

FIG. 4   ExoIII analysis on *in vitro* reconstituted nucleosomal core particles. The particles were reconstituted on a 305-bp *Eco*RI–*Hind*III fragment of the pBBIIIF plasmid [B. R. Braun, D. L. Riggs, G. A. Kassavetis, and E. P. Geiduschek, *Proc. Natl. Acad. Sci. U.S.A.* **86**, 2530 (1989)] containing a copy of the yeast 5 S rRNA repeat gene, labeled in the upper strand at the *Eco*RI site. Reconstitution with c.e. nucleosomal core particles was according to a standard salt dilution protocol (10, 11). ExoIII treatment (40 min) was performed according to Protocol 5. The position of the borders produced by 13 nucleosomal core particles is indicated (from NCP 4 to 16) on the right-hand side of the gel autoradiogram; 16 nucleosomal core particles form on this DNA (10). The borders of NCP 1 to 3 on the upper strand are not detected here owing to their innermost position relative to the entry site of ExoIII (indicated at right in the map). Lanes B show reconstitutions at two different histone to DNA ratios, with the similarity of the patterns obtained showing that saturation is reached.

phase was shown on the 5 S rDNA with several different techniques: DNase I digestion, hydroxyl radical footprinting, MN restriction digestion, and MN–primer extension mapping (10). The coherence of the data with the results obtained with ExoIII rules out any suspicion of ExoIII strand invasion, that is, penetration of the enzyme even along DNA complexed with proteins. The risk of strand invasion is real, but it relates only to conditions of overdigestion.

The pattern of borders reported in Fig. 4 does not change as a function of increasing time of digestion (not shown). The borders are stable, contrary to what would be observed in the case of strand invasion. Protocol 6 describes how to obtain the quantitative evaluation of strand invasion by ExoIII.

Protocol 6: Quantitative Evaluation of Strand Invasion with Exonuclease III

1. Around 500 ng of homogeneously labeled DNA fragment (specific activity 0.2–1.0 μCi/pmol) is reconstituted as in Protocol 5. Reconstitution is performed at a molar ratio of donor to acceptor DNA of approximately 10.
2. Add 3 μl of 50 mM CaCl$_2$ (final concentration of 1.5 mM) and transfer the sample to a 30°C water bath.
3. Add 1 μl of 0.2 U/μl MN (final concentration 2 U/ml) and incubate 12 min at 37°C.
4. Stop with 1.2 μl of 250 mM EGTA (3 mM final concentration). Add 5.5 μl of 50 mM MgCl$_2$ (2.5 mM final concentration).
5. Divide the sample into four aliquots (~27 μl each).
6. Add to the aliquots 0, 1.35, 4.05, and 12.15 U of ExoIII (final concentration 0, 50, 150, and 450 U/ml, respectively).
7. Incubate 20 min at 30°C.
8. Stop the reaction with 3.0 μl of stop mix (25 mM EDTA, 2% SDS) and add 1 μg proteinase K. Incubate 30 min at 56°C.
9. Phenol extract, ethanol precipitate, wash the pellet with 70% ethanol, and dry.
10. Resuspend in loading buffer and load $10–50 \times 10^3$ cpm on a 6% polyacrylamide denaturing gel with size markers.

Figure 5 shows an example of this quantitative evaluation. An internally labeled 326-bp DNA encompassing the *S. cerevisiae* 5 S rRNA gene was reconstituted and extensively digested with MN (Fig. 5, lane 1). The pattern obtained shows that most of the DNA strands range between 146 and 125 bp (as described previously; see Ref. 12). The MN-treated DNA was digested with increasing amounts of ExoIII (Fig. 5, lanes 2 to 4). Quantitative evaluation shows that ExoIII invasion under the conditions described in the legend to Fig. 5 is a minor effect, is dose-dependent, and is sensitive to temperature (higher temperature gives higher invasion) and ionic strength (lower strength gives higher invasion). Evaluation by scanning densitometry of the occurrence of the band which is the product of 1 helical turn strand invasion (Fig. 5, arrowhead) recommends reaction conditions of 30°C, 50 U/ml ExoIII, and 100 mM NaCl. At higher dosage (150 U/ml), ExoIII invades 1 helical turn in 8.3% of the molecules from each end.

References

1. S. Venditti and G. Camilloni, *Mol. Gen. Genet.* **242,** 100 (1994).
2. C. Wu, *Nature (London)* **286,** 854 (1980).
3. J. Sambrook, E. F. Fritsch, and T. Maniatis, "Molecular Cloning: A Laboratory Manual." Cold Spring Harbor Laboratory, Cold Spring Harbor, New York, 1989.

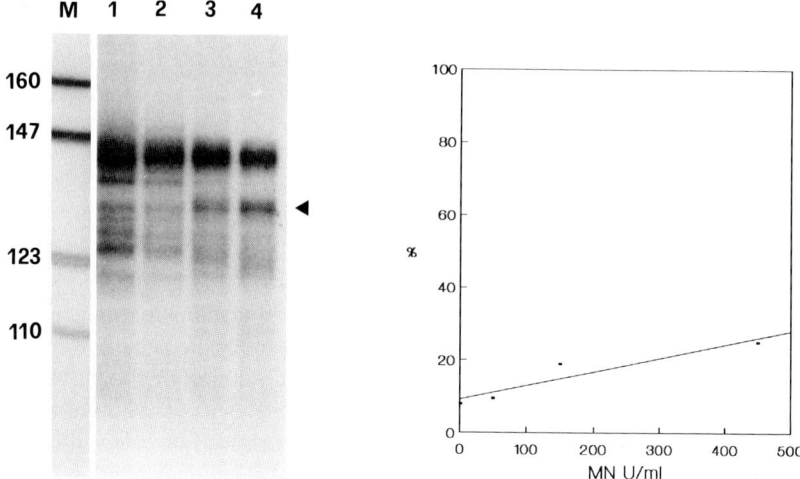

FIG. 5   Quantitative evaluation of strand invasion by ExoIII. *Left:* the 6% denaturing acrylamide gel shows size markers (lane M) and results of ExoIII treatments [20 min in 20 mM Tris-HCl, pH 8.0, 100 mM NaCl, 2.5 mM MgCl$_2$, 0.1% Nonidet P-40, 2% glycerol, 30°C, with 50 (lane 2), 150 (lane 3), and 450 U/ml ExoIII (lane 4)] of the sample shown in lane 1. This sample was produced by extensive MN digestion of nucleosomal core particles reconstituted on a 326-bp internally labeled DNA fragment. *Right:* The graph shows the evaluation (%) of the band which is the product of 1 per helical turn strand invasion (indicated by an arrowhead), relative to the full-sized monomer band.

4. T. E. Shrader and D. M. Crothers, *Proc. Natl. Acad. Sci. U.S.A.* **86,** 7418 (1989).
5. P. Venditti, G. Costanzo, R. Negri, and G. Camilloni, *Biochim. Biophys. Acta* **1219,** 677 (1994).
6. H. R. Drew and C. R. Calladine, *J. Mol. Biol.* **195,** 144 (1987).
7. F. Thoma, *Biochim. Biophys. Acta* **1130,** 1 (1992).
8. A. Almer and W. Hörz, *EMBO J.* **5,** 2681 (1986).
9. F. Dong, J. C. Hansen, and K. E. van Holde, *Proc. Natl. Acad. Sci. U.S.A.* **87,** 5724 (1990).
10. M. Buttinelli, E. Di Mauro, and R. Negri, *Proc. Natl. Acad. Sci. U.S.A.* **90,** 9315 (1993).
11. H. R. Drew and A. A. Travers, *J. Mol. Biol.* **186,** 773 (1985).
12. D. Riley and H. Weintraub, *Cell (Cambridge, Mass.)* **13,** 281 (1978).

[10] Novel Assays for Ligation and Cleavage Activities of Site-Specific Recombinases

Paul D. Sadowski, Guohua Pan, and Roland Brousseau

Introduction

Conservative, site-specific recombinases bring about rearrangements of DNA at specific chromosomal target sequences. They are called conservative because recombination occurs by concerted breakage and reunion of the DNA duplex without the removal or resynthesis of DNA. The mechanisms of action of several recombinases have been extensively characterized biochemically (reviewed in Refs. 1 and 2). All of the enzymes carry out a concerted breakage and reunion of DNA strands that involves covalent attachment of the recombinase to the DNA via a phospho–amino acid linkage. The covalent intermediate is thought to conserve the energy of the phosphodiester bond and thereby obviate the need for an external energy source to promote sealing of the phosphodiester backbone.

The recombinases are grouped into two broad families: (a) the invertase/resolvase group and (b) the integrase group. Members of the former group use a serine residue to promote covalent attachment of the recombinase to the 5′-phosphoryl group at the site of the nick. Members of the integrase family use a conserved tyrosine residue that covalently attaches to the 3′-phosphoryl group (3).

The conservative site-specific recombinases perform a variety of DNA rearrangements depending on the location and disposition of the target sequences (e.g., inversion of the DNA between inverted targets, excision of the DNA between directly oriented targets, intermolecular recombination between two DNA molecules each of which bears a target sequence). Furthermore, the topological requirements and the need for accessory proteins vary with the recombinase. Two members of the integrase family, the Cre protein of bacteriophage P1 and the FLP protein of the 2-μm circle plasmid of yeast, are being extensively used to modify the genomes of eukaryotic cells and organisms (4–6). They are the recombinases of choice because they have simple target sequences, show relaxed topological requirements, and have no need for accessory protein cofactors.

In this chapter, we describe the development of novel assays that measure the ligation activity of the FLP protein and other enzymes. The assays make use of activated substrates that bear a 3′-phosphoryltyrosine; the tyrosine acts as a leaving group and permits the assay of ligation activity independently of the ability of the protein to cleave the DNA. While studying the requirements for ligation, we also developed an assay that measures the ability of the FLP protein to cleave a phosphodiester bond and to covalently attach to the 3′-PO_4 end.

Methods in Molecular Genetics, Volume 6

Mechanism of Cleavage and Ligation by Integrase Family Members

All members of the integrase family of recombinases share four absolutely conserved residues that are involved in breakage and reunion of the DNA strands (3, 7). A conserved tyrosine (amino acid 343 of FLP) is the nucleophile that promotes strand breakage and covalently attaches to the 3′-phosphoryl group at the site of nicking (8, 9). The other terminus at this nick is a 5′-hydroxyl group. Two arginines (amino acids 191 and 308 of FLP) and a histidine (amino acid 305) are also conserved in all integrase family members and are involved in the ligation step of the reaction (10–12). During this step, the 5′-OH group of an incoming DNA strand acts as the nucleophile that breaks the 3′-phosphotyrosine bond and establishes the 5′–3′ phosphodiester bond of the DNA backbone with the liberation of free FLP recombinase. The tyrosine acts as a leaving group in this nucleophilic attack.

Principle of Ligation Assay

Because of the close coupling between the cleavage and ligation steps of the FLP recombinase, it was difficult to isolate a large amount of the FLP–DNA covalent intermediate and to separate clearly the cleavage step from the ligation step of the reaction. We have taken advantage of a mutation in one of the conserved catalytic residues of FLP to generate a large amount of the cleaved, covalent FLP–DNA intermediate (13). Digestion of the complex with protease yielded a substrate that bore a 3′-phosphotyrosine moiety and that was an active substrate for FLP-mediated ligation (10). We then synthesized oligonucleotides that had a 3′-phosphoryltyrosine residue and showed that such substrates were also active for the ligation by the FLP recombinase (11). Analogous substrates for the phage λ integrase and phage P1 Cre proteins as well as for mammalian topoisomerase I were also active as ligation substrates.

Principle of Cleavage Assay

While studying the ligation assay, we unexpectedly discovered a specific assay that measures strand cleavage and covalent attachment of FLP to DNA (11). The substrate consists of a FLP target sequence with a nick at one of the natural cleavage sites. The nick contains a 5′-PO$_4$ group and an extra protruding 3′-deoxynucleoside. The former group blocks ligation, and the latter is removed with covalent attachment of the FLP protein to the 3′-phosphoryl group.

Materials and Methods

Enzymes

Wild-type and mutant FLP proteins are prepared from *Escherichia coli* cells bearing a plasmid containing the appropriate FLP gene under the control of the T7 promoter (14). The mutant proteins are named as previously described, for example, FLP R191K denotes a FLP protein in which amino acid 191 has been changed from an arginine to a lysine residue (13). The proteins are purified via BioRex 70 chromatography, and some are subjected to an additional Sephacryl S-300 step. Purity ranges from 15 to 50% for the mutant proteins. The wild-type FLP is 85% pure. FLP proteins are stored in small aliquots (10–20 μl) at $-70°$C and are subjected to freeze–thawing no more than twice.

Pronase is obtained from Boehringer Mannheim (Indianapolis, IN) and is dissolved in 10 mM Tris-HCl, pH 7.4, 10 mM NaCl at 200 mg/ml. It is heated to 37°C for 1 hr and is stored at $-20°$C. Proteinase K is purchased from Boehringer Mannheim and is stored at $-20°$C at a concentration of 10 mg/ml in water.

DNA Substrates

A 142-bp DNA fragment containing the FLP recombination target (FRT) is excised from plasmid pGP25 (15) with *Eco*RI and *Hind*III (Fig. 1). Oligonucleotides of defined sequence are synthesized on an Applied Biosystems (Foster City, CA) Model 380B DNA synthesizer using standard phosphoramidite chemistry. A 3'-phosphoryl group is attached to an oligonucleotide using 2-[2-(4,4'-dimethoxytrityl-oxy)ethylsulfonyl]ethyl-2-(2-cyanoethyl)-(*N,N*-diisopropyl)phosphoramidite, purchased from Glen Research (Sterling, VA). The synthesis is done by the Carbohydrate Research Centre, Faculty of Medicine, University of Toronto.

Preparation of Oligonucleotides Bearing 3'-Phosphotyrosine

See Fig. 2 for additional information.

1. Removal of the dimethoxytrityl protecting group from the Teflon-based solid support: Approximately 1 μmol (30 mg) of Teflon-based Oligo-Affinity Support (OAS; Glen Research; product number 20-40000-10) is treated with an excess (2 ml) of 3% trichloroacetic acid in dichloromethane for 10 min at room temperature to remove the dimethoxytrityl protecting group from the 5'-hydroxyl of the

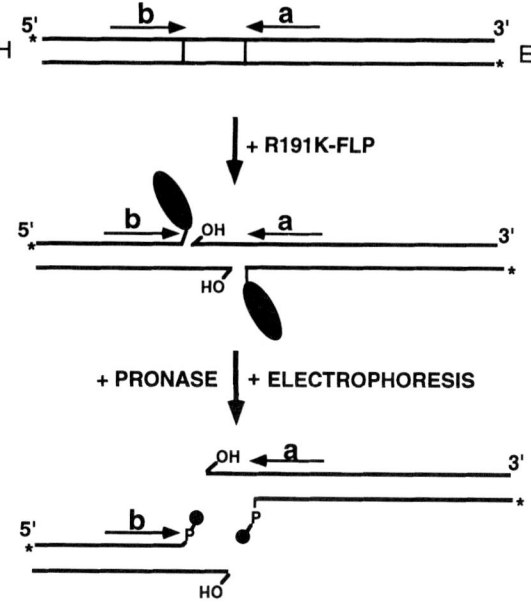

FIG. 1   Schematic representation of preparation of half-site ligation substrates. A 142-bp DNA fragment was excised from plasmid pGP25 (14) with HindIII (H) and EcoRI (E), and the 5' ends were labeled with [32]P (asterisks, top). The fragment contains the FRT (FLP recognition target) that is represented by two inverted horizontal arrows (a and b) surrounding an open box. The arrows represent 13-bp, inverted symmetry elements to which a FLP molecule binds site-specifically. The open box represents the 8-bp core that separates symmetry elements a and b. The fragment is incubated with the variant FLP protein R191K which nicks the top and bottom strands at the margins of the core region (middle) and becomes covalently attached to the 3'-PO$_4$ end (solid ovals); the apposing 5' end bears an OH group. After Pronase treatment, the two half-sites (bottom) contain a 3'-PO$_4$-tyrosine residue (solid dot) and can be separated by preparative electrophoresis.

adenosine residue bound on the support. The support is then washed several times with anhydrous acetonitrile and allowed to air dry at room temperature.

2. Coupling of the tBu-Tyr-Fmoc to the deprotected 5'-hydroxyl of the Teflon-based solid support: The deprotected support is reacted with a 100-fold excess of tBu-Tyr-Fmoc (Institut Armand Frappier, Laval, Quebec; 45.9 mg) in the presence of 1-hydroxybenzotriazole (HOBT; 16 mg) and 1,3-dicyclohexylcarbodiimide (DCC; 25 mg) in 1 ml of dry pyridine. The slurry is stirred overnight at room temperature with the help of a micromagnetic stirring bar. After the reaction the solid support is washed with pyridine, followed by dimethylformamide (DMF).

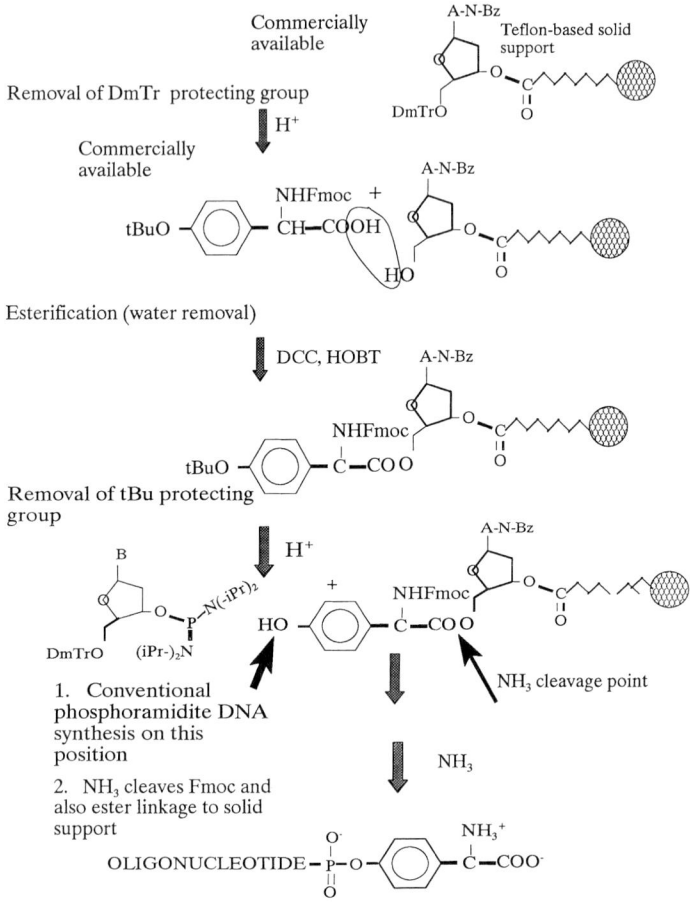

FIG. 2   Schematic representation of synthesis of $3'$-PO_4-tyrosine-terminated oligonucleotides (see text).

3. Removal of the *tert*-butyl protecting group: The tBu protecting group is removed through treatment with 1 ml of 50% trifluoroacetic acid (TFA) in DMF for 1 hr at room temperature.[1] The solid support is washed with DMF, aqueous pyridine, DMF and acetonitrile, and dried *in vacuo* at room temperature.

4. Phosphoramidite synthesis of the oligonucleotide on the immobilized tyrosine

[1] Incomplete removal of the tBu protecting group may cause failure of the first coupling step (step 4). This can be avoided by prolonging TFA/DMF treatment to 12 hrs or using TFA in CH_2Cl_2 for 1 hr (B.-P. Zhao, to be published).

residue: The tyrosine-derivatized solid support is placed within a standard Applied Biosystems synthesis column and used to synthesize the oligonucleotides on an Applied Biosystems 380A DNA synthesizer using the standard phosphoramidite 1 μmol cycle.

5. Removal of the Fmoc and nucleoside protecting groups and cleavage from the solid support: At the end of the synthesis, cleavage with concentrated ammonia at 55° C for 5 hr releases the 3'-phosphotyrosine oligonucleotide from the adenosine residue bound to the solid support and also removes the Fmoc protecting group and the nucleoside protecting groups.

6. Purification of the 3'-phosphotyrosine oligonucleotide: The 3'-phosphotyrosine oligonucleotide is purified by reversed-phase high-performance liquid chromatography (HPLC) [RP-18 column; solvent A, 0.1 M triethylamine acetate (TEAA) in water; solvent B, 75% acetonitrile in 0.1 M TEAA; linear gradient from 0% solvent B to 100% solvent B in 100 min]. The elution time is 60 min. The presence of tyrosine in the conjugate is confirmed by amino acid analysis after hydrolysis in 6 N HCl at 150° C for 1 hr.

Preparation of Activated Substrate by FLP R191K and Pronase Digestion

The procedure is diagrammed in Fig. 1. A 5'-labeled DNA fragment containing an FRT site is incubated with FLP R191K followed by Pronase treatment. The mutant FLP protein covalently attaches to the 3'-PO$_4$ group at the sites of cleavage via tyrosine 343. Pronase removes most of the protein but leaves the phosphotyrosine linkage intact. The substrate undergoes double-strand breakage, and the two half-sites can be separated by acrylamide gel electrophoresis and used as ligation substrates. We refer to these substrates as half-site substrates.

1. About 1 pmol of 5'-^{32}P-labeled DNA fragment from plasmid pGP25 is incubated in an 80-μl reaction that contains 50 mM Tris-Cl buffer (pH 7.4), 33 mM NaCl, 1 mM EDTA, and 8 μg sonicated, denatured calf thymus DNA with approximately 50 pmol FLP R191K protein.

2. After 90 min at 25° C, 20 μg Pronase is added, and incubation is continued at 37° C for 120 min.

3. The DNA is extracted with phenol–chloroform (1 : 1, v/v) and precipitated with ethanol.

4. The DNA is subjected to at least two more cycles of FLP R191K and Pronase treatment. Depending on the FLP preparation, a total of three or four cycles of treatment may be required.

5. The cleaved DNA fragments (half-sites a and b) are then run on an 8% polyacrylamide gel to separate the half-sites from uncleaved or singly nicked DNA. Gen-

erally, about 80% of the full-site containing fragment is converted to half-sites. The half-sites which contain a 3'-phosphoryltyrosine group are located by autoradiography and isolated by electroelution. Incidentally, the remaining full-length molecules contain a high proportion of singly nicked substrates. The presence of tyrosine on the 3'-termini is confirmed by acid hydrolysis and high voltage electrophoresis, but we have been unable to determine whether additional amino acids are still attached to the tyrosine (16).

Assembly of Synthetic Activated Substrates

The synthetic ligation substrates are generally assembled from three synthetic oligonucleotides (Fig. 3). A 40-nucleotide-long top strand is hybridized to two complementary oligonucleotides that constitute a nicked bottom strand. The nick is positioned precisely at the normal site of nicking by FLP, and the nick is bounded by the 3'-PO_4-tyrosine and a 5'-OH group. The 3'-PO_4-tyrosine-terminated oligonucleotide is 5'-labeled with ^{32}P. Other structures for the substrates are possible: for example, a synthetic half-site analogous to those created by FLP R191K and Pronase digestion can be assembled.

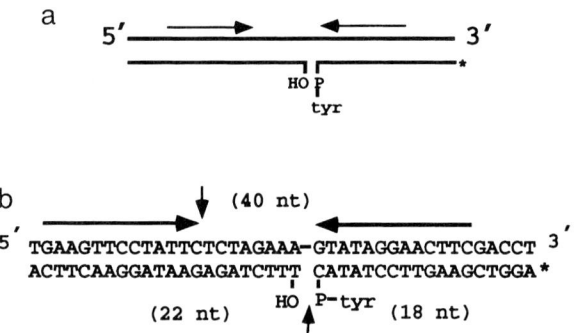

FIG. 3   Structure of synthetic ligation substrate. (a) Schematic representation. The top strand is annealed to two oligonucleotides, one of which bears 3'- and 5'-OH ends (bottom left) and the other bears a 3'-PO_4-tyrosine residue and a 5'-$^{32}PO_4$ group. The horizontal arrows represent inverted symmetry elements of the FRT site. (b) Actual sequence of oligonucleotides. The 13-bp symmetry elements are demarcated by the horizontal arrows surrounding the 8-bp, AT-rich core. The short vertical arrows indicate the sites of phosphodiester bond cleavage normally carried out by FLP. The bottom strands surround a nick bearing 5'-OH and 3'-PO_4-tyrosine ends. Ligation of the two bottom oligonucleotides is readily detected on a sequencing gel.

1. The 3'-PO$_4$-tyrosine oligonucleotide is 5'-^{32}P-labeled with polynucleotide kinase.
2. The labeled oligonucleotide is then annealed with equivalent quantities (50 pmol of each molecule) of the two unlabeled oligonucleotides. The annealing mixture (15 μl) contains 0.1 M NaCl and 5 mM MgCl$_2$.
3. After heating to 70°C for 10 min, the mixture is cooled to 65°C over 25 min, kept at 65°C for 4 min, and then cooled to 20°C (at a rate of 1°C per 5 min) in a thermal cycling apparatus.
4. The substrates are purified by preparative polyacrylamide gel electrophoresis, electroelution, and ethanol precipitation.

Assembly of Nicked Substrates to Measure Ligation and Covalent Attachment

The substrates used to detect cleavage and covalent attachment of the FLP protein are shown schematically in Fig. 4. All consist of synthetic oligonucleotides designed to create a substrate with a nick on the bottom strand of the FRT site. The nick contains a 3'-hydroxyl and a 5'-hydroxyl group (Fig. 4a), a 3'-PO$_4$ group and a 5'-PO$_4$ group (Fig. 4b), a protruding 3'-deoxyadenosine residue and a 5'-OH group (Fig. 4c), or a 3'-deoxyadenosine and a 5'-PO$_4$ group (Fig. 4d). The sequences of the oligonucleotides are otherwise the same as for the synthetic ligation substrates, and the oligonucleotides are annealed as described for the ligation substrates.

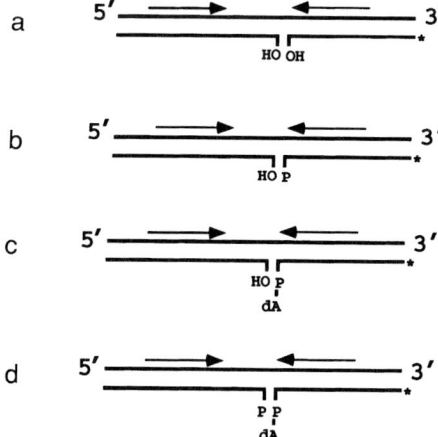

FIG. 4   Structure of nicked synthetic substrates. The substrates have the same structure as the synthetic ligation substrate shown in Fig. 3 but differ in the nature of the terminus of the site of the nick, namely, (a) 3'-OH, 5'-OH; (b) 3'-PO$_4$, 5'-OH; (c) 3'-PO$_4$-dA, 5'-OH; (d) 3'-PO$_4$-dA, 5'-PO$_4$. The asterisk represents the 5'-^{32}P label.

Intramolecular ligation   **Intermolecular ligation**

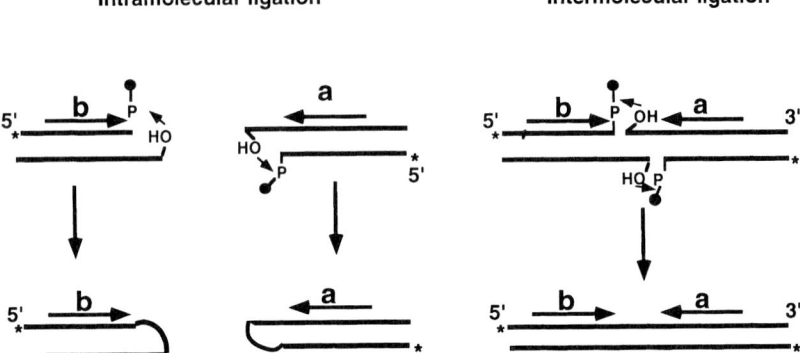

F𝒾ɢ. 5   Schematic representation of ligation of half-site substrates. The a- and b-half-sites were prepared as in Fig. 2. For intramolecular ligation (left and middle), FLP catalyzes a nucleophilic attack by the 5'-OH group (small arrow) on the 3'-PO₄-tyrosine (solid dot) linkage to yield a hairpin structure caused by ligation of the top and bottom strands. The half-site hairpins (bottom, left and middle) run ahead of the substrate on a nondenaturing acrylamide gel. Intermolecular ligation (right) between an a- and a b-half-site results from two nucleophilic attacks by the 5'-OH groups on the 3'-PO₄-tyrosine linkages. A full-length DNA fragment (bottom) is reconstituted.

Ligation Assay: Half-Site Substrates

The half-site substrates arise after digestion with FLP R191K and Pronase (Fig. 1). As can be seen in Fig. 5, ligation can yield three potential products: each half-site can undergo an intramolecular ligation reaction to form either an a-half-site hairpin or a b-half-site hairpin (bottom left, middle). An intermolecular ligation reaction between the two activated half-sites will reconstitute the starting substrate (bottom, right). The three products can be resolved from the substrates by either nondenaturing or denaturing polyacrylamide gel electrophoresis.

1. The a-half-site, the b-half site, or a mixture of the two (0.01 pmol each) is incubated with FLP protein (0.1 μg) in a 30-μl reaction containing 50 mM TAPS buffer, pH 8.0, 10% polyethylene glycol 4000, 200 mM NaCl, 1.5% glycerol, 1 mM EDTA, and 1 μg sonicated denatured calf thymus DNA.
2. After 30 min at 25°C, 1 μl of 0.1% sodium dodecyl sulfate (SDS) and 5 μg proteinase K are added, and the incubation is continued for 30 min at 37°C.
3. Reactions are analyzed on both native 8% polyacrylamide gels and 8% thin DNA sequencing gels. The half-site hairpin products run slightly faster than the half-site substrates on the native gel but much more slowly than the substrate on the sequencing gel.

Ligation Assay: Synthetic Activated Substrates

The ligation assays with synthetic activated substrates have the same composition as described above except they contain 0.05 pmol (of molecules) of substrate. Incubation is for 45 min at 25° C. Reactions are terminated with SDS and proteinase K and analyzed on 8% DNA sequencing gels.

Ligation Assay: Nicked Substrates

Ligation assays with nicked substrates are identical to those using the synthetic activated substrates except the various nicked substrates (Fig. 4) are used.

Assay for Covalent Attachment of FLP Protein to Substrate

The assay conditions for covalent attachment of FLP to substrates are identical to the ligation assays using nicked substrates (see above). After incubation, the reactions are adjusted to 10% (w/v) glycerol, 3% SDS, 60 mM Tris-HCl (pH 6.8), and 5% (w/v) 2-mercaptoethanol (final concentrations). The samples are then boiled for 2 min and run on a 7.5% SDS–polyacrylamide gel according to the method of Laemmli (17).

Results

Measurement of Ligation Using Half-Site Substrates

Incubation of activated half-site substrates with wild-type FLP protein showed efficient production of the three expected products, namely, the two half-site hairpins (HaL and HbL, Fig. 6, lanes 1–5 and 6–10) as well as the reconstituted full-site (Ha + HbL, Fig. 6, lanes 11–15). The yields of product(s) were high, frequently exceeding 50%. The presence of the two half-sites in the reaction (which contain complementary 8-nucleotide, 5′ single strands) seemed to favor the intermolecular ligation over the formation of (intramolecular) hairpins (Fig. 6, lanes 12 and 13). Similar efficiencies of ligation were obtained when the wild-type protein was replaced by FLP Y343F (Fig. 6, lanes 3, 8, and 13), a mutant protein that is incapable of DNA strand cleavage because it lacks the nucleophilic tyrosine hydroxyl group (8). This observation provides clear evidence that the ligation reaction does not require the cleavage function of the enzyme. This assay also permitted us to show that other mutant proteins (notably R191K and H305L) are defective in ligation activity (Fig. 6, lanes 4 and 5, 9 and 10, and 14 and 15). Extensive *in vitro* complementation

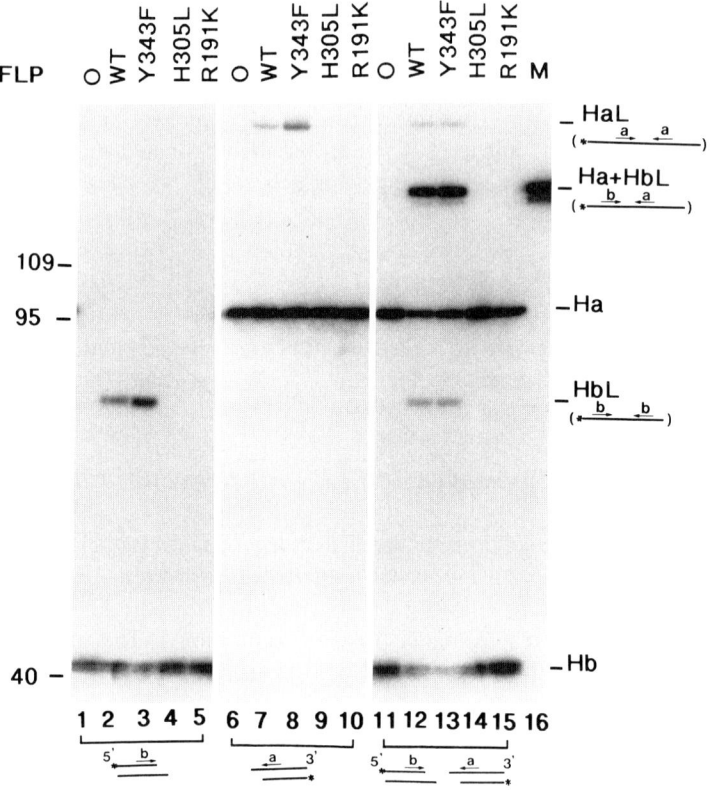

FIG. 6   FLP-mediated strand ligation of half-site substrates. Reactions were carried out as described in the text and analyzed on an 8% denaturing acrylamide gel. The FLP proteins used are indicated at the top of each lane: 0, no protein; WT, wild-type FLP; Y343F, H305L, R191K, mutant FLP proteins as defined in the text. Lanes 1–5, half-site b; lanes 6–10, half-site a; lanes 11–15, mixture of both half-sites a and b (diagrammed at bottom). Lane 16 (M) shows intact marker substrate. Numbers at left indicate the length (nucleotides) of marker fragments. Ha, half-site a; Hb, half-site b; HaL, intramolecular ligation product of half-site a; HbL, intramolecular ligation product of half-site b; Ha + HbL, intermolecular ligation product between the two half-sites. The products are diagrammed at right. Asterisks represent the 5′-^{32}P label. The horizontal arrows with letters a or b represent the 13-bp symmetry elements of the FRT site. [From G. Pan and P. D. Sadowski, *J. Biol. Chem.* **267,** 12397 (1992); with permission.]

analyses as well as direct biochemical studies have shown that the cleavage and ligation functions of FLP are spatially separated in the protein (12, 18).

These substrates showed the absolute requirement for the nucleophilic 5′-hydroxyl group, as phosphorylation of it with polynucleotide kinase completely abolished ligation (11). The activated half-sites were also used to show that a single-stranded oligonucleotide that contained only eight 5-terminal nucleotides complementary to the 8-nucleotide protrusion of the substrate could be efficiently ligated (10). We have subsequently found (G. Pan, unpublished) that only three complementary nucleotides suffice for efficient ligation of a single-stranded oligonucleotide. Thus, it appears that only a single FLP molecule bound adjacent to the site of ligation is necessary to catalyze the reaction (10).

Use of Synthetic Ligation Substrates

Synthetic ligation substrates use synthetic oligonucleotides, one of which bears a 3′-phosphoryl tyrosine. This oligonucleotide is also 5′-^{32}P-labeled and is then hybridized to a complementary oligonucleotide (unlabeled) so as to reconstruct a binding site for the FLP recombinase (Fig. 3). Ligation is measured on a DNA sequencing gel. These substrates were constructed to show that the only requirement for the leaving group on the 3′-PO$_4$ terminus was, in fact, tyrosine. They are also much easier to prepare than the half-site substrates because no enzymatic digestion is required. The results showed that ligation of the synthetic substrates was just as efficient as with the half-site substrates (see Fig. 7, compare lanes 2–4 versus lanes 9, 11–13, and 15–17). Furthermore, the synthetic substrates showed the same enzymatic properties as the half-site substrates, namely, FLP Y343F gives an efficient ligation and FLP R191K is ligation-defective (Fig. 7, lanes 3 and 5).

The synthetic substrates were also useful to show that the assay had applicability beyond the FLP protein (11). FLP is a member of the integrase family of recombinases, all of which share the conserved R-H-R-Y tetrad of amino acids (3, 7). This tetrad is thought to underlie a common catalytic mechanism of breakage and reunion of DNA strands. It was therefore of interest to learn whether other integrases could utilize activated tyrosine-containing substrates for ligation. Accordingly, substrates analogous to the FLP ligation substrate (Fig. 3) were assembled to contain binding sites for either the phage λ integrase protein or the phage P1 Cre protein surrounding a nick that bore 5′-OH and 3′-PO$_4$-tyrosine termini. These substrates served as active ligation substrates for their respective proteins, although neither protein worked as efficiently as FLP did on its substrate. Furthermore, the λ integrase mutant analogous to FLP Y343F also catalyzed ligation.

Integrase proteins resemble topoisomerases in that members of both groups attach covalently to the DNA via a phosphotyrosyl linkage (1, 19). Therefore, to broaden the applicability of the ligation assay further, we also designed a synthetic substrate to assay ligation promoted by mammalian topoisomerase I and found that the substrate was successfully ligated by the enzyme (11).

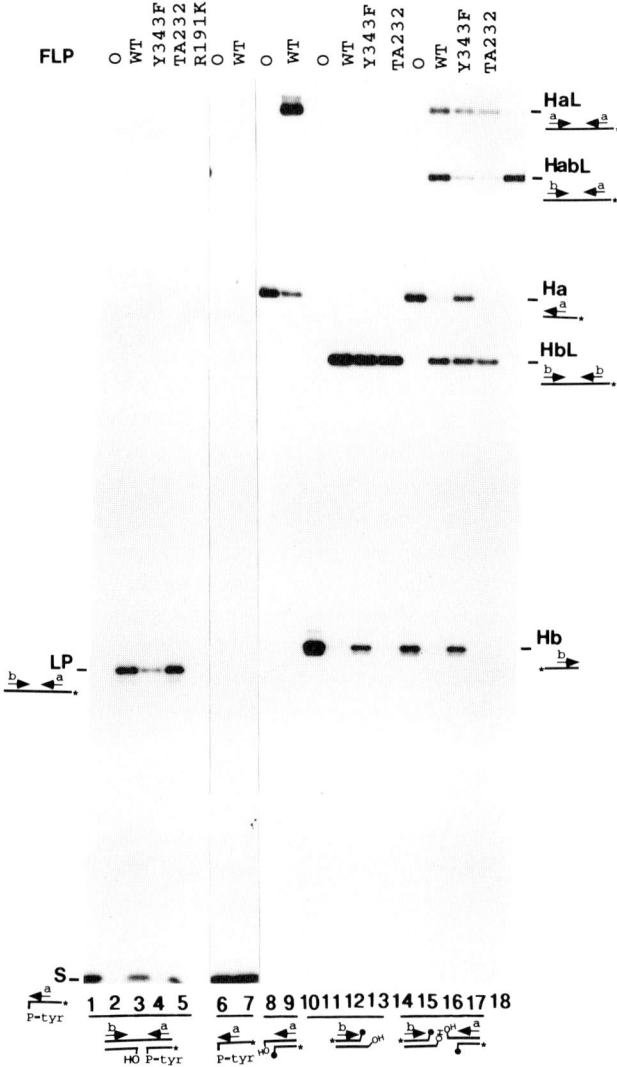

FIG. 7   Ligation of synthetic activated substrates by FLP proteins. Lanes 1–5 contained the synthetic activated substrate. Lanes 6 and 7 contained only the 5'-labeled, 3'-phosphotyrosine-terminated oligonucleotide. Lanes 8–17 contained the activated ligation substrates prepared after FLP cleavage as described previously by Pan and Sadowski (10). Lane 18 contained a full-sized DNA fragment containing two symmetry elements. The synthetic activated substrate (lanes 1–5), the half-site a (lanes 8 and 9), the half-site b (lanes 10–13), or a mixture of the two half-sites (lanes 14–17) were incubated with partially purified FLP proteins (0.5 pmol for WT, 0.4 pmol for TA232, 0.2 pmol for Y343F, 0.9 pmol for R191K). The reactions were then analyzed on an 8% denaturing polyacrylamide gel. The FLP proteins in the reactions are in-

Use of Synthetic Nicked Substrates to Assay Ligation and Covalent Attachment

We examined the influence of the termini at site of the nick on the ligation activity (Fig. 8A). Ligation did not occur if both the 3' and 5' termini bore hydroxyl groups (Fig. 8A, lanes 1–4). This shows the necessity for the 3'-PO$_4$-tyrosine to act as a leaving group for the ligation reaction. A minute amount of ligation product was detectable when the nick contained a 3'-phosphoryl group (lanes 5–8). This product was seen only with the wild-type protein, however (Fig. 8A, lane 6), and not with the cleavage-incompetent Y343F protein (Fig. 8A, lane 7). A much larger amount of ligation occurred if the nick contained a protruding 3'-PO$_4$-dA residue (Fig. 8A, lanes 9–12). Again, however, only the wild-type protein was active (Fig. 8A, lane 10).

This result suggested that ligation of the nicked substrates required covalent attachment of the FLP protein. This supposition was reinforced by the finding that the covalent FLP–substrate intermediate was detectable on an SDS–polyacrylamide gel (Fig. 8B, lane 2) using a substrate with a protruding 3'-PO$_4$-dA at the nick. Furthermore, the amount of FLP–DNA covalent intermediate was substantially increased if the 5' end at the nick bore a phosphate group (Fig. 8B, lane 7). We conclude that accumulation of the covalent intermediate occurs because the 5'-PO$_4$ group blocks ligation which would in turn release the covalently attached FLP protein. We propose that the ligation of the nicked substrate occurs by the mechanisms portrayed in Fig. 9. FLP cleaves the 3'-terminal phosphodiester bond and attaches covalently to the 3'-PO$_4$ terminus and liberates free dA (Fig. 9a). The 5'-OH group mounts a nucleophilic attack on the 3'-phosphotyrosine linkage with ligation of the nick and liberation of FLP (Fig. 9b).

The nicked substrate with a 5'-phosphoryl group can undergo cleavage and covalent attachment but not ligation. Hence the covalent protein–DNA intermediate accumulates. Thus, this substrate measures the cleavage potential of the wild-type FLP

---

dicated at the top of each lane. The substrates, the ligation products, and their structures are indicated on the right- or left-hand sides. The arrows labeled a or b represent the symmetry elements. The asterisks indicate the radioactively labeled 5' ends of the DNA fragments. The black dots at the 3' terminus of the activated substrate represent the phosphotyrosine and possibly other amino acid(s) retained after Pronase digestion. Ha, Half-site a; Hb, half-site b; HaL, intramolecular ligation product of half-site a; HbL, intramolecular ligation product of half-site b; HabL, intermolecular ligation product between two half-sites. S, Synthetic oligonucleotide bearing a 3'-phosphotyrosine; LP, ligation product of the synthetic activated substrate. 0, No FLP protein in the reaction; WT, wild-type FLP protein; Y343F, R191K, mutant proteins defined as in text; TA232, 4-amino acid insertion at position 115. The DNA substrates in the reactions are shown at the bottom. [From G. Pan, K. Luetke, C. D. Juby, R. Brousseau, and P. Sadowski, *J. Biol. Chem.* **268,** 3683 (1993); with permission.]

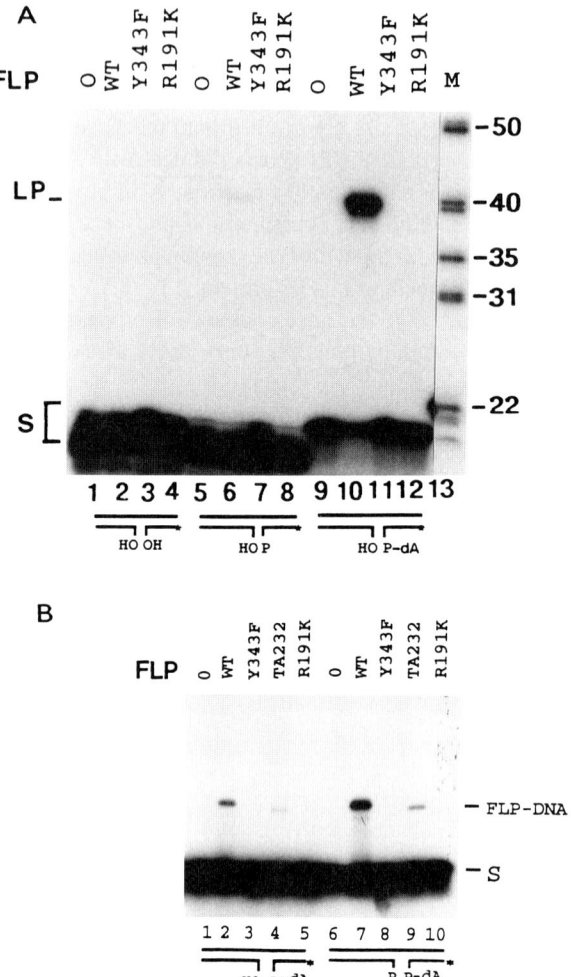

FIG. 8   Ligation and covalent attachment of the FLP proteins using different nicked substrates. (A) The various nicked substrates were incubated with FLP proteins (indicated at top). Ligation was detected by denaturing polyacrylamide gel electrophoresis. Lanes 1–4, Substrate bears 5'-OH and 3'-OH at the nick; lanes 5–8, substrate bears 5'-OH and 3'-PO$_4$ at the nick; lanes 9–12, substrate with 5'-OH and 3'-PO$_4$-dA at the nick. S, Substrate; LP, ligation product. The numbers down the right-hand side represent the position of migration of size markers (in nucleotides, lane 13). (B) Covalent attachment of FLP protein revealed by 7.5% SDS–polyacrylamide gel electrophoresis. Lanes 1–5, Nicked substrate with 5'-OH and 3'-PO$_4$-dA termini at the nick; lanes 6–10, nicked substrate with 5'-PO$_4$ and 3'-PO$_4$-dA termini. Proteins added are indicated at the top of each lane. S, Substrate; FLP–DNA, FLP–DNA covalent intermediate. The substrates are also diagrammed. The asterisks represent the 5'-^{32}P label. [From G. Pan, K. Luetke, C. D. Juby, R. Brousseau, and P. Sadowski, *J. Biol. Chem.* **268**, 3683 (1993); with permission.]

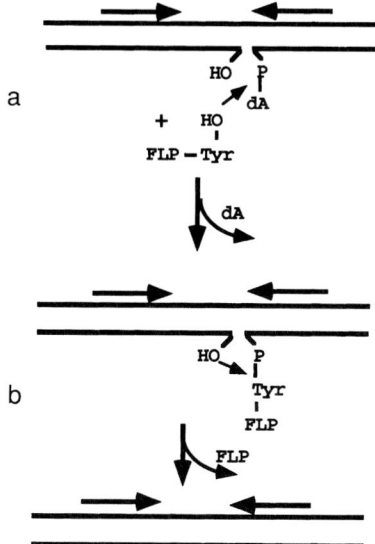

FIG. 9 FLP-mediated cleavage and ligation of nicked substrate bearing a 5'-OH and 3'-PO$_4$-dA. FLP removes the dA from the 3' terminus at the nick and attaches to the 3'-PO$_4$ via a phosphotyrosine linkage (a). In the ligation, FLP catalyzes a nucleophilic attack (small arrow) by the 5'-OH on the phosphotyrosine to reform the phosphodiester bond (b). The horizontal arrows represent the symmetry elements of the FRT site. If the 5' end of the nick bears a PO$_4$ group, no nucleophilic attack can occur and the FLP–DNA intermediate accumulates. A mutant FLP protein that lacks the nucleophilic tyrosine (FLP Y343F) cannot carry out this reaction. [From G. Pan, K. Luetke, C. D. Juby, R. Brousseau, and P. Sadowski, *J. Biol. Chem.* **268,** 3683 (1993); with permission.]

protein in the absence of ligation. The general applicability of the assay to other site-specific recombinases and topoisomerases is being studied.

Discussion

In this chapter, we have described a new class of substrates that measures the ligation activity of site-specific recombinases and topoisomerase I. In most instances, the ligation activity was independent of cleavage activity of the protein, and the assays allowed the mapping of the ligation and cleavage functions in the FLP protein.

The assay also clarifies the mechanism of ligation by the enzymes and confirms that tyrosine acts as the leaving group for a nucleophilic attack by the adjacent 5'-OH group. Somewhat surprisingly, the active site for ligation by FLP is apparently able to accommodate the 3'-phosphoryltyrosine on the substrate in addition to the tyrosine (or phenylalanine) present at position 343. This assay also confirms the postulate that the energy of the phosphodiester bond of the DNA is conserved in the phosphotyrosine linkage. Unlike other polynucleotide ligases which require DPN or ATP as an energy source, site-specific recombinases require none. It has long been known that bacterial and phage ligases promote the formation of a 5′-phosphoryl-AMP intermediate (20). The AMP moiety presumably fulfils the same function as the 3′-PO$_4$ tyrosine does for FLP, namely, to act as the leaving group during phosphodiester bond closure.

It will be of interest to broaden these assays to include members of the invertase/resolvase family of recombinases and other topoisomerases. The former group uses a serine residue to attach to the 5′-phosphoryl end of the DNA, whereas topoisomerase II uses a tyrosine residue to form a 5′-PO$_4$–tyrosine linkage (19). It will likewise be interesting to broaden the cleavage/attachment assay described here to include other recombinases and topoisomerases. These assays may be useful in dissecting the mechanism of action of nicking-closing enzymes and in isolating new inhibitors for them.

Acknowledgments

Work in the Sadowski laboratory is supported by the Medical Research Council of Canada. We thank Karen Luetke and Carl Juby for technical assistance. We thank Frieda Chan for typing the manuscript.

References

1. P. D. Sadowski, *FASEB J.* **7,** 760 (1993).
2. A. Landy, *Curr. Opin. Genet. Dev.* **3,** 699 (1993).
3. P. Argos, A. Landy, K. Abremski, J. B. Egan, L. E. Haggard, R. H. Hoess, M. L. Kahn, B. Kalionis, S. V. Narayana, L. S. Pierson III, N. Sternberg, and J. M. Leong, *EMBO J.* **5,** 433 (1986).
4. K. G. Golic and S. Lindquist, *Cell (Cambridge, Mass.)* **59,** 499 (1989).
5. K. G. Golic, *Science* **252,** 958 (1991).
6. M. Lakso, B. Sauer, B. J. Mosinger, E. J. Lee, R. W. Manning, S. H. Yu, K. L. Mulder, and H. Westphal, *Proc. Natl. Acad. Sci. U.S.A.* **89,** 6232 (1992).
7. K. E. Abremski and R. H. Hoess, *Protein Eng.* **5,** 87 (1992).
8. B. R. Evans, J. W. Chen, R. L. Parsons, T. K. Bauer, D. B. Teplow, and M. Jayaram, *J. Biol. Chem.* **265,** 18504 (1990).
9. C. A. Pargellis, S. E. Nunes-Duby, L. Moitoso de Vargas, and A. Landy, *J. Biol. Chem.* **263,** 7678 (1988).
10. G. Pan and P. D. Sadowski, *J. Biol. Chem.* **267,** 12397 (1992).
11. G. Pan, K. Luetke, C. D. Juby, R. Brousseau, and P. Sadowski, *J. Biol. Chem.* **268,** 3683 (1993).

12. G. Pan, K. Luetke, and P. D. Sadowski, *Mol. Cell. Biol.* **13,** 3167 (1993).

13. H. Friesen and P. D. Sadowski, *J. Mol. Biol.* **225,** 313 (1992).

14. H. Pan, D. Clary, and P. D. Sadowski, *J. Biol. Chem.* **266,** 11347 (1991).

15. G. Proteau, D. Sidenberg, and P. Sadowski, *Nucleic Acids Res.* **14,** 4787 (1986).

16. G. Pan, Ph.D. Thesis, University of Toronto (1993).

17. U. K. Laemmli, *Nature (London)* **227,** 680 (1970).

18. G. Pan and P. D. Sadowski, *J. Biol. Chem.* **268,** 22546 (1993).

19. J. J. Champoux, *in* "DNA Topology and Its Biological Effects" (N. R. Cozzarelli and J. C. Wang, eds.), p. 217. Cold Spring Harbor Laboratory, Cold Spring Harbor, New York, 1990.

20. B. M. Olivera, Z. W. Hall, Y. Anraku, J. R. Chien, and I. R. Lehman, *Cold Spring Harbor Symp. Quant. Biol.* **33,** 27 (1968).

[11]   Physical Monitoring of Mitotic and Meiotic Recombination in *Saccharomyces cerevisiae*

James E. Haber and Neal Sugawara

Introduction

Much of what we understand about recombination between homologous DNA comes from genetic studies in fungi such as *Saccharomyces cerevisiae*. From the meiotic segregation of closely linked markers, one can deduce a sequence of steps in the exchange of DNA strands, including the formation of heteroduplex DNA, the repair of mismatches, and the eventual position of crossing-over. Similar studies can be carried out in mitotic cells, where one can also explore the effects of initiating higher levels of recombination by using ultraviolet light or X-irradiation to create lesions in DNA. Finally, highly efficient site-specific recombination events both at the yeast mating type locus and in the mobile ω^+ intron located in mitochondria have permitted a detailed examination of a set of events where the site of initiation of recombination is known.

A variety of genetic studies have led to elaboration of several molecular models of recombination (1). The paradigms for these models are the single-strand nick model proposed by Meselson and Radding (2) (Fig. 1A) and the double-strand break (DSB) repair model described by Szostak *et al.* (3) (Fig. 1B). More recently, attention has also been paid to a third type of recombination, termed single-strand annealing by Lin *et al.* (4) (Fig. 1C). This kind of recombination differs from the other mechanisms in that it is inherently nonreciprocal, such that DNA located between homologous sequences will inevitably be lost.

A full understanding of the molecular events that actually occur in a particular type of recombination demands that one be able to define each of the steps in recombination and to identify the proteins that are essential (or important) for catalyzing these events. This chapter focuses on a number of different ways in which discrete steps in recombination, and the identity of gene products that are important for those steps, can be analyzed in *S. cerevisiae*. The first section outlines the important results that have been obtained using these approaches to study mitotic and meiotic recombination. The second section provides detailed descriptions of many of the methods, especially those that have been used in our laboratory.

Methods in Molecular Genetics, Volume 6

Mechanisms of Mitotic and Meiotic Recombination in *Saccharomyces cerevisiae*

Mitotic Recombination Induced by HO Endonuclease

Recombination can be initiated in mitotic cells by the expression of a site-specific endonuclease to deliver a double-strand cleavage of chromosomal DNA. The most thoroughly studied system is the genetically programmed switching of the yeast mating-type (*MAT*) genes [reviewed most recently by Haber (5)]. *MAT*a and *MAT*α encode different gene products because they each contain about 700 bp of different DNA sequences (Fig. 2). On the same chromosome, but at a distance, are two unexpressed donor loci, one (*HMR*) containing **a**-specific sequences and the other (*HML*) containing α-specific sequences. Cells can switch from one mating-type gene to the other by a gene conversion event that is catalyzed by a site-specific cleavage of *MAT* DNA by the HO endonuclease. The altered chromatin structure of the silent copies makes them inaccessible to the HO endonuclease, so that only *MAT* is cleaved. The sequence of molecular steps that occur during this recombination event can be identified by isolating DNA from cells in which switching is initiated at the same time in most of the cells in the population. Such synchrony is easy to achieve by placing the HO endonuclease gene under the control of a galactose-inducible promoter (6, 7).

The overall switching process is easily followed on a Southern blot of *Sty*I-digested DNA taken at intervals after *HO* induction (Fig. 3). The most striking observation is that switching takes approximately 60 min to complete, from the time of the first appearance of the double-strand cleavage of *MAT* to the first appearance of the opposite mating-type product. This long time is characteristic of all of the DSB-induced recombination events that have been followed in yeast, even when the recombining DNA segments are very close together. This makes the analysis of recombination intermediates feasible.

Several intermediate steps in *MAT* recombination have been identified. After cleavage by HO endonuclease, the DSB is processed by a 5' to 3' exonuclease that leaves a single-stranded 3' tail (Figs. 2 and 3) that can be easily detected by several methods, described below. The 3'-ended single-strand can then invade into an intact donor locus, and the end can then be used as a primer to initiate new DNA synthesis. This step, too, can be detected by an assay based on the polymerase chain reaction (PCR) in which the PCR product can only be synthesized when a Yα-specific and MAT_{distal} primer are covalently joined together (Fig. 2). This step occurs approximately 30 min after the appearance of the DSB and about 30 min before the original Y**a** sequences are removed from *MAT* and replaced with Yα sequences (8). A second PCR assay reveals the completion of the process (Fig. 2C, primers pC and pD). These physical monitoring assays have also shown that the Y region DNA is somehow protected from exonucleolytic digestion (8). This protection depends on the presence of the donor sequences, although the basis for such protection is unknown. A com-

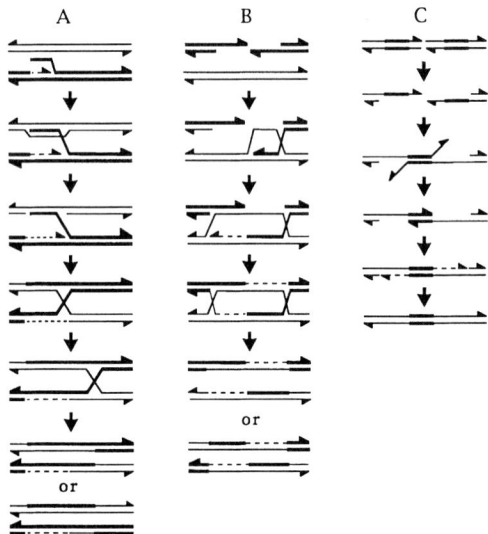

Fig. 1   Models of DNA recombination. (A) Meselson–Radding model. A nicked DNA strand is displaced by new DNA synthesis from the 3′ end and invades a homologous sequence. During the extension of heteroduplex DNA, the branched structure may isomerize to create a symmetrical Holliday structure, so that branch migration can occur. Cleavage of the Holliday junction can result in both crossovers and noncrossovers containing regions of heteroduplex DNA. Mismatch correction can lead to gene conversion. (B) The double-strand break repair model. A double-strand break (DSB) is enlarged by 5′ to 3′ exonuclease digestion to have two long 3′-ended single strands of DNA that can both invade a homologous sequence. If there is also 3′ to 5′ exonuclease activity a gap may be created. New DNA synthesis primed from the two 3′ ends can copy donor sequences and fill in any gap. A symmetric and an asymmetric Holliday junction can be cleaved to yield both crossover and noncrossover products containing heteroduplex DNA. Repair of mismatches may result in gene conversion of markers near the DSB. (C) Single-strand annealing model. A DSB created between two regions of flanking homology is acted on by 5′ to 3′ exonucleases exposing the complementary single strands of the flanking homologous regions. These sequences can anneal with one another. Subsequent removal of nonhomologous single-stranded tails and initiation of new DNA synthesis lead to the formation of a deletion. This process is inherently intrachromosomal and is nonconservative in that the sequences between the homologous regions are degraded and lost.

bination of genetic and physical studies have also been used to demonstrate that there is little, if any, 3′ to 5′ exonuclease activity that would shorten the 3′-ended single-strand when a homologous donor is present (9). Additional steps in recombination are likely to include the resolution of branched structures such as Holliday junctions, but these intermediates have not yet been identified.

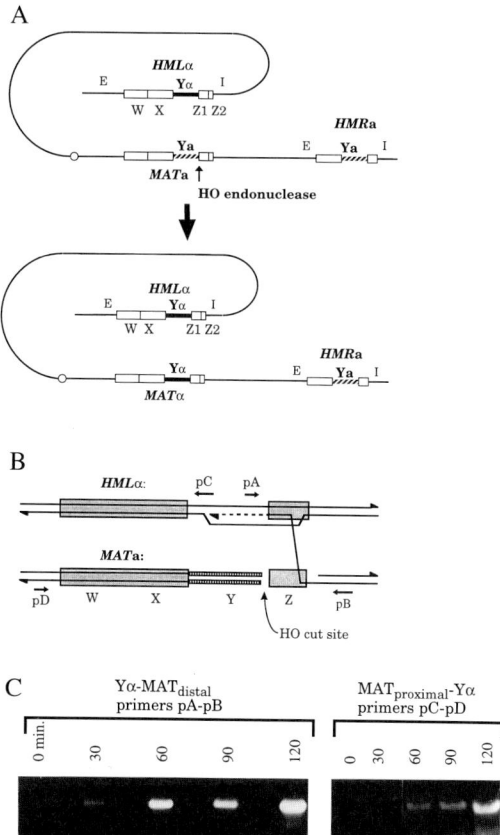

FIG. 2 Switching of the yeast mating-type (*MAT*) gene. (A) The *MAT* locus may contain either Y**a** or Yα sequences that in turn determine the mating type of the cell. A *MAT***a** cell can switch to *MAT*α by a gene conversion process in which the Y**a** sequences are removed and replaced by a copy of Yα DNA derived from the silent locus, *HML*α, located 185 kb away on the left arm of the same chromosome. *HML*α is maintained in an unexpressed state by the interaction of several gene products with two cis-acting sites, designated E and I. A similar gene conversion process can occur in which *MAT*α pairs with the unexpressed donor *HMR***a**, located 95 kb more distal on the right arm of the same chromosome. *MAT* shares more homology (regions W and Z2) with *HML* than with *HMR*. (B) On expression of a galactose-inducible *HO* gene, the *MAT* locus is cleaved a few base pairs to the right of the Y/Z junction. After 5′ to 3′ exonuclease digestion of the Z region DNA (and beyond) the 3′-ended single strand can invade the donor locus and initiate new DNA synthesis. These initial steps can be detected by a PCR assay dependent on the creation of a covalent DNA strand between Yα (initially only in the donor) and sequences distal to *MAT* using primers pA and pB. A second PCR assay using primers pC and pD reveals the joining of Yα DNA to the proximal side of the *MAT* locus. (C) A PCR assay detects the time of initial strand invasion and primer extension (Yα-MAT$_{distal}$; primers pA and pB). The completion of switching is revealed by a second PCR product (MAT$_{proximal}$-Yα; primers pC and pD). The second PCR product appears approximately 30 min after the Yα-MAT$_{distal}$ strand invasion intermediate.

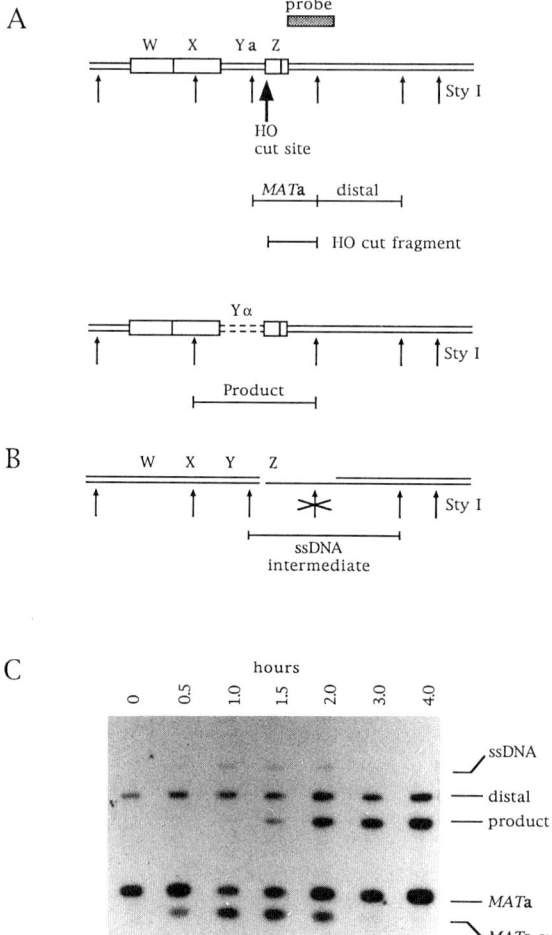

FIG. 3   Southern hybridization analysis of mating-type switching induced by HO endonuclease. A *MAT***a** strain was grown in YP–lactate medium and induced to switch by induction of the *GAL:HO* gene. (A) Digestion by *Sty*I gives two bands on a Southern blot that hybridize to a *MAT*–distal probe (shaded box) and that correspond to the *MAT***a** fragment and the MAT_{distal} fragment. When the HO endonuclease cuts *MAT***a** at the Y/Z junction, a new fragment is formed (0.7 kb) (labeled *MAT***a** cut fragment). The *MAT*α gene conversion product lacks a *Sty*I site in the Yα-specific sequence and hence results in a larger restriction fragment (8) that is distinguishable from the *MAT* Y**a** *Sty*I fragment. (B) After DSB formation an exonucleolytic activity degrades the 5′ strand on the distal side of the DSB leaving a 3′ tail. *Sty*I is incapable of cutting this single-stranded DNA, resulting in higher molecular weight restriction fragments. Degradation of the proximal side of the DSB is not observed on the Southern blot in strains where the *HML* cassette is present (8). (C) DNA was isolated at the designated times after *HO* induction, digested with *Sty*I, and electrophoresed on an alkaline denaturing gel. The Southern blot was probed with a MAT_{distal} probe (shaded box in A).

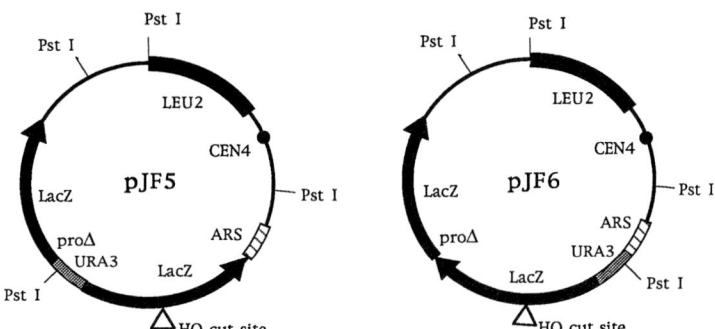

FIG. 4   Structure of the *LacZ* plasmids. pJF5 and pJF6 contain two mutant *LacZ* sequences in inverted or direct orientations, respectively. One *LacZ* sequence contains a 117-bp *HO* cut site. The other *LacZ* sequence lacks a promoter (proΔ). *LEU2* and *URA3* are selectable markers. CEN4 (centromere) and ARS (autonomous replicating sequence) allow the plasmids to be replicated and segregated in yeast.

Alternative Pathway of Break Repair Independent of Gap Repair: Single-Strand Annealing between Homologous Sequences Flanking Double Strand Breaks

To extend these studies to more general DSB-mediated recombination, several laboratories have also constructed other substrates in which the HO endonuclease can be used to create a double-strand break (10–12). For example, a 117-bp fragment of the *MATa* locus has been inserted into the *Escherichia coli β*-galactosidase (*LacZ*) gene in centromere-containing plasmids that also carry a promoterless *LacZ* gene that can act as a donor for repair of the DSB (Fig. 4). Plasmids in which the two repeats are in inverted and direct orientation have been constructed. HO-mediated recombination is quite efficient, with more than 50% of the cells becoming functionally Lac$^+$. Similar results are obtained when the same pair of *LacZ* sequences are integrated into a chromosome (10).

When an analysis was carried out using *LacZ* sequences that are in direct orientation on a plasmid, it was easy to detect both Lac$^+$ products that have undergone a gene conversion and those that have experienced a deletion. There were several aspects to these results that suggested that the two events did not result from the simple alternative resolution of a common intermediate containing a symmetrical Holliday junction intermediate. First, the proportion of deletions was much greater than gene conversions without exchanges (87 versus 13%) (10). Second, despite the abundance of the deletion product, there was no evidence of the expected reciprocal exchange product, a 4.5-kb circular product that contains a single *Pst*I site. Finally, the kinetics of appearance of the gene conversion product and the deletion were different (13); if they were alternatively resolved outcomes of the same intermediate, they should have the same time of appearance.

The explanation for these discrepancies seems to be that the deletions arise by an alternative mechanism known as single-strand annealing (SSA) or resection/reannealing. As first proposed by Lin *et al.* (4), efficient recombination can occur between homologous regions flanking a DSB after extensive 5′ to 3′ exonuclease digestion exposes complementary single-stranded regions. Evidence to support the existence of this mechanism in yeast has come from physical monitoring studies that show the creation of such long single-stranded regions (14). Moreover, the timing of the deletions in the *LacZ* direct repeat plasmids is delayed by about 60 min when 4.4 kb of additional DNA is inserted between the DSB and one of the two homologous regions (13). There was no effect on the timing of gene conversion events, although the proportion of gene conversions rose significantly when the appearance of deletions was retarded. The 1-hr delay caused by the insertion of 4.4 kb of DNA is consistent with movement of a 5′ to 3′ exonuclease at about 1 nucleotide/sec.

The SSA mechanism in yeast is also highly efficient when the DSB is introduced on the chromosome. When flanking regions of about 1 kb are separated by as much as 15 kb, essentially all the cells that experience a DSB are able to repair the break by this deletion mechanism. The efficiency with which such repair occurs is dependent on the size of the flanking regions. Between 30 and 90 bp of homology is sufficient for these events to occur at a low frequency (14; N. Sugawara and J. E. Haber, unpublished data).

Strand Repair Initiated by Other Endonucleases and by Ionizing Radiation

The I-*Sce*I endonuclease was identified as a site-specific mitochondrial enzyme that promoted the spread of an intron in the ribosomal RNA gene (which encodes this endonuclease) to rRNA genes that lack the intron. In all formal respects I-*Sce*I-promoted gene conversions of the intron appeared to be similar to those events initiated by the HO endonuclease. A more direct demonstration of the equivalence of the events initiated by the two endonucleases became possible when a version of the gene that could be expressed in the yeast nucleus was constructed (15). A comparison of HO- and I-*Sce*I-mediated DSBs in *LacZ* plasmids containing the respective cleavage site showed that they were apparently identical.

Ionizing radiation such as X-rays creates DSBs at apparently random sites in DNA. The repair of such breaks can be followed using chromosome-separating gels in which any random break of a chromosome will produce smaller fragments. In a diploid, where each break is likely to be covered by an intact sequence, DSBs can be repaired by gene conversions. When samples are taken after X-irradiation and separated by electrophoresis in the contour-clamped homogeneous electric field (CHEF) mode, it becomes clear that the kinetics of repair are roughly similar (about 1 hr) to the repair of single lesions created by HO endonuclease (16).

Analysis of Specific Gene Functions during Recombination

One of the other significant advantages of physical monitoring of recombination is that it is possible to pinpoint the molecular steps that are impaired by different recombination-defective mutations. Because one does not demand the recovery of any viable products, it is possible to examine the extent to which recombination can occur even in cells that are rendered inviable. Several important functions have been studied in this way. The most successful of these studies demonstrated that a deletion of the *RAD1* gene prevented the completion of recombination when the HO-cleaved ends of the DSB were not homologous to the donor locus. Thus, whereas *RAD1* was not required for *MAT* switching or for X-ray induced DSB, the gene was essential to complete HO-induced recombination of the *LacZ* plasmids (13). Recombination could be restored by inserting into the donor locus a nearly identical *HO* cutting site (but containing a single base-pair mutation that prevented cutting). In this way the ends of the DSB were once again homologous to the donor and recombination could proceed. *RAD1* is required both for gap-filling and single-strand annealing events.

Similar types of analysis have been carried out for deletions of several known X-ray sensitive genes including *RAD50, XRS2, RAD51, RAD52, RAD54, RAD55,* and *RAD57* (13–15, 17–19). The results of these studies show that different recombination processes are affected in different ways by a number of these mutations. Deletions of both *RAD50* and *XRS2* slow down, but do not prevent, *MAT* switching, *LacZ* conversions, and single-strand annealing. Both genes have a marked effect on the rate of 5′ to 3′ exonuclease activity (though there is no evidence that either gene encodes an exonuclease). The double mutation *rad50 xrs2* slows recombination even more; *MAT* switching under these conditions takes as long as 10 hr to complete (18). It should be possible by following the kinetics of other intermediates to determine if these gene products are also required for subsequent steps in recombination.

RAD51, RAD52, RAD54, RAD55, and *RAD57* are all required at or before strand invasion and primer extension during *MAT* switching. There is an apparent increase in the amount of 5′ to 3′ exonucleolytic activity in all of these mutant strains (e.g., compare Figs. 3C and 5A). This may represent an uncoupling of an exonuclease from the rest of a strand-exchange "machine" owing to a block in the recombination pathway.

Further investigation of the *RAD* genes in gene conversion events has been undertaken using the *LacZ* plasmids containing inverted repeats. A *rad52* deletion prevented all recombination; however, other mutants in the series (*rad51, rad54, rad55, rad57*) proved not to be required (19). Both gene conversions with and without crossing-over were obtained at nearly wild-type levels, although some of the mutants appeared to cause delays in the appearance of the final products. This is especially surprising for the deletion of *RAD51*, as this protein appears to have some of the structural and biochemical properties of the *E. coli recA* protein (20, 21). However, despite the fact that the Rad51 protein will form filaments with double-stranded

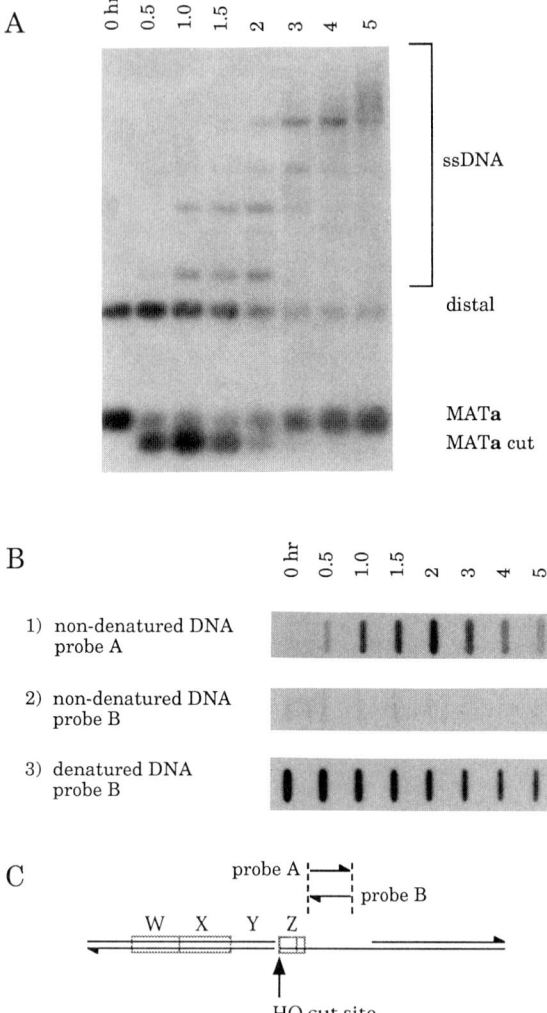

FIG. 5   Production of single-stranded structures in a *rad51* strain. (A) DNA was extracted from a *rad51* strain after *HO* induction at the times shown. Samples were digested with *Sty*I and electrophoresed on an alkaline denaturing gel. The Southern blot was probed with a double-stranded DNA probe (shown in Fig. 3A). A product band is not seen because *rad51* mutants are defective in mating-type switching. (B) Native time-course samples were applied to a nylon membrane using a slot-blot apparatus and hybridized with RNA probes. Row 1 shows native DNA samples probed with RNA probe A shown in (C) that hybridizes to the 3′ single-stranded tail formed after *HO* induction. The nondenatured samples in row 2 were hybridized with the RNA probe of the opposite strand, probe B in (C). Samples in row 3 were denatured prior to loading onto the membrane and were subsequently hybridized with probe B. Slot-blot analyses may detect single-stranded structures not apparent using Southern blots of denaturing gels. (C) Probes A and B indicate the location and directions of the RNA probes used in (B).

DNA, it has not yet been shown to function as a strand exchange protein in yeast. The physical monitoring results argue that although *RAD52* is essential for gene conversion and crossing-over, *RAD51, RAD54, RAD55,* and *RAD57* are not needed for some of these events. The possibility that these genes are only needed when the amount of homology between the donor and recipient is small (as it is at *MAT*) can be easily tested.

The requirements for single-strand annealing are different from those for *MAT* switching. Unlike mating-type switching, *RAD51, RAD54, RAD55,* and *RAD57* are apparently not required for single-strand annealing (E. L. Ivanov, J. Fishman-Lobell, N. Sugawara, and J. E. Haber, unpublished). The situation with *RAD52* is more complicated. *Rad52* mutants have very inefficient SSA when the flanking homologous substrates are 1 kb (22). They show about 10% of wild-type levels of deletions when the flanking regions are about 2.5 kb (13, 15), but viability is essentially normal when the *HO* cut site is placed in an array of tandem rDNA repeats (17). This seems to suggest that *RAD52* is needed to stabilize shorter regions of heteroduplex DNA, but not longer ones. The data are consistent with earlier genetic studies of *RAD52*-independent spontaneous recombination events (23, 24).

Dependence of Recombination on Cell Cycle and Other Physiological Conditions

Physical monitoring can also be used to examine the ability of cells to complete recombination when they are unable to grow or to carry out certain essential processes such as chromosomal DNA replication. For example, *MAT*a cells can be arrested in the G_1 stage of the cell cycle with the mating pheromone α-factor, and then HO endonuclease can be induced. *MAT* switching, *LacZ* conversions, and SSA can all occur under these conditions, although the proportion of *LacZ* conversions relative to SSA deletions decreased from 17 to 5% (13). *MAT* switching has been shown to occur with similar kinetics in cells synchronously growing at different stages of the cell cycle (25).

Cells carrying a temperature-sensitive allele of various essential genes important for DNA replication and repair can also be arrested at their restrictive temperature and then tested for the ability to carry out HO-mediated recombination. We have shown in this way that topoisomerase II is not needed for recombination; nor is topoisomerase I (J. Fishman-Lobell and J. E. Haber, unpublished results). A similar analysis could be carried out using various temperature-sensitive mutations in different DNA polymerase genes or other possible "players" such as DNA ligase and helicases.

We do know that some new protein synthesis is required for the completion of recombination. Rudin and Haber (22) added cycloheximide to cells soon after HO induction and blocked most of the completion of SSA. The need for new protein synthesis may account for the inherent slowness of DSB-mediated recombination. One interesting experiment would be to induce one DSB, with I-*Sce*I, and then induce an HO-mediated event some time thereafter. If the first DSB induces the synthesis of rate-limiting components, the second event should now occur more quickly.

Meiotic Recombination

Physical monitoring of the kinetics of meiotic recombination was first accomplished by Borts *et al.* (26) by looking for the time of appearance of novel restriction fragments created by reciprocal crossing-over. Reciprocal exchange appeared to occur at least an hour after meiotic DNA synthesis and the commitment of cells to undergo meiotic recombination. These initial studies have been followed by more detailed investigations into the timing of meiotic events by using strains that show more rapid, though not much more synchronous, meiosis in which it is possible to monitor the time of crossing-over versus both biochemical events such as new DNA synthesis and cytological events such as chromosome condensation and the formation of the synaptonemal complex (27). Interestingly, crossing-over appears to be delayed until the synaptonemal complex begins to dissolve.

Early Intermediates of Meiotic Recombination: Double-Strand Breaks

Double-strand breaks were first detected at the meiotic "hot spot" in the promoter region of the *ARG4* gene (28). The level of DSB appears to correlate with the frequency with which meiotic recombination occurs near this site (29). Meiotic DSBs are similar to those created by HO or I-*Sce*I in that they create overhanging 3′ ends; moreover, they are processed by a 5′ to 3′ exonucleolytic activity (30). A more detailed analysis of the formation of DSB was facilitated by the isolation of a very useful "separation of function" mutation of *RAD50* known as *rad50-S*. This mutation permits the formation of DSB but does not allow their subsequent exonucleolytic processing (31); thus the DSBs created at a locus can be stabilized. In *rad50-S* backgrounds, several important experiments have been carried out. First, by examining the integrity of whole chromosomes on chromosome separating gels, Simchen and co-workers have been able to demonstrate that there are several distinctive DSB hot spots along several different chromosomes (32). This is a very important step in understanding the relation between DSB formation, chromosome pairing, and chromosome segregation (e.g., see Refs. 33 and 34).

A second important study, by Wu and Lichten (35), makes use of *rad50-S* to investigate the basis of the 40- to 50-fold range in recombination that is seen when the same *LEU2* or *ARG4* genes are placed in different chromosomal locations (36). It appears that the position effects are reflected in changes in the level of the DSB at *ARG4* and in surrounding sequences. An extension of these studies has been to demonstrate that the level of a DSB at any one chromosomal position is correlated with the degree of hypersensitivity of those same regions to DNase I (37). These types of approaches may help define the nature of the as yet unidentified meiotic endonuclease that creates breaks at sites that otherwise show no DNA sequence identity.

As another way to follow the metabolism of DSBs in meiosis, M. Hoekstra (personal communication) has placed the *HO* gene under the control of a meiotic-specific promoter, *SPO13*. Thus, HO-mediated events can theoretically be followed even in mutant strains where normal meiotic DSB may be prevented. We have found that many, if not all, of the DSBs are created at the appropriate time in meiosis (after DNA replication) (L. O. Ross, A. L. Malkova, D. Dawson, and J. E. Haber, unpublished).

Still another DSB that can be studied physically in meiosis is that created by the "intron homing" endonuclease encoded by the VDE endonuclease that is excised by a unique protein splicing event from the yeast vacuolar ATPase protein VMA1 (38–40). The VDE endonuclease creates a DSB that promotes a gene conversion event between a target lacking the DNA segment that encodes the VDE intron and the intact gene. Recently, L. Gilbert and F. Stahl (personal communication) have inserted a VDE cutting site into the *ARG4* gene and shown that it stimulates very high levels of gene conversion. This should prove to be a useful comparison to the events initiated by the normal *ARG4* meiotic DSB. Its high level of cleavage should make physical monitoring studies more sensitive.

Heteroduplex-Containing Intermediates of Recombination

Lichten *et al.* (41) made use of partially denaturing gel techniques to identify the formation of single base-pair mismatches created by DNA strand exchange during meiosis. These studies confirmed genetic results that C/C mismatches were not efficiently corrected but that G/G mismatches were. The kinetics of G/G mismatch correction could be inferred to be rapid, as no such mismatches could be detected unless the cells were defective for the mismatch correction gene *PMS1*. A more recent study (42) has raised interesting problems in understanding the timing of recombination events by revealing that stable heteroduplex-containing DNA cannot be recovered until about the time that reciprocal recombination is seen, long after the formation and apparent disappearance of the meiotic DSB. Further work will be needed to determine if the absence of heteroduplex DNA soon after the formation of 3'-ended single-stranded DNA is a reflection of the formation initially of very short regions of heteroduplex DNA or of some other unstable structure.

Another means to follow heteroduplex formation *in vivo* employs the use of mutations composed of two short palindromic sequences with either a *Bam*HI or *Pst*I restriction site in the center. Nag and Petes (43) observed that a heteroduplex formed between these palindromes is resistant to digestion with *Pst*I or *Bam*HI, consistent with the proposal that a cruciform structure is formed. They confirmed the timing of the heteroduplex formation during meiosis and showed that *rad50* but not *rad52* mutants are defective in meiotic heteroduplex formation. The ability of *rad52* mutants to form heteroduplex DNA in meiosis is consistent with earlier physical monitoring studies that found that *rad52* diploids could complete crossing-over at a reduced but significant level (26). This is quite different from what we have observed in HO-induced mitotic cells (13, 22). Possibly there may be a second, meiotic-specific *RAD52* homolog, as has been found for *RAD51* and its meiotic-specific relative *DMC1*.

Branched Intermediates of Recombination

Bell and Byers (44) and then Brewer and Fangman (45) have developed two-dimensional gel electrophoresis techniques to resolve intermediates in DNA replication and recombination. The approaches should be applicable to look for Holliday junctions or other branched intermediates of recombination. Indeed, Symington (46) has reported such structures in *in vitro* studies of recombination using yeast extracts. More recently two laboratories have reported success in detecting joint, interhomolog molecules from cells undergoing meiosis (47, 48). Several observations indicate that these molecules are recombination intermediates. These include their molecular mass and position on two-dimensional gels, the timing of their presence, and their requirements for *SPO11* and *RAD50* products. Schwacha and Kleckner (48) utilized a psoralen crosslinking reagent to enhance the detection of joint molecules. When these crosslinks were removed they showed that DNA strands recombinant for flanking markers were absent. This led to the proposal that the molecules contain paranemic joints, double Holliday junctions, or more complex structures.

Roles of Specific Genes during Meiotic Recombination

Many meiotic-defective mutations completely prevent the formation of viable spores or even of viable recombinants when cells are pulled back from meiosis and returned to vegetative conditions. Nevertheless, much can be learned by physical analysis of the formation of recombination products and intermediates. In the first of these studies, Borts *et al.* (26) showed that diploids homozygous for *rad50* failed to complete reciprocal recombination. More recently, Cao *et al.* (31) confirmed this result and showed that *rad50* mutants also fail to create meiotic DSBs. As mentioned before, the *rad50-S* mutation allows the formation of the DSB but prevents further processing. Given that a null allele of *rad50* reduces but does not eliminate 5′ to 3′ exonuclease activity in mitotic cells (14), it is not clear if there are other exonucleases in mitotic cells missing in meiotic cells or if *rad50-S* mutants are particularly blocked in exonuclease activity.

Meiotic DSBs are also not formed in strains carrying other mutations that are also blocked early in meiosis, including *mre11* (49), *spo11* (31), and *xrs2* (50). An indication that *XRS2* is required before or simultaneously with *RAD50* is that an *xrs2 rad50-S* double mutant does not accumulate DSBs.

In contrast, several meiotic-defective mutations that genetically appear to be unable to complete recombination actually can complete the majority of physically monitored crossing-over events. Thus, both *rad57* and *rad6* mutants show nearly wild-type levels of crossing-over despite failing to make viable spores (26). Another class of mutants are unable to complete sporulation and fail to recombine by physical assay, yet they show substantial gene conversion and (in some cases) crossing-over when returned to growth conditions. This class includes deletions of *ZIP1* (51) and *SEP1* (52). Whether some of these mutations change the appearance and/or stability of heteroduplex-containing DNA will be an interesting question to investigate.

Analysis of Intermediates of Recombination

Formation of Single-Stranded DNA by Exonucleolytic Digestion

Approximately 15 min after the appearance of the DSB, one can begin to detect the appearance of the first intermediate in recombination: the creation of long 3′-ended single-stranded tails by the action of a 5′ to 3′ exonuclease. The extent of such exonucleolytic digestion can be determined by use of alkaline denaturing Southern blots, in which one can see the appearance of a set of apparently partially digested restriction fragments that possess a higher molecular weight than the completely digested DNA. These larger fragments are the reflection of single-stranded DNA, adjacent to the HO cleavage site, that cannot be digested by the restriction enzyme (Figs. 3 and 5A). These newly created fragments are visible only as a faint smear of heterogeneously sized fragments when neutral gels are used, presumably because their dual single/double-stranded nature does not permit them to migrate as discrete bands. The exonucleolytic activity may also initiate digestion in an asynchronous manner or in a processive manner, resulting in a population of degraded molecules of varying sizes. What is observed on the denaturing gel, therefore, is not the degraded strand but, rather, the opposite full-length single-stranded DNA.

The single-stranded structure of possible intermediates can be established by the use of strand-specific hybridization probes. Strand-specific probes can consist of RNA probes made by cloning template sequences into readily available vectors possessing transcription promoters like those from the pGEM (Promega, Madison, WI) or pBluescript (Stratagene, La Jolla, CA) collections. Labeled run-off transcripts are prepared by cleaving the template by restriction digestion and polymerizing from the appropriate promoter flanking the template (53). Strand-specific probes can also be obtained by preparing labeled oligonucleotides (53).

In an alternative procedure, the DNA is run on a neutral gel and transferred to a nylon membrane without denaturing the DNA (28, 30). The principal components that bind to the membrane are the single-stranded DNA, although some double-stranded DNA also binds. The membrane is probed with a labeled single-stranded RNA probe that binds to the single-stranded DNA bound to the membrane. Residual amounts of double-stranded DNA bound to the membrane do not hybridize with the probe at a significant level. A RNA probe of the opposite strand is generally used as a negative control.

A more direct method for quantitative purposes employs the use of slot-blots or dot-blots (Fig. 5B). As above, native DNA is applied directly to the nylon membranes without denaturation and then hybridized with single-stranded probes. The single-stranded probes will detect DNA that has become single-stranded but will not detect double-stranded DNA. Denatured DNA samples can also be applied to membranes and probed with single-strand-specific probes. These samples can reveal degradation kinetics that may be obscured by electrophoretic migration on gels. This approach circumvents difficulties in analyzing single-stranded components that do not migrate as discrete bands on gels. It also avoids the difficulty of transferring large restriction fragments out of gels without DNA nicking and denaturation as in the native DNA transfers described above.

Strains

All strains here were derived from *S. cerevisiae* DBY745, a strain of the S288C background. Strains were screened for the ability to grow on galactose medium under anaerobic conditions. The strains possess a GAL10:HO fusion maintained on a centromere plasmid [either pFH800 (12, 54) or pGal-HO (7, 55), a YCp50-based vector].

Media and Solutions

YP–lactate medium is 3% (w/v) lactic acid, 1% (w/v) yeast extract, and 2% (w/v) peptone, neutralized to pH 5.5 using NaOH. A 20% galactose solution is prepared by dissolving galactose at room temperature followed by filter sterilization. Glass beads (0.5 μm in diameter) are acid and base washed (56). DNA extraction buffer is 100 mM Tris, pH 8, 50 mM EDTA, 2% (w/v) sodium dodecyl sulfate (SDS). Other reagents are prepared as described by Sambrook *et al.* (53). Selective media are described by Sherman *et al.* (57).

Induction of Recombination and DNA Extraction

1. Grow cells from a single colony overnight in 5 ml at 30°C. Selective medium should be used if a plasmid such as a GAL::HO plasmid needs to be maintained.
2. Spin down the cells in a centrifuge and discard the supernatant. Resuspend the pellet in 10 ml of YP–lactate medium and shake the cells at 30°C for 5–10 hr or until the cells begin to grow again. Strains undergo a variable lag phase on the shift to lactate medium during which they do not divide. The pregrowth in lactate medium is an optional step intended to overcome the lag phase and make it easier to obtain the proper cell density of the final culture (see next step). Glucose is not used as a carbon source because it represses the *GAL10* promoter.
3. Inoculate YP–lactate medium with enough cells to give 10^7 cells/ml the next day. Shake the culture overnight at 30°C using a flask large enough to provide sufficient aeration (usually 0.5–1.0 liter in a 4-liter flask with a foam plug).

4. When the cells have grown to a density of 10^7 cells/ml, take a sample (usually 50 ml) for the uninduced control and then add $\frac{1}{10}$th volume of 20% galactose (~2% final concentration) to the remaining culture. Continue shaking the culture at 30°C and take 50-ml samples at the desired time points. For experiments studying mating-type switching, add $\frac{1}{10}$th volume of 20% glucose after 30 min to repress the expression of the HO endonuclease. *HO* expression is stopped so that HO endonuclease will not cleave the product of the switching event. Samples of the cell culture may be diluted and plated onto YPD or selective media to follow the reaction phenotypically if possible and to measure the stability of the GAL::HO plasmid.

5. At each time point, centrifuge the samples until the cells have sedimented. Discard the supernatant and resuspend the cells in 400 μl of extraction buffer. Transfer the cells to a microcentrifuge tube containing approximately 500 μl of glass beads (0.7 g) and 400 μl of phenol. The glass bead/phenol mixture may be prepared beforehand and stored at 4°C or on ice in the dark.

6. Vigorously vortex the tube for 1 min. During this time remove the tube from the vortexer once or twice and briefly shake the tube. Store the sample on ice until finished with all the time-course samples. Add enough extraction buffer to the samples to equalize the volumes.

7. Allow the tubes to sit on ice for several minutes or longer prior to centrifuging them in a microcentrifuge for 10–15 min at 4°C. Chloroform is not used in this procedure.

8. Carefully remove the top aqueous layer to a new tube. Add 400 μl of phenol to the new tube. Mix by inverting the tubes. Leave on ice 1–2 min, centrifuge samples 10–15 min at 4°C, and save the aqueous layer in new tubes.

9. Add 50 μl of 3 M sodium acetate (pH 5.2) and 600 μl of 2-propanol. Mix.

10. Centrifuge the samples in a microcentrifuge for 1 min at 4°C or room temperature. Discard the supernatant.

11. Add 300 μl TE containing 10 μg of RNase to each tube. Incubate at 37°C with occasional, minimal vortexing for 30–60 min or until the pellet has dissolved. Most of the RNA has been digested when the pellet has dissolved.

12. Add 30 μl of 3 M sodium acetate and 300 μl of 2-propanol. Spin for 2 min at room temperature or 4°C. Discard the supernatant. Rinse with 95% (v/v) ethanol. Dry the pellet. Resuspend the pellet in TE and digest 10–20% of each sample for use on Southern blots.

Denaturing Gels

Denturing gel electrophoresis is carried out according to the procedure of McDonnel *et al.* (58) as modified by Maniatis *et al.* (59). Agarose is melted in 50 mM NaCl, 1 mM EDTA, poured into a gel tray, and allowed to solidify. After solidification, the gel is submerged in a gel box in 50 mM NaOH, 1 mM EDTA and allowed to equilibrate for 30 min or longer. This procedure is used because agarose will decompose if boiled under alkaline conditions. Alternatively, the agarose can be melted in water and allowed to cool to 60°C. At that temperature the agarose can be adjusted to 50 mM NaOH, 1 mM EDTA by addition of concentrated NaOH and EDTA (53). Ethidium bromide is omitted because it does not bind to DNA under alkaline conditions.

The DNA samples are prepared by adjusting the solution to 0.3 M sodium acetate and 5 mM EDTA (pH 8.0) followed by the addition of 2 volumes of ethanol to precipitate the DNA. After chilling the samples on dry ice (10 min) and centrifuging in a microcentrifuge (10 min), the supernatant is discarded and the pellet is rinsed in 95 or 70% (v/v) ethanol. As much of the supernatant as possible should be removed before drying the pellet. Resuspend the pellet in alkaline gel loading buffer [1× buffer: 50 mM NaOH, 1 mM EDTA, 2.5% (w/v) Ficoll (Type 400), and 0.025% (w/v) bromocresol green]. Size markers can be denatured by directly adding alkaline gel loading buffer to a small volume of plasmid or phage DNA digested by restriction enzymes.

After loading the DNA on the gel, place a glass plate on the gel to prevent the dye from diffusing from the agarose during the course of the run. After the DNA has migrated far enough, remove the gel and, if desired, stain the gel with ethidium bromide (0.5 μg/ml) in 1× TAE electrophoresis buffer. The DNA will be faint because the DNA is single-stranded. Soak the gel in 0.25 N HCl for 6–7 min with gentle agitation. Rinse the gel with water and soak the gel in 0.5 N NaOH, 1.5 M NaCl for 30 min with gentle agitation. Rinse briefly with water and apply the gel to the DNA transfer apparatus. The DNA can be blotted onto either neutral or positively charged nylon membranes using 2× or 10× SSC by any of various methods. We generally employ a capillary transfer table or a vacuum blotter.

Southern hybridizations can be carried out according to different protocols. We hybridize according to the protocol of Church and Gilbert (60) at 65°C in 7% SDS, 0.5 M sodium phosphate (pH 7.2), and 1 mM EDTA, after cross-linking the DNA to the nylon membrane using UV light. We prepare RNA probes according to Melton *et al.* (61) as modified by Promega, and DNA probes according to Feinberg and Vogelstein (62).

Native Gels

For nondenatured DNA transfers, gels are run in 1× TAE and the DNA is transferred in 10× SSC by capillary action or by vacuum blotting. By either means, a fraction of medium to high molecular weight DNA is not transferred efficiently because the DNA has not been nicked and denatured. For this reason it is desirable to choose restriction enzymes that yield restriction fragments less than 5 kb in size if possible. Higher molecular weight bands are usually visible on Southern blots but are not quantitatively transferred. The extent of transfer should be estimated after the transfer by soaking the gel in 1× gel running buffer plus ethidium bromide (0.5 μg/ml) for 30–40 min and visualizing the gel under UV light.

Demonstration that a band on a Southern blot represents a single-stranded entity relies on the ability of the band to hybridize to a RNA probe but not to the RNA probe of the opposite sense. Because of this, two blots are generally prepared from DNA samples electrophoresed in parallel, one for each probe. Positive controls are essential and may be provided by denaturing (boiling) linearized plasmid DNA (~0.05 ng of hybridizable sequence per band per well). In addition, a third blot is prepared in parallel in which the DNA is acid-nicked and alkaline-denatured prior to the DNA transfer onto the nylon membrane (see above). This Southern blot may be probed with one of the RNA probes or with a double-stranded DNA probe.

Slot-Blots

Three sets of samples are loaded in a typical slot-blot experiment described below.

1. Native samples: Dilute 1 μg of yeast genomic DNA from the time-course samples to 200 μl, adjusting the final concentration to 10$\times$ SSC.
2. Denatured samples: Denature 0.1 μg of genomic DNA by adding NaOH to a final concentration of 0.4 N NaOH. Neutralize with 3 M sodium acetate, pH 5.7. The amount needs to be determined empirically for each new solution of sodium acetate. In our experiments the reaction induced by the HO endonuclease produces only a small amount of single-stranded DNA in the native samples. Therefore, we use more native DNA (10$\times$) than denatured DNA in order to obtain comparable signal strengths.
3. Quantitation samples and other controls: Digest a plasmid containing the target sequence with a restriction enzyme and prepare samples ranging in amount from 0.05 to 1.0 pg of hybridizable sequence. Denature the samples with 0.4 N NaOH and then neutralize the samples using sodium acetate as described in the previous step. These samples will be useful if the amount of single-stranded DNA formed is going to be quantitated (63). For other controls include a sample with nondenatured plasmid DNA (0.1 ng of hybridizable sequence), samples of nonradioactive RNA transcripts complementary to the RNA probes that will be used (0.001 ng of hybridizable sequence), and, if possible, denatured genomic DNA lacking hybridizable sequences as a negative control.

One blot is prepared for each RNA probe. A small amount of signal can sometimes be detected using the noncomplementary RNA probe. For this reason, single-stranded probes are preferred to denatured double-stranded DNA probes. Load the samples into the wells of the slot-blot or dot-blot apparatus and transfer the DNA to two nylon membranes, one for each RNA probe. After the samples have been transferred to the membrane, rinse the sides of each well with 500 μl of 10$\times$ SSC. The DNA is then cross-linked to the membrane with UV light and hybridized according to the protocol of Church and Gilbert (60).

Strand Invasion and Heteroduplex-Containing Intermediates

Polymerase Chain Reactions

The PCR can be used to amplify strand invasion and crossover products if the primers are positioned in unique regions on opposite sides of the shared homologous sequences (see Fig. 2B). The PCR product can be used to detect strand invasion prior to the formation of a mature recombination product. In the case of the *MAT* assay (Fig. 2), the PCR measures strand invasion and strand extension by new DNA synthesis. Many of the *rad52* series of mutants (*rad51, rad52, rad54,* and *rad57*), for example, are deficient in creating this intermediate as assayed by PCR (8, 19). The PCR assay is a very sensitive method that can be employed when only small amounts of recombined product are present. Furthermore, PCR creates a PCR product of a discrete length and hence overcomes the difficulty of detecting a strand invasion structure that may be heterogeneous in size.

Primers are chosen to hybridize in the unique sequences outside the regions of shared homology such that a PCR product will be created only on formation of a recombination product or a strand invasion intermediate. This method appears to be most successful in detecting early intermediates of recombination when the amount of homology shared between the recombining substrates is small ($\sim$300 bp), as it is in the *MAT* locus. Otherwise, a high background may be detected because of template switching by DNA polymerase during the PCR cycles. With longer substrates it may be possible to insert a small nonhomology to which a primer will hybridize in the middle of one of the homologous regions that will undergo strand invasion and recombination. The usual precautions in selecting primers also apply. The primers should be 15–20 nucleotides long or long enough so that they hybridize to unique sequences and give only one end product that is less than 2 kb. Care should also be taken to avoid primers that anneal to one another or themselves. When selecting primers it may be useful to also select primers to serve as positive controls. This can be accomplished by choosing primers that anneal to the starting template prior to the induction of recombination.

If the PCR product is to be quantitated, it is prudent to establish whether there is a linear relationship between the amount of recombination product template and the final amplified PCR product. To determine this, mix the genomic DNAs from the unrecombined sample and from the product samples in various proportions. Amplified PCR products should increase in a linear manner with respect to the amount of the template. We have shown that the PCR reactions we have used are linear over a 10-fold range of product with a constant amount of total DNA (8).

The procedure for carrying out the PCR varies according to the polymerase used, the primers chosen, and the PCR instrument. Here we describe a protocol in which we used Pyrostase and a Techne (Cambridge, U.K.) PCR instrument (PHC-1) to amplify the PCR product of the pA and pB primers.

1. Aliquot the genomic DNA to 0.5-ml microcentrifuge tubes. For time-course DNA we use about 1% of the total in a 10-μl volume.
2. Prepare a master mix of buffer, deoxynucleoside triphosphates (dNTPs), and primers in 80 μl according to the manufacturer's specifications. A master mix is used to minimize pipetting errors. Pyrostase buffer conditions are 50 mM Tris, pH 9.0 (at 25°C), 1.5 mM MgCl$_2$, 20 mM (NH$_4$)$_2$SO$_4$, 0.2 mM each of dATP, dCTP, dGTP, and TTP, and 50 μg/ml bovine serum albumin (BSA).
3. Heat the samples at 100°C for 2–5 min. Transfer the tubes to a 72°C heating block.
4. Add 10 μl of diluted Pyrostase per tube in a 1× reaction buffer.
5. Overlay the samples with 100 μl of oil. Subject the samples to 25 cycles of 1.5 min at 95°C, 3 min at 48°C, and 3 min at 72°C followed by a final incubation of 7 min at 72°C.
6. After amplification, the products are extracted with equal volumes of phenol and chloroform, precipitated with 0.3 M sodium acetate and 2 volumes of ethanol, and run on an agarose gel. The gel is subsequently stained with ethidium bromide (0.5 μg/ml) for visualization of the PCR products.

Detection of Heteroduplexes by Denaturing Gel Electrophoresis

The following method is based on the denaturing gradient gel method developed by Fischer and Lerman (64) as modified by Lichten *et al.* (41). Short fragments of DNA (300–500 bp) containing single base pair mismatches will denature more easily than the same DNA sequences that are perfectly matched. Thus, when both heteroduplex and homoduplex DNAs migrate in gels containing a high concentration of urea, heteroduplex-containing DNA will be retarded relative to homoduplex DNA. An algorithim (MELT) to predict the effects of different mismatches on the denaturation of a known DNA sequence has been developed by Silverstein and Lerman; the program is available from Michael Lichten (Bldg. 37, Room 4D14, NIH, Bethesda, MD). This program is useful because the position of the mismatch within a particular restriction fragment strongly influences the course of denaturation. Thus, not all mismatches are easy to resolve from homoduplexes. Moreover, in only some cases can the two alternative heteroduplex molecules, formed by annealing strands of two homoduplexes differing by a single base pair, be resolved both from one another and from the homoduplex parent molecules.

For routine analysis of a particular heteroduplex, it is first necessary to determine the optimum concentration of denaturant that will lead to the maximum resolution of heteroduplex and homoduplex molecules. This is accomplished by pouring a polyacrylamide gel containing a gradient of urea, ranging from approximately 15 to 45% of a 7 M urea solution dissolved in 40% (w/v) formamide. The polyacrylamide gels can contain between 6.5% (37.5:1 ratio of monomer to bisacrylamide) and 11% acrylamide (49:1). The gradient gel is constructed by setting a Hoefer (San Francisco,

CA) SE600 gel pouring assembly on its side and inserting a syringe needle into a space at the (now upper) edge of what would normally be the top spacer. Once the gradient gel is cast, the gel is placed in its normal position, the spacer is removed, and a nondenaturing acrylamide loading gel is poured. The samples are then loaded, containing a synthetic mixture of denatured and reannealed DNA from two parent restriction fragments differing by the selected base pair. The gel is then run at 60° C and 150 V. Gels can be either dried or transferred to Nytran membranes (Schleicher & Schuell, Keene, NH) by electroblotting.

An autoradiograph of the gradient gel will establish the optimal concentration of urea to best separate heteroduplex-containing molecules from the homoduplex parents. Once this concentration is established, subsequent analysis can be carried out on DNA samples collected at intervals after initiating recombination, using nondenaturing gels with the selected concentration of urea.

Acknowledgments

We thank Susan Lovett and Evgeny Ivanov for comments on the manuscript. The work described in this review was supported primarily by National Institutes of Health Grants GM20056 and GM29736.

References

1. T. D. Petes, R. E. Malone, and L. S. Symington, *in* "The Molecular and Cellular Biology of the Yeast *Saccharomyces*" (J. R. Broach, J. R. Pringle, and E. W. Jones, eds.), Vol. 1, p. 407. Cold Spring Harbor Laboratory, Cold Spring Harbor, New York, 1991.
2. M. S. Meselson and C. M. Radding, *Proc. Natl. Acad. Sci. U.S.A.* **72,** 358 (1975).
3. J. W. Szostak, T. L. Orr-Weaver, R. J. Rothstein, and F. W. Stahl, *Cell (Cambridge, Mass.)* **33,** 25 (1983).
4. F.-L. Lin, K. Sperle, and N. Sternberg, *Mol. Cell. Biol.* **4,** 1020 (1984).
5. J. E. Haber, *Trends Genet.* **8,** 446 (1992).
6. R. E. Jensen and I. Herskowitz, *Cold Spring Harbor Symp. Quant. Biol.* **49,** 97 (1984).
7. I. Herskowitz and R. E. Jensen, *in* "Methods in Enzymology" (C. Guthrie and G. R. Fink, eds.), Vol. 194, p. 132. Academic Press, San Diego, 1991.
8. C. I. White and J. E. Haber, *EMBO* **9,** 663 (1990).
9. B. L. Ray, C. I. White, and J. E. Haber, *Mol. Cell. Biol.* **11,** 5372 (1991).
10. N. Rudin, E. Sugarman, and J. E. Haber, *Genetics* **122,** 519 (1989).
11. A. Ray, I. Siddiqi, A. L. Kolodkin, and F. W. Stahl, *J. Mol. Biol.* **201,** 247 (1988).
12. J. A. Nickoloff, J. D. Singer, M. F. Hoekstra, and F. Heffron, *J. Mol. Biol.* **207,** 527 (1989).
13. J. Fishman-Lobell, N. Rudin, and J. E. Haber, *Mol. Cell. Biol.* **12,** 1291 (1992).
14. N. Sugawara and J. E. Haber, *Mol. Cell. Biol.* **12,** 563 (1992).
15. A. Plessis, A. Perrin, J. E. Haber, and B. Dujon, *Genetics* **130,** 451 (1992).

16. J. C. Game, M. Bell, J. S. King, and R. K, Mortimer, *Nucleic Acids Res.* **18,** 4453 (1990).
17. B. A. Ozenberger and G. S. Roeder, *Mol. Cell. Biol.* **11,** 1222 (1991).
18. E. L. Ivanov, N. Sugawara, C. I. White, F. Fabre, and J. E. Haber, *Mol. Cell. Biol.* **14,** 3414 (1994).
19. N. Sugawara, E. L. Ivanov, J. Fishman-Lobell, B. L. Ray, X. Wu, and J. E. Haber, *Nature* **373,** 84 (1994).
20. A. Shinohara, H. Ogawa, and T. Ogawa, *Cell (Cambridge, Mass.)* **69,** 457 (1992).
21. T. Ogawa, Y. Xiong, A. Shinohara, and E. H. Egelman, *Science* **259,** 1896 (1993).
22. N. Rudin and J. E. Haber, *Mol. Cell. Biol.* **8,** 3918 (1988).
23. J. E. Haber and M. Hearn, *Genetics* **111,** 7 (1985).
24. M. F. Hoekstra, T. Naughton, and R. E. Malone, *Genet. Res.* **48,** 9 (1986).
25. B. Connolly, C. I. White, and J. E. Haber, *Mol. Cell. Biol.* **8,** 2342 (1988).
26. R. H. Borts, M. Lichten, and J. E. Haber, *Genetics* **113,** 551 (1986).
27. R. Padmore, L. Cao, and N. Kleckner, *Cell (Cambridge, Mass.)* **66,** 1239 (1991).
28. H. Sun, D. Treco, N. P. Schultes, and J. W. Szostak, *Nature (London)* **338,** 87 (1989).
29. B. de Massy and A. Nicolas, *EMBO J.* **12,** 1459 (1993).
30. H. Sun, D. Treco, and J. W. Szostak, *Cell (Cambridge, Mass.)* **64,** 1155 (1991).
31. L. Cao, E. Alani, and N. Kleckner, *Cell (Cambridge, Mass.)* **61,** 1089 (1990).
32. D. Zenvirth, T. Arbel, A. Sherman, M. Goldway, S. Klein, and G. Simchen, *EMBO J.* **11,** 3441 (1992).
33. M. Goldway, T. Arbel, and G. Simchen, *Genetics* **133,** 149 (1993).
34. M. Goldway, A. Sherman, D. Zenvirth, T. Arbel, and G. Simchen, *Genetics* **133,** 159 (1993).
35. T.-C. Wu and M. Lichten, *in* "Meiosis II: Contemporary Approaches to the Study of Meiosis" (F. P. Haseltine and S. Heyner, eds.), pp. 19–36. AAAS, Washington, D.C., 1993.
36. M. Lichten, R. H. Borts, and J. E. Haber, *Genetics* **115,** 233 (1987).
37. T.-C. Wu and M. Lichten, *Science* **263,** 515 (1994).
38. A. A. Cooper, Y. J. Chen, M. A. Lindorfer, and T. H. Stevens, *EMBO J.* **12,** 2575 (1993).
39. M. C. Bremer, F. S. Gimble, J. Thorner, and C. L. Smith, *Nucleic Acids Res.* **20,** 5484 (1992).
40. F. S. Gimble and J. Thorner, *Nature (London)* **357,** 301 (1992).
41. M. Lichten, C. Goyon, N. P. Schultes, D. Treco, J. W. Szostak, J. E. Haber, and A. Nicolas, *Proc. Natl. Acad. Sci. U.S.A.* **87,** 7653 (1990).
42. C. Goyon and M. Lichten, *Mol. Cell. Biol.* **13,** 373 (1993).
43. D. K. Nag and T. D. Petes, *Mol. Cell. Biol.* **13,** 2324 (1993).
44. L. Bell and B. Byers, *Anal. Biochem.* **130,** 527 (1983).
45. B. J. Brewer and W. L. Fangman, *Cell (Cambridge, Mass.)* **51,** 463 (1987).
46. L. S. Symington, *EMBO J.* **10,** 987 (1991).
47. I. Collins and C. S. Newlon, *Cell* **76,** 65 (1994).
48. A. Schwacha and N. Kleckner, *Cell* **76,** 51 (1994).
49. M. Ajimura, S.-H. Leem, and H. Ogawa, *Genetics* **133,** 51 (1992).
50. E. L. Ivanov, V. G. Korolev, and F. Fabre, *Genetics* **132,** 651 (1992).
51. M. Sym, J. A. Engebrecht, and G. S. Roeder, *Cell (Cambridge, Mass.)* **72,** 365 (1993).
52. D. X. Tishkoff, B. Rockmill, G. S. Roeder, and R. D. Kolodner, *Genetics* **139,** 489–494 (1995).

53. J. Sambrook, E. F. Fritsch, and T. Maniatis, *in* "Molecular Cloning: A Laboratory Manual." Cold Spring Harbor Laboratory, Cold Spring Harbor, New York, 1989.

54. J. A. Nickoloff, E. Y. Chen, and F. Heffron, *Proc. Natl. Acad. Sci. U.S.A.* **83,** 7831 (1986).

55. D. W. Russell, R. Jensen, M. J. Zoller, J. Burke, B. Errede, M. Smith, and I. Herskowitz, *Mol. Cell. Biol.* **6,** 4281 (1986).

56. R. F. Schleif and P. C. Wensink, *in* "Practical Methods in Molecular Biology." Springer-Verlag, New York, 1981.

57. F. Sherman, G. R. Fink, and J. B. Hicks, *in* "Methods in Yeast Genetics." Cold Spring Harbor Laboratory, Cold Spring Harbor, New York, 1983.

58. M. W. McDonnel, M. N. Simon, and F. W. Studier, *J. Mol. Biol.* **110,** 119 (1977).

59. T. Maniatis, E. F. Fritsch, and J. Sambrook, *in* "Molecular Cloning: A Laboratory Manual." Cold Spring Harbor Laboratory, Cold Spring Harbor, New York, 1983.

60. G. M. Church and W. Gilbert, *Proc. Natl. Acad. Sci. U.S.A.* **81,** 1991 (1984).

61. D. A. Melton, P. A. Krieg, M. R. Rebagliati, T. Maniatis, K. Zinn, and M. R. Green, *Nucleic Acids Res.* **12,** 7035 (1984).

62. A. P. Feinberg and B. Vogelstein, *Anal. Biochem.* **137,** 266 (1984).

63. S. Swillens, P. Cochaux, and R. Lecocq, *Trends Biochem. Sci.* **14,** 440 (1989).

64. S. G. Fischer and L. S. Lerman, *Proc. Natl. Acad. Sci. U.S.A.* **80,** 1579 (1983).

[12] Genetic Dissection of Complex Heteromultimeric Enzymes: The Case of Yeast DNA-Dependent RNA Polymerases

Pierre Thuriaux, Michel Werner, Sophie Stettler, and Dominique Lalo

Introduction

This chapter discusses some genetic methods currently used in *Saccharomyces cerevisiae* to analyse the organization of complex heteromultimeric yeast enzymes. On the basis of studies of yeast DNA-dependent RNA polymerases (E.C. 2.7.7.6), we consider the construction of null alleles, the isolation of conditional mutations from mutagenized allele libraries, the characterization of extragenic suppressors, and the use of genetic tests to monitor protein–protein association *in vivo.*

The properties of yeast RNA polymerases I, II, and III have been reviewed (1, 2). The enzymes dissociate into 14, 12, and 16 distinct subunits, respectively, after denaturing gel electrophoresis. They transcribe nonspecific DNA matrices *in vitro,* but depend on a variety of general transcription factors for faithful gene transcription. Almost all of the RNA polymerase genes have now been cloned (see Refs. 1, 2, and references therein). In agreement with earlier biochemical and immunological data, sequence comparison showed that yeast RNA polymerases have a closely related subunit organization. The two large subunits are similar in each of the three enzymes, and are related, respectively, to the β' and β subunits of bacterial enzymes. There are two small polypeptides shared by enzymes I and III (AC40 and AC19),[1] and five common to all three enzymes (ABC27, ABC23, ABC14.5, ABC10α, and ABC10β). Finally, each enzyme has a small number of specific subunits.

Host Strains and Vectors

Yeast Strains

Table I lists some polyauxotrophic strains bearing the *ura3, trp1, leu2, ade2, ade3, his3,* and *lys2* markers that are complemented by commonly used yeast/*Escherichia*

[1] Subunits are designated by the prefix A, B, or C (corresponding to enzyme I, II, or III) followed by a number giving their molecular mass as estimated by sodium dodecyl sulfate–polyacrylamide gel electrophoresis (SDS-PAGE). AC and ABC are used as prefixes for subunits shared by enzymes I and III, and by all three enzymes.

Methods in Molecular Genetics, Volume 6

TABLE I   Useful Host Strains of *Saccharomyces cerevisiae*

Strain	Genotype	Ref.
YNN281 (YPH52)	*MAT*a *ura3-52 his3-Δ200 trp1-Δ1 lys2-801 ade2-101*	*a*
YNN282 (YPH54)	*MAT*α *ura3-52 his3-Δ200 trp1-Δ1 lys2-801 ade2-101*	*a*
YPH102	*MAT*α *ura3-52 his3-Δ200 lys2-801 ade2-101 leu2-Δ1*	*a*
W301-1A	*MAT*a *ura3-1 his3-11,15 trp1-Δ1 lys2-801 ade2-101 leu2-3,112 can1-100*	*b*
W301-1B	*MAT*α *ura3-1 his3-11,15 trp1-Δ1 lys2-801 ade2-101 leu2-3,112 can1-100*	*b*
YPH499	*MAT*a *ura3-52 his3-Δ200 trp1-Δ63 lys2-801 ade2-101 leu2-Δ1*	*a*
YPH500	*MAT*α *ura3-52 his3-Δ200 trp1-Δ63 lys2-801 ade2-101 leu2-Δ1*	*a*
FL100	*MAT*a	*c*
FL200	*MAT*α	*c*
X2180-1A	*MAT*a *gal2*	*c*
S288C	*MAT*α *gal2*	*c*
Σ1278b	*MAT*a	*c*
A364A	*MAT*a *ade1 ade2 ura1 his7 lys2 tyr1 gal1*	*c*
Y367	*MAT*a *ade2 ade3 ura3 leu2-3,112 lys2 trp1*	*d*

[a] P. Hieter, C. Mann, M. Snyder, and R. W. Davis, *Cell (Cambridge, Mass.)* **40**, 381 (1985).
[b] B. J. Thomas and R. Rothstein, *Cell (Cambridge, Mass.)* **56**, 619 (1989).
[c] R. K. Mortimer and J. R. Johnston, *Genetics* **113**, 35 (1986).
[d] A. Bender and J. R. Pringle, *J. Mol. Cell. Biol.* **11**, 1295 (1991).

coli shuttle vectors. These strains have a complex pedigree, with a great deal of anisogeny (3). The ochre allele *ade2-101* (also called *ade2-1*) and the amber allele *lys2-801* occasionally lead to pseudoreversion by nonsense suppression, but all other markers are deletions (e.g., *trp1-Δ63*, *trp1-Δ1*, *his3-Δ200*, *leu2-Δ1*), insertions (*ura3-52*), or multiple point mutations (*his3-11,15*, *leu2-3,112*). The *trp1-Δ1* deletion (also called *trp1-Δ901*) extends over the upstream activating sequence (UAS) elements of the adjacent *GAL3* gene, which may affect experiments requiring a fast transcriptional induction in the presence of galactose. The *his3-Δ200* marker should also be avoided as it considerably increases the spontaneous rate of mitochondrial "petite" mutation, presumably to some interference with the neighbor gene *PET56*. *ade2*⁻ colonies accumulate a red pigment (whereas *ade2 ade3* double mutants are white), which forms the basis of a useful visual detection test to monitor plasmid loss (see plasmid shuffling assay below).

Vectors

A variety of shuttle vectors bearing *URA3*, *LYS2*, *LEU2*, *HIS3*, *ADE2*, or *TRP1* selectable markers (sometimes mutagenized to eliminate unappropriate restriction sites) have been constructed (4–6). Centromeric vectors are spontaneously lost at a rate of about 10% per generation (5). Consequently, a clone grown under nonselective growth conditions will be practically plasmid-free after 10 generations. Copy numbers range between 1 and a few molecules per cell. Episomal vectors based on the indigenous 2 μm plasmid have an average copy number of 10 to 40 per cell, with a stability comparable to centromeric plasmids. A variety of expression vectors allow a high and/or regulatable transcription (7).

Gene Inactivation

Gene inactivation determines whether a gene product is an essential cell component. It is also a prerequisite to the functional testing of mutagenized libraries and to the construction of conditional mutant strains. An inactivation protocol is given in Fig. 1 for the *RPC40*-encoded RNA polymerase subunit AC40, shared by enzymes I and III (8). Briefly, a selectable marker (*HIS3* in this case) is inserted by recombination *in vitro* to create the null allele *RPC40-Δ∷HIS3*. This allele is transferred by homologous recombination to one of the two *RPC40* copies of a suitable *his3⁻* diploid host. Heterozygous transformants, selected by prototrophy for uracil, are induced to sporulate, and the corresponding asci are microdissected (9). A 2:2 segregation of germinating versus nongerminating spores (the former cosegregating with histidine auxotrophy) indicates a gene encoding an essential gene product. As a control, a heterozygous diploid is transformed by a plasmid-borne copy of the wild-type gene, using *TRP1* as selectable marker, which yields viable haploids containing the interrupted chromosomal allele rescued by the plasmid-borne wild-type gene, and thus displaying a double prototrophy for histidine and tryptophan. These strains are unable to lose the *TRP1* plasmid bearing the wild-type *RPC40* gene, as this would be a lethal event. This result forms the basis of the plasmid shuffling assay described in the next section.

Comments

Inactivation is ideally achieved by deleting the coding sequence between two ad hoc restriction sites located (or created by oligonucleotide mutagenesis) at the amino-terminal and carboxyl-terminal ends, and replacing it with an insert bearing an appropriate selective marker. Natural restriction sites can also be used to create a partial deletion, or even a mere insertion interrupting the coding sequence at a unique restriction site, located as near as possible to the amino-terminal end. In the latter case, it is advisable to use inactivating cartridges flanked by nonsense triplets put in all possible frames (10). The possibility of a spurious viable phenotype arising from a partly functional truncated gene product cannot be ignored. For example, a deletion replacing the amino-terminal half of *RPC53* (encoding the 53-kDa subunit of enzyme III) by a *HIS3* insert also containing the amino-terminal end of *PET56*

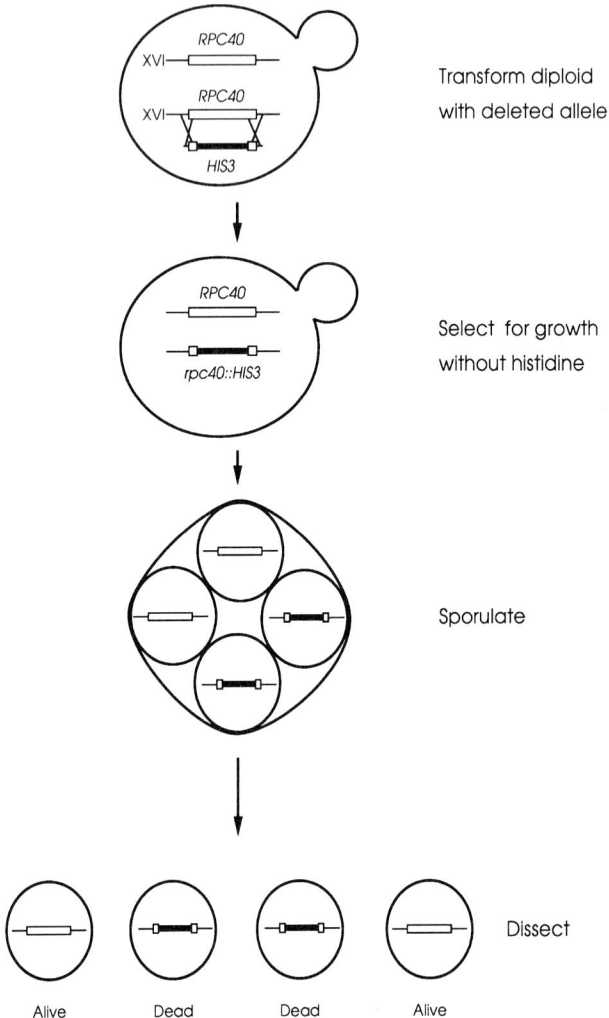

FIG. 1   Deletion of *RPC40* gene. A diploid *his3⁻/his3⁻* strain is transformed with a linear fragment bearing an allele of *RPC40* in which the deleted part of the gene has been replaced by the selectable marker *HIS3*. The heterozygosity of the His⁺ transformants at the *RPC40* locus is checked by Southern blotting. The *RPC40/rpc40::HIS3* diploids are sporulated, asci are dissected, and individual spores are germinated. The presence of two viable spores, which are invariably His⁻, and two dead spores indicate that the deleted gene is essential.

happened to fuse the first 75 codons of the latter gene to the last 165 codons of *RPC53*. This in-frame fusion had partial gene activity leading to a thermosensitive (ts) phenotype, whereas a complete deletion was lethal (11).

Many gene-lethality reports are merely based on the systematic segregation of two viable, auxotrophic spores in the meiotic offspring, without a control showing the complementation of that lethal phenotype by a plasmid-borne wild-type copy of the disrupted gene. Yet, this control is important to rule out spurious cis-acting effects on the neighbor gene (especially when using deletions that extend somewhat beyond the target coding sequence itself) and to identify possible cases where a null allele prevents spore germination but not vegetative growth.

Targeting the inactivated allele to the chromosome requires a homologous gene conversion event between the chromosomal gene (acting as recipient DNA) and a linear fragment bearing the interrupted gene (acting as donor DNA). Successful conversion requires the donor DNA fragment to share about 250 bp with the target gene on each side of the interrupting cartridge. However, the few base pairs of terminal nonhomology often generated by the presence of additional restriction sites on the donor DNA do not prevent homologous recombination. Some constructions are prone to integrate elsewhere than at the target gene (despite the conventional wisdom that recombination is mostly due to homologous exchanges in *S. cerevisiae*), and the structure of the heterozygous diploid should therefore be confirmed by Southern genomic hybridization.

Most of the commonly used yeast vectors (4–6) bear suitable inactivating cartridges that are often cured of unsuitable internal restriction sites, and some are endowed with nonsense triplets inserted in all frames to prevent read-through effects (10). The choice of the inactivating cartridge will affect subsequent genetic strategies because the wild-type gene marking the disruption prevents further use of the corresponding auxotrophic marker. *URA3* cartridges flanked by directly repeated DNA segments can be eliminated from the target gene by unequal crossovers, thus leaving a single repeat to inactivate the target gene and allowing reuse of the *ura3* auxotrophic marker in further genetic constructions (12).

Application to Yeast RNA Polymerases

The 25 RNA polymerase subunit-encoding genes isolated so far in *S. cerevisiae* were all inactivated. In most cases, null alleles had a lethal phenotype, indicating that the corresponding subunit is strictly required for growth and, by extension, plays an essential role in the transcription process [conditional mutations provide a definitive genetic proof of the latter point if they can be shown to affect specifically the relevant RNA polymerase(s)]. Full inactivation of B32 and B12.6 in enzyme II, and of A49 and A12.2 in enzyme I, led to a conditionally lethal phenotype, implying that the corresponding subunits have but an optional role in the transcription machinery.

Moreover, inactivating subunit A34 has little effect on growth. In contrast, all the components of RNA polymerase III cloned so far encode an essential gene product (see Refs. 1, 2 and references therein).

Allele Libraries and Conditional Mutants

Figure 2 describes the functional screening of an allele library prepared by *in vitro* mutagenesis of an *RPC40* plasmid, the isolation of conditional alleles, and their conversion into a chromosomal allele. The library, borne on a yeast *URA3* centromeric vector, is transferred to a haploid host where the chromosomal *rpc40-Δ::HIS3* null allele (8) is complemented by a wild-type *RPC40* gene borne on a *TRP1, SUP11-1* plasmid. Because the *SUP11-1* marker (encoding an ochre-specific informational suppressor) confers a white pigmentation to the normally red *ade2⁻* host, a "plasmid shuffle" assay (4) can be applied to individual transformants to determine whether a given *in vitro*-generated mutant allele can support growth in the absence of the wild-type gene. The ability of a given transformant to lose the *SUP11-1* plasmid carrying the wild-type *RPC40* gene is monitored by its ability to form red-pigmented sectors in a white colony at various temperatures. A failure to lose the *SUP11-1* plasmid indicates that this $rpc40^{ts}$ mutation (borne on the remaining *URA3* plasmid) is nonfunctional at the temperature considered. This mutation can be converted into a chromosomal allele by spontaneous gene conversion directed toward the chromosomal *rpc40-Δ::HIS3* null allele. As shown in Fig. 3, this conversion replaces the chromosomal *HIS3* deletion by the $rpc40^{ts}$ allele, therefore relieving the host cell from the necessity of keeping the initial *URA3* plasmid, and thus allowing the formation of spontaneous subclones that are auxotrophic for uracil (owing to plasmid loss) and for histidine (owing to the initial conversion event). The corresponding strains are easily selected by streaking on FOA medium (4) and checking for histidine auxotrophy.

Comments

Randomly mutagenized allele libraries of a given gene can be produced by hydroxylamine (NH_2OH)-mediated *in vitro* mutagenesis of that gene on an appropriate shuttle vector, which essentially generates G to A transitions (see Ref. 4 for a detailed protocol). Typically, about 5% of the defective mutants have a ts phenotype, but no clear-cut ts mutation was obtained in some genes in spite of intensive screening (13). A highly effective $NaHSO_3$ mutagenesis of single-stranded DNA was reported for the AC40 subunit common to RNA polymerases I and III, but the alleles obtained were almost exclusively due to multisite substitutions (8). Linker scanning (14) and multisite replacement of acidic residues by alanine (15) have been used with some

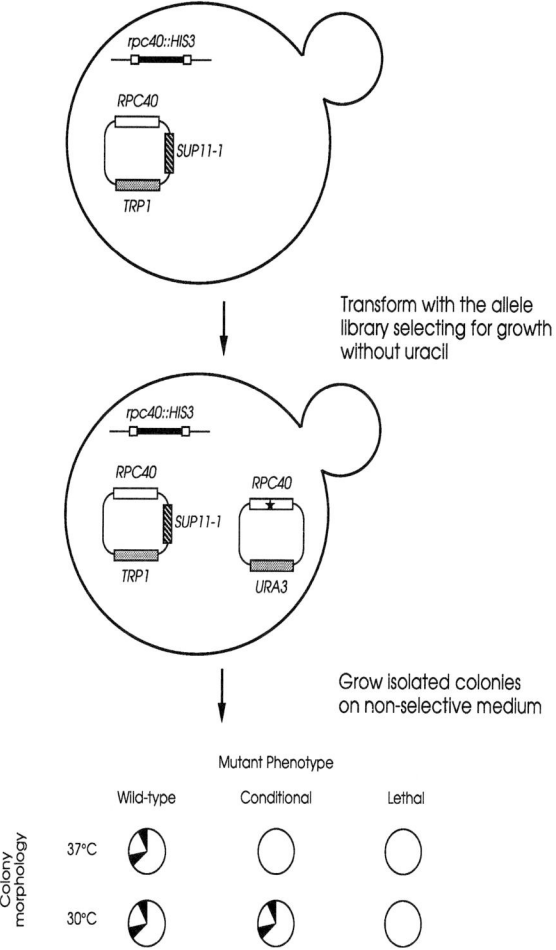

Fig. 2   Shuffle assay of allele libraries. A recipient strain (DLY11: *ura3-52 lys2-801 trp1-Δ1 ade2-1 his3-D200 rpc40-Δ::HIS3 [DLp01: CEN4 RPC40 TRP1 SUP11-1]*) is transformed with mutagenized *RPC40 URA3* plasmids. Transformants containing the plasmid bearing the wild-type *RPC40* gene are white because they also bear a *SUP11-1* ochre suppressor. *SUP11-1* suppresses the Ade⁻ phenotype and red color of *ade2* colonies. When transformants containing both plasmids are grown on nonselective medium, the wild-type *RPC40 TRP1 SUP11-1* plasmid can be lost if the mutagenized *RPC40* allele can complement the *rpc40-Δ::HIS3* chromosomal deletion. In this case white colonies with red sectors will appear. Thermosensitive *rpc40ᵗˢ* alleles will lead to sectored colonies at the permissive temperature only, whereas lethal alleles will lead only to white colonies.

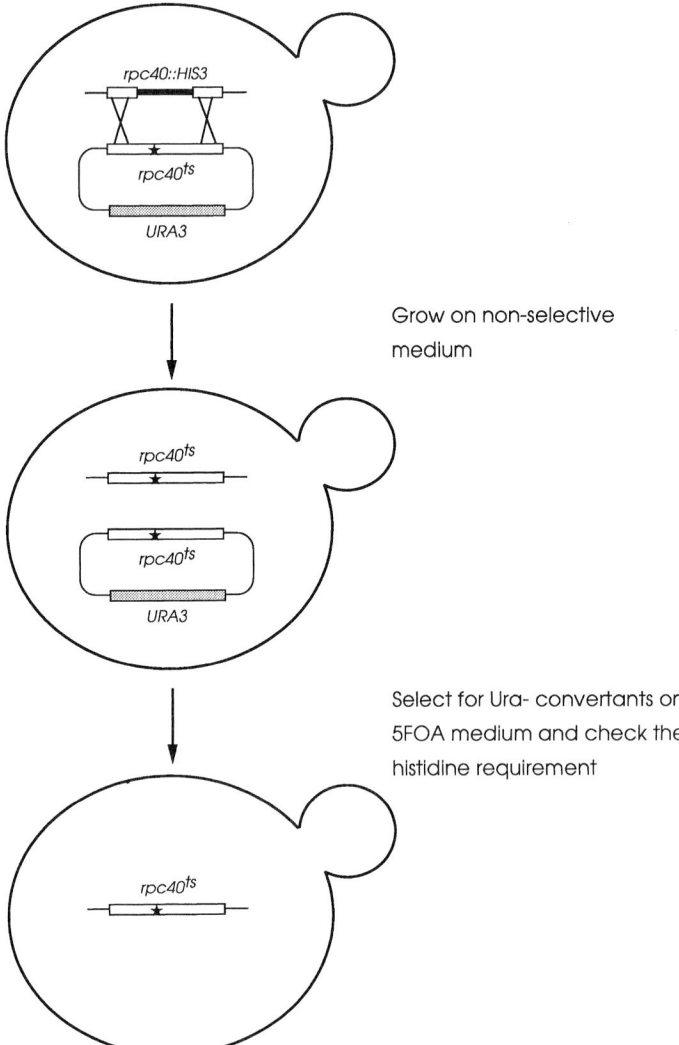

Grow on non-selective
medium

Select for Ura- convertants on
5FOA medium and check the
histidine requirement

FIG. 3   Replacement of a deleted *rpc40 :: HIS3* chromosomal allele by a mutant *rpc40ts* allele.
A chromosomal *rpc40 :: HIS3* deletion can be converted to a thermosensitive *rpc40ts* allele by
growing, on nonselective medium at the permissive temperature, a strain in which the latter
allele is borne on a plasmid. In a small proportion of the cells, the chromosomal allele will
undergo gene conversion. Such events can be selected on FOA medium because only clones
that have lost the *URA3* plasmid and have a functional allele of *RPC40* can grow. Such convertants will be His⁻ because the *rpc40 :: HIS3* deletion has been replaced by the *rpc40ts* allele
lacking the *HIS3* gene.

success to generate ts mutations. Site-directed mutagenesis of conserved domains also yielded conditional mutations, which were often due to conservative substitutions at or near invariant residues.

Temperature-sensitive mutations are not exclusively associated with base-pair substitutions. As noted above, gene deletions removing the B32 subunit of RNA polymerase II and the A49 or A12 subunits of RNA polymerase I led to a conditional growth phenotype. The partial truncation of the carboxyl-terminal end of the B220 subunit of RNA polymerase II (16, 17) and the C31 subunit of RNA polymerase III (17a) was successfully used to generate ts mutants for these subunits. In both cases, the C-terminal fragment was highly repetitive and negatively charged (owing to phosphorylation of B220, and to a high proportion of acidic residues in the case of C31). Its partial deletion therefore altered the charge distribution at the surface of the cognate enzyme. A dramatic case of deletion with a ts phenotype was obtained by removing the entire amino-terminal half of the C53 subunit of RNA polymerase III (11). This phenotype was strongly dependent on the host genetic background (18).

In Fig. 2, the loss of the plasmid bearing the wild-type gene is monitored by the simple and inexpensive assay based on the white pigmentation of normally red $ade2^-$ colonies when they harbor a plasmid with the $ADE2^+$ allele or the ochre suppressor allele $SUP11-1$. However, white colonies can also arise from spontaneous mutants blocked in one of the five enzymatic steps leading to pigment accumulation, or from mitochondrial petites (the latter are quite frequent in certain strains but are easily identified by the failure to grow on media with glycerol as carbon source). One way of overcoming this problem is to use $ade2$, $ade3$ host strains that are white but acquire a red pigmentation when complemented by an $ADE3^+$ plasmid. A convenient but more expensive screen is based on the capacity to lose a $URA3$ vector: $ura3^-$ clones having lost the $URA3$ plasmid are selected as resistant colonies after replica plating transformants to (FOA) medium (4). Mitochondrial petite clones can also lead to spurious 5-Fluoro-orotic acid FOA resistance. Other screens are based on resistance phenotypes conferred by plasmids bearing $CAN1$, $CYH2$, or $LYS2$ inserts (4, 8).

Conditional mutations affecting a heteromultimeric enzyme may affect enzyme function, assembly from individual subunits, or the stability of quaternary structures. These three levels are not mutually exclusive. In the case of stable enzymes, mutations that affect assembly with little effect on activity will give rise to mutants with delayed kinetics of inactivation, since the preformed enzyme retains its activity until it is diluted out by growth. In contrast, functionally defective mutants (or mutants generating a highly unstable enzyme) should have a rapid phenotypic response. They are, however, conspicuously rare for RNA polymerase genes, which might well be a common pattern in the case of heteromultimeric protein complexes.

Plasmid-borne alleles are heterogeneous in terms of copy number per cell, even in the case of centromeric plasmids. In some cases, the copy number heterogeneity leads to phenocopies, that is, to pseudorevertant clones that do not breed true on further reisolation, which may reflect a gene-dosage dependency of conditional alleles producing proteins defective in assembly and/or stability of the enzyme quaternary structure.

In yeast, conditional mutants are typically isolated at 37° C, but somewhat higher (38° C) or lower (35° C) temperatures have also been used. Our experience with osmotic remediality or sensitivity to deuterated water (see Ref. 4 and references therein) was frustrating, owing to a large number of false-positive isolates. Selecting for cold-sensitive (cs) mutants (unable to grow at 16°–20° C) was quite effective, but about half of the mutants also had a ts phenotype. It is sometimes assumed that cs mutants are preferentially altered in the assembly of multiprotein complexes, but this property is not exclusive to cs mutants, as there are several examples of ts RNA polymerase mutants specifically affecting enzyme assembly (8, 19). Finally, a simple and useful way of conditionally expressing a given RNA polymerase subunit is to put its transcription under the control of a tightly regulatable promoter such as the *GAL1-10* promoter (7). Unlike the use of ts mutations, this provides no information on the structure–function relationship of the protein considered.

Application to Yeast RNA Polymerases

Conditional or leaky mutations have been isolated for most of the RNA polymerase genes cloned so far, and all are invariably defective (*in vitro* and/or *in vivo*) in the cognate RNA polymerase I, II, and/or III activity. Thus, the corresponding subunits functionally belong to these enzymes, in keeping with earlier biochemical and immunological evidence (1, 2). Such conditional mutations were used to demonstrate that the 35 S rRNA is the only transcript (or at least the only essential transcript) requiring RNA polymerase I activity for its synthesis (20). Similarly, conditional RNA polymerase III mutations were used to establish that U6 and the RNA component of RNase P are RNA polymerase III transcripts (see Refs. 1, 2, and references therein).

Some of the constructions obtained by *in vitro* mutagenesis of RNA polymerase subunits were analyzed in more detail. Mutations affecting lysine residues of domains **H** and **I** on the second largest subunit of RNA polymerase II (B150) allowed us to identify which lysyl groups are affinity-labeled by nucleotide derivatives and established that the two domains are in close spatial contact and participate in the active site (21). A mutation in a nonconserved region of the corresponding subunit of enzyme III specifically affected the termination process (22). Mutants of the largest subunit (B220) displaying an increased sensitivity to 6-azauracil (thus possibly defective in transcript elongation) were clustered in a region immediately downstream of the conserved domain **g** (23). As many of the conditional RNA polymerase mutants appear to be defective in the assembly or stability of the cognate enzyme(s), their detailed characterization should help to clarify the assembly pathway(s) of the three enzymes. In the case of enzyme II, epitope-tagged subunits were used to show that the B44 subunit combines with B150 and then with B220 as part of the assembly pathway (19). Finally, a mutation leading to an increased lability of enzyme III mapped in the conserved zinc-binding domain of the largest subunit, suggesting an essential role of that domain in stabilizing the quaternary structure of RNA polymerases (24).

Extragenic Suppression

Extragenic suppressors can be isolated from conditional mutants as spontaneous sub-clones, after *in vivo* mutagenesis, or by transformation with a genomic library borne on a high copy number vector. A symmetric approach, not dealt with in this chapter, is to isolate secondary mutations enhancing the deleterious phenotype of the conditional allele by a colethal effect. These techniques are powerful ways to identify proteins that "genetically interact" with the product of the conditional allele, keeping in mind that such interactions can be due to a direct protein–protein association (thus providing a tool to analyze the spatial organization of heteromultimeric complexes) but may also reflect less direct functional relationships.

Suppressor mutations can be isolated from a mutagenized cell population as rare colonies growing at the restrictive temperature (genuine revertants appear first, whereas suppressor mutants may be delayed owing to a partial suppressor effect). UV mutagenesis is generally used because of its minimal health hazard, although chemical mutagenesis with ethyl methane sulfonate is more efficient (see Ref. 25 for mutagenesis protocols with discussion of safety aspects). Spontaneous suppressors should be derived from carefully reisolated subclones of the suppressible mutant strain, to ensure their independent origin. Backcrosses to an isogenic wild-type strain will distinguish between extragenic suppressors and intragenic ones or true revertants, taking advantage of tetrad analysis based on the microdissection of meiotic asci (see Ref. 11 for experimental details). Tetrad analysis also allows the identification of pseudosuppressor effects (26) due to duplications of the mutant alleles, which lead to a partial correction of the mutant phenotype by increased gene dosage.

Multicopy suppressors are isolated by transformation of the mutant yeast strain with an appropriate multicopy library followed by plating on selective medium at the restrictive temperature. Transformation must be efficient enough to reconstitute several times the total yeast genome. The yield in suppressors may be improved by incubating the plates overnight at the permissive temperature immediately after the transformation procedure. Several genomic libraries are available on multicopy vectors of *S. cerevisiae* (see Ref. 18 for experimental details of a library construction). Yeast libraries put under the control of a strong inducible promoter (7) are also available.

Comments

In some cases, up to one-quarter of the putative multicopy suppressor clones initially selected at the restrictive temperature were false-positive. That could not be confirmed by recovering the corresponding plasmids in *E. coli* and transforming back to the yeast ts mutant, nor by isolating spontaneous subclones cured from the transforming plasmid and checking for the restoration of the initial ts phenotype. Both controls are therefore crucial before concluding that a bona fide multicopy suppressor has been identified.

Extragenic suppression may be due to cryptic suppressor alleles harbored in some laboratory strains and revealed by crosses or genetic transformation involving anisogenic transfer of DNA. Thus, a presumptive multicopy suppressor of RNA polymerase mutants turned out to operate independently of its copy number and to be caused by the $SS\Delta1$-v allele present in the FL100 strain used to prepare the multicopy library (18). This allele suppresses a surprisingly large number of ts mutations, possibly because of an altered phosphorylation pattern of the transcription machinery.

Extragenic suppressor mutations usually have no selectable phenotype in their own right, and they must therefore be cloned via a genomic library prepared from the suppressor strain itself (see Ref. 6 for experimental details) and screened for the capacity to restore growth on transformation in the ts mutant. This only allows the cloning of dominant or partially dominant suppressor alleles. Given the rapid progress made in systematic chromosome sequencing, it may be advisable to invest some time in standard genetic mapping of the suppressor allele, as this could allow the direct cloning of its wild-type gene from preexisting libraries (see Ref. 27 for mapping strategies of mutations corresponding to uncloned genes).

Because mutational suppressors alter the sequence of the suppressor gene, they may operate by compensating the structural change in a given subunit (initially introduced in a given subunit by the suppressible allele) by altering another subunit of the same complex. Accordingly, extragenic suppression may identify which subunits contact one another in a given heteromultimeric edifice. Moreover, it may provide clues about the spatial organization of the two gene products, if, as is arguable, the residues corresponding to the suppressed and suppressible alleles are in spatial contact (28, 29). Multicopy suppressors, which act by overexpression of their wild-type gene product rather than by altering its structure, may reflect a direct protein–protein interaction by a mechanism involving the correction of an assembly defect via a mass action effect favored by the local increase in the concentration of a partner subunit (see Table II).

Suppression may also operate by functional rescue, for example, in cases where the suppressor allele allows the cell to recruit another gene product with a similar function, instead of directly correcting the defect of the suppressible allele. Similarly, the physiological consequences of the initial defect may be bypassed by the suppressor allele, even though the initial cellular defect would not be affected. Thus,

TABLE II   Extragenic Suppression of Conditional Mutations Affecting Yeast RNA Polymerases

Suppressible allele	Suppressor allele	Hypothetical suppression mechanism	Ref.
Mutations affecting two largest subunits			
rpa190-1, rpa190-5	rap135-C1127R	Interaction between two largest subunits through zinc-binding domains	a
	SRP1-1	Interaction of RNA polymerase I with nucleolar structure	b
rpb1-1, rpb1-5	rpb2-510, rpb2-513	Domain-to-domain interaction between large subunits of RNA polymerase II	c
rpo21-4	RPB6 (multicopy)	B220–ABC23 association during RNA polymerase II assembly	d
rpo21-23 (and others)	PPR2 (multicopy)	Correction of elongation defect by over-expression of general transcription factor SII	e
rpb1-Δ101	SRB2-1	Dominant suppressor, specific to CTD domain; possibly encoding new (general) transcription factor of RNA polymerase II	f
rpb1-Δ131 and rpb1-Δ128	spt2-Δ (SPT2 = SIN1)	Weak interaction with CTD domain of B220 (note: allele specificity versus other rpb1 mutations unknown)	g
rpc160-41 (and others)	SSD1-v	Modified phosphorylation pattern of cognate RNA polymerase	h
rpb1-1	kex2-1 (SRB1 = KEX1)	Nonspecific suppression due to gener-ally improved thermoresistance of kex2⁻ strains at 38°C	i
rpc160-209 (and rpc31-236)	FLH1	Bypass of slower rRNA maturation in RNA polymerase III mutants	j
Mutations affecting common subunits			
rpc40-V78R	RPC19, RPB10 (multicopy)	AC40–AC19–ABC10β association during RNA polymerase I and III assembly	k
rpc40-V78R	SRS1 (multicopy)	Interaction with serine-rich protein, pos-sibly nucleolar	k
rpc19-G73D	RPC40, RPB10 (multicopy)	AC40–AC19–ABC10β association during RNA polymerase I and III assembly	k
Mutations affecting enzyme-specific subunits			
rpc82-6	RPC31 (multicopy)	C31–C82 association during enzyme assembly	l
rpc31-236	RPC160 (multicopy)	C31–C60 association during enzyme assembly	l
rpc31-236	DED1 (multicopy)	Functional bypass by overexpression of RNA or DNA helicase activity	h
rpc31-236	UBI4 (multicopy)	Modification of chromatin structure	h
rpa12-1::LEU2	RPA190 (multicopy)	A12–A190 association during enzyme assembly	m

continued

TABLE II   *(continued)*

[a] R. Yano and M. Nomura, *Mol. Cell. Biol.* **11**, 754 (1991).

[b] R. Yano, M. Oakes, M. Yamigishi, J. A. Dodd, and M. Nomura, *Mol. Cell. Biol.* **12**, 5640 (1992).

[c] C. Martin, S. Okamura, and R. A. Young, *Mol. Cell. Biol.* **10**, 1908 (1990).

[d] K. F. Wertman, D. G. Drubin, and D. Botstein, *Genetics* **132**, 337 (1992).

[e] J. Archambault, F. Lacrourte, A. Ruet, and J. D. Friesen, *Mol. Cell. Biol.* **12**, 4142 (1992).

[f] A. J. Koleske, S. Buratowski, M. Nonet, and Young, *Cell (Cambridge, Mass.)* **69**, 883 (1992).

[g] C. L. Peterson, W. Kruger, and I. Herskowtiz, *Cell (Cambridge, Mass.)* **64**, 1135 (1991).

[h] S. Stettler, N. Chiannilkulchai, S. Hermann-Le Denmat, D. Lalo, F. Lacroute, A. Sentenac, and P. Thuriaux, *Mol. Gen. Genet.* **239**, 169 (1993).

[i] C. Martin and R. A. Young, *Mol. Cell. Biol.* **9**, 2341 (1989).

[j] S. Hermann-Le Denmat, M. Werner, A. Sentenac, and P. Thuriaux, *Mol. Cell Biol.* **14**, 2905 (1994).

[k] D. Lalo, C. Carles, A. Sentenac, and P. Thuriaux, *Proc. Natl. Acad. Sci. U.S.A.* **90**, 5524 (1993).

[l] P. Thuriaux and A. Sentenac, *in* "The Molecular Biology of the Yeast *Saccharomyces*" (J. R. Broach, J. R. Pringle, and E. W. Jones, eds.), Vol. 2, p. 1. Cold Spring Harbor Laboratory, Cold Spring Harbor, New York, 1992.

[m] Y. Nogi, R. Yano, J. Dodd, C. Carles, and M. Nomura, *Mol. Cell. Biol.* **13**, 114 (1993).

mutations reducing the synthesis of a given cellular component may be suppressed by increasing the metabolic stability of that component.

Application to Yeast RNA Polymerases

Table II summarizes the mutational or multicopy suppressor effects observed on conditional RNA polymerase I, II, or III mutants. Several of the suppressors define RNA polymerase subunits, suggesting direct subunit interactions. Thus, extragenic suppression between the conserved zinc-binding domains on the N terminus of the largest subunit and the C terminus of the second largest subunit in enzyme I (29) would be consistent with a close spatial contact of these two domains through a "head-to-tail" connection between the two subunits.

The occurrence of multicopy suppression between various RNA polymerase subunits suggests that the corresponding polypeptides are associated during the assembly of the cognate enzyme(s) (2, 30). In keeping with results obtained by the two-hybrid method (see below), the data point to an association between the AC40 and AC19 subunits (common to enzymes I and III), themselves interacting with ABC10β, one of the five subunits shared by all three enzymes (30). Moreover, the C82, C31, and C34 subunits of enzyme III appear to be closely associated with one another and to interact with C160, the largest subunit of that enzyme (31). Because RNA polymerases combine with a variety of general transcription factors to achieve faithful transcription, extragenic suppression effects may also occur between some of the factors and RNA polymerase subunits. Thus, the *PPR2* gene, which encodes an elongation factor of RNA polymerase II, operates as a multicopy suppressor of 6-azauracil-sensitive mutations of the B220 subunit (23). Similarly, a conditional mutation affecting the TBP transcription factor was sensitive to multicopy suppression by the 70-kDa subunit of the general transcription factor TFIIIB of RNA polymerase III (32, 33).

Several extragenic suppressors isolated from conditional RNA polymerase mutations define genes that do not code for another subunit. They are still poorly characterized, but some of them, at least, may directly interfere with the properties of the transcriptional machinery. The dominant *SSΔ1*-v allele (18) may affect RNA polymerase phosphorylation. The strong multicopy suppressor effect of the *DED1* gene (encoding a putative RNA or DNA helicase of unknown function) might reflect a functional bypass of an elongation phenotype associated with the *rpc31-236* allele (18), even though the *DED1* gene product does not appear to be itself a functional component of the RNA polymerase III-dependent transcription machinery. The dominant *SRB2-1* allele (34) and the *SPT2* gene (35) suppressed truncated forms of the CTD domain of the RNA polymerase largest subunit (B220), and they presumably encode RNA polymerase II transcription factors. On the other hand, the weak suppressor effect of the *FHL1* gene (belonging to the fork-head protein family of transcriptional activators and stimulating the maturation of ribosomal RNA) is presumably by functional rescue, as RNA polymerase III mutants are themselves slightly defective in rRNA maturation (36). Other weak extragenic suppressors listed in Table II presumably have but an indirect effect on transcription, such as the ubiquitin-encoding gene *UBI4* (a multicopy suppressor of *rpc31-236;* Ref. 28) or *SRP40,* coding for an exceptionally serine-rich protein with a multicopy suppressor effect on *rpc40-V78R* (30). Moreover, the suppressor effect of *kex2*⁻ on ts RNA polymerase II mutations at 38°C might well be due to a rather unspecific effect, because *kex2*⁻ mutations also improve the thermoresistance of a wild-type control (37).

Two-Hybrid System for Studying Protein Complexes

The two-hybrid system (38) provides a convenient tool to study the *in vivo* physical interaction between two proteins. The method relies on the observation that RNA polymerase II transcription activators are composed of an activation domain and a DNA-binding domain operating independently of one another. The domains need not be on the same polypeptide for activation to take place, provided that they are fused to proteins which interact, creating a heterodimeric activator. This observation forms the basis of an *in vivo* assay where the activation of a reporter gene is used as a tool for assessing the interaction between two proteins fused, respectively, to the DNA-binding and to the activation domains of a transcription activator.

The system is based on fusions with domains of the Gal4p yeast transcription activator. Amino acids 1 to 147 of Gal4p define a domain able to bind to the UAS of galactose-regulated genes. Amino acids 768 to 881 define an autonomous activation domain. The proteins under study are fused at the C terminus of Gal4p (1–147) in pMA424 (39) or pGBT9 (P. L. Bartel and S. Fields, personal communication) and Gal4p (768–881) in pGAD2F (40) or pGAD424 (Bartel and Fields, personal communication). pGBT9 and pGAD424 are smaller, provide more unique sites usable for protein fusions, and are produced at a higher copy number in *E. coli*. The transcription of the fusion genes is driven by the strong constitutive yeast *ADH1* promoter. An *ADH1* terminator is placed downstream of the multiple cloning site to avoid interference with transcription of the yeast markers. Because transcription factors have to be targeted to the nucleus, the fusion proteins need a nuclear localization signal (NLS). The DNA-binding domain of Gal4p bears such an NLS, whereas the SV40 NLS had to be N-terminally added to the Gal4p activation domain (40). When the two constructions are harbored by an appropriate host strain (e.g., Y526) bearing *lacZ* under the control of the *GAL1* promoter, the association of the two fusion proteins generates a functional trans-activator, as monitored by a colony color assay.

Comments

The most convenient way to generate the protein fusions is to prepare a polymerase chain reaction (PCR) fragment encompassing the whole open reading frame (ORF). The 5′ restriction site can be placed just upstream of the start codon of the ORF by amplification with oligonucleotides carrying the appropriate sites. For ORFs too large or too difficult to amplify, oligonucleotide-directed mutagenesis may be used to introduce convenient sites for the construction of translation fusions. The constructions are then transferred into yeast by any standard transformation protocol selecting for the yeast marker present on the fusion vector. The simple lithium acetate protocol works well with the tester strain Y526 (Table I) and allows cotransformation of two test plasmids using a double selection strategy. However, the higher transformation efficiencies attainable with electroporation or spheroplasting methods may be required when screening fusion banks (see below). The *GAL4* gene of the tester strain has been deleted because Gal4p would otherwise activate the reporter gene. The *GAL80* gene has also been deleted because it prevents the induction of the reporter gene on media lacking galactose. We have tested various carbon sources in solid media for induction of the reporter gene and found that raffinose gives the best results while allowing fast growth.

β-Galactosidase induction can be monitored on plates by overlaying the transformants grown on raffinose with 10 ml X-Gal agar [0.5% agar, 500 mM potassium phosphate buffer, pH 7.0, 0.1% (w/v) sodium dodecyl sulfate (SDS), 7% *N,N*-dimethylformamide, 0.4 mg/ml 5-bromo-4-chloro-3-indolyl-β-D-galactopyranoside] and incubating the plates for 24 hr at 30°C. It is commonly observed that when two proteins interact, 5 to 10% of the transformants remain white, probably owing to the excision of the reporter gene by homologous recombination between the tandem *URA3* genes. In some cases where the strong expression of the fusion protein is toxic (e.g., with the TATA-binding protein), the number of false negatives may reach 50%.

Interactions deduced from observation of plates can be quantitatively assessed by measuring β-galactosidase activity in cellular extracts. Strains bearing the wild-type *GAL4* activator give specific activities of about 10,000 units, whereas the background is less than 1 unit β-galactosidase per milligram protein. Constructions giving a distinct but weak blue color have specific activities as low as 3 units, which probably correspond to weak or transient associations of disputable biological relevance. In contrast, interactions between RNA polymerase III subunits give rise to specific activities in the range of 50–400 units. Constructions giving lower β-galactosidase levels should therefore be taken with some caution (especially since the vectors used are multicopy and thus may lead to protein overexpression). They may nevertheless correspond to transient but biologically relevant associations between two protein partners.

A protein tested in the two-hybrid system must be checked for its ability to activate transcription by itself, as there are several examples of proteins that, when fused to the Gal4p DNA-binding domain, activate transcription because they bear an activation domain. Conversely, a protein fused to the Gal4p activation domain might have some affinity for the *GAL1* promoter and thus activate *lacZ* transcription. A Gal4p activation domain fusion protein may also directly interact with the Gal4p DNA-binding or activation domains rather than with the polypeptide fused to these domains. Therefore, Gal4p DNA-binding domain fusions must be tested against an unfused Gal4p activation domain, and conversely, before concluding that a genuine protein interaction between the two fusion polypeptides has been observed.

A failure to generate a functional heterodimer may reflect the instability of the corresponding fusion proteins. This problem may be circumvented by testing the interactions in the two reciprocal constructions, which may also detect potential problems stemming from steric hindrance by one of the two Gal4p domains. However, constructions giving a negative result in both reciprocal assays must be tested for the expression of the fused protein before concluding that no interaction occurs.

Genetic interactions revealed by multicopy suppression analysis was not necessarily detected by the two-hybrid system, and vice versa. In the case of multisubunit enzymes such as RNA polymerases, this may arise because the corresponding associations occur between intermediary subcomplexes, rather than by a pairwise association of two individual subunits. Conversely, multicopy suppression presumably acts by displacing association/disassociation equilibria toward the formation of multisubunit complexes, and it might therefore operate in an indirect way rather than via direct protein–protein contact.

In a derivative of the original method, the DNA-binding domain comes from the bacterial repressor lexA (41). The important feature of this system is that two reporter genes, *LEU2* and *lacZ,* are used to avoid false positives. Moreover, interactions can be selected by their activation of the *LEU2* gene, allowing growth of yeast transformants in the absence of leucine.

One of the major uses of the two-hybrid method is to allow the identification of previously unknown partners of a given protein. This application is possible by taking advantage of appropriate genomic DNA or cDNA libraries fused to activation domains (40, 41). The libraries are searched by cotransformation with the test protein fused to the GAL4 (1–147) or lexA DNA-binding domain.

Application to Yeast RNA Polymerase III

We have used the two-hybrid method to study protein–protein interactions between 8 of the 13 yeast RNA polymerase III subunits (30, 31). Our results confirmed and extended the data obtained from the biochemical study of conditional mutants or from multicopy suppression experiments. For example, the interaction between C82 and C31 suggested by the multicopy suppression of an *rpc82* mutation by *RPC31*, and by the simultaneous loss of C82, C34, and C31 subunits from a mutant RNA polymerase III affected in C160, was further substantiated by the two-hybrid method. In the case of the C34 subunit it has been possible to study the interaction with three different partners, as this polypeptide appears to contact subunits C31 and C82, and to interact with the 70-kDa subunit of the general initiation factor TFIIIB. The data suggest that C34 (presumably associated with C31 and C82) has an important role in the specific association between RNA polymerase III and cognate initiation factors (31).

Conclusions and Perspectives

Early attempts to isolate yeast RNA polymerase mutants by classic approaches have been unsuccessful, and genetic studies of RNA polymerases therefore only started with the cloning of RNA polymerase subunit genes. Most of the genes have now been cloned. Sequence comparison among the three eukaryotic enzymes, and relative to the bacterial and archaebacterial enzymes, provided important insights into the subunit organization. In parallel, the isolation of null alleles and conditional mutations established that most of the subunits are essential components of the transcription machinery. The conditional mutants are being analyzed in terms of functional defects or impaired subunit assembly. Moreover, the isolation of a large number of suppressor alleles by multicopy suppression or by *in vivo* mutagenesis has proved to be a fruitful approach, in particular by revealing patterns of interaction between the subunits of a given RNA polymerase. This information is complemented by that obtained using the recently developed two-hybrid method, which directly demonstrates the capacity of two given subunits to combine *in vivo*, and by data based on

immunological and biochemical studies, including three-dimensional image recon-
structions of yeast RNA polymerase II (42) and I (43). A coherent picture of the
spatial and functional organization of eukaryotic RNA polymerases will doubtless
emerge from these converging approaches.

Acknowledgments

We thank André Sentenac and Sylvie Hermann for helpful discussions, and Cathy Jackson for careful reading of the manuscript. Recent work done in the authors laboratory was supported by Grant SCI-CT91-0702 from the European Union.

Note added in proof. Since the writing of this paper, a helpful comparison of the various vectors used in the two-hybrid assay has been published (44).

References

1. A. Sentenac, M. Riva, P. Thuriaux, J. M. Buhler, I. Treich, C. Carles, M. Werner, A. Ruet, J. Huet, C. Mann, N. Chiannilkulchai, S. Stettler, and S. Mariotte, *in* "Transcriptional Regulation" (K. R. Yamamoto and S. L. McKnight, eds.), Vol. 1, p. 24. Cold Spring Harbor Laboratory, Cold Spring Harbor, New York, 1992.
2. P. Thuriaux and A. Sentenac, *in* "The Molecular Biology of the Yeast *Saccharomyces*" (J. R. Broach, J. R. Pringle, and E. W. Jones, eds.), Vol. 2, p. 1. Cold Spring Harbor Laboratory, Cold Spring Harbor, New York, 1992.
3. R. K. Mortimer and J. R. Johnston, *Genetics* **113**, 35 (1986).
4. R. S. Sikorski and J. D. Boeke, *in* "Methods in Enzymology" (C. Guthrie and G. R. Fink, eds.), Vol. 194, p. 302. Academic Press, San Diego, 1991.
5. N. Bonneaud, O. Ozier-Kalogeropoulos, G. Li, M. Labouesse, L. Minvielle-Sebastia, and F. Lacroute, *Yeast* **7**, 609 (1991).
6. M. D. Rose and J. R. Broach, *in* "Methods in Enzymology" (C. Guthrie and G. R. Fink, eds.), Vol. 194, p. 195. Academic Press, San Diego, 1991.
7. J. C. Schneider and L. Guarente, *in* "Methods in Enzymology" (C. Guthrie and G. R. Fink, eds.), Vol. 194, p. 373. Academic Press, San Diego, 1991.
8. C. Mann, J. M. Buhler, I. Treich, and A. Sentenac, *Cell (Cambridge, Mass.)* **48**, 627 (1987).
9. F. Sherman and J. Hicks, *in* "Methods in Enzymology" (C. Guthrie and G. R. Fink, eds.), Vol. 194, p. 21. Academic Press, San Diego, 1991.
10. G. Berben, J. Dumont, V. Gilliquet, P. A. Bolle, and F. Hilger, *Yeast* **7**, 457 (1991).
11. N. Chiannilkulchai, A. Moenne, A. Sentenac, and C. Mann, *J. Biol. Chem.* **267**, 23099 (1992).
12. E. Alani, L. Cao, and N. Kleckner, *Genetics* **116**, 541 (1987).
13. C. Mosrin, M. Riva, M. Beltrame, E. Cassar, A. Sentenac, and P. Thuriaux, *Mol. Cell. Biol.* **10**, 4737 (1990).

14. J. Archambault, K. T. Schappert, and J. D. Friesen, *Mol. Cell. Biol.* **10,** 6123 (1990).
15. K. F. Wertman, D. G. Drubin, and D. Botstein, *Genetics* **132,** 337 (1992).
16. M. Nonet, D. Sweetser, and R. A. Young, *Cell (Cambridge, Mass.)* **50,** 909 (1987).
17. L. A. Allison, J. K. Wong, V. D. Fitzpatrick, M. Moyle, and C. J. Ingles, *Mol. Cell. Biol.* **8,** 321 (1988).
17a. V. Thuillier, S. Stettler, A. Sentenac, P. Thuriaux, and M. Werner, *EMBO J.* **14**, 351 (1995).
18. S. Stettler, N. Chiannilkulchai, S. Hermann-Le Denmat, D. Lalo, F. Lacroute, A. Sentenac, and P. Thuriaux, *Mol. Gen. Genet.* **239,** 169 (1993).
19. P. A. Kolodziej and R. A. Young, *Mol. Cell. Biol.* **11,** 4669 (1991).
20. Y. Nogi, L. Vu, and M. Nomura, *Proc. Natl. Acad. Sci. U.S.A.* **88,** 7026 (1991).
21. I. Treich, C. Carles, A. Sentenac, and M. Riva, *Nucleic Acids Res.* **20,** 4721 (1992).
22. P. James, S. Whelen, and B. D. Hall, *J. Biol. Chem.* **266,** 5616 (1991).
23. J. Archambault, F. Lacroute, A. Ruet, and J. D. Friesen, *Mol. Cell. Biol.* **12,** 4142 (1992).
24. M. Werner, S. Hermann-Le Denmat, I. Treich, A. Sentenac, and P. Thuriaux, *Mol. Cell. Biol.* **12,** 1087 (1992).
25. C. W. Lawrence, *in* "Methods in Enzymology" (C. Guthrie and G. R. Fink, eds.), Vol. 194, p. 273. Academic Press, San Diego, 1991.
26. J. H. McCusker, M. Yamagishi, J. M. Kolb, and M. Nomura, *Mol. Cell. Biol.* **11,** 746 (1991).
27. F. Sherman and P. Wakem, *in* "Methods in Enzymology" (C. Guthrie and G. R. Fink, eds.), Vol. 194, p. 38. Academic Press, San Diego, 1991.
28. C. Martin, S. Okamura, and R. A. Young, *Mol. Cell. Biol.* **10,** 1908 (1990).
29. R. Yano and M. Nomura, *Mol. Cell. Biol.* **11,** 754 (1991).
30. D. Lalo, C. Carles, A. Sentenac, and P. Thuriaux, *Proc. Natl. Acad. Sci. U.S.A.* **90,** 5524 (1993).
31. M. Werner, N. Chaussivert, I. M. Willis, and A. Sentenac, *J. Biol. Chem.,* **268**, 20721 (1993).
32. S. Buratowski and H. Zhou, *Cell (Cambridge, Mass.)* **71,** 221 (1992).
33. T. Colbert and S. Hahn, *Genes Dev.* **6,** 1940 (1992).
34. A. J. Koleske, S. Buratowski, M. Nonet, and R. A. Young, *Cell (Cambridge, Mass.)* **69,** 883 (1992).
35. C. L. Peterson, W. Kruger, and I. Herskowitz, *Cell (Cambridge, Mass.)* **64,** 1135 (1991).
36. S. Hermann-Le Denmat, M. Werner, A. Sentenac, and P. Thuriaux, *Mol. Cell Biol.* **14**, 2905 (1994).
37. C. Martin and R. A. Young, *Mol. Cell. Biol.* **9,** 2341 (1989).
38. S. Fields and O. K. Song, *Nature (London)* **340,** 245 (1989).
39. J. Ma and M. Ptashne, *Cell (Cambridge, Mass.)* **51,** 113 (1987).
40. C. T. Chien, P. L. Bartel, R. Sternglanz, and S. Fields, *Proc. Natl. Acad. Sci. U.S.A.* **88,** 9578 (1991).
41. A. S. Zervos, J. Gyuris, and R. Brent, *Cell (Cambridge, Mass.)* **72,** 223 (1993).
42. S. A. Darst, A. M. Edwards, E. W. Kubalek, and R. D. Kornberg, *Cell (Cambridge, Mass.)* **66,** 121 (1991).
43. P. Schultz, H. Célia, M. Riva, A. Sentenac, and P. Oudet, *EMBO J.* **12**, 2601 (1993).
44. P. Legrain, M. C. Dokhelar, and C. Transy, *Nucleic Acids Res.* **22**, 3241 (1994).

[13]   Pheromone Procedures in Fission Yeast

John Davey, Richard Egel, and Olaf Nielsen

Introduction

Comparative analyses of signal transduction in mammalian cells and lower eukaryotes such as yeast have revealed a striking conservation of modular components
and regulatory cascades (reviewed in Ref. 1), and it is likely that further investigations into the yeast systems could provide valuable new information into the molecular interactions involved in cell-to-cell communication. The pheromone response
pathway of yeast cells has received particular attention, and recent reviews reveal the
similarities between the budding yeast *Saccharomyces cerevisiae* (2) and the fission
yeast *Schizosaccharomyces pombe* (3). Although the two yeast are remarkably similar, they are sufficiently different to prevent the direct transfer of methods from one
to the other, and the well-characterized methodology of *S. cerevisiae* must often be
modified for *S. pombe*. Here we review several current methods for studying pheromones and signal transduction in *S. pombe* and describe the biological background
for these procedures.

Mating Types

Haploid cells of the fission yeast *S. pombe* exist in one of two mating types, *Minus*
and *Plus*, which differ only in the allele carried at the mating-type locus *mat1*, encoding two different subfunctions in each mating type. The genes have no function
during mitotic growth and, in wild-type cells, are only expressed when the nutrient
supply becomes limiting, especially for a nitrogen source. Under starvation conditions the cells become potential mating partners and communicate through the production of mating pheromones. These activities are controlled by the early acting
mating-type subfunctions. *P* cells release P-factor and respond to M-factor by
expressing the M-factor receptor, whereas *M* cells release M-factor and express the
receptor for P-factor. The pheromones act over a relatively short distance and induce
mating-specific changes in cells of the opposite mating type. Responding cells elon-
gate and bend toward the pheromone source (4, 5); other changes include a G_1 arrest
of cell division (6, 7), an increase in cell agglutinability, and altered patterns of gene
transcription (reviewed in Ref. 3). This includes several genes involved in zygote
formation and also the late acting mating-type subfunctions which are needed for
meiosis.

Methods in Molecular Genetics, Volume 6

Pheromones

P-factor is an unmodified peptide of 23 amino acids that is released by *P* cells (7). It is encoded by the *map2* gene, and the primary translation product contains an amino-terminal signal sequence and four repetitive units of the pheromone. There are potentially three different species of P-factor, but one apparently dominates the others in quantity (Fig. 1) and has been demonstrated to have pheromone activity (7).

The dominant species is encoded by the second and fourth repeats in the precursor. In the other two units, the N-terminal threonine is replaced by a serine residue, and the glutamine residue at position 10 is replaced by a histidine in one, but not the other, repeat. The fates and activities of the two minor species have not been investigated in any detail. Each repeat is separated by a spacer region that contains pairs of basic residues which could act as possible cleavage sites. The structure of the P-factor precursor is analogous to that of α-factor in *S. cerevisiae* and is likely to be processed in a similar way. This involves three enzymes: a dibasic endopeptidase that cleaves on the C-terminal side of -Lys-Arg- sites, a carboxypeptidase that removes basic residues from the C terminus, and an aminopeptidase that trims the N terminus. These processing events are likely to occur as the P-factor is transported along the secretory pathway.

M-factor, the pheromone released by *M* cells, is a nonapeptide in which the C-terminal cysteine residue is carboxymethylated and S-farnesylated (8) (Fig. 2). It is encoded by three genes, *mfm1, mfm2,* and *mfm3* (8, 9), which, although similar, map to different regions in the genome and are unlinked (10). Although the biogenesis of M-factor is not fully characterized, it is likely to resemble that of the **a**-factor pheromone from the budding yeast *S. cerevisiae.* The two pheromones are similar in structure, and each appears to be synthesized as a precursor which undergoes proteolytic processing and posttranslational modification to produce the active pheromone (reviewed in Ref. 11). Briefly, the primary translation product contains a -CAAX box at the C terminus which serves as the recognition sequence for the addition of a farnesyl residue to the cysteine residue. The last three amino acids are then removed, and the exposed carboxyl group is protected by methylation. Proteolytic processing at the amino terminus yields the mature pheromone, which is likely to be released into the medium following direct transport across the plasma membrane via an ATP-dependent transporter similar to the *STE6* gene product that transports **a**-factor in *S. cerevisiae.*

Thr-Tyr-Ala-Asp-Phe-Leu-Arg-Ala-Tyr-Gln-Ser-Trp-Asn-Thr-Phe-Val-Asn-Pro-Asp-Arg-Pro-Asn-Leu

FIG. 1   Structure of P-factor.

Tyr-Thr-Pro-Lys-Val-Pro-Tyr-Met-Cys(S-farnesyl)-OCH$_3$

Fig. 2   Structure of M-factor.

Events at the Target Cell

Both pheromone receptors are seven-transmembrane proteins (12, 13) which interact with a heterotrimeric G protein so that stimulation releases the Gα subunit (encoded by *gpa1*) that plays a positive role in generating an intracellular signal (14). Propagation of the intracellular signal involves a series of protein kinases (*byr2, byr1,* and *spk1*) that are functionally homologous to the mitogen-activated protein (MAP) kinases involved in the control of proliferation and differentiation in many eukaryotes (reviewed in Ref. 1). Signal transduction is also influenced by the *ras1*-associated pathway (15), although ras1 is not a signal transmitter and its precise role remains unclear.

Degradation of P-factor

M cells exposed to P-factor secrete a protease that efficiently degrades P-factor. The protease is encoded by the *sxa2* gene, and mutants are hypersensitive to P-factor (7, 16). Sxa2 is a serine carboxypeptidase which appears to be functionally analogous to the *BAR1* aspartyl protease that degrades α-factor in *S. cerevisiae* (17). Transcription of the *sxa2* gene is induced in response to P-factor, and the secreted protease is likely to degrade the extracellular pheromone as part of an adaptation process. The degradation of P-factor by Sxa2 is very efficient, and many of the pheromone-induced changes expected when cells are exposed to pheromone cannot be detected in wild-type cells. It is therefore routine to use mutant *M* cells that are defective in *sxa2* when monitoring the activity of P-factor.

Nutritional Concerns and Media

Investigations into pheromone-associated events in *S. pombe* are complicated because all functions of sexual differentiation are repressed during mitotic growth. Therefore, and unlike the situation in *S. cerevisiae,* there is no response when mating pheromones are added to mitotically growing cells. One approach to overcome this nutritional problem is to grow the cells in a nitrogen-limited medium until the nitrogen source has been consumed. Alternatively, the cells can be grown in a defined medium and the mating response induced by transferring to the same medium but lacking a nitrogen source.

TABLE I   Media for *Schizosaccharomyces pombe*

Medium	Component	Amount
MSL[a]	Glucose	10 g
	Arginine monochloride	2 g
	KH_2PO_4	1 g
	$MgSO_4 \cdot 7H_2O$	200 mg
	NaCl	100 mg
	$CaSO_4 \cdot 2H_2O$	100 mg
	Vitamins (1000× stock)	1 ml
	Minerals (10,000× stock)	100 μl
MSA	As MSL but include 20 g agar per liter	
PM[a]	Glucose	20 g
	NH_4Cl	5 g
	Potassium hydrogen phthalate	3 g
	Na_2HPO_4	2.2 g
	Salts (50× stock)	20 ml
	Vitamins (1000× stock)	1 ml
	Minerals (10,000× stock)	100 μl
SSL[a]	Glucose	10 g
	KH_2PO_4	2 g
	$MgSO_4 \cdot 7H_2O$	500 mg
	Aspartic acid	200 mg
	Na_2HPO_4	199 mg
	$CaCl_2 \cdot 2H_2O$	100 mg
	Vitamins (1000×)	1 ml
	Minerals (10,000×)	100 μl
Stock solution of salts (50×)[a]	$MgCl_2 \cdot 6H_2O$	52.5 g
	KCl	50 g
	Na_2SO_4	2 g
	$CaCl_2 \cdot 2H_2O$	735 mg
Stock solution of minerals (10,000×)[b]	Citric acid	1 g
	Boric acid	500 mg
	$MnSO_4 \cdot H_2O$	500 mg
	$ZnSO_4 \cdot 7H_2O$	400 mg
	Molybdic acid	305 mg
	$FeCl_3 \cdot 6H_2O$	200 mg
	KI	100 mg
	$CuSO_4 \cdot 5H_2O$	40 mg
Stock solution of vitamins (1000×)[b]	Nicotinic acid	1 g
	Inositol	1 g
	Pantothenic acid	100 mg
	Biotin	1 mg
Yeast extract (YE)[a,c]	Glucose	30 g
	Yeast extract	5 g

[a] Recipes give amounts to prepare 1 liter.
[b] Recipes give amounts to prepare 100 ml.
[c] Used for normal vegetative growth.

The compositions of several media are given in Table I. Yeast extract (YE) is used for the routine culturing of cells, and the mating reaction is induced by growth in synthetic sporulation liquid (SSL) after the low amount of aspartic acid has been used (18) or by a shift from MSL to MSL−N (19, 20). MSL is a nitrogen-enriched modification of SSL, whereas MSL−N lacks a nitrogen source. These media are phosphate-buffered. Phthalate-buffered media such as PM (21, 22) or EMM (23) have also been used; however, the rapid succession of events induced by a shift from MSL to MSL−N (19) is much retarded in PM, and the efficiency of conjugation and/or sporulation is often rather low in this medium.

The process by which *S. pombe* monitors the level of extracellular nitrogen is not understood, but the reduction in the intracellular concentration of cAMP observed when cells are starved of nitrogen may be the trigger for sexual differentiation. High levels of cAMP repress the sexual pathway by activating a cAMP-dependent protein kinase that inhibits transcription of genes required for differentiation. A key enzyme in the biosynthesis of cAMP is adenylate cyclase (encoded by the *cyr1* gene; 24, 25), and mutants lacking the enzyme have no detectable cAMP, are derepressed for nutritional signals, and undergo sexual differentiation in the presence of a nitrogen source (26–28). These mutants respond to mating pheromone in rich medium and undergo an arrest of division at the G_1 stage of the cell cycle (6, 7).

Strains

Many of the methods described in this chapter involve the use of strains with specific characteristics. These strains are available from the authors (Table II).

TABLE II   *Schizosaccharomyces pombe* Strains

Strain	Genotype
EG-432	*mat1-P Δmat2,3 :: LEU2 leu1-32 ura4-D18*
EG-544	*mat1-M Δmat2,3 :: LEU2 leu1⁺*
EG-545	*mat1-P Δmat2,3 :: LEU2 leu1⁺*
EG-571	*mat1-M int-H1 :: ura4⁺ Δmat2,3 :: LEU2 leu1-32 ura4-D18 ade6-M26*
EG-575	*mat1-Mc :: ura4⁺ Δmat2,3 :: LEU2 leu1-32 ura4-D18 ade6-M26*
EG-670	2n: EG-432/EG-575
EG-699	*mat1-M Δmat2,3 :: LEU2 sxa2-563*
EG-794	*mat1-P Δmat2,3 :: LEU2 cyr1 :: ura4⁺*
EG-796	*mat1-M Δmat2,3 :: LEU2 cyr1 :: ura4⁺ sxa2-563*
EG-812	*mat1-M int-H1 :: ura4⁺ Δmat2,3 :: LEU2 leu1-32 ade6-M26 sxa2-563*
EG-817	*mat1-Pc :: int-1 Δmat2,3 :: LEU2 leu1-32 ura4-D18*
EG-818	*mat1-Pc :: int-1 Δmat2,3 :: LEU2 leu1-32 ura4-D18 sxa2-563*
EG-819	2n: EG-571/EG-817
EG-820	2n: EG-812/EG-818

Preparation of Mating Pheromones

Both of the *S. pombe* mating pheromones have now been isolated and characterized, and active preparations of each can be obtained either by purification of the native pheromone from culture medium or by chemical synthesis (8, 7, 29). This section describes the purification and chemical synthesis of both M-factor and P-factor.

Purification of M-Factor

The basic principles of the purification scheme (8, 30) are (1) batch extraction of M-factor from culture medium by adsorption to polystyrene beads; (2) elution of M-factor from the beads using propan-1-ol; (3) fractionation on a column of Sephadex LH-60; and (4) reversed-phase high-performance liquid chromatography (HPLC) on a C_{18} column. In most purification schemes it is advisable to perform repeated assays for the material being purified so that the correct decisions concerning yield and enrichment can be made at each step. Several assays for pheromone activity have now been developed and are described elsewhere in this chapter, but all are relatively time-consuming to perform and slow to provide a result. Given these problems, and in view of the reliability of the purification scheme, we no longer routinely assay for M-factor activity at each stage of the purification.

Strains and Culture Conditions

The most suitable strain to use for the routine production of M-factor is a nonswitching, haploid *M* strain such as EG-544. Homozygous diploid strains do produce slightly more M-factor than haploid cells (30), but the increase is not sufficiently significant to overcome the difficulties that can be associated with diploid strains. An important consideration when choosing an alternative strain is its nutritional requirements, and it is considerably more straightforward to use strains that do not require additional nutrients for growth. As with almost all aspects of the mating response in *S. pombe,* the production of M-factor is repressed during normal mitotic growth and is induced when the nitrogen source becomes limiting. Additives such as leucine or adenine can be utilized as a nitrogen source and thus repress mating-related activities. The fine balance between growth and starvation is difficult, but not impossible, to achieve. Culturing in SSL to stationary phase yields sufficient pheromone for purification.

Preparing Beads

Amberlite XAD-2 is a polymeric adsorbent supplied as white insoluble beads for use in columns or in batch operations to adsorb soluble organic compounds from aqueous solutions. The adsorption forces involved are primarily of the van der Waals type, and adsorption/elution can be affected by the hydrophobicity of the solvent. Organic compounds will therefore adsorb to the beads in an aqueous environment and can be released by an organic solvent. The beads are commercially available from Sigma Chemical Co. (St. Louis, MO). It is necessary to first treat the beads with an organic solvent to ensure removal of any prebound material.

1. Transfer about 600 ml of beads to a 2-liter measuring cylinder and wash extensively with distilled water. Fill the cylinder with water, mix by inversion, and allow the beads to settle before removing the excess water.
2. Remove the excess water and add about 500 ml of propan-1-ol to the beads while still in the cylinder. Mix by inverting the cylinder and allow the beads to settle before removing the excess liquid and repeating with another 500 ml of propan-1-ol.
3. Load the beads into a column (5 × 30 cm) and wash with 2 liters of propan-1-ol at a flow rate of 100 ml/hr.
4. Transfer the beads from the column to a 2-liter measuring cylinder, wash them twice with distilled water, and then transfer them back into the column.
5. Wash the beads with 10 liters of distilled water at a flow rate of 100 ml/hr.
6. Remove the beads from the column and sterilize by autoclaving before storing at room temperature. Ensure that the beads are immersed in water when being autoclaved and during storage.

Preparing Cells

The standard preparation involves ten 1-liter cultures. This is more a reflection of the equipment available in our laboratory than an optimized protocol, and preparations have ranged from 50-ml to 20-liter cultures. All cultures are incubated with constant shaking to aerate the medium and keep the cells and beads in suspension.

1. Use an overnight culture of cells to inoculate 400 ml of YE and incubate with shaking at 30° C for 48 hr.
2. Collect the cells by centrifugation (2500 g for 10 min); wash twice in SSL before resuspending in ten 1-liter cultures at 10^6 cells/ml. Incubate at 30° C for another 24 hr.
3. The cells should now be in an early stationary phase ($\sim 2 \times 10^7$ cells/ml). Add 30 ml (30%, v/v) of washed Amberlite XAD-2 resin to each 1-liter culture and continue incubation at 30° C for a further 72 hr.
4. Separate the beads from the cells. Because the beads are larger than the yeast cells they can be separated by repeated resuspending and decanting the mixture using distilled water. Once most of the cells have been removed, pour the mixture into a column (2.5 × 50 cm). Any remaining cells pass through the column, but the beads are retained and can be washed with distilled water. Gently tap the column to ensure that the beads are well packed and the amount of water in the column is minimized.

5. Recover the adsorbed material by eluting with propan-1-ol at a flow rate of 50 ml/hr and collecting 5-ml fractions. The elution can be followed using an in-line spectrophotometer (280 nm); alternatively, a clear brown ring is visible at the front of the propan-1-ol as it passes down the column.

6. Ideally, the amount of M-factor in each fraction should now be assayed (see below); however, the procedure is very reliable, and almost all of the M-factor activity is contained in the first 7 fractions of eluted material. These fractions can be identified as the first 7 containing material to absorb at 280 nm or as the first 7 fractions containing brown-colored material. Fractions containing M-factor should be pooled and dried by lyophilization. Add an equivalent volume of water to the column eluate to aid freezing, and freeze as a shell in a 250-ml round-bottom flask immersed in liquid nitrogen.

7. Resuspend the dried sample in methanol at 100 μl per liter of original culture medium. There is likely to be a considerable amount of insoluble material which should be removed by centrifugation for 5 min at 10,000 g in a microcentrifuge. The sample should now be stored at $-20°C$ for 3 days, during which time additional material precipitates. This material should be removed by centrifugation; it is not M-factor, and its precipitation at $-20°C$ provides an effective purification step.

Fractionation on Sephadex LH-60

Most Sephadex columns separate molecules according to size, and substances are eluted in order of decreasing molecular size. However, such molecular sieving is not the only factor that influences separation on Sephadex LH-60 in organic solvents. There are a number of gel–solute interactions that affect the separation, and although these cannot be predicted they often improve the usefulness of such columns.

1. Prepare a column (1.5 × 90 cm) of Sephadex LH-60 in methanol.

2. The material from the original 10 liters of culture medium is resuspended in approximately 1 ml of methanol. This is carefully loaded onto the column and eluted at the rate of about 10 ml/hr with 3-ml fractions being collected.

3. Elution is monitored by absorbance at 280 nm. M-factor is eluted in the first of two relatively broad peaks of absorbing material. The appropriate fractions are pooled to provide approximately 15 ml of material which contains about 45% of the activity present in the crude material. This sample should not be concentrated prior to further purification by HPLC.

Reversed-Phase Chromatography

Column details are as follows. A Spherisorb C_{18} ODS2 column (4.6 mm × 25 cm) (Phase-Sep, Clwyd, UK) is used with a precolumn of C-130B (Uptight) with Pellicular ODS (Whatman, 4102-0101). The eluent is 80% (v/v) methanol, 0.1% (v/v) trifluoroacetic acid at an elution rate of 1 ml/min. The sample volume is 2 ml (equivalent to 2 liters of culture medium).

1. The material from the Sephadex LH-60 column (in 100% methanol) should be diluted to 80% (v/v) methanol prior to loading on the HPLC column.
2. Monitor the elution by absorbance at 220 nm. M-factor has a retention time of approximately 30 min and is easy to identify since it is the penultimate product to be eluted from the column. Material eluting after about 20 min is an oxidized form of M-factor with a greatly reduced specific activity. The ratio of oxidized to native pheromone varies from one preparation to another.

Some Useful Figures

One unit of M-factor is defined as the amount of pheromone that induces a 4% response in the cell volume change assay described previously (30).

The specific activity of M-factor is approximately 440 units/nmol (330 units/μg).
The concentration of M-factor in culture medium is 8 units/ml.
Recovery after separation on Sephadex LH-60 is approximately 36,000 units (45%).
Recovery after separation on HPLC is approximately 5000 units (6%).

Chemical Synthesis of M-Factor

The farnesylation and carboxymethylation of M-factor are essential for its activity, but the modifications complicate chemical synthesis. Hence, the production of synthetic M-factor is time-consuming and inefficient. A detailed description of the synthesis can be found elsewhere (29). The main problem is that the farnesyl group on the C-terminal cysteine residue of the M-factor contains three double bonds and is not stable to strong acids, and so the pheromone cannot be synthesized directly by solid-phase methods. An alternative strategy involving minimum side-chain protection was developed based on a strategy similar to that used for synthesizing the **a**-factor from *S. cerevisiae* (31). It involves the coupling of an N-terminal hexapeptide to a C-terminal farnesylated tripeptide methyl ester. The protected hexapeptide can be prepared using standard solid-phase synthesis, whereas synthesis of the C-terminal tripeptide is achieved by solution-phase chemistry. Coupling of the two peptides generates the nonapeptide, which is purified by reversed-phase HPLC.

Preparation of P-Factor

Being an unmodified peptide, P-factor can be synthesized by standard solid-phase methods on an automated synthesizer. In addition, HPLC-purified P-factor is available from Dr. John Fox (Alta Bioscience, School of Biochemistry, The University of Birmingham, Birmingham B15 2TT, England).

If necessary, P-factor can be purified from culture medium using a scheme that is broadly similar to that described for M-factor. The secreted P-factor is first adsorbed onto Amberlite XAD-2 and the eluted material fractionated on Sephadex LH-20 prior to purification by reversed-phase chromatography. Further details are described in the original report of purification (7). It was found to be important to use a strain that overproduces P-factor because rather little pheromone is secreted by wild-type *P* cells that have not been exposed to the M-factor pheromone.

Assays for Pheromone Activity

The first assays for monitoring pheromone activity in *S. pombe* utilized chambers in which cells of one mating type were placed on agar slabs close to, but not touching, cells of the opposite type. The responding cells undergo a mating-specific elongation ("shmooing") toward the source of the pheromone. Such assays were the first to demonstrate the diffusible nature of the pheromones (4, 5) and are sufficiently quantitative to allow comparisons of pheromone production and pheromone response in different strains (32, 33). Unfortunately, the assays require a relatively high degree of experimental skill, and the results are often complicated by excessive rounds of residual cell division. Furthermore, although comparisons among different strains are possible, the results are only semiquantitative, and it would be difficult to measure accurately pheromone activity by these methods. Several assays have therefore been developed, and each can be used to monitor either M-factor or P-factor.

Cell Elongation

The elongation of cells in response to mating pheromones is accompanied by an increase in cell volume. The increase is most easily monitored using a Coulter Channelyser, although the size distribution of the population can be determined by performing repeated counts at different threshold settings on a Coulter Counter. Even if there is no access to a machine suitable for monitoring the change in cell volume, microscopic examination of the cells can provide a semiquantitative estimate of the pheromone activity.

Many strains can be used in the assay, but, as mentioned earlier when describing the production of pheromones, it is most convenient if the cells do not require additional nutrients for growth. The inclusion of additives such as leucine or adenine can complicate the nutritional conditions required for a response. If the response is to be monitored visually, then the normally elongated shape of wild-type cells can make it difficult to detect and quantitate pheromone-induced elongation. The use of strains possessing activated alleles of the *ras1* gene should be considered in such cases. When these cells are grown under conditions of nitrogen limitation, they possess a short and swollen cell shape, and the mating-specific elongation is more marked than that in wild-type cells (4). Strains defective in *gap1* (34) (also called *sar1;* 35) possess the same phenotype as those with an activated *ras1* gene. *M* cells are used to monitor P-factor activity must lack the *sxa2* protease that efficiently degrades the pheromone. A strain suitable for monitoring M-factor activity would be EG-545, whereas EG-699 would be suitable for monitoring P-factor activity.

1. Use an overnight culture of cells to inoculate 50 ml of YE and incubate with shaking at $30°$ C for 24 hr.
2. Collect the cells by centrifugation (2500 *g* for 10 min) and wash once in SSL before resuspending in SSL at 10^6 cells/ml. Incubate at $30°$ C for another 24 hr.
3. Collect the cells by centrifugation and resuspend in SSL at 10^6 cells/ml.
4. Dispense 900-μl aliquots of cells into each well of a 24-well culture plate.
5. Add the sample to be assayed in a final volume of 100 μl of SSL.
6. Incubate with shaking in a humidified chamber at $30°$ C for 48 hr.
7. Monitor the pheromone-dependent change in the size distribution of the culture.

Nonsynchronized cultures contain cells with a range of sizes, and exposure to pheromone not only increases cell size but broadens the size distribution of the culture. These features can complicate the comparison of treated and untreated cultures. A simple and reliable comparative measurement is to determine the median volume of each culture, where the median volume is the midpoint of the culture such that 50% of the cells are larger than and 50% are smaller than the median. The median volume of the culture increases following exposure to pheromone. An alternative method for comparison is to determine the percentage of cells in each treated culture that are larger than the median cell size of cells not exposed to pheromone (30). There is a linear relationship between this value and the logarithm of the pheromone concentration used to treat the cells.

Although these measurements provide a reliable comparison between different cultures, they do require the use of a Coulter Channelyser or Coulter Counter. If these are unavailable, it is still possible to make semiquantitative estimates of pheromone activity by microscopic examination of the treated cultures. One possibility is to make serial 2-fold dilutions of the pheromone and to determine the lowest concentration that can cause a visible response in the tester cells.

G_1 Arrest of Cell Division

In the budding yeast *S. cerevisiae,* addition of mating pheromone to mitotically growing cells of the complementary mating type causes an arrest in the G_1 phase of the cell cycle. It is a transient arrest which ensures that conjugation occurs between cells containing a 1*C* complement of chromosomes. Such an arrest does not occur when the experiment is performed with *S. pombe,* and mitotically growing cells exhibit no response when exposed to pheromone. The difference is due to the coupling of sexual differentiation and nutritional starvation in *S. pombe,* and, as described in the introduction, it is necessary to perform the experiment in strains that are defective for the *cyr1* gene that encodes adenylate cyclase. These cells are derepressed for sexual differentiation and respond to pheromones in the presence of a nitrogen source. *M* cells used to monitor P-factor activity must again lack the sxa2 protease that efficiently degrades the pheromone. A strain suitable for monitoring M-factor activity would be EG-794, whereas EG-796 would be suitable for monitoring P-factor activity.

1. Inoculate an appropriate volume of cells from a preculture into a defined minimal medium so that the cell density following overnight growth at $30°$ C will be about 3×10^6 cells/ml.
2. Add pheromone to a final concentration of $2 \ \mu M$. Pheromone stocks are kept at $2 \ mM$ in methanol at $-20°$ C. An equivalent volume of methanol can be added to a second culture of cells as control.
3. At appropriate time intervals (probably every hour), determine the cell density and remove and fix 3×10^6 cells for analysis by flow cytometry (see below).

A second mutation that bypasses the nutritional requirements for pheromone response is the temperature-sensitive *pat1-114* allele. The *pat1* (or *ran1*) gene encodes a protein kinase that couples sexual differentiation to nutritional starvation (21, 36, 37). The *pat1-114* kinase is completely inactivated at $35°$ C, and the cells attempt to undergo meiosis regardless of ploidy or nutritional conditions. At $30°$ C the enzyme is only partially inactivated, and cells transcribe pheromone-dependent genes in the absence of pheromone and mate in the presence of a nitrogen source (21, 38). Even at the apparently permissive temperature of $23°$ C, however, the *pat1-114* kinase is not fully active, and the cells display an increased sensitivity to pheromones and exhibit a G_1 arrest of the cell cycle in response to pheromone (6). Apart from the temperature considerations of using the *pat1-114* mutation, there is no apparent difference between the growth arrest in these mutants and that seen in cells lacking *cyr1.*

Flow Cytometry

Flow cytometry allows the DNA content of individual cells to be determined. In a mitotically growing culture of *S. pombe,* most cells will have a $2C$ complement of DNA because the G_1 phase of the cycle is much shorter than the G_2 phase and newly separated cells have already undergone DNA replication (39). The percentage of cells with a $1C$ DNA content increases as the culture becomes arrested in G_1. A G_1 arrest can be effected in a number of ways: as a consequence of a mutation in a gene required to progress through G_1, after transfer to nitrogen-free medium, or following pheromone treatment of *cyr1⁻* cells or *pat1-114* cells at 23°C.

1. Collect approximately 3×10^6 cells by centrifugation and wash once with cold sterile water. Using too many cells can lead to incomplete staining and artifactual results.
2. Resuspend the cells in 300 μl of water.
3. While constantly mixing the cells, slowly add 700 μl of cold ethanol. The cells at this point can be stored almost indefinitely at $-20°$C.
4. Harvest the cells and resuspend in 1 ml of 50 mM sodium citrate (unbuffered, pH 7–8).
5. Harvest the cells and resuspend in 500 μl of 50 mM sodium citrate containing 0.1 mg/ml RNase A.
6. Incubate at 37°C for at least 2 hr.
7. The fixed cells can be harvested and stained by resuspending in 200 μl of 50 mM sodium citrate containing 2 μg/ml propidium iodide or in 200 μl of 10 mM Tris-HCl (pH 7.4), 10 mM MgCl$_2$ containing 50 μg/ml mitramycin and 10 μg/ml ethidium bromide (40).

Plate Assay for Pheromone Activity

A particular problem of studying pheromone activity in *S. pombe* is the lack of a simple plate assay, and it would be useful to develop an assay where the pheromone-induced G_1 arrest could be monitored as a halo of growth inhibition. The tester strain in such an assay would obviously need to possess a defective *cyr1* gene (or the *pat1-114* allele) to allow a pheromone response during mitotic growth, and the *M* cells used to detect P-factor would also need to be defective for *sxa2*. Such an assay has not yet been successfully performed. The tester cells do respond to the pheromone by shmooing, but they appear to recover quickly from the stimulation, with no noticeable delay in colony formation. The solution might be to isolate a strain that is either supersensitive to the pheromones or one that is defective in its ability to recover from stimulation, and we are currently in the process of isolating such strains.

As an alternative to pheromone-induced haloes of growth inhibition, we have developed a plate assay (20) which exploits the observation that pheromone stimulation is necessary for *S. pombe* to undergo meiosis and sporulate (32). The pheromones induce the meiosis-specific subfunctions of the mating-type locus, *mat1-Pm* and *mat1-Mm*. A heterozygous diploid cell will therefore sporulate in nitrogen-free medium because it produces both pheromones and both receptors and autostimulates the response pathway. In contrast, a mutant strain defective in early subfunctions of the mating-type locus will sporulate only on exposure to the appropriate pheromone.

Thus, a mutant lacking *mat1-Pc* is unable to produce either P-factor or the M-factor receptor and will sporulate only if supplied with P-factor. Similarly, disruption of *mat1-Mc* produces a strain that sporulates only on addition of M-factor. The assay is assessed by the positive iodine reaction of the sporulating cells, and the extent of the stained halo provides a semiquantitative measure of the amount of pheromone being produced. Further details are described elsewhere (20).

The diploid tester strains used in the assay are designed to be homogeneous, non-switching, and reasonably stable. They are heterozygous for *mat1-M/mat1-P* and are mutated in one of the early subfunctions, either Mc^- or Pc^-. The silent mating-type cassettes have been deleted, and at least one of the *mat1* cassettes is unable to receive the double-strand break usually associated with mating-type switching and mitotic recombination (41). On rare occasions, the strains can become homozygous by proximal mitotic recombination to produce nonsporulating derivatives which might interfere with the assay. A suitable strain for the detection of M-factor is EG-670, whereas the strain EG-819 can be used to detect P-factor. The haloes produced by EG-819 in response to P-factor are smaller than those produced by EG-670 in response to M-factor. This correlates well with previous observations. Larger haloes are formed by strain EG-820, which lacks the sxa2 protease (20).

1. Grow the diploid tester strain and the pheromone source strain on MSA at 30°C.
2. Suspend a loopful of each strain into separate tubes containing 1 ml of sterile water.
3. Transfer 7-μl droplets of the tester strain to a fresh MSA plate. This provides a control of the tester strain without added pheromone.
4. Add 1 μl of source cells to the tube containing the tester strain, mix, and transfer 7-μl droplets to the MSA plate.
5. Add 2 μl of source cells to the tube containing the tester strain, mix, and transfer 7-μl droplets to the MSA plate.
6. Add 7 μl of source cells to the tube containing the tester strain, mix, and transfer 7-μl droplets to the MSA plate. Steps 4–6 provide an increasing concentration of source cells to form microcolonies within the patches of diploid tester cells.
7. Incubate the MSA plate at 30°C for 3 days.
8. Expose the now confluent patches on the MSA plate to iodine vapor (*caution:* use a fume hood) and inspect the haloes with a dissection microscope.

Pheromone-Controlled Transcription

Northern blot analysis of pheromone-induced genes is a powerful technique for studying the signal transduction pathway. The most useful genes are those whose expression has an absolute requirement for a pheromone signal. This is true for the *mat1-Pm* gene in *P* cells (15) and the *sxa2* gene in *M* cells (7). Several other genes are induced by nitrogen starvation but show enhanced expression when stimulated by pheromone. The three *mfm* genes (9) and the *fus1* gene (J. Petersen, personal communication) belong to this class, whereas genetic analysis suggests that *mam2* (12, 42), *map3* (13), and *map2* (7) may also be enhanced by pheromone. A Northern blot of the *gpa1* gene detects a pheromone-induced band in addition to the constitutively expressed transcript (14). It is obviously necessary to use strains possessing the *sxa2*⁺ gene when monitoring P-factor-induced expression of this gene, but we suggest that strains lacking the gene should be used when other P-factor induced genes are being studied. Induction of *sxa2* occurs rapidly following exposure to P-factor, but the induction is transient as the sxa2 protease degrades the extracellular P-factor. The P-factor-dependent transcription of other genes may be increased in strains defective in *sxa2*.

The method detailed here has been developed for the preparation of RNA from 30 ml of cells (38). Gloves should be worn to prevent degradation by ribonucleases, and all tips, tubes, and solutions should be treated appropriately.

1. Inoculate an appropriate volume of cells from a preculture into a defined minimal medium so that the cell density following overnight growth at 30°C will be 5–8 × 10⁶ cells/ml. Cultures with lower densities will give low yields of RNA, whereas the sexual differentiation program is not induced efficiently if the culture density is above 1 × 10⁷ cells/ml before shifting to nitrogen-depleted medium.

2. Harvest the cells and resuspend at the same density in minimal medium lacking a nitrogen source.

3. Add pheromone to a final concentration of 20 n*M* and continue incubating for at least another 5 hr to induce the pheromone response.

4. Immediately chill the cells by adding 30 ml of culture to a 40-ml centrifuge tube containing crushed ice. All subsequent steps should be performed at 0°C.

5. Harvest the cells and ensure the complete removal of the supernatant.

6. Resuspend the cells in 500 μl of ice-cold LETS buffer [0.1 *M* LiCl, 10 m*M M* EDTA, 10 m*M M* Tris-HCl, pH 7.4, 0.02% (w/v) sodium dodecyl sulfate (SDS)] and transfer to a siliconized 15-ml Corex tube containing 2.2 g glass beads (1 mm diameter). Add 600 μl of ice-cold phenol–chloroform previously equilibrated with LETS buffer. Vortex briefly. Samples at this point can be stored at −20°C and should be thawed on ice before continuing.

7. Vortex for a total of 3 min. Avoid excessively heating the sample by alternating between 30 sec of vortexing and 30 sec of incubation on ice.

8. Add 500 μl of ice-cold LETS buffer and vortex briefly.

9. Centrifuge at 12,000 *g* for 10 min at 0°C and transfer 800 μl of supernatant to a microcentrifuge tube.

10. Add 500 μl of ice-cold phenol–chloroform (equilibrated with LETS buffer), vortex, and centrifuge at 10,000 g for 5 min at 0°C. Repeat the extraction until the interphase is clear (usually requires two extractions).
11. Extract the supernatant twice with 500 μl of chloroform.
12. Transfer the supernatant to a fresh tube and add $\frac{1}{10}$ volume of 5 M LiCl. Vortex briefly and leave at -20°C for at least 3 hr.
13. Collect the RNA by centrifugation at 20,000 g for 45 min at 0°C, wash with 70% (v/v) ethanol, and resuspend in 100 μl of water. The sample should be stored at -20°C, and 10 μg of total RNA should be used per track for Northern blot analysis.

Acknowledgments

Work done in our groups was supported by the Cancer Research Campaign, UK (J.D.), and the Danish Center for Microbiology (R.E. and O.N.)

References

1. B. Errede and D. E. Levin, *Curr. Opin. Cell Biol.* **5,** 254 (1993).
2. J. Kurjan, *Annu. Rev. Genet.* **27,** 147 (1993).
3. O. Nielsen, *Trends Cell Biol.* **3,** 60 (1993).
4. Y. Fukui, Y. Kaziro, and M. Yamamoto, *EMBO J.* **5,** 1991 (1986).
5. U. Leupold, *Curr. Genet.* **12,** 543 (1987).
6. J. Davey and O. Nielsen, *Curr. Genet.* **26,** 105 (1994).
7. Y. Imai and M. Yamamoto, *Genes Dev.* **8,** 328 (1994).
8. J. Davey, *EMBO J.* **11,** 951 (1992).
9. S. Kjaerulff, J. Davey, and O. Nielsen, *Mol. Cell. Biol.* **14,** 3895 (1994).
10. R. Egel, *Curr. Genet.* **26,** 187 (1994).
11. S. Michaelis, *Semin. Cell Biol.* **4,** 17 (1993).
12. K. Kitamura and C. Shimoda, *EMBO J.* **10,** 3743 (1991).
13. K. Tanaka, J. Davey, Y. Imai, and M. Yamamoto, *Mol. Cell. Biol.* **13,** 80 (1993).
14. T. Obara, M. Nakafuku, M. Yamamoto, and Y. Kaziro, *Proc. Natl. Acad. Sci. U.S.A.* **88,** 5877 (1991).
15. O. Nielsen, J. Davey, and R. Egel, *EMBO J.* **11,** 1391 (1992).
16. Y. Imai and M. Yamamoto, *Mol. Cell. Biol.* **12,** 1827 (1992).
17. V. L. MacKay, S. K. Welch, M. Y. Insley, T. R. Manney, J. Holly, G. C. Saari, and M. L. Parker, *Proc. Natl. Acad. Sci. U.S.A.* **85,** 55 (1988).
18. R. Egel, *Planta* **98,** 89 (1971).
19. R. Egel and M. Egel-Mitani, *Exp. Cell Res.* **88,** 127 (1974).
20. R. Egel, M. Willer, S. Kjaerulff, J. Davey, and O. Nielsen, *Yeast* **10,** 1347–1354 (1994).
21. D. Beach, L. Rodgers, and J. Gould, *Curr. Genet.* **10,** 297 (1985).
22. M. Kelly, J. Burke, M. Smith, A. Klar, and D. Beach, *EMBO J.* **7,** 1537 (1988).

23. S. Moreno, A. Klar, and P. Nurse, *in* "Methods in Enzymology" (C. Guthrie and G. R. Fink, eds.), Vol. 194, p. 795. Academic Press, San Diego, 1991.

24. Y. Yamawaki-Kataoka, T. Tamaoki, H.-R. Choe, H. Tanaka, and T. Kataoka, *Proc. Natl. Acad. Sci. U.S.A.* **86,** 5693 (1989).

25. D. Young, M. Riggs, J. Field, A. Vojtek, D. Broek, and M. Wigler, *Proc. Natl. Acad. Sci. U.S.A.* **86,** 7989 (1989).

26. T. Maeda, N. Mochizuki, and M. Yamamoto. *Proc. Natl. Acad. Sci. U.S.A.* **87,** 7814 (1990).

27. M. Kawamukai, M. Ferguson, M. Wigler, and D. Young, *Cell Regul.* **2,** 155 (1991).

28. A. Sugimoto, Y. Iino, T. Maeda, Y. Watanabe, and M. Yamamoto, *Genes Dev.* **5,** 1990 (1991).

29. S.-H. Wang, C.-B. Xue, O. Nielsen, J. Davey, and F. Naider, *Yeast* **10,** 595 (1994).

30. J. Davey, *Yeast* **7,** 357 (1991).

31. C.-B. Xue, J. M. Becker, and F. Naider, *Int. J. Pept. Protein Res.* **37,** 476 (1991).

32. U. Leupold, O. Nielsen, and R. Egel, *Curr. Genet.* **15,** 403 (1989).

33. U. Leupold, M. Sipiczki, and R. Egel, *Curr. Genet.* **20,** 79 (1991).

34. Y. Imai, S. Miyake, D. A. Hughes, and M. Yamamoto, *Mol. Cell. Biol.* **11,** 3088 (1991).

35. Y. Wang, M. Boguski, M. Riggs, L. Rodgers, and M. Wigler, *Cell Regul.* **2,** 453 (1991).

36. Y. Iino and M. Yamamoto. *Mol. Gen. Genet.* **198,** 416 (1985).

37. P. Nurse, *Mol. Gen. Genet.* **198,** 497 (1985).

38. O. Nielsen and R. Egel, *EMBO J.* **9,** 1401 (1990).

39. C. J. Bostock, *Exp. Cell Res.* **60,** 16 (1970).

40. K. Skarstad, H. B. Steen, and E. Boye, *J. Bacteriol.* **163,** 661 (1985).

41. R. Egel, *Cold Spring Harbor Symp. Quant. Biol.* **45,** 1003 (1981).

42. H.-P. Xu, M. White, S. Marcus, and M. Wigler, *Mol. Cell. Biol.* **14,** 50 (1994).

Section III

Bacterial Gene Structure and Regulation

[14] Use of DNA Methylation Deficient Strains in Molecular Genetics

Lene Juel Rasmussen and M. G. Marinus

Introduction

Instead of the protective immune system present in higher organisms, bacteria possess restriction–modification systems to protect themselves from the intrusion of foreign DNA. A restriction–modification system is a two-component system consisting of a restriction endonuclease that catalyzes the cleavage of DNA and a modification methylase which protects the endogenous DNA by methylation of bases in the recognition sequence of the endonuclease. In *Escherichia coli,* two general restriction systems are present to exclude foreign DNA: (1) the K-type restriction–modification system (*hsdRMS*) and (2) the methylation-dependent system (MDRS) (Table I).

DNA from a wide variety of higher eukaryotes and bacteria contains methylated cytosines and adenines which, in *E. coli,* are restricted by the *mcrABC* and the *mrr* gene products, respectively. Cloning of the genomic DNA from other organisms into *E. coli* requires a strain deficient in all of the MDRS genes to prevent degradation of the incoming foreign DNA, and the highest methylation tolerance is obtained using a strain mutated in *mcrA* and deleted for the *mrr, hsdRMS,* and *mcrBC* genes. *Escherichia coli* strains which carry a mutation in the *hsdR* gene are very useful for the construction of cDNA libraries of unmethylated non-*E. coli* DNA. After transformation of the foreign DNA into a *hsdR* mutant strain, the incoming DNA will be methylated by the *hsdM*-encoded restriction methylase; thereafter, the DNA can be transformed into any *E. coli* host strain.

Not only does cloning of foreign DNA require restriction–modification deficient strains, but these strains are also valuable for plasmid and phage recovery from eukaryotic cells. Bacterial phage or plasmid which is integrated into the eukaryotic genome would be methylated by host-encoded methylases. The phage or the plasmid is recovered from the genome together with a piece of the flanking eukaryotic DNA by digestion with appropriate restriction endonucleases and transfected into a bacterial strain. To avoid exclusion of the foreign DNA, the bacterial host should be restriction–modification deficient.

Escherichia coli contains two methylases, DNA adenine methylase (Dam) and the DNA cytosine methylase (Dcm), that are not part of any restriction–modification system (Table II). The Dam methylase plays a role in the initiation of chromosome replication, transposition, mismatch repair and gene expression. In contrast, the biological function of the Dcm methylase is not clear, though it has been implicated to play a role in very short patch repair.

TABLE I   The Two General Restriction Systems
in *Escherichia coli*

Gene	Gene product	Function
K-type restriction–modification system		
hsdR	Restriction endonuclease	Cuts unmethylated DNA
hsdM	Modification methylase	Protects *E. coli* DNA from degradation
Methylation-dependent system		
mcrA	Restriction endonuclease	Cuts DNA that is methylated at cytosines (foreign cytosine methylation)
mcrBC	Restriction endonuclease	Cuts DNA that is methylated at cytosines (foreign cytosine methylation)
mrr	Restriction endonuclease	Cuts DNA that is methylated at adenines and cytosines (foreign adenine and cytosine methylation)

The primary use of *dam* and *dcm* methylation deficient strains in DNA technology is to obtain DNA molecules which lack methylation at either or both Dam and Dcm recognition sites. This allows for digestion with methylation-sensitive restriction enzymes. These enzymes are inactive if their recognition sequences contain or overlap with Dam or Dcm recognition sites. A detailed list of endonucleases affected by DNA methylation is given by Kessler and Manta (1) and by Nelson and McClelland (2). An example of the use of methylation to create very rare cleavage sites is to prevent the action of *Cla*I at overlapping *dam* sites (ATCGATC) but not at nonoverlapping

TABLE II   Dam and Dcm Methylases from *Escherichia coli*

Gene	Gene product	Function
dam	Dam methylase	Methylates adenine in -GATC- sequences in DNA
dcm	Dcm methylase	Methylates cytosine in -CC(A/T)GG- sequences in DNA

sites (ATCGATG/A/T), and other examples are described by Grimes *et al.* (3) and by Kobb and Szybalski (4). DNA isolated from strains deficient in Dcm-mediated cytosine methylation has been shown to give better results in Maxam–Gilbert DNA sequencing. This is probably because 5-methylcytosine is resistant to chemical attack by hydrazine and, therefore, 5-methylcytosine positions are absent in Maxam–Gilbert sequence patterns. In contrast, DNA isolated from *dam* mutant strains is not suitable for the Sanger dideoxy chain termination method of DNA sequencing owing to single-strand DNA breaks present in the DNA, presumably as the result of nicking by the MutH protein (5).

Transformation of Methylation Deficient Strains

Escherichia coli is still the most commonly used microorganism for cloning and expression of foreign DNA. However, other microorganisms can also be used for this purpose. A *Salmonella* strain that carries mutations in all three restriction–modification systems has been constructed for cloning of heterologous DNA (6). Also, mutant strains of gram-positive *Bacillus* species have been developed that are deficient in the 6GM restriction-modification system and can, therefore, be used for the cloning of foreign DNA (7).

The transformation efficiency in several microorganisms is higher when the transforming DNA lacks Dam and Dcm methylation [*Streptomyces avermitilis* (8) and *Bacillus thuringiensis* (9)]. Shuttle-vector DNA isolated from a *dam dcm* double-mutant *E. coli* strain increased the transformation efficiency of *Streptomyces avermitilis* 400 to 10,000-fold. Plasmid DNA isolated from *B. thuringiensis, B. megaterium,* or a *dam dcm E. coli* strain efficiently transformed several *B. thuringiensis* strains. It was also found that DNA isolated from one *B. thuringiensis* species transformed another species very poorly, indicating that *B. thuringiensis* strains differ in their endogenous DNA modification and restriction systems.

Methods for Transformation

Preparation of Plasmid DNA from Methylase-Deficient Escherichia coli Strains

1. Grow bacterial strain overnight at 37° C in L broth (Luria broth, see recipe below).
2. Transfer 1.5 ml of cell culture to a centrifuge tube and centrifuge for 3 min at 10,000 rpm.
3. Remove supernatant and resuspend cells in 0.1 ml of cold solution I [1% (w/v) glucose, 10 mM EDTA, pH 8.0, 40 mM Tris-HCl, pH 8.0].
4. Add 0.2 ml of freshly prepared solution II [200 mM NaOH, 1% (w/v) sodium dodecyl sulfate (SDS)] and mix gently.

5. Add 0.15 ml of 3 M sodium acetate, pH 4.8, and mix gently.
6. Centrifuge for 15 min at 10,000 rpm at 4° C.
7. Transfer the supernatant to a new Eppendorf tube and add 0.2 ml of 96% (v/v) ethanol. Incubate for 20 min at −20° C.
8. Centrifuge for 15 min at 15,000 rpm at 4° C.
9. Wash pellet once with 1 ml of 70% ethanol and once with 1 ml of 96% ethanol. Dry the DNA and resuspend the dry pellet in 50 μl TE buffer (10 mM Tris-HCl, pH 8.0, 0.1 mM EDTA, pH 8.0).

There is no significant reduction in the yield of plasmid DNA from a *dam* mutant strain.

Transformation in Salmonella typhimurium

1. Grow bacterial strain to a cell density of about 1 × 10⁸ cells/ml in L broth.
2. Transfer 1 ml of cell culture to a centrifuge tube and centrifuge for 1 min at 15,000 rpm.
3. Remove the supernatant, resuspend cells in 0.1 ml of cold 100 mM $CaCl_2$, and incubate on ice for 5 min.
4. Add 2–20 μg plasmid DNA, heat-shock the cells immediately for 1 min at 42° C, and incubate the cells in L broth for 90 min to allow for expression of the drug resistance marker on the plasmid.
5. Plate cells on plates containing the appropriate antibiotic to select for the plasmid and incubate at 37° C.

The above procedure is from Ryu and Hartin (6). For transformation with foreign DNA, strain JR501 can be used because it is restriction but not modification deficient (10).

Transformation in Escherichia coli

1. Grow bacterial strain to a cell density of about 5 × 10⁸ cells/ml in L broth.
2. Transfer 10 ml of cell culture to a centrifuge tube and centrifuge for 3 min at 12,000 rpm.
3. Remove the supernatant and resuspend cells in 10 ml of ice-cold CP buffer (50 mM $CaCl_2$, 10 mM PIPES, pH 6.8) and incubate on ice for 20 min.
4. Collect the cells by centrifugation for 5 min at 8000 rpm.
5. Resuspend the cells in 10 ml of fresh ice-cold CP buffer. The cells at this point can be stored at 4° C for several days, depending on the strain, without any significant loss in transformation efficiency.
6. For transformation, 1–2 μg of plasmid DNA and 40 μl of cold PCM buffer (10 mM PIPES, 10 mM $CaCl_2$, 10 mM $MgSO_4$, pH 6.8) are added to 100 μl of competent cells.

7. The transformation mixture is incubated on ice for 5 min, followed by heat-shock for 5 min at 37°C.
8. Add 350 μl L broth and incubate for 90 min to allow expression of the drug-resistance marker on the plasmid.
9. Plate and incubate the cells at 37°C.

The above procedure is adapted from Beckingham and White (11). For transformation with foreign DNA, strain GM2299 can be used because it is an easily transformable strain and lacks both the MDRS and the K-type systems (12). Other suitable *dam* and *dcm* strains have been described by Palmer and Marinus (12).

Transformation in Bacillus thuringiensis

1. Grow *B. thuringiensis* culture overnight to saturation with shaking at 30°C in BHIG [brain–heart infusion (see below) supplemented with 0.5% (v/v) glycerol]. Dilute the overnight culture 1:20 in BHIG and incubate for 1 hr at 30°C with shaking.
2. Wash the cells once in EB (0.625 M sucrose, 1 mM MgCl$_2$) and resuspend the pellet in $\frac{1}{2}$ volume of EB.
3. Mix 0.8 ml of cells with less than 1–2 μg of DNA in a 0.4-cm cuvette and chill the mixture on ice for 5 min.
4. Set a 5-Ω resistor in series between the cuvette and a Bio-Rad (Richmond, CA) Gene Pulser and use a single discharge (2500 V, 25 μF) for electroporation.
5. Incubate the cells on ice for 5 min.
6. Dilute the cells into 1.6 ml of BHIG medium and incubate with shaking at 30°C for 1 hr.
7. Plate and incubate the cells at 30°C.

The protocol given is adapted from MacNeil (8). Some alternative methods for transformation of *Bacillus* strains are given by Belliveau and Trevors (13), Bone and Ellar (14), Lereclus *et al.* (15), and Mahillon *et al.* (16), and transformation of *B. subtilis* has been described by Khanna and Stotzky (17).

Culture Media

L broth is used for growth of *S. typhimurium* and *E. coli.* To prepare L broth, dissolve 10 g Bacto-tryptone (Difco, Detroit, MI), 5 g Bacto-yeast extract (Difco), and 10 g NaCl in water to 1000 ml. Adjust to pH 7.0–7.2 with 10 N NaOH. Sterilize by autoclaving and store at room temperature.

For growth of *B. thuringiensis* and *E. coli,* it is necessary to make up brain–heart infusion (BHI) medium. Dissolve 20 g Bacto brain–heart infusion (Difco) in water to 1000 ml. Adjust to pH 7.0–7.2 with 10 N NaOH. Sterilize by autoclaving. Supplement with 0.5% (v/v) glycerol to obtain BHIG.

Recombinant DNA Technology

Construction of DNA Libraries

Some important factors governing the choice of a strain for the construction of cDNA and genomic libraries are (1) the transformation efficiency of the strain, (2) the restriction of foreign DNA by host-encoded restriction–modification systems, and (3) the stability of the vector containing the inserts.

In contrast to genomic libraries, the preparation of cDNA libraries does not result in the restriction of incoming foreign DNA by the K-type restriction–modification or the methylation-dependent restriction systems in *E. coli*. However, some cDNA cloning procedures described in Sambrook *et al.* (18) involve methylation of DNA to create cleavage sites for certain restriction enzymes. Because most eukaryotic and prokaryotic genomic DNAs are methylated, it is important to use a restriction deficient host strain to construct genomic DNA libraries. In the case of *E. coli* as host, the preferred strain should carry a mutation in the *mcrA* gene and also be deleted for the *hsdR, mcrBC,* and *mrr* genes.

Plasmid and phage recovery from eukaryotic genomes involve methods that are useful for studying mutagenesis in eukaryotic cells. A plasmid or a phage is randomly integrated into a genomic or a cDNA library to create mutant clones. The mutagenized library is then allowed to integrate into the eukaryotic genome using gene replacement techniques. Cells which show the desired phenotype are selected, the plasmid or the phage is recovered together with a piece of genomic DNA by restriction digestion and transformed into a bacterial strain for further characterization. The mutagenized clones become methylated by the host-encoded methylases after their integration into the host genome. To avoid exclusion of the heterologous eukaryotic DNA, it is advisable to use a bacterial strain which is deficient in restriction and modification for recovery of the integrated plasmid or phage. Two methods for plasmid (19) and phage recovery (20) are given below, but other methods can be found in Burns *et al.* (21).

Methods for Plasmid and Phage Recovery

Plasmid Recovery from Eukaryotic Cells

1. Isolate chromosomal DNA and digest the DNA with a restriction enzyme which is unique for the plasmid to be rescued.
2. Ligate the restricted genomic DNA at a concentration of 10 μg/ml to favor circularization.
3. Analyze the ligation mixture on an agarose gel to determine the ligation efficiency.

4. Transform 50 μg of the ligated DNA into a competent methylation deficient *E. coli* strain, using the method described earlier for transformation of *E. coli*.
5. Plate the cells on plates containing a drug which selects for the plasmid.

Phage Recovery from Eukaryotic Cells

1. Isolate chromosomal DNA and digest the DNA with an enzyme which does not cut the phage vector.
2. Purify the vector from a preparative agarose gel.
3. Treat the DNA with an *in vitro* λ phage packaging extract (*in vitro* packaging extracts are commercially available from several biotechnology companies).
4. Analyze the DNA on an agarose gel to verify the presence of vector monomers.
5. Transfect the *in vitro* packaged phage particles into a methylation deficient *E. coli* strain.

The vectors isolated from eukaryotic cells would be in a hypermethylated state that could lead to host restriction during *in vitro* packaging or after recovery of the phage particles in bacteria. Highly methylated vectors can be rescued with high efficiency by using *E. coli* C-derived packaging extracts and propagation of the phage in *E. coli* C, which naturally lacks the *dcm* gene responsible for cytosine methylation. Alternatively, one could use any *E. coli* strain which carries a mutation in the *dcm* gene.

Site-Directed Mutagenesis

Site-directed mutagenesis is a widely used method to study the structure and function of DNA sequences. The strategy for site-directed mutagenesis is to clone the target DNA into a vector which allows DNA to be obtained in a single-stranded form. An oligonucleotide primer which is complementary, except for the region to be altered, is hybridized to the single-stranded DNA, and the primer is extended using DNA polymerase. After the primer extension step, DNA ligase is used to circularize the heteroduplex, which is then transformed into *E. coli*. The transformants would contain either the nonmutated plasmid or plasmids that contain the oligonucleotide-directed mutation.

The efficiency of site-directed mutagenesis depends on the quality of each step in the procedure and on the *E. coli* strain used as the host. If the heteroduplex contains either a mismatch or 1- to 3-bp insertions or deletions, mismatch repair of the unannealed bases in the primer can lead to loss of the desired mutant species. Preparation of the single-stranded DNA template from a Dam methylase-deficient strain alleviates this problem because the template strand is unmethylated and is no longer resistant to mismatch repair. As the *mutHLS* system would fail to distinguish between the mutant and the wild-type strands, random repair to either strand will occur. Alternatively, one can methylate the heteroduplex molecule *in vitro* using commercially available Dam methylase before transformation into the *E. coli* host. This prevents preferential repair of the mutant strand because both strands of the heteroduplex become resistant to mismatch repair.

Methods for Methylation of DNA in Vitro

The *E. coli* Dam methylase can be used to methylate GATC sites *in vitro*. For this purpose, one can use either the purified enzyme, which is commercially available, or a cell extract from a Dam methylase proficient strain.

Preparation of Crude Extract for in Vitro Dam Methylation

1. Grow cells to a density of 2×10^8 cells/ml in a 1-liter flask containing 100 ml BHI (brain–heart infusion) medium (for preparation of BHI medium, see above).
2. Transfer the cell culture to a centrifuge tube and centrifuge for 5 min at 10,000 rpm.
3. Wash the cell pellet once with TE buffer (10 mM Tris-HCl, pH 8.0, 0.1 mM EDTA, pH 8.0).
4. Freeze the cell pellet (tube upside down) at $-70°$C.
5. Thaw the pellet and resuspend in 2 ml buffer (60 mM Tris-HCl, pH 7.6, 1 mM MgSO$_4$, 20% glycerol).
6. Add 0.2 ml lysozyme (from a 10 mg/ml stock solution).
7. Incubate on ice for 10 min.
8. Sonicate the cells 6 times for 10 sec duration each; chill the tube on ice between sonications.
9. Centrifuge at 4°C for 10 min at 10,000 rpm.
10. Add 50 μl of RNase and DNase, each from 1 mg/ml stock solutions.
11. Incubate at room temperature for 30 min.
12. Add EDTA to a final concentration of 4 mM and dispense 200 μl of cell extract into Eppendorf tubes. Freeze the tubes at $-70°$C.

Methylation of DNA Template Using a Crude Extract

1. Mix in an Eppendorf tube 2 μl crude extract, 200 μl methylase assay buffer [50 mM HEPES–KOH, pH 7.8, 1 mM EDTA, 1 mM dithiothreitol (DTT), 200 mM potassium glutamate, 5% glycerol] (22), 5 μM S-adenosylmethionine, and 20 μg DNA.
2. Incubate for 20 min at 37°C (any temperature between 30° and 42°C can be used with satisfactory results).
3. Transfer the reaction mixture onto ice and add 200 μl of buffer-saturated phenol (the phenol should be saturated with 0.5 M Tris-HCl, pH 8.0, at room temperature).

4. Mix for 10 sec on a vortex mixer.
5. Centrifuge for 30 sec at room temperature at 10,000 rpm.
6. Remove 200 μl of the aqueous solution (top layer). Be careful not to transfer the interphase or the phenol phase.
7. Confirm the methylation status of the DNA by testing for susceptibility to cleavage by *Dpn*I.

At this point the DNA should be fully Dam methylated, and the heteroduplex molecule can be transformed into any mismatch repair proficient *E. coli* strain.

Methylation of DNA Template Using Purified Dam Methylase

1. Add 20 units Dam methylase (New England Biolabs, Beverly, MA) to 2 μg purified heteroduplex DNA, 5 μM *S*-adenosylmethionine, and 100 μl methylase assay buffer (50 mM HEPES–KOH, pH 7.8, 1 mM EDTA, 1 mM DTT, 200 mM potassium glutamate, 5% glycerol) (22).
2. Incubate at 37°C for 1 hr.
3. Heat the reaction mixture to 70°C for 5 min.
4. Extract the mixture with phenol (as above) to inactivate the methyltransferase.
5. Confirm the methylation status of the DNA by testing for susceptibility to cleavage by *Dpn*I.

Transform the fully methylated heteroduplex DNA into any mismatch repair proficient *E. coli* strain.

Studying Protein–DNA Interactions and Understanding Chromosome Structure

Genomic DNA sequences are being determined so fast that a complete nucleotide composition of several organisms can be expected within the near future. This has created a need for efficient techniques to determine the regions of the genomes that are bound by proteins and to understand the structural organization of chromosomes.

Dam methylase can be utilized for identifying both DNA–protein interactions and for probing chromosome structure and function. Both Dam and Dcm methylases can be expressed in other organisms such as *Saccharomyces cerevisiae* which do not naturally harbor the enzymes. Methods have been developed which use endogenous or artificially introduced methylases to methylate all genomic targets except those protected by protein or nonprotein factors interfering with methylase action (12, 23, 24). The protected targets are identified using methylation-sensitive restriction endonucleases. The extent of *dam* and *dcm* methylation can be determined using type II restriction enzymes that cleave the cognate recognition sequences but vary in their sensitivity to methylation. For example, *Mbo*I cleaves only unmethylated GATC *dam* recognition sites, whereas *Dpn*I is selective for methylated GATC sites and

*Sau*3AI cleavage is unaffected by methylation. Similarily, cleavage by *Eco*RII is inhibited by methylation of *dcm* recognition sites, whereas *Bst*NI cleavage is unaffected by cytosine methylation.

Analysis of DNA Sequences Using Methylation-Sensitive Restriction Endonucleases

1. Grow strain in L broth (see above) to OD_{600} of 1.0. Harvest 3 ml bacteria by centrifugation at 6000 *g* for 10 min.
2. Wash twice with SE (75 m*M* NaCl, 25 m*M* EDTA, pH 7.4).
3. Resuspend the pellet in 480 μl water and add an equal volume of 2% (w/v) low melting point agarose.
4. Load onto an agarose block mold ($\sim$240 μl per block).
5. Cut each block into four pieces. Incubate overnight at 56°C in a 1.5-ml Eppendorf tube containing 1 ml of 0.5% (w/v) proteinase K, 0.5 *M* EDTA, pH 7.4, 1% (w/v) *N*-lauroylsarcosine, pH 9.6.
6. Transfer to a new Eppendorf tube containing 1 ml of 1 m*M* PMSF (phenylmethylsulfonyl fluoride) in TLE buffer (10 m*M* Tris-HCl, 0.1 m*M* EDTA, pH 7.5) and rotate gently for 2 hr at room temperature. Replace with fresh PMSF in TLE buffer and repeat this step.
7. Wash three times with an additional 1 ml of TLE buffer without PMSF. Blocks can be stored at this stage for several weeks at 4°C.
8. Cut each block into thirds or halves for restriction enzyme digestion. Equilibrate in 150 μl restriction enzyme buffer on ice for 30 min. Replace with fresh buffer, add restriction enzyme (5 units for overnight incubation, 20 units for a 4-hr incubation), and incubate at 37°C.
9. Replace the restriction buffer with 1 ml of 0.5× TBE buffer (1× TBE is 89 m*M* Tris–borate, 89 m*M* boric acid, 1 m*M* EDTA, pH 8.0) and hold on ice for 30 min before gel electrophoresis.
10. Analyze the DNA by pulsed-field gel electrophoresis in a 1% agarose gel using a Bio-Rad CHEF Mapper and the following parameters: run time 26 hr 40 min at 6 V/cm, initial switch time 11.75 sec, final switch time 1 min 33.69 sec, included angle 120°, and running buffer 0.5× TBE buffer.
11. Stain the gel with ethidium bromide in water (final concentration 0.5 μg/ml) for 30 min.
12. Destain the gel in water for 1 hr.

The above procedure is from Palmer and Marinus (12). Plasmids containing the *E. coli dam* and *dcm* genes have been constructed from which these genes can be moved into shuttle vectors for expression of the methylases in organisms other than *E. coli*. Plasmids containing *E. coli* replicons and methylase genes are described in Table III.

TABLE III   Plasmids for Overexpression of *dam*
and *dcm* Genes from *Escherichia coli*

Plasmid	Gene	Vector/replicon	Ref.
pTP166	*dam*	pBR322	*a*
pMQ191	*dam*	pACYC184	*a*
pALO160	*dam*	pJEL109	*b*
pYin	*dam*	ColE1 derivative	*c*
pDCM1	*dcm*	pBR322	*d*

[a] M. G. Marinus, A. Poteete, and J. A. Arraj, *Gene* **28**, 123 (1984).
[b] A. Lobner-Olesen, E. Boye, and M. G. Marinus, *Mol. Microbiol.* **6**, 1841 (1992).
[c] V. U. Nwosu, *Biochem. J.* **283**, 745 (1992).
[d] A. S. Bhagwat, A. Sohail, and R. J. Roberts, *J. Bacteriol.* **166**, 751 (1986).

Conclusions

The identification and characterization of restriction–modification systems in bacteria have been of immense value to modern recombinant DNA technology. Both in basic and in medical sciences, the interest in exploring eukaryotic organisms has increased considerably since the 1980s. The complexity and the lack of knowledge of how to handle higher eukaryotes as genetic systems have made it difficult to use those organisms in modern DNA technology for purposes such as cloning and expression of genes, purification of proteins, and mutagenesis.

Bacteria, and in particular *E. coli,* are convenient systems for most recombinant DNA work. *Escherichia coli* grows very fast, and reliable methods for DNA purification and transformation (some are described in this chapter) have been developed, which makes the organism ideal for manipulating recombinant DNA. *Escherichia coli* strains which are deficient in all of the restriction–modification systems have been constructed. Furthermore, *dam* and *dcm* mutant derivatives of *E. coli* exist and are available from the authors.

Despite the advantages mentioned earlier, there are some problems with the *dam* mutant strains. One problem is the low transformation frequency of the strains. The Dam methylase plays a role in the initiation of chromosome replication as well as in the initiation of plasmid replication for several commonly used cloning vectors. In *E. coli,* a fully Dam methylated origin of replication is able to initiate, whereas a hemimethylated origin fails to do so. Transformation of the *dam* mutant strains with fully methylated plasmids leads to the hemimethylation of the plasmid origin after the first round of replication, and this is believed to lower the number of transformants.

Instability of the vector containing certain inserts can, in some cases, cause problems owing to hyperrecombination. This is usually circumvented by using strains deficient in RecA activity. In this context, it is important to note that the viability of *dam* mutants requires the function of the recombination pathways involving the RecA, RecB, RecC, and RecJ gene products, whereas the RecF, RecN, and RecQ proteins are not essential for viability (25).

It has been reported that the yield of plasmid DNA from a *dam* mutant strain is lower than that obtained from an isogenic wild-type strain. In our hands, however, there is no significant reduction in the yield of plasmid DNA from a *dam* mutant strain. The discrepancy could be due to the difference in the genetic backgrounds of the strains used. We as well as other workers have found that the transformation efficiency of restriction–modification deficient *E. coli* strains is high and similar to that of the isogenic parental strains.

Methods such as footprinting and gel-shift assays have been developed to study protein–DNA interactions *in vitro*. There has been a growing need to study these interactions *in vivo*. In this chapter, we have described a method, pulsed-field gel electrophoresis, that can be used to analyze the regions of DNA protected by proteins *in vivo*. The Dam methylase sensitive restriction enzymes *Dpn*I and *Mbo*I can also be used to study DNA–protein interactions *in vivo* with the requirement that the organism under study expresses the Dam methylase. The *E. coli dam* gene has been cloned and can be expressed in other organisms. One approach for the detection of DNA–protein interactions using methylation sensitive restriction enzymes is to compare the restriction pattern of the DNA in a strain expressing the DNA-binding protein with that in a strain lacking this protein. If the recognition sequence for the DNA-binding protein contains GATC sites, these will exist unmethylated or hemimethylated in the wild-type strain owing to the protection from methylation by the DNA-binding protein. GATC sites will then be sensitive to both *Mbo*I and *Sau*3AI but resistant to *Dpn*I digestion. In contrast, the sites will remain unprotected in a mutant strain lacking the DNA-binding protein and hence will be fully methylated. The DNA from the mutant will then be susceptible to both *Dpn*I and *Sau*3AI digestion but will be resistant to *Mbo*I. If the DNA sequence to be studied does not contain GATC sites, other restriction methylases and enzymes of different sequence specificities can be used.

The use of methylation deficient strains in modern DNA technology is gaining wider applications, and the future prospects for their use in studying DNA–protein interactions *in vivo* are promising. Understanding how Dam methylase affects bacterial gene expression could serve as a model for studying methylation control of gene expression in eukaryotes. Also, exploiting Dam methylation to study DNA–protein interactions as mentioned above may help define the precise nucleotide contacts of the DNA-binding proteins.

Acknowledgments

A part of this work was supported by a grant from the Danish Natural Science Research Council to L.J.R. We thank D. RayChaudhuri and U. von Freiesleben for critical reading of the manuscript and for helpful suggestions.

References

1. C. Kessler and V. Manta, *Gene* **92,** 1 (1990).
2. M. Nelson and M. McClelland, *Nucleic Acids Res.* **20,** 2145 (1992).
3. E. Grimes, M. Koob, and W. Szybalski, *Gene* **90,** 1 (1990).
4. M. Kobb and W. Szybalski, *in* "Nucleic Acids and Molecular Biology" (F. Eckstein and D. M. J. Lilley, eds.), Vol. 8, Springer-Verlag, Berlin, 1994.
5. M. Carraway and M. G. Marinus, unpublished data (1990).
6. J.-I. Ryu and R. J. Hartin, *BioTechniques* **8,** 43 (1990).
7. P. Haima, S. Bron, and G. Venema, *Mol. Gen. Genet.* **209,** 335 (1987).
8. D. J. MacNeil, *J. Bacteriol.* **170,** 5607 (1988).
9. A. Macaluso and A.-M. Mettus, *J. Bacteriol.* **173,** 1353 (1991).
10. S. P. Tsai, R. J. Hartin, and J.-I. Ryu, *J. Gen. Microbiol.* **135,** 2561 (1989).
11. K. Beckingham and R. White, *J. Mol. Biol.* **137,** 349 (1980).
12. B. R. Palmer and M. G. Marinus, *Gene* **143,** 1 (1994).
13. B. H. Belliveau and J. T. Trevors, *Appl. Environ. Microbiol.* **55,** 1649 (1989).
14. E. J. Bone, and D. J. Ellar, *FEMS Microbiol. Lett.* **58,** 171 (1989).
15. D. Lereclus, O. Arantes, J. Chaufaux, and M.-M. Lecadet, *FEMS Microbiol. Lett.* **60,** 211 (1989).
16. J. Mahillon, W. Chungjatupornchai, J. Decock, S. Dierickx, F. Michiels, M. Peferoen, and H. Joos, *FEMS Microbiol. Lett.* **60,** 205 (1989).
17. M. Khanna and G. Stotzky, *Appl. Environ. Microbiol.* **58,** 1930 (1992).
18. J. Sambrook, E. F. Fritsch, and T. Maniatis, "Molecular Cloning: A Laboratory Manual." Cold Spring Harbor Laboratory, Cold Spring Harbor, New York, 1989.
19. S. G. N. Grant, J. Jessee, F. R. Bloom, and D. Hanahan, *Proc. Natl. Acad. Sci. U.S.A.* **87,** 4645 (1990).
20. J. A. Gossen, W. J. F. De Leeuw, C. H. T. Tan, E. C. Zwarthoff, F. Berends, P. H. M. Lohman, D. L. Knook, and J. Vijg, *Proc. Natl. Acad. Sci. U.S.A.* **86,** 7971 (1989).
21. N. Burns, B. Grimvalde, P. B. Ross-Macdonald, E.-Y. Choi, K. Finberg, G. S. Roeder, and M. Snyder, *Genes Dev.* **8,** 1087 (1994).
22. K.-H. Hulsmann, A. Bergerat-Couland, and U. Hahn, *Nucleic Acids Res.* **18,** 7189 (1990).
23. M. X. Wang and G. M. Church, *Nature (London)* **360,** 606 (1992).
24. S. Ringquist and C. L. Smith, *Proc. Natl. Acad. Sci. U.S.A.* **89,** 4539 (1992).
25. K. R. Peterson and D. W. Mount, *J. Bacteriol.* **175,** 7505 (1993).

[15]   Identifying Ends of Infrequent RNA Molecules in Bacteria

Vincent J. Cannistraro and David Kennell

Introduction

Procedures to identify the sequence of end nucleotides in a population of RNA molecules can be valuable for many studies. The most common application has been the determination of the site of transcription initiation or termination from the 5′ or 3′ ends of the RNA. A less common application has been the identification of the sites of cleavages of an RNA. The number of ends generated by cleavages in growing bacteria is far more than the number from initiation or termination. RNA can be grouped into two metabolically distinguishable classes: the (relatively) stable and the unstable. The former includes the ribosomal RNA (rRNA) which in *Escherichia coli* can comprise as much as 80%, or more, of the total RNA mass and the transfer RNA (tRNA) which is about 15%. The messenger RNA (mRNA) is the unstable component which, as a result of its instability, is approximately 50% of the RNA being synthesized but only about 3% of the mass (1). Even the stable RNA is a product of several cleavages. The rRNAs in *E. coli* are synthesized from seven operons as single long transcripts of approximately 5500 nucleotides that include the 16 S, 23 S, and 5 S rRNAs, one or more tRNAs, as well as spacer regions, each of which are cut out before further processing to the final products (2–4). The RNases responsible for converting precursor transcripts into the final functional products are called processing RNases.

The 5′ and 3′ ends of mRNA are even more diverse in the cell as a result of its mechanism of degradation. As first shown for the metabolism of *lac* mRNA (5) and subsequently for other mRNAs (e.g., 6–8), the degradation includes many endonucleolytic cleavages from the 5′ to 3′ end. Since the mass decay of the resulting oligonucleotides is extremely rapid [almost as fast as the rate of synthesis (9)], and given the very low total mass of any mRNA, it was not a certainty that cleavage ends would have a half-life sufficiently long to allow their detection and identification. The concentration of RNA ends in a cell population vary at least 10,000-fold between cleavage ends of the more abundant mRNAs (<1 end/cell) (10) and ends of stable RNAs and another 1000-fold for the least abundant mRNAs. Thus, identification procedures must be chosen that are sufficiently sensitive.

Procedures to Map the Ends of RNA Molecules

The two most commonly used mapping procedures in the current literature are indirect. To map a 5′ end, both of them use a 5′-^{32}P-labeled DNA (or RNA) probe to

Methods in Molecular Genetics, Volume 6

hybridize to an appropriate length of the RNA. In one case the 3' end of the probe extends beyond the 5' end of the RNA to be determined. The hybrid is then treated with S1 nuclease, which degrades only the single-stranded "tail" at the 3' end of the DNA. Knowing the exact position of the 5'-^{32}P end of the DNA then allows locating the 5' end of the RNA from the size of the ^{32}P-labeled DNA determined in a gel. The other procedure is conceptually related in that, instead of degrading an overlapping end of the probe, the 5' end is extended by reverse transcriptase to meet the 5' end of the RNA. Both techniques, termed S1 nuclease mapping (11) and primer extension (12, 13), have been extremely useful tools but have limitations. S1 nuclease can cut at mismatched base pairs, but more importantly, in terms of accuracy, it can terminate one or more nucleotides from the correct 5' end, especially when the end is rich in A or T residues which give less stable hybrid bonds (14). The latter consideration is significant for identification of the 5' transcription starts of RNA since a large fraction of them initiate with 5'-pppAp-.

Reverse transcriptase can pause or terminate prematurely at double-stranded regions of the RNA. The probability of such stops cannot be predicted from the calculated free energy of a presumptive stem–loop structure, as apparently other factors are also important (12). Alternatively, the predictive accuracy for estimating a given stem–loop free energy may be deficient. It has been shown that RNA nucleotide interactions can be more complex than a conventional base pairing (15); for example, the sequence of the loop nucleotides can affect the stability of the stem (16). On the basis of conventional base pairing, the probability that there will be one or more stable secondary structures of 5 bp or more is very high in any random sequence of 40 or more nucleotides (17). For this reason, the primer should be within 100 nucleotides from the 5' end (13). It is recommended that S1 nuclease mapping be done simultaneously to verify the correct 5' end (12). The same probe used for S1 mapping can be degraded to provide sequence ladders run in parallel lanes. In matching a primer-extended band to a given nucleotide in the sequencing lanes, allowance has to be made for the somewhat faster migration of the primer-extended band with its 3'-OH end as opposed to the chemically cleaved bands with 3'-P ends. Since it is recommended that the 5'-^{32}P-labeled primer should be 75–150 nucleotides, if double-stranded, or 30–40 nucleotides long if single-stranded (13), the total length of the final product will probably be greater than 80 nucleotides, so that a precise alignment to the sequence ladders is crucial.

A variation of primer extension uses T4 DNA polymerase instead of reverse transcriptase (18). The RNA is annealed to a single-stranded DNA containing the complementary strand to the 5' region of the RNA, for example, from M13 phage DNA that contained the cloned gene sequence. A 5'-^{32}P-labeled primer is annealed upstream, and T4 DNA polymerase then fills in the gap and stops at the 5' end of the RNA. The length of the ^{32}P-labeled DNA then gives the end nucleotide of the DNA or one nucleotide before the 5' end of the RNA. The problem of interfering secondary structures is lessened somewhat because a DNA duplex is less stable than an RNA duplex.

It has been concluded that all three procedures can be in error one or more nucleotides when identifying an RNA end nucleotide (12, 14, 18). The importance of a method that allows identification of the exact end nucleotides with a high degree of accuracy is especially critical in cases in which a distribution of different ends in a population is being monitored. For example, the first observation that bacterial mRNA degradation includes many endonucleolytic cleavages included S1 nuclease mapping (5), but exact cleavage specificity could only be determined by direct sequencing of the resultant ends (5). If only one of every five ends had been incorrect by being off by one nucleotide, the actual sequence specificity would have been obscured. The following sections outline methods for the direct determination of RNA ends.

Measuring Relative Frequencies of Each Nucleotide on the Ends of an RNA Population

In Vivo Transcription Starts of Total Cell RNA

The preferred starting nucleotides for RNA synthesis can be identified by labeling growing bacteria with $^{32}P_i$ and completely digesting the purified ^{32}P-labeled RNA to mononucleotides with alkali. The initiating nucleotides have 5'-triphosphate ends which can be separated easily from all other nucleotides with the A, C, G, and U tetraphosphates separated from one another for quantitation.

1. Bacteria are grown in a minimal salts–sugar medium, centrifuged at room temperature in mid-log phase ($\sim 3 \times 10^8$ cells/ml for *E. coli*) and the pellet rinsed twice (at 37°C if growth is at 37°C), and then resuspended in the same medium but lacking phosphate and with another buffer such as 10 mM Tris-HCl, pH 7.2, plus $^{32}P_i$. The cells are shaken for 5 to 8 min before bringing to 0°C and pelleting. Collecting and washing cells on a large filter can be faster and more gentle and has been used often (e.g., Ref. 19). Any abrupt changes in growth conditions should be avoided, such as chilling the cells or changing the growth medium (e.g., a "shift-down" from broth to mininal salts media), as they can cause a significant lag before growth resumes with an ill-defined physiology. For example, the ratio of total mRNA to stable RNA synthesis could be changing. The choice of labeling time is governed by a combination of physiological considerations. Exponential growth during the labeling period can be verified in a separate culture without $^{32}P_i$ by following the A_{600}. The bacteria should continue growing exponentially during this short time, but the labeling time must be long enough to label the precursor nucleotide pools uniformly and incorporate sufficient label for analyses. At the same time the labeling time should be as short as possible (e.g., 5 sec) in order to avoid enriching for ends of stable molecules. However, this latter condition is met

if the sizes of RNA molecules chosen to study are all small, for example, between the bromphenol blue (BPB) and xylene cyanol (XC) dye markers on a 20% polyacrylamide gel (~7 to 27 nucleotides/molecule); only transcripts initiated, or cleaved, during the last second of labeling will be analyzed.

2. After labeling, the bacteria can be opened by any of a number of procedures. However, an important consideration is to inhibit the alkaline phosphatase and RNase I that are released from the periplasm. We increase the EDTA concentration to 20 mM using the EDTA–lysozyme–sodium dodecyl sulfate (SDS) method of lysis and detect no phosphatase or RNase activity in extracts (20). We routinely monitor for phosphatase activity by incubating [γ-^{32}P]ATP with an aliquot of the sample, since the released ^{32}P$_i$ can be separated easily from the ATP by thin-layer chromatography (21). An irreversible inhibitor of RNase activity, such as diethyl pyrocarbonate (22), can be added on lysis (23).

3. Protein can be removed by three phenol extractions in 0.2 M NaCl followed by ether extraction of the aqueous phase and then ethanol precipitation in 0.2 M NaCl. In separate experiments we have found that virtually all the oligonucleotides, even the smaller ones and mononucleotides, are present in the ethanol precipitate with the volumes of cells and concentrations used (20).

4. After vacuum drying, the pellet is resuspended in 0.5 ml of 10 mM Tris-HCl, pH 7.2, 1 mM EDTA, 10% glycerol, plus BPB and XC dyes, and applied to several small wells or to one large well (~12 cm) in a 20% polyacrylamide gel (1.5 mm × 20 cm × 40 cm) containing 7 M urea, as described (20). The thick gel is used simply for ease of subsequent manipulations and to accommodate larger sample volumes. Electrophoresis is monitored by the position of the dye markers. The ^{32}P bands are visualized by autoradiography, and one or more bands of a given size, or size range, are cut out for analysis. If the ends of mRNA are being analyzed, the gel samples can be limited to those with molecules less than about 60 nucleotide lengths in order to avoid any stable RNA species which are higher in the gel (with the exception of smaller nascent transcripts of stable RNA or decaying spacer fragments).

5. The gel band is crushed and the RNA eluted from it by gentle agitation in water overnight at 4°C. The sample is then applied to a small DE52 (Whatman, Clifton, NJ) column and salts eluted in 0.3% triethylamine carbonate (TEC) followed by elution of the oligonucleotides in 30% TEC (21). The TEC is removed by evaporation in a Speed-Vac (Savant, Holbrook, NY), and the sample is resuspended in 10 μl of 0.5 N KOH, incubated for at least 4 hr at 37°C, then neutralized with 7% perchloric acid; the salt is removed by centrifugation, and HCl is added to 0.1 N to eliminate 2′,3′ cyclic bonds (24).

6. The mononucleotides are separated by the two-dimensional Sanger system (24) with the first-dimension electrophoresis on a 55-cm cellulose acetate strip (Sartorius, Edgewood, NY) followed by homochromatography on a poly(ethylenimine) (PEI) sheet (Macherey-Nagel, Duren, Germany). In this system the 5′ ends from transcription initiation are the fastest migrating nucleotides in electro-

phoresis but do not migrate in homochromatography because they contain four phosphates, pppNp. The pNp nucleotides are separated in both directions (21), whereas 5'-OH-Np hardly move in the first dimension but run with the solvent front in homochromatography. The latter molecules contain the bulk of the ^{32}P from an *in vivo* labeling because they include all the internal nucleotides in the uniformly labeled oligonucleotides.

In an experiment in which molecules between the dye markers were pooled, the results showed that 70% or more of the 5'-^{32}P in 5'-labeled nucleotides was in pppGp, which had at least four times more label than did pppAp (Fig. 1). There were much

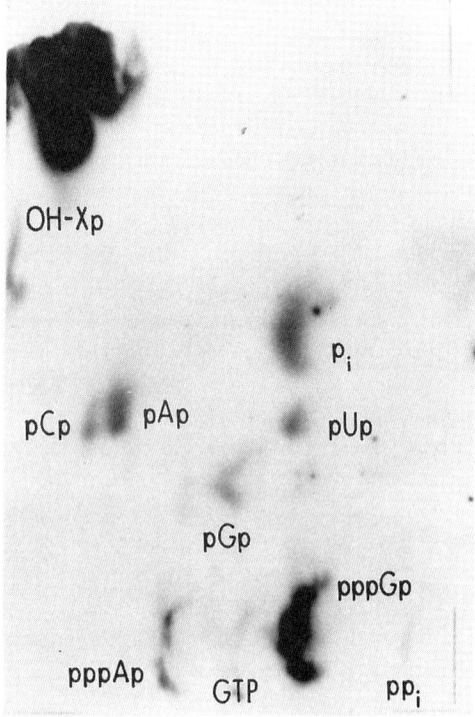

FIG. 1 The 5'-end nucleotides on RNA oligonucleotides in growing *E. coli.* Strain DK533, which lacks RNase I or I* activity, was labeled *in vivo* with ^{32}P$_i$, and the label in the 5' nucleotide of oligonucleotides was determined as described. Most of the ^{32}P is in the 5'-OH-Np molecules which were derived from all the internal nucleotides released by alkali plus the 5' end nucleotides on oligonucleotides that had a 5'-OH group. These molecules migrate with the solvent front in homochromatography, and part of that pool is seen in the top left corner. Other species are also identified. Electrophoresis was from left to right and homochromatography bottom to top. [Reprinted from *Eur. J. Biochem.* **213,** 285 (1993), with permission of the publisher.]

smaller amounts of ^{32}P in pNp molecules, and control experiments showed that most of those molecules probably arose by unavoidable hydrolysis of pppGp and, to a lesser extent, pppAp (20). These results indicate that almost all of the RNA oligonucleotides in the bacteria with a phosphate on the 5' end are nascent RNA molecules in the process of being synthesized. Of these, about 80% start with G and 20% with A; C and U starts were too low to detect in this kind of global analysis. This ratio is in contrast to the predicted starting nucleotide for approximately 70 E. coli mRNAs; in those tabulations the ratio of A to G starts was about 2 to 1 (25, 26).

It is unlikely that the results from in vivo labeling differ from the tabulated mRNA starts because the former also include starts of stable RNAs, as well as mRNAs. The transcripts from the seven rrn operons, for rRNA and several tRNAs, incorporate 40 to 45% of the total pulse-labeled RNA. However, they are more than 6 times longer than the average mRNA, and the number of cleavages per RNA length is much lower for a stable RNA precursor. Furthermore, most of them start with A or G with no preference reported for one or the other nucleotide (3). The total starts for the tRNA molecules not synthesized from the rrn operons is not known but is probably only about 5% of the total RNA initiations. Thus, the preponderance of pppGp- starts must result from the frequency of G starts in the total mRNA population. Possible explanations for the apparent disagreement with the tabulated results have been discussed (20).

The End Nucleotides from Degradative Cleavages of mRNA

The preceding analysis of in vivo ends by in vivo labeling could not be used to determine the frequencies of each nucleotide at the 5' ends of oligonucleotides derived by cleavages of mRNA. The reason is dictated by the enzymology of E. coli endoribonucleases (endoRNases) and chemistry of alkaline hydrolysis of RNA. All of the known major E. coli processing endoRNases (RNases E, H, P, and III) generate 5'-P_i and 3'-OH ends, whereas the known degradative endoRNases (RNases M, R, I, and I*) generate 5'-OH and 3'-P_i ends (20). The 5'-OH ends generated by degradative cleavages in the cell would be indistinguishable from the 5'-OH ends generated from all the internal nucleotides by the in vitro alkaline hydrolysis, which also gives 5'-OH-, 3'-P_i ends. To circumvent this problem, the oligonucleotides can be labeled in vitro so that only the 5' end nucleotide has ^{32}P.

7. The above steps 1 through 3 are the same except that exponential cells are centrifuged directly without washing or growth in the presence of ^{32}P$_i$.
8. After ethanol precipitation, the dried pellet is resuspended in as small a volume of 10 mM Tris-HCl, pH 7.2, as possible, and 20 μl is 5'-end-labeled with [γ-^{32}P]ATP in the T4 polynucleotide 5'-hydroxyl-kinase (kinase) reaction, using RNase-free enzyme (e.g., from BRL, Gaithersburg, MD, or USB, Cleveland, OH).

9. To label oligonucleotides that have phosphate(s) on the 5′ ends, another aliquot is first treated with alkaline phosphatase. A convenient procedure, which saves having to remove the alkaline phosphatase by phenol extractions of a very small volume, is to titer the enzyme first using [γ-^{32}P]ATP and then use so little phosphatase that it takes 2 or 3 hr for the reaction to be complete. As the subsequent kinase reaction is over in 10 to 15 min, the dephosphorylations during the reaction are negligible when the sample is immediately loaded onto a polyacrylamide gel for electrophoresis.

10. Gel bands are eluted, and the oligonucleotides are alkali-digested and run in the first dimension on cellulose acetate strips (see steps 4–6 above). The mononucleotides are transferred by blotting onto PEI strips but are not run in the second homochromatography dimension.

The 5′ end nucleotides from different size classes of oligonucleotides in a typical experiment are shown in Fig. 2. The following results were obtained. First, in the wild-type parent strain, about 50% of ends with a 5′-OH were 5′-OH-Ap- for all sizes of oligonucleotides except the very small ones, with the mononucleotide pool having a distribution of the four nucleotides not too different from that in total cell RNA. The next most abundant was 5′-OH-Gp- (about half as abundant as 5′-OH-Ap-), and the least abundant was 5′-OH-Up- (barely detectable).

Second, RNase I* has been identified as a form of the periplasmic endoRNase I

Length (~nt)	RNAse I$^+$						RNAse I$^-$						RNAse I$^-$ (+Alkaline Phosphatase)					
	%A	C	A	G	U	%G	%A	C	A	G	U	%G	%A	C	A	G	U	%G
>80	53					28	67					14	56					29
50	55					13	70					14	58					23
30	49					27	63					10	58					30
3	48					26	58					12	54					33
1	33					30	59					8	55					27

FIG. 2 Abundance of each 5′-nucleotide in RNA oligonucleotides from growing *E. coli*. The oligonucleotides were separated by size and the relative amounts of each of the four nucleotides at the 5′ ends determined by the procedures outlined. The approximate lengths in nucleotides is given on the far left-hand side, and the approximate percentages of A and G for each size are indicated. RNase I$^+$ are representative oligonucleotides from wild-type strain DK 390; RNase I$^-$ denotes samples from the *rna* mutant, DK533. The far right-hand part shows 5′-nucleotides for oligonucleotides from DK533 first treated with alkaline phosphatase before labeling in the kinase reaction. [Reprinted from *Eur. J. Biochem.* **213,** 285 (1993), with permission of the publisher.]

that is in a different cell component (27); it is not present in an *rna* gene mutant that makes no RNase I. In such a mutant the fraction of oligonucleotides starting with 5'-OH-Ap- is dramatically increased to close to 70% of the total and is close to that level even in the small molecules and mononucleotide pool. These results do not reflect a preference of the kinase for a particular end nucleotide, in particular 5'-OH-Ap-. As a control there is approximately equal labeling of ends derived from RNase I cleavage, which has no obvious nucleotide specificity. Moreover, early studies concluded that there may even be a slight preference for pyrimidine over purine nucleotides, whereas 5'-OH-Ap- and 5'-OH-Gp- ends labeled equally well (28).

Third, after dephosphorylation, the level of subsequent phosphorylation increases. The important result is that almost all of the increase is due to the phosphorylation of 5'-OH-Gp- molecules. The level of [5'-^{32}P]Gp- increased 3- to 4-fold over its level in the absence of prior dephosphorylation. There was a much smaller increase in [5'-^{32}P]Ap- and no detectable increase in [5'-^{32}P]Cp- or [5'-^{32}P]Up-.

In conclusion, the large increase in [^{32}P]Gp- that results from dephosphorylating the oligonucleotides before end-labeling is consistent with the conclusions from *in vivo* experiments that almost all of the 5' ends with a phosphate are ends from transcription initiation, since the bulk start with 5'-pppGp-, and, at least for the processing endoRNases known, there is no evidence that their activity generates a great abundance of 5'-pGp- ends compared to the other nucleotides (20). Also, as noted, oligonucleotides resulting from cleavages of stable RNA precursors would be a small fraction of the total. The results indicated that most of the cleavages in bulk mRNA generate a 5'-OH-Ap- end and the least abundant are those with a 5'-OH-Up- end. Although some cleavages by the processing endoRNases do occur, they represent too small a fraction of the total to be detected in the analyses. The other conclusion, derived from the results with a strain lacking RNase I*, is that RNase I* must contribute some cleavages to mRNA and is especially important for the breakdown of the smaller oligonucleotides. The cleavage specificity was defined further by sequence determinations of the ends of the mRNA oligonucleotides.

Sequence Determination of the Ends of Infrequent RNA Molecules

Identifying the sequence of 5' or 3' ends of a specific RNA that is of very low abundance in the cell is much more difficult than is measurement of frequencies of each end nucleotide in a larger population. The fully induced mRNA from the *lac* operon of *E. coli* accounts for perhaps 0.1% of the RNA mass of the cell. Degradation starts as soon as the 5' end of the mRNA is synthesized (29), with a constant fraction of the surviving molecules being inactivated per unit time. The degradation progresses by cleavages that occur throughout the entire molecule at any of the vulnerable sites (5). The released oligonucleotides are then degraded almost as fast as they were synthesized (~40 nucleotide/sec at 37° C) (9). It will be a challenging task to determine the average distance between cleavages on a single message molecule, since what has been observed are the specific cleavage products from populations of millions of molecules.

On the basis of the broad specificity of the cleavage site (see below), it is highly probable that a given molecule is cleaved at only a fraction of the many sites observed in the greater population. If correct, a given 5′ end might only occur from a cleavage in, say, 1 of every 5 molecules. If the total length of an oligonucleotide were 120 nucleotides, its total lifetime expectancy would only be 3 to 4 sec (120/40). *LacZ* is the first message in the operon and has a functional half-life of 80 to 100 sec at 37° C. This means that one-half of the original 5′ initiation end is still present after 80 to 100 sec. Thus, from those relative life expectancies and the probability that an oligonucleotide with a given 5′ end is produced in only a fraction of the decaying molecules, it would seem reasonable that the concentration of a given 5′ cleavage end might be in the range of 1000 times lower than is the original 5′ initiation end. Possibly the only reason they can be detected at all is that in *E. coli* there is no evidence for a 5′-exoRNase activity and thus the 5′ end may be lost last.

Procedures to Sequence RNA

Specific endoRNases can be used to sequence RNA (24, 30–33) and are available commercially in kits (e.g., Pharmacia, Piscataway, NJ; USB). The activity of each enzyme has to be titered carefully because some enzymes introduce lower preference cuts at higher activities. Chemical procedures have also been developed which are analogous to the chemical sequencing of DNA (34). In both methods, some of the cleavages are not specific to one base recognition but favor one base over another, which can give some ambiguity to the reading. However, the main problem with both enzymatic and chemical methods for sequencing the very infrequent RNAs in a cell is that quite large amounts of the RNA are required to run the five or six lanes of enzyme-digested RNA plus a lane for the partially alkali-digested marker "ladder" and the undigested control.

We have used the wandering spot procedure (24, 35, 36), which is limited in terms of the length of RNA that can provide an unambiguous sequence but requires much less RNA sample. The following examples illustrate determinations of end sequences of specific messages transcribed *in vivo* from the chromosomal DNA in wild-type *E. coli*. Larger amounts of specific mRNAs could be obtained per cell from messages transcribed from plasmids or from specially constructed overproducing strains.

Preparation of Specific mRNA for End Sequencing

Most published transcription initiation sites have relied on analyses of the products of *in vitro* transcription, with some cases verified by S1 nuclease or primer extension mapping of the *in vivo* mRNA. It should be recognized that those results were not derived from a direct sequencing of the *in vivo* RNA. As far as we know, the only determination of the *in vivo* initiation end of any mRNA by its direct sequencing is that of the *E. coli lac* mRNA, and the general procedure is outlined. The methods, developed more than a decade ago (21, 37, 41), could be modified or improved with introduction of newer technologies and simplified by using cells that overproduce the message being sequenced.

11. The message to be analyzed must be purified free from all other RNA in the cell with resultant unavoidable losses. We have purified a specific region of a message by two cycles of hybridization to the complementary DNA trapped on a nitrocellulose filter (Schleicher & Schuell, Keene, NH) (37). In the original experiments, a large volume of growing bacteria (~6 liters when ends from cleavages were being identified) were pulse-labeled with [5-^{3}H]uridine for 2 min at 7 min after inducing the *lac* operon. The ^{3}H label is present only for monitoring the amount of hybrid RNA and should be kept as low as possible, as the incorporated ^{3}H causes RNA breaks with time of storage at any temperature as a function of the amount incorporated. A few critical considerations of the more detailed purification procedures (37) are mentioned.

12. If possible, it is best to avoid using commercial enzymes; if unavoidable, keep the amounts used to a minimum, to avoid any contamination with pancreatic RNase (RNase A) or related activities, which are widespread in plant and animal tissues. The lysate is sonicated briefly to reduce viscosity and treated with a minimum amount of commercial DNase (RNase free) that has been further purified by a column fractionation and tested by gel band assay for products of its incubation with RNA.

13. The T4 polynucleotide kinase reaction must be maximized in order to analyze the ends of extremely infrequent RNA molecules. Unfortunately, the kinase reaction is very sensitive to a number of inhibitors derived from the RNA preparation. The double-stranded DNA, for the region of RNA to be hybridized, is diluted in 1 M NaCl and filtered through a 0.2-μm nitrocellulose filter presoaked in 1 M NaCl; the DNA is collected in the filtrate. This quick step eliminates components in the DNA preparation that would bind to the filter and, after elution, interfere with the kinase reaction. The DNA is denatured by incubating in 0.5 N NaOH for 3 hr at room temperature. This is a useful denaturing step that also eliminates any RNA and nuclease activities as well as other alkali-labile contaminants from the DNA (38). The sample is then brought to 50 mM Tris-HCl, pH 8.2, and neutralized with 6 N HCl. If the DNA fragment is small, special precautions must be taken to avoid renaturation before the single strands are trapped on the filter (37). The DNA can be diluted to 0.5 μg/ml and placed in a boiling bath before plunging the tubes into ice water and adding 20× SSC (1× SSC is 15 mM sodium citrate plus 0.15 M NaCl, pH 6.0) to give 6× SSC and filtering through a filter, prewashed with 6× SSC. The baked filter is preincu-

bated in the hybridization mix for 1 hr at 52°C, which removes loosely bound DNA which would decrease the yield of filter-bound hybrid.

14. Hybridization is performed in 200 μl of 3× SSC plus 50% formamide and 2 mM EDTA (39). For a $\frac{1}{16}$ sector of a 24-mm filter with about 1 μg/sector of a DNA fragment containing the first 479 bp of the *lacZ* mRNA, the maximum level of hybrid (37) is attained in 20 hr at 52°C or in more than twice that time at 37°C (37). Hybridization at 45°C for 3 days followed by elution results in an average of 1 break in the RNA per 2000 nucleotides (37), which is sufficiently low to be insignificant compared to the number of ends of mRNA generated *in vivo*.

15. Nonspecific RNA could be removed from the filter by RNase A treatment, but controls have shown that the concentration could be decreased 100-fold below that originally recommended (40) to only 0.2 μg/ml. RNase T1 could be used in place of RNase A, but it has to be present at the high concentration of 10,000 units/ml (5) (~50 times more than RNase A on a microgram basis) (41). No RNase is used after the second hybridization.

16. As noted, a major consideration is to obtain an intact specific RNA that contains no inhibitors for the sensitive kinase reaction which follows as soon as possible afterward. The procedure recommended is to treat the washed filter with iodo-acetate to inhibit any RNase activity, then wash and elute the hybrid RNA in 400 μl of Tris-HCl, pH 8.0, plus 0.1% SDS at 100°C for 45 sec, followed by occasional shaking at 0°C for 5 min. The RNA is purified further by a second hybrid reaction, but the SDS is omitted in that elution, as it is a potent inhibitor of kinase activity (0.001% SDS inhibits the kinase reaction by ~30%) (21). Other denaturing agents, such as formamide or urea, or longer treatments at 100°C in water release unacceptable amounts of DNA, which inhibits the reaction, and an inhibitor of the kinase from the nitrocellulose filter itself. The DNA filter sector is kept as small as possible to minimize the amount of inhibitor released.

17. If the RNA is to be dephosphorylated, it is advantageous to treat with heat-labile calf intestine alkaline phosphatase before the last hybridization reaction rather than after it (21).

18. The picomoles of 5′ ends in the recovered RNA in a 400-μl kinase reaction could probably be 1000 times below the K_m value for the kinase–RNA association (21). It has been found that the incorporation increases with increasing [γ-^{32}P]ATP concentration to at least 7 μM, which is about 20 mCi of [γ-^{32}P]ATP made from carrier-free ^{32}P$_i$. Lower amounts of [^{32}P]ATP could be used, especially with higher yields of RNA.

19. In such a reaction approximately 99.7% of the ATP could still be unreacted even if all the ends were labeled. It is important that more than 99.9% of the [γ-^{32}P]ATP be removed before proceeding. Repeated ethanol precipitations are inadequate, but molecular sieving spin columns are effective in removing about 99% of the

ATP with each centrifugation when the sample is brought to 200 mM sodium citrate (or chloride) plus 10 mM EDTA. We generally run four such columns using BioGel P-4 polyacrylamide gel beads (Bio-Rad, Richmond, CA) and one BioGel P-10 column to eliminate small oligonucleotides that could have arisen from breakdown before labeling (21).

Separating and Sequencing the 5' Ends from the Mix of Oligonucleotides

20. Ideally, the next step would be to separate the oligonucleotides, which would range in size up to the length of the DNA probe. This might be feasible if the region being examined were sufficiently limited in size by the probe size. Another approach is to treat the eluted RNA with RNase T1, which is very specific for cleavage only after G residues, to give a mix of smaller oligonucleotides that can be fractionated more easily. The main shortcoming of this approach is that oligonucleotides that start with G or have G in the second position can be missed. To correct for this, another RNase with different specificity can be used in a separate aliquot. For example, RNase A that cleaves only after pyrimidines has been used to help determine the initiation start(s) of *lac* mRNA.

21. After RNase T1 digestion, the resulting mix of ^{32}P-labeled oligonucleotides can be partially separated by electrophoresis through a polyacrylamide gel. Samples in the resulting bands are eluted, concentrated, and then separated by electrophoresis on cellulose acetate strips to give quite well-separated oligonucleotides (see steps 4–6 above). Other separation procedures could be used.

22. Each ^{32}P-labeled oligonucleotide can be sequenced by the wandering spot technique outlined above. They are resuspended in 5 μl of 0.5 N KOH in a capillary, and hydrolysis is limited to 30 min at 37°C. The conditions had previously been found to give a good distribution of ^{32}P in the resulting sizes of products from mononucleotide to full-length substrate that ended with a G.

The Initiation Start of lac mRNA

The first identification of the *in vivo* 5' ends of *E. coli* mRNA used a DNA probe that contained 479 bp of the first message of the *lac* operon (*lacZ*) plus about 300 bp that preceded it. It was used to identify the 5' ends of RNA at the start of the *lac* mRNA (42). The results demonstrated the enormous difference in abundance of the initiation start end(s) and any ends derived from endonucleolytic cleavages. The bulk of the ^{32}P was in two oligonucleotides: 5'-A-A-U-U-G- accounted for 70 to 80% of all the label and 5'-A-U-U-G- for almost all of the remainder (Fig. 3). There was also a much weaker oligonucleotide 5'-A-U-U-A-G- that could originate from a weak initiation at position −52 (the first A in A-A-U-U-G- is defined as position +1) or, less

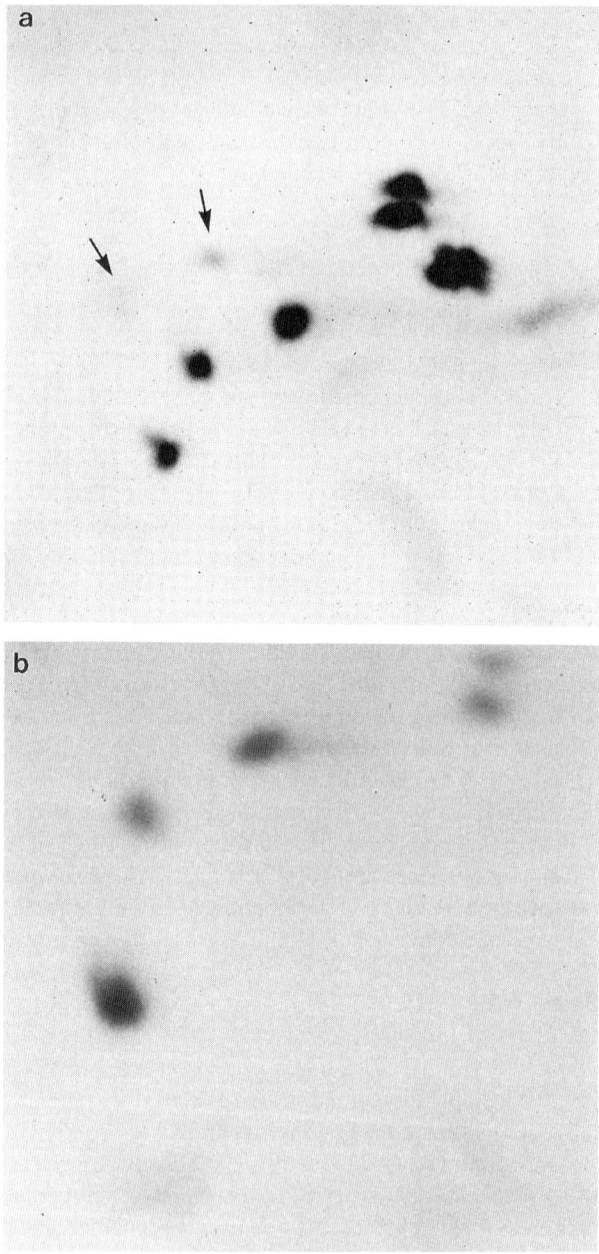

FIG. 3   End oligonucleotides that identify the 5′ transcription ends of *lac* mRNA in growing *E. coli* cells. The *lac* mRNA was purified by two rounds of hybridization to a 789-bp fragment of DNA containing the end of *lacI* to +479 bp of *lacZ*. The *lac* RNA was 5′-end-labeled,

likely, cleavage at the $-53/-52$ bond in a transcript reading from the *lacI* gene (42). As noted above, the ^{32}P in the 5′ end of an oligonucleotide derived from cleavage is probably hundreds of times less abundant than the ^{32}P in a 5′ initiation end. We did not try to extend the resolution sufficiently to identify the major cleavage ends in this region, but we did observe a probable family of very weakly labeled ends that included 5′-A-C-A-C-A-G-, 5′-A-C-A-G-, 5′-C-A-G-, and 5′-A-G-. The first of these would be derived by cleavage between nucleotides 23 and 24 and terminate at the first G in the presumed -A-G-G-A- ribosome binding site. Such cleavages could be initial inactivating events for decay of the *lacZ* message.

Comparison to Other Studies of the Initiation Start of lac mRNA

The original *in vitro* experiments of Maizels with *lac* UV5 DNA (43) found that 5′-A-A-U-U-G- was the major start but that about 15% of the starts were 5′-G-A-A-U-U-G-. Later *in vitro* studies with wild-type DNA reported about equal starts of G- and A- (44); still later *in vitro* experiments found that about 6% of the starts were 5′-A-U-, and the investigators favored the idea that it represented starts of 5′-A-U-U-G- (45).

The results of a careful S1 nuclease mapping to identify the *in vivo* start(s) of *lac* mRNA showed the limitations in accuracy of the method. The major bands protected by *lac* mRNA corresponded to four 5′ ends of approximately equal intensity at positions +1 through +4. All the residues in those positions are either A or U and would have given a weak hybrid. The authors noted that this weak duplex could have resulted in variable penetration of the nuclease to give the cluster of four consecutive end nucleotides (46). However, there was no band corresponding to the 5′-G- at -1, in agreement with the results from direct sequencing. With the use of RNase T1, the direct sequencing experiments would have missed 5′-G- starts. Therefore, an experiment was done using RNase A in place of RNase T1. The possible start at position -1 would have given an oligonucleotide with 5′-G-A-A-U, which would have been

---

digested with T1 RNase, and the resulting oligonucleotides separated, as described. Each oligonucleotide was partially digested with alkali and the resulting products separated in two dimensions: electrophoresis (left to right) and homochromatography (bottom to top). (A) The predominant oligonucleotide is 5′-A-A-U-U-G. The arrows point to the minor contamination of that oligonucleotide by the oligonucleotide shown in (B). (B) The oligonucleotide 5′-A-U-U-G is 4 to 5 times less abundant than the one in (A). The sequence is read from top to bottom. Note that pAp has 2′ and 3′ isomers that separate mostly in homochromatography. pCp isomers separate mostly in the electrophoresis dimension. Because pAp and pCp have the most similar migration rates in both dimensions, the isomer separations provide a useful criterion to distinguish between them. pUp runs very slowly in electrophoresis and pGp very slowly in homochromatography. [From *J. Mol. Biol.* **182,** 241 (1985), with permission of the publisher.]

readily observed in the analysis. However, it was not detected in such experiments (42).

Various explanations can be proposed to account for differences between the *in vitro* and *in vivo* observations. Most basic, of course, is that the reaction conditions in the cell may be different in ways significant for this reaction. For example, the *in vitro* reactions generally used equimolar concentrations of the nucleoside triphosphates, but the level of ATP in the cell is about three times higher than that of GTP (47, 48). It was shown in the later *in vitro* reactions that varying the ratio of these substrates did change the ratio of the initiation starts in the same direction (45). Another explanation has to do with the fact that *in vivo* experiments are conducted with actively metabolizing systems, whereas the *in vitro* studies are with relatively static systems. It seems less likely, but the terminal 5′-G may also be synthesized *in vivo* but then rapidly removed. The same could be possible for the terminal 5′-A to account for the presence of 5′-A-U-U-G. These reactions would be missing in a relatively pure *in vitro* synthesis reaction. Further studies will be necessary to identify the correct answer, but we think that direct sequencing of the ends from molecules taken from growing cells is probably the most reliable method that has been used. Unfortunately, it has not been applied to other messages.

5′ and 3′ Ends Resulting from Degradative Cleavages of mRNA

As noted, the first demonstration that *E. coli* mRNA is cleaved in many positions as a part of the degradation mechanism was shown for a region of the *lac* mRNA. S1 nuclease mapping showed the presence of discrete 5′- and 3′-ended intermediates (5), and cleavage ends were actually sequenced and provided an indication of the cleavage specificity of the endonucleolytic cuts. The preceding studies of the initiation ends of *lac* mRNA demonstrated the degree of sensitivity that would be required for their detection, let alone identification by sequencing. Because earlier analyses of the changing size distributions of a population of decaying *lac* mRNA had indicated that the *lac* mRNA is cleaved between the first *lacZ* message and the second *lacY* message (23, 49), an effort was made to sequence the expected cleavage 5′ and 3′ ends at that site. The DNA probe was 380 bp containing 53 bp of the end of *lacZ*, the first 273 bp of *lacY*, and the 54 bp of intervening sequence. Using the methods outlined above, it was a surprise to find that not only was there cleavage in the intervening sequence, but there were cleavages throughout the entire region being probed.

About 20 of the more abundant oligonucleotides were 5′ end sequenced. The results were quite consistent. In all but two cases, the cleavages occurred after a pyrimidine residue, and in all but two of these the cleavage was at a pyrimidine–adenosine bond (Table I). The wandering spots of a few oligonucleotides are shown in Fig. 4. At this level of resolution it is important to consider every possible source of error and perform adequate controls. The results suggested an enzyme specificity

TABLE I   The 5′ Ends of Abundant *lac* mRNA Molecules[a,b]

Oligonucleotide end	Position(s)[c]	S₁ band[d]	Base cleavage[e]
AAUAAU	52, 55	55	U↓A
AUUUCG	86	Yes	U↓A
AAG	95, 233	95	U↓A
AUUAUG	105	Yes	C↓A
UAUG	107	Yes	U↓U
AUG	7 positions	108	U↓A
AAACUUUUG	128	Weak	C↓A
ACUUU	131, 163	163	U↓A
UUUAUCAUG	168	Yes	U↓U
AUCAUG	171	Yes	U↓A
ACUUCCCG	184	Yes	U↓A
AUUUG[f]	201	No	G↓A
AUUUUUG	249	Yes	U↓A
AUUUCUCUG	261	Yes	U↓A
AACCG	61, 286	286	C↓A
ACCG	62, 287	287	A↓A
AAACUCG	312	Yes	C↓A
AAAUACCUG	327	Yes	C↓A
AUCUUCG	7[g]	Yes	U↓A

[a] From *J. Mol. Biol.* **192**, 257 (1986), with permission of the publisher.

[b] The *lac* mRNA oligonucleotides were from the 380-bp region of the operon between the *Eco*RI site near the end of *lacZ* and the first *Mbo*II site in *lacY*, except for the last 5′ end sequence which was from an adjacent downstream 296-bp DNA fragment. The sequences of more abundant oligonucleotides are shown.

[c] The position from the first nucleotide in the 380 nucleotide region.

[d] "Yes" means that after S1 nuclease mapping with 5′-³²P-labeled DNA, a band was seen in a gel that would correspond to an RNA end within a few nucleotides of the position indicated by direct sequencing.

[e] The 3′-nucleotide is deduced from the known nucleotide sequence of the DNA.

[f] Oligonucleotide was observed only in an experiment in which RNase T₁ was used to eliminate nonspecific RNA bound to the hybrid filter.

[g] The position from the first nucleotide in the 296 nucleotide region.

that has a preference for pyrimidine–adenosine cuts but can also cut pyrimidine–pyrimidine bonds. This specificity is the same as that of RNase A. Since that enzyme had been used to eliminate nonspecific RNA from the first filter hybridization, a major concern was the possible survival of a residual amount of enzyme activity even though presumably it had been inactivated by iodoacetate and SDS–phenol extractions. The experiments were repeated using RNase T1 in place of RNase A (see step 15 above). Although not so many oligonucleotides were sequenced using RNase T1, the three that were sequenced agreed with those observed using RNase A,

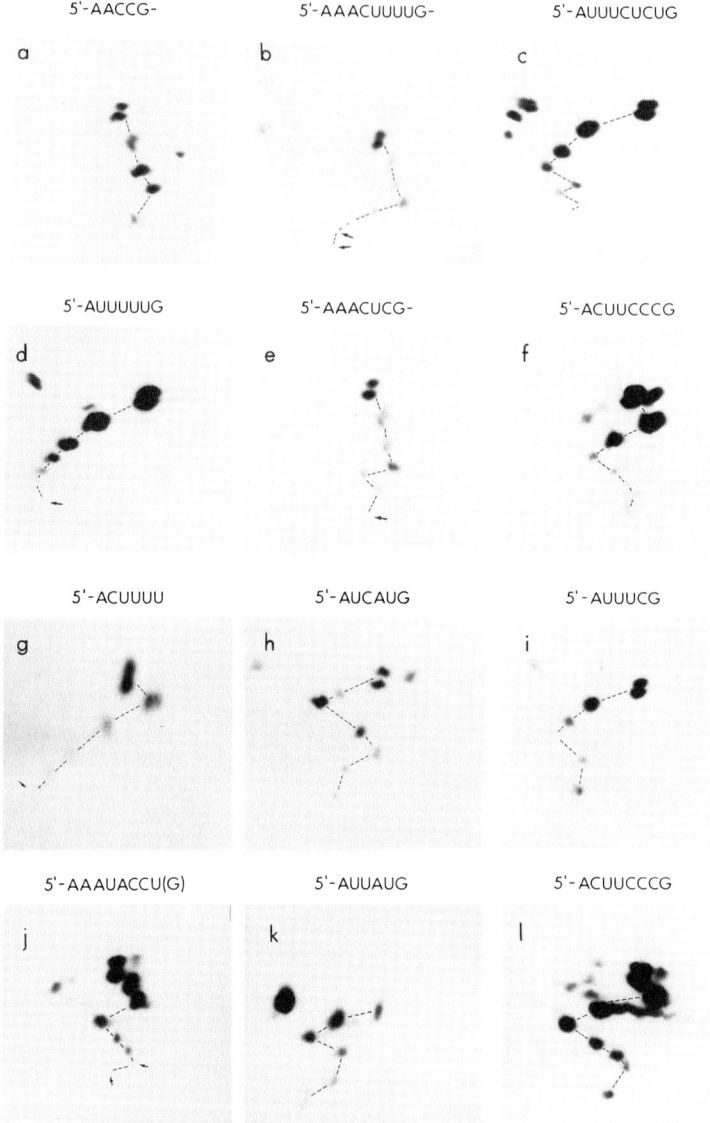

FIG. 4   Sequences of oligonucleotides derived from the 5′ ends of the more abundant *in vivo* molecules of *lac* mRNA from the region of the mRNA defined by a DNA fragment of 380 bp that spans the end of *lacZ*, the first 273 bp of *lacY*, and the intervening 54 bp. The *lac* mRNA was purified by two rounds of hybridization, 5′-end-labeled, and digested with RNase T1, and the resulting 5′-end oligonucleotides were sequenced, as described. The arrows point to ^{32}P-labeled oligonucleotides that required longer exposures than shown to be made visible. Oligonucleotides in (a) through (i) were derived from RNA–DNA hybrids that were treated with RNase A to remove nonspecifically bound RNA. Parts (j) through (l) were from hybrids treated with RNase T1 instead of RNase A. [Reprinted from *J. Mol. Biol.* **192,** 257 (1986), with permission of the publisher.]

including the large 5'-A-C-U-U-C-C-C-G. Thus, the specificity could not be an artifact of the RNase A treatment.

Sequencing the 3' Ends of RNA

Sequencing the 3' ends from cleavages in the mRNA is more difficult technically than is sequencing the 5' ends. The 3' ends can be ^{32}P-labeled using [5'-^{32}P]Cp in the RNA ligase reaction (50), but many fewer counts are incorporated than are incorporated in the kinase reaction. To increase the incorporation, the purification was limited to one hybrid reaction which gave 5–10 times greater yield of *lac* mRNA (41). The increased nonspecific RNA contamination could be eliminated by RNase A treatment if an excess of denatured heterologous DNA had been immobilized on the filter with the DNA probe. However, this resulted in the release of amounts of DNA during elution of the RNA hybrid, and the DNA inhibited the subsequent ligase reaction. Digesting the DNA with DNase I increased the inhibition, but it was found that treatment of the DNase products with alkaline phosphatase eliminated that inhibition (see Ref. 41 for more details).

The ligation reaction generates a new 3' end with a terminal C residue and an internal ^{32}P. Mild alkali digestion then gives an unusual pattern of two sets of wandering spots. The first set is produced from the expected ladder of single hits, but in each case there is a small fraction of molecules that experienced a second hydrolysis that removed the terminal C residue but would leave the internal ^{32}P. This "shadow" set of spots were observed adjacent to the primary single hit series. Four oligonucleotide 3' ends were presented, and they all ended (started) with 3'-U residues (5) (Fig. 5). The 380-nucleotide region of the RNA happens to be rich in U residues, which may explain why no 3'-C was observed, but the results are consistent with the specificity for pyrimidine–adenosine cleavages.

Mechanism of mRNA Degradation

These experiments remain the only ones that have sequenced directly the 5' and 3' ends which result from the endonucleolytic cleavages of mRNA that are occurring in the growing cell and, presumably, are the primary events in the degradation of this unstable component. There were many technical problems to overcome, and there was no guarantee that the search would (or could) be successful. The results showed a preference for hydrolysis of pyrimidine–adenosine bonds. Although limited to the mRNA from a small segment of the total genome, the generality of the results was evident from the analysis, described in the preceding sections, of the 5'-end nucleotides on oligonucleotides derived from cleavages of the total mRNA in the cell. Most of those ends terminated with 5'-OH-Ap-, which is consistent with most cleavages

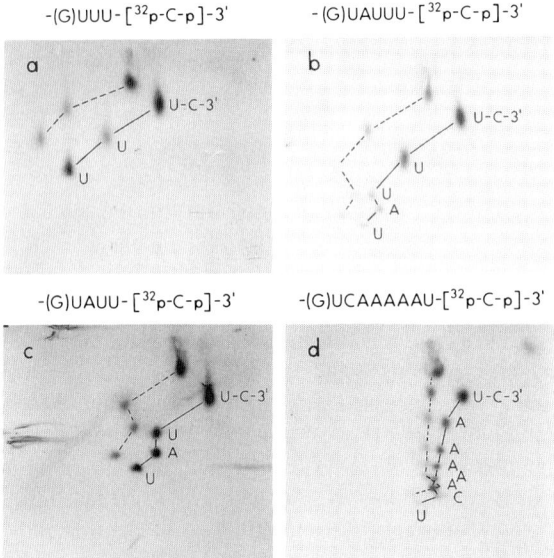

FIG. 5   Sequences of oligonucleotides derived from the 3′ ends of the more abundant *in vivo* molecules of *lac* mRNA from the same 380-bp region described in Fig. 4. The same procedures were followed as in Fig. 4, except that the molecules were 3′-end-labeled with [5′-^{32}P]Cp. The alkali digestion produces two sets of ^{32}P-labeled oligonucleotides. The major set results from one cleavage per molecule, and this set is connected by the continuous line. The minor set results from two cleavages per molecule, with one of those eliminating the 3′-terminal Cp; those products are connected by the broken line. Because the oligonucleotides were derived by RNase T1 digestion, each is assumed to be adjacent to a 5′-Gp. [Reprinted from *J. Mol. Biol.* **192**, 257 (1986), with permission of the publisher.]

between pyrimidine and adenosine residues from a degradative endoRNase. These observations suggested that there must be an endoRNase activity in *E. coli* with that preference. Subsequently, such an enzyme (RNase M) was identified (51).

Conclusions

The cases described here demonstrate the value of direct sequence determinations of RNA ends from growing cells. They were first used to identify the *in vivo* transcription start(s) of an mRNA. More challenging was their application to identify the sequence specificity for the *in vivo* endonucleolytic cleavages in the degradation of mRNA. Other applications could include the identification of cleavages and cleavage specificities for the processing of stable RNA transcripts. Sequencing the 3′ ends of

RNA could be applied to identify transcription termination, or final processed, $3'$ ends. Finally, although the procedures have been used to study RNA metabolism in *E. coli,* they should be applicable to any cell type.

References

1. D. Kennell, *J. Mol. Biol.* **34,** 85 (1968).
2. D. Apirion and A. Miczak, *BioEssays* **15,** 113 (1993).
3. S. Jinks-Robertson and M. Nomura, *in "Escherichia coli* and *Salmonella typhimurium"* (F. Neidhardt, ed.), p. 1328. American Society for Microbiology, Washington, D.C., 1987.
4. T. C. King and D. Schlessinger, *in "Escherichia coli* and *Salmonella typhimurium"* (F. Neidhardt, ed.), p. 703. American Society for Microbiology, Washington, D.C., 1987.
5. V. J. Cannistraro, M. N. Subbarao, and D. Kennell, *J. Mol. Biol.* **192,** 257 (1986).
6. M. N. Subbarao and D. Kennell, *J. Bacteriol.* **170,** 2860 (1988).
7. C. M. Arraiano, S. D. Yancey, and S. R. Kushner, *J. Bacteriol.* **170,** 4625 (1988).
8. D. J. Ebbole and H. Zalkin, *J. Biol. Chem.* **263,** 913 (1988).
9. E. Schneider, M. Blundell, and D. Kennell, *Mol. Gen. Genet.* **160,** 121 (1978).
10. D. Kennell and H. Riezman, *J. Mol. Biol.* **114,** 1 (1977).
11. A. J. Berk and P. A. Sharp, *Cell (Cambridge, Mass.)* **12,** 721 (1977).
12. W. R. Boorstein and E. A. Craig, *in* "Methods in Enzymology" (J. Dahlberg and J. N. Abelson, eds.), Vol. 180, p. 347. Academic Press, San Diego, 1989.
13. J. Sambrook, E. F. Fritsch, and T. Maniatis, "Molecular Cloning: A Laboratory Manual," 2nd Ed. Cold Spring Harbor Laboratory, Cold Spring Harbor, New York, 1989.
14. P. A. Sharp, A. J. Berk, and S. M. Berget, *in* "Methods in Enzymology" (L. Grossman and K. Moldave, eds.), Vol. 65, p. 750. Academic Press, San Diego, 1980.
15. J. A. Jaeger and I. Tinoco, Jr., *Biochemistry* **32,** 12522 (1993).
16. C. Tuerk, P. Gauss, C. Thermes, D. R. Groebe, M. Gayle, N. Guild, G. Stormo, Y. D'Aubenton-Carafa, O. C. Uhlenbeck, I. Tinoco, Jr., E. N. Brody, and L. Gold. *Proc. Natl. Acad. Sci. U.S.A.* **85,** 1364 (1988).
17. D. Kennell, *in* "Maximizing Gene Expression" (W. Reznikoff and L. Gold, eds.), p. 101. Butterworth, Boston, 1986.
18. M. C.-T. Hu and N. Davidson, *in* "Methods in Enzymology" (R. Wu, ed.), Vol. 217, p. 446. Academic Press, San Diego, 1993.
19. D. Kennell and C. Simmons, *J. Mol Biol.* **70,** 451 (1972).
20. V. J. Cannistraro and D. Kennell, *Eur. J. Biochem.* **213,** 285 (1993).
21. V. J. Cannistraro, B. M. Wice, and D. E. Kennell, *J. Biochem. Biophys. Methods* **11,** 163 (1985).
22. I. Fedorcsak and L. Ehrenberg, *Acta Chem. Scand.* **20,** 107 (1966).
23. M. Blundell and D. Kennell, *J. Mol. Biol.* **83,** 143 (1974).
24. G. G. Brownlee, "Determination of Sequences in RNA." North-Holland, Amsterdam, and Elsevier, New York, 1972.
25. D. K. Hawley and W. R. McClure, *Nucleic Acids Res.* **11,** 2237 (1983).
26. U. Siebenlist, R. B. Simpson, and W. Gilbert, *Cell (Cambridge, Mass.)* **20,** 269 (1980).
27. V. J. Cannistraro and D. Kennell, *J. Bacteriol.* **173,** 4653 (1991).

28. A. Novogrodsky, M. Tal, A. Traub, and J. Hurwitz, *J. Biol. Chem.* **241,** 2933 (1966).
29. V. J. Cannistraro and D. Kennell, *J. Bacteriol.* **161,** 820 (1985).
30. H. Donis-Keller, H. Maxam, and W. Gilbert, *Nucleic Acids Res.* **4,** 2527 (1977).
31. A. Simoncsits, G. G. Brownlee, R. S. Brown, J. R. Rubin, and H. Guilley, *Nature (London)* **269,** 833 (1977).
32. M. S. Boguski, P. A. Heiter, and C. C. Levy, *J. Biol. Chem.* **255,** 2160 (1980).
33. H. Donis-Keller, *Nucleic Acids Res.* **8,** 3133 (1980).
34. D. A. Peattie, *Proc. Natl. Acad. Sci. U.S.A.* **76,** 1760 (1979).
35. M. Silberklang, A. Rochiantz, A.-L. Haenni, and U. L. Rajbhandary, *Eur. J. Biochem.* **72,** 465 (1977).
36. C. P. E. Tu and R. Wu, *in* "Methods in Enzymology" (L. Grossman and K. Moldave, eds.), Vol. 65, p. 620. Academic Press, San Diego, 1980.
37. V. J. Cannistraro, M. B. Strominger, B. M. Wice, and D. E. Kennell, *J. Biochem. Biophys. Methods* **11,** 153 (1985).
38. D. Kennell and A. Kotoulas, *J. Mol. Biol.* **34,** 71 (1968).
39. S. Gillespie and D. Gillespie, *Biochem. J.* **125,** 481 (1971).
40. D. Gillespie and S. Spiegelman, *J. Mol. Biol.* **12,** 829 (1965).
41. V. J. Cannistraro, P. Hwang, and D. E. Kennell, *J. Biochem. Biophys. Methods* **14,** 211 (1987).
42. V. J. Cannistraro and D. Kennell, *J. Mol. Biol.* **182,** 241 (1985).
43. N. Maizels, *Nature (London)* **215,** 647 (1973).
44. J. Majors, *Proc. Natl. Acad. Sci. U.S.A.* **72,** 4394 (1975).
45. A. J. Carpousis, J. E. Stefano, and J. D. Gralla, *J. Mol. Biol.* **157,** 619 (1982).
46. M. L. Peterson and W. S. Reznikoff, *J. Mol. Biol.* **185,** 525 (1985).
47. A. S. Bagnara and L. R. Finch, *Eur. J. Biochem.* **41,** 421 (1974).
48. J. D. Friesen, N. P. Fiel, and K. von Meyenberg, *J. Biol. Chem.* **250,** 304 (1975).
49. L. W. Lim and D. Kennell, *J. Mol. Biol.* **135,** 369 (1979).
50. T. E. England and O. C. Uhlenbeck, *Nature (London)* **275,** 560 (1978).
51. V. J. Cannistraro and D. Kennell, *Eur. J. Biochem.* **181,** 363 (1989).

[16] Identification of *rcs* Genes in *Escherichia coli* O9:K30:H12 and Involvement in Regulation of Expression of Group IA K30 Capsular Polysaccharide

Chris Whitfield, Wendy J. Keenleyside, P. Ronald MacLachlan, Padman Jayaratne, and Anthony J. Clarke

Introduction

Escherichia coli strains produce a variety of cell surface polysaccharides. Some of the polymers are serotype-specific, such as the lipopolysaccharide O-antigen and the K-antigen capsular polysaccharide (CPS). The more than 70 K-antigens recognized (1) are subdivided into groups IA, IB, and II on the basis of a variety of chemical and genetic criteria (2, 3). In addition, there are approximately 160 O-antigens (1). Other cell surface polysaccharides are not considered serotype-specific, because they are widely distributed among *E. coli* strains with different O- and K-serotypes. Examples include the slime exopolysaccharide known as colanic acid (4) and enterobacterial common antigen (5). Research interest in the O- and K-polysaccharides stems from their diversity and the origin of the different forms, as well as their role as potential virulence determinants. Lipopolysaccharide O-antigens and capsular K-antigens are involved in resistance against complement-mediated serum killing (6) and phagocytosis (7), respectively. In contrast, there is no obvious role in pathogenesis for the nonspecific polysaccharides. Colanic acid, for example, is generally only produced at low temperature, on nitrogen-limited minimal medium with excess carbon (4), or in strains with mutations which result in defective or altered regulatory genes (8). The polymer may be involved in protection during desiccation (9).

The biosynthetic and regulatory relationships between the different cell surface polysaccharides in *E. coli* are poorly understood. Although relatively little is known about the regulation of K-antigenic CPS, the regulation of colanic acid in *E. coli* has received considerable attention, primarily in the laboratory of Gottesman. The production of colanic acid is regulated by three positive regulators (RcsA, RcsB, and RcsF, where Rcs stands for regulator of capsule synthesis) and by two negative regulators (RcsC and Lon). The postulated regulatory scheme is shown in Fig. 1. We are interested primarily in the group IA CPSs of *E. coli,* and our prototype strain is *E. coli* serotype O9:K30:H12. The K30 CPS shares superficial structural features with colanic acid (Fig. 2), and with other group I-like extracellular polysaccharides (EPSs) produced by the human pathogen *Klebsiella pneumoniae* and by the plant pathogen *Erwinia*. The production of EPSs of *K. pneumoniae* and *Erwinia* spp. is

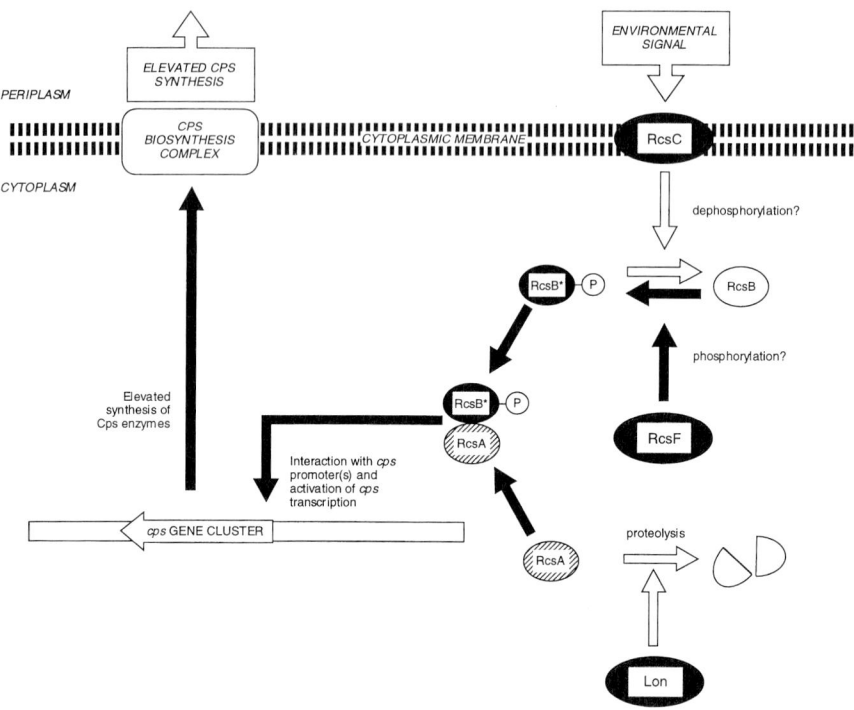

FIG. 1   Regulation of colanic acid production in *Escherichia coli* K-12. The model is modified from that initially described by Gottesman and Stout (8). Reactions having a positive
effect on CPS production are indicated by filled arrows; those with a negative effect are highlighted by hollow arrows. RcsB and RcsC are believed to form an environmentally responsive
two-component regulatory system, where RcsC provides the sensor and RcsB the effector [V.
Stout and S. Gottesman, *J. Bacteriol.* **172,** 659 (1990)]. RcsB is thought to be activated by
phosphorylation based on homology with other related, and better characterized, systems. The
activation of RcsB may involve RcsF, and it has been proposed that RcsF is a protein kinase
[F. G. Gervais and G. R. Drapeau, *J. Bacteriol.* **174,** 8016 (1992)]. However, RcsF lacks the
consensus features expected of a kinase and its effect may therefore be indirect. RcsC probably
acts as a phosphatase, to dephosphorylate RcsB. In its activated form, RcsB forms a heterodimer with RcsA, an auxiliary factor [8; V. Stout, A. Torres-Cabassa, M. R. Maurizi, D. Gutnick,
and S. Gottesman, *J. Bacteriol.* **173,** 1738 (1991)]. The heterodimer activates transcription of
chromosomal *cps* (colanic acid biosynthesis) genes. In *E. coli* K-12, RcsA is present in low
quantities, because it is a substrate for the ATP-dependent protease, Lon [V. Stout, A. Torres-
Cabassa, M. R. Maurizi, D. Gutnick, and S. Gottesman, *J. Bacteriol.* **173,** 1738 (1991)]. This
provides one explanation for the observation that *E. coli* K-12 strains are generally not mucoid
under routine laboratory conditions.

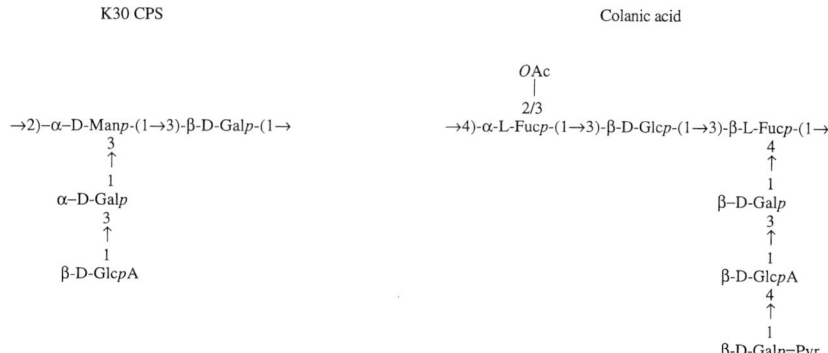

K30 CPS   Colanic acid

FIG. 2   Structures of the repeating units of the K30 CPS [A. K. Chakraborty, H. Friebolin, and S. Stirm, *J. Bacteriol.* **141,** 971 (1980)] and colanic acid [P. J. Garegg, B. Lindberg, T. Onn, and I. W. Sutherland, *Acta Chem. Scand.* **25,** 2103 (1971)] produced by *E. coli* K-12. The pyruvate (Pyr) substituent is present in colanic acid from *E. coli* K-12, but the residue at this site varies in colanic acids isolated from different bacteria.

also regulated by Rcs proteins (10). In this chapter we review strategies used to isolate and characterize the role of *rcs* regulatory genes involved in the expression of the group IA CPS in *E. coli* O9 : K30 : H12.

Isolation of *Escherichia coli* Mutants with Elevated Synthesis of K30 Capsular Polysaccharide

To simplify analysis of CPS in a strain with multiple cell surface polysaccharides, a mutant strain, *E. coli* CWG44 O⁻ : K30 : H12, is used routinely (11). This strain produces wild-type amounts of K30 CPS but has a deletion in the *rfb* genes, whose products are responsible for the expression of the lipopolysaccharide O9 antigen (12). Mutants of *E. coli* CWG44 with altered formation of CPS were identified by their colony phenotypes after ethyl methanesulfonate mutagenesis. A number of highly mucoid colonies were identified. Because many *E. coli* strains can produce colanic acid (4), mucoidy could be due to overexpression of either the K30 CPS or colanic acid. The presence of K30 CPS was confirmed using the K30-specific monoclonal antibody MAb4-15A (13) and the K30-specific bacteriophage coliphage K30. However, these approaches indicate only that K30 CPS is present and do not provide information regarding potential simultaneous production of additional polymers (e.g., colanic acid). The cell surface polysaccharides were therefore further characterized by analysis of their composition by high-performance liquid chromatography (HPLC). The K30 and colanic acid structures are distinct (Fig. 2) and give rise to markedly different HPLC profiles (Fig. 3). All of the mutants tested were found to overproduce K30 CPS without any detectable colanic acid.

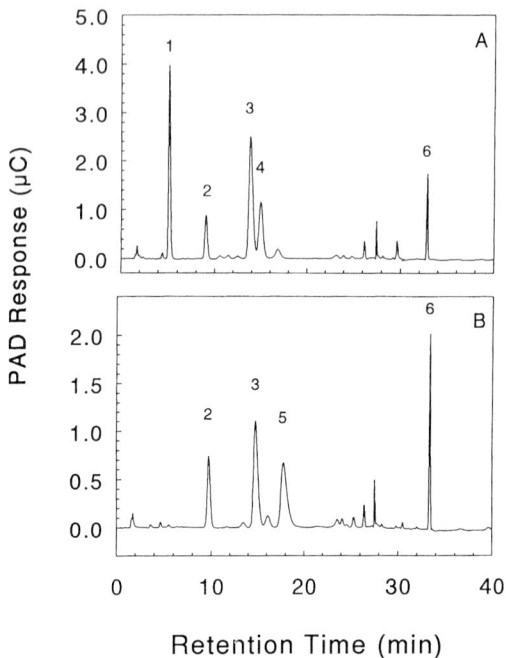

FIG. 3   HPLC profiles of the constituent monomers of colanic acid (A) and K30 CPS (B). Chromatography was performed using a Dionex series 4500 HPLC system, equipped with a pulsed-electrochemical detector (PED). The PED, operating in the pulsed-amperometric detector (PAD) mode, had electrode potentials set as follows: E_1, 0.1 V, 500 msec applied duration; E_2, 0.6 V, 10 msec; and E_3, −0.6 V, 5 msec. The system was calibrated using standards comprising 10 nmol amounts of monosaccharides, dissolved in distilled water. Hydrolyzed polysaccharide samples in 25 μl were injected onto a Dionex CarboPAc PA-1 pellicular anion-exchange column, previously equilibrated in 15 mM NaOH. The flow rate was set at 1 ml/min. Separation of neutral and basic monosaccharides was achieved isocratically using 15 mM NaOH. After 20 min, a linear gradient to 100 mM NaOH and 150 mM sodium acetate was applied over 40 min to elute acidic sugars. The column was maintained in 100 mM NaOH/150 mM sodium acetate for 15 min, then washed in 150 mM NaOH for 4 min, prior to equilibration in 15 mM NaOH. Column effluents were mixed with 300 mM NaOH, pumped under helium pressure at 0.3–0.4 ml/min, prior to PAD detection. The response factors of the PAD were determined for each relevant monosaccharide [A. J. Clarke, V. Sarabia, W. Keenleyside, P. R. MacLachlan, and C. Whitfield, *Anal. Biochem.* **199,** 68 (1991)]. The numbers above each peak correspond to L-fucose (1), 2-deoxy-D-glucose internal standard (2), D-galactose (3), D-glucose (4), D-mannose (5), and D-glucuronic acid (6). The structures of the polysaccharides (Fig. 2) predict compositions of 1 Man : 2 Gal : 1 GlcA (K30) and 2 Fuc : 1 Gal : 2 Glc : 1 GlcA (colanic acid).

In *E. coli* K-12, defects affecting the Lon protease result in activation of colanic acid and a mucoid phenotype, owing to stabilization of RcsA (14, 15). Other characteristics of *lon* mutants include UV sensitivity and filamentation. Examination of the CPS overproducing mutants in *E. coli* K30 led to the identification of *E. coli* CWG55, which showed all of the characteristics expected of a *lon* mutant (16). The nature of the defect in a second K30 CPS overproducer is addressed below. These results indicate that (i) regulatory mutations affecting K30 CPS formation can be isolated easily and (ii) at least some mutants are analogous to those affecting colanic acid expression in *E. coli* K-12 and may involve components of the *rcs* system.

Experimental Procedures

Mutagenesis

Escherichia coli CWG44 (O^- : K30) is mutagenized by treatment with ethyl methanesulfonate (17). Bacteria are grown to mid-exponential phase in Luria–Bertani (LB) broth (per liter: 10 g tryptone (Difco, Fisher Scientific, Unionville, Ontario), 5 g yeast extract (Difco, Fisher Scientific), 10 g NaCl, adjusted to pH 7.0 using 5 M NaOH) (17). The cells are harvested, washed, and resuspended in M9 minimal salts (per liter: 6 g Na_2HPO_4, 3 g KH_2PO_4, 0.5 g NaCl, 1 g NH_4Cl, and 10 ml of 10 mM $CaCl_2$ added after autoclaving) (17). Ethyl methanesulfonate stock is prepared by mixing 0.5 ml ethyl methanesulfonate (Sigma, St. Louis, MO) with 12 ml of M9 minimal salts. For mutagenesis, 5 ml of mutagen stock is mixed with 5 ml of culture. The mixture is then incubated at 37°C with shaking. Samples are taken at various times, and mutants are selected from incubation times sufficient to give over 70% killing. After mutagenesis, samples of mutagenized cells are washed in M9 salts, resuspended, and incubated at 37°C for 4.5 hr without shaking. Samples are then diluted into LB broth and incubated at 37°C overnight with shaking. Serial dilutions are made and plated on LB plates. Mutants are selected by visible differences in colony morphology after 18 hr of growth at 37°C.

Determination of K30 Capsular Polysaccharide Expression

Mutants are screened for the ability to produce K30 CPS by a slide agglutination reaction with MAb-15A (13). Bacteria from a single colony, grown on an LB plate, are resuspended in 0.1 ml of 0.9% (w/v) saline. A 10-μl sample is then placed on a glass microscope slide, and 1 μl of MAb4-15A ascites fluid is added and mixed by gentle rocking. A flocular precipitate indicating a positive agglutination reaction is visible within 30 sec. The presence of K30 CPS is also determined by a bacteriophage sensitivity test using K30 CPS-specific coliphage K30 (18). Using a glass serological pipette, a drop of an overnight LB broth culture is spread as a bar on a plate of LB medium and allowed to dry. A 10-μl sample of bacteriophage lysate [10^9–10^{10} plaque-forming units (pfu)/ml stock] is inoculated on the bar and allowed to dry. The plate is incubated at 37°C for 4–6 hr and then examined for visible clearing due to bacteriophage-mediated lysis.

Isolation of Capsular Polysaccharide

A small-scale extraction process has been developed for screening large numbers of strains. Bacteria are harvested from colonies grown on petri plates containing LB medium for 18 hr at 37°C. Plates are inoculated to give separated colonies, as lower yields of CPS are obtained from confluent lawns of bacterial growth. Cells are scraped from five plates per strain into 20 ml of 0.9% (w/v) saline and resuspended in a vortex mixer. The cells are collected by centrifugation (10,000 *g* for 10 min at 4°C) and both the cell pellet and the supernatant are retained. The supernatant contains cell-free CPS. The cell pellet is then extracted with hot aqueous phenol to liberate cell-bound CPS, following a modification of the method originally described by Westphal and Jann (19). The cell pellet is resuspended in 10 ml of 0.9% (w/v) saline, and an equal volume of 90% aqueous phenol at 65°C is added. The mixture is incubated at 65°C for 15 min and stirred frequently. The phases are then separated by centrifugation (20 min at 14,000 *g*, 4°C) and the aqueous phase removed. The phenol phase is reextracted with an equal volume of distilled water, and the combined aqueous phase fractions, containing cell-bound CPS, are dialyzed for 2 days against several changes of distilled water. $MgCl_2$ (final concentration 10 m*M*) is added to both the cell-free and cell-bound CPS samples, and both samples are then treated with deoxyribonuclease (10 µg/ml final concentration) and ribonuclease (10 µg/ml) for 1 hr at 37°C. Finally, proteinase K (10 µg/ml final concentration) is added, and the samples are incubated at 65°C for 1 hr to degrade residual protein. Because both lipopolysaccharide and CPS are extracted into the aqueous phase in the hot-phenol method, contaminating lipopolysaccharide is removed as a pellet by an ultracentrifugation step at 105,000 *g* for 16 hr. The lipopolysaccharide-free supernatant is collected, dialyzed against distilled water, and lyophilized.

Composition of Capsular Polysaccharide

For composition analysis, polysaccharide samples of 0.2–0.5 mg are hydrolyzed to constituent monosaccharides in trifluoroacetic acid at 100°C for 12 hr, *in vacuo*. The samples are lyophilized and redissolved in 10 m*M* NaOH. As an internal standard, 12.5 nmol of 2-deoxy-D-glucose is added immediately prior to analysis. The components of acid-hydrolyzed polysaccharide samples are characterized by HPLC, using a Dionex series 4500 HPLC system (Dionex, Mississauga, Ontario) equipped with a pulsed-electrochemical detector (PED) (Fig. 3). Individual components are identified by their retention times, established using the appropriate monosaccharide standards. Standards are run before each unknown sample.

Cloning Genes Involved in Regulation of K30 Capsular Polysaccharide Expression

A pVK102 cosmid-based gene library from *E. coli* O9:K30:H12 DNA is constructed in *E. coli* LE392. Among the collection of recombinant *E. coli* LE392 strains

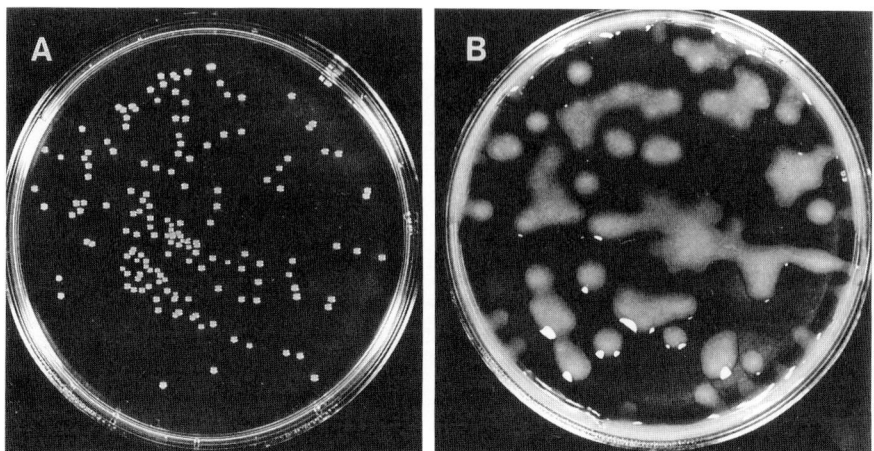

FIG. 4   Mucoid phenotype of *E. coli* LE392 containing a recombinant plasmid (pWQ600) carrying $rcsB_{K30}$. (A) *Escherichia coli* LE392. (B) The mucoid polysaccharide evident in colonies of *E. coli* LE392[pWQ600] was purified and characterized by HPLC analysis. Colanic acid was the only polymer detected.

are several which give rise to highly mucoid colonies (Fig. 4). Two classes of recombinants are identified expressing visibly different amounts of polysaccharide. Highly mucoid colonies are represented by *E. coli* LE392[pWQ600] and less mucoid colonies by *E. coli* LE392[pWQ499]. The physical maps of the two plasmids, pWQ499 and pWQ600, are different (Fig. 5), and DNA cloned in these plasmids does not cross-hybridize. The polysaccharide produced by these colonies is examined by HPLC, and, in all cases, the product is identified as colanic acid. Consequently, plasmids giving rise to mucoid colonies are considered to contain genes encoding regulatory elements which interact with the colanic acid system in *E. coli* K-12.

   The regulation of colanic acid is well established, and the genes for two positive regulators, *rcsA* (15) and *rcsB* (20), have been well characterized. Defined mutations are available in each gene in *E. coli* K-12, affording the identification of cloned homologous genes by complementation. Complementation of *E. coli* VS20186 and VS20187 with pWQ499 and pWQ600 (Table I) demonstrates that the plasmids contain functional homologs of *rcsA* and *rcsB*, respectively. By testing subcloned fragments for the ability to confer a mucoid phenotype in VS20186 and VS20187, the genes are localized on each of the cosmids. Subclones containing $rcsB_{K30}$ but not $rcsC_{K30}$ cause a dramatic increase in mucoid phenotype, when compared to pWQ600. The DNA sequence for each region is determined, unambiguously confirming that the plasmids contain $rcsA_{K30}$ (pWQ499) (16) and $rcsB_{K30}$–$rcsC_{K30}$ (pWQ600) (21). The Subscripts K30 and K-12 are used to discriminate between the homologs from *E. coli* K30 and *E. coli* K-12. The predicted $RcsA_{K30}$ protein differs in only 2 of 107 amino acid residues when compared to its counterpart from *E. coli* K-12 (16).

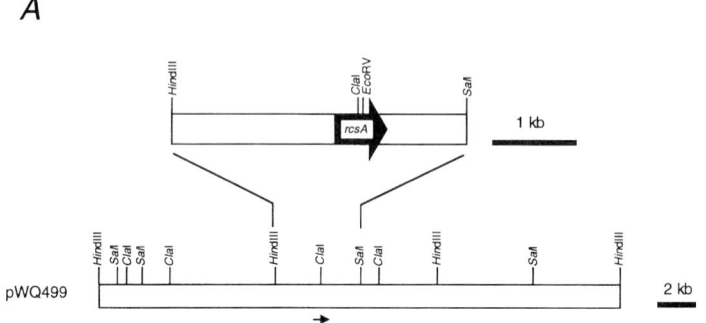

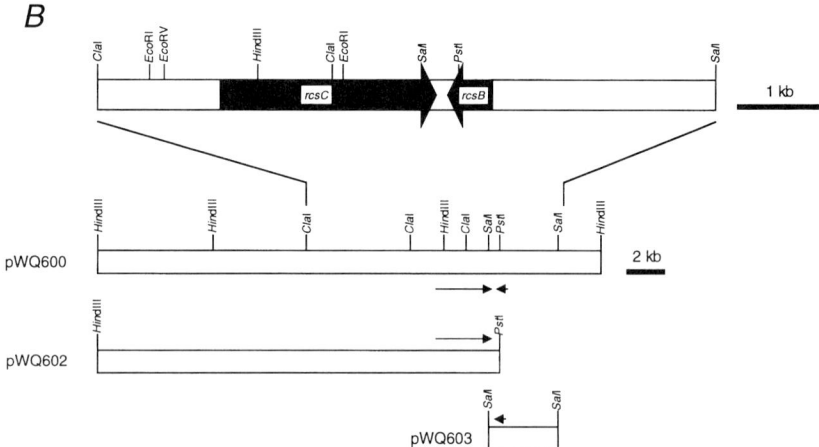

FIG. 5   Physical maps of plasmids carrying the $rcsA_{K30}$ and $rcsB_{K30}-rcsC_{K30}$ regions cloned from *E. coli* O9:K30:H12. Plasmids pWQ499 and pWQ600 were isolated from mucoid recombinants which produced colanic acid. The identity and position of the *rcs* genes were established by complementation analyses using *E. coli* K-12 derivatives with defined *rcsA* and *rcsB* defects (Table I). The identity of the genes was confirmed by nucleotide sequence analysis. Plasmids pWQ499, pWQ600, and pWQ602 were constructed using the vector pVK102 and are therefore mobilizable. The $rcsB_{K30}$-containing fragment is cloned in pUC19 to give pWQ603. [Data from P. Jayaratne, W. J. Keenleyside, P. R. MacLachlan, C. Dodgson, and C. Whitfield, *J. Bacteriol.* **175,** 5384 (1993), and Keenleyside *et al.* (16).]

$RcsB_{K30}$ and $RcsB_{K-12}$ are identical, and $RcsC_{K30}$ contains 6 amino acid changes when compared to its K-12 homolog (21). Potential promoters and regulatory features associated with the genes are also conserved.

TABLE I   Complementation of *rcsA* and *rcsB* Mutations in
Escherichia coli K-12 with Plasmids Containing
DNA Fragments from *Escherichia coli*
O9:K30:H12[a]

	Mucoid phenotype in	
Plasmid	*E. coli* VS20186 (*lon rcsA*)	*E. coli* VS20187 (*lon rcsB*)
None	−	−
pWQ499 (*rcsA*$^+$)	+++[c]	−
pWQ600 (*rcsB*$^+$, *rcsC*$^+$)	+[b]	+[b]
pWQ602 (*rcsC*$^+$)	−	−
pWQ603 (*rcsB*$^+$)	+++[c]	+++[c]

[a] To identify strains harboring homologs of *rcsA* and *rcsB*, plasmids were
introduced by electroporation into *E. coli* strains with defined mutations
rcsA and *rcsB*. *Escherichia coli* VS20186 [F⁻ *araD139* Δ(*argF-lac*)U169
rpsL150 flbB5301 deoC1 relA1 ptsF25 rbsR rcsA51 ∷ Δkan lon-100] and
VS20187 [F⁻ *araΔ139* Δ(*argF-lac*)U169 *rpsL150 flbB5301 deoC1 relA1
ptsF25 rbsR rcsB62 ∷ Δkan lon-100*] were generously provided by Dr. V.
Stout (Arizona State University). These strains are nonmucoid in the absence
of wild-type copies of *rcsA* and *rcsB*, respectively. However, in the presence
of *rcsA*$^+$ and *rcsB*$^+$ complementing plasmids, the *lon* defect ensures colanic
acid production and a mucoid colony phenotype. The mucoid phenotype was
assessed visually after 18 hr of growth on LB plates at 37°C.
[b] Multicopy *rcsB* can overcome the *rcsA* defect.
[c] Highly mucoid phenotypes were scored +++.

To begin investigation of the role of the *rcs* genes in *E. coli* K30, plasmids
pWQ499 (*rcsA*$^+$) and pWQ600 (*rcsB*$^+$ *rcsC*$^+$) were introduced into *E. coli* CWG44
O⁻:K30. These strains are relatively refractory to standard procedures for making
competent cells (e.g., CaCl₂ treatment), and their electroporation efficiencies are very
low with large plasmids. However, pVK102 is a broad host range mobilizable plasmid and can be transferred to other gram-negative bacteria if the transfer functions
are supplied in trans. Consequently, pWQ499, pWQ600, and pWQ602 were introduced by using a helper plasmid, pRK2013, in triparental mating experiments. Electroporation was used to introduce the smaller construct, pWQ603, in the vector
pUC19. Highly mucoid colonies resulted from CWG44 exconjugants containing either *rcsA*$_{K30}$ or *rcsB*$_{K30}$ in multicopy (Table II). In each case, HPLC analysis showed
that the only polymer present was K30 CPS; no colanic acid was detected. These
results indicated that, at least in multicopy, *rcsA*$_{K30}$ and *rcsB*$_{K30}$ elevated K30 CPS
expression, providing further support for the possibility that the K30 CPS and colanic
acid share a common regulatory system.

TABLE II   Ability of Multicopy *rcsA* and *rcsB* to Stimulate an Increase in Mucoid
Phenotype Due to Overproduction of K30 Polysaccharide in *Escherichia coli*
K30 Derivatives[a]

	Overproduction of K30 CPS in				
Plasmid	E69 (*rcsA*[+], *rcsB*[+])	CWG120 (*rcsA*)	CWG121 (*rcsB*)	CWG123 (*rcsC*)	CWG131 (*rcsC rcsB*)
None	−	−	−	+	−
pWQ499 (*rcsA*[+])	+	+	−	+	−
pWQ600 (*rcsB*[+], *rcsC*[+])	+	+	+	NT	NT
pWQ603 (*rcsB*[+])	+	+	+	+	+
pWQ602 (*rcsC*[+])	−	−	−	+	NT

[a] Mucoid phenotypes were assessed by colony morphology and visible increase in CPS production after growth on LB
for 18 hr at 37°C. Each transconjugant was compared to the parental strain. The identity of K30 CPS was determined
by HPLC analysis. NT, Not tested. Modified from Jayaratne *et al.* (21).

Experimental Procedures

Construction of Gene Library from Escherichia coli O9 : K30 : H12

A gene library of chromosomal DNA from *E. coli* O9 : K30 : H12 is constructed in
the broad host range cosmid vector pVK102 (22). High molecular weight chromosomal DNA is purified by a modification of the method of Hull *et al.* (23). Mid-
exponential phase bacteria from 10 ml of LB broth culture are harvested and washed
once in 0.9% (w/v) saline. The cells are then resuspended in 2 ml ice-cold 50 mM
Tris-HCl/1 mM EDTA/25% (w/v) sucrose, pH 8.0. Two milligrams lysozyme is
then added, and the mixture is incubated on ice. After 5 min, 10 μl of 20 mg/ml
stock proteinase K is added. Then 0.5 M EDTA (0.4 ml) is added with gentle mixing,
and the cells are lysed by the addition of 0.25 ml of 10% (w/v) *N*-lauroylsarcosine.
The suspension is maintained on ice for 1 hr. Lysis is not complete at this stage,
and the tubes are therefore incubated at 50°C for 4 hr to complete lysis. Instead of
further purification by CsCl-gradient centrifugation, the lysate is extracted with TE-
equilibrated phenol and phenol–chloroform. The phases are separated by centrifugation (8000 g for 10 min). DNA is then precipitated from the aqueous phase using
ethanol, spooled on a glass rod, and washed in 70% (v/v) ethanol. After drying in air,
the DNA is redissolved in TE overnight at 4°C. DNA is partially digested using
*Hin*dIII under conditions that favor fragments of 20–50 kbp. The digests are then
fractionated on 38-ml 10–40% sucrose gradients prepared in 20 mM Tris-HCl/1 M

NaCl/5 m*M* EDTA, pH 8.0, using a Beckman SW28 rotor (26,000 rpm, 24 hr, 15° C), essentially as described by Sambrook *et al.* (24). Samples from alternate fractions are examined on 0.8% agarose gels, and fractions containing 20–30 kbp fragments are pooled and the DNA collected by ethanol precipitation. DNA fragments are ligated to *Hin*dIII-digested pVK102 vector, and ligation mixtures are packaged into λ phages *in vitro* using Packagene packaging extracts (Promega, Madison, WI), according to the manufacturer's instructions. The resulting recombinant phages are transduced into *E. coli* LE392 [F$^-$ *hsdR514*(r$_K^-$ m$_K^-$)*supE44 supF58 lacY1* or Δ(*lac-proAB*)6 *galK2 galT22 metB1 trpR55* λ$^-$] (24). Strains resistant to tetracycline (15 μg/ml final concentration) are selected.

Complementation Experiments to Identify Plasmids Carrying Functional Homologs of rcsA and rcsB

To identify strains harboring homologs of *rcsA* and *rcsB*, plasmids are introduced into *E. coli* strains VS20186 [F$^-$ *araD139* Δ(*argF-lac*)U169 *rpsL150 flbB5301 deoC1 relA1 ptsF25 rbsR rcsA51*::Δ*kan lon-100*] and VS20187 [F$^-$ *ara*Δ*139* Δ(*argF-lac*)U169 *rpsL150 flbB5301 deoC1 relA1 ptsF25 rbsR rcsB62*::Δ*kan lon-100*] (Table I). The strains have defective minitransposon insertions in *rcsA* and *rcsB* and suppress the normally mucoid phenotype of *lon* derivatives. However, in the presence of *rcsA$^+$* and *rcsB$^+$* complementing plasmids, the *lon* defect causes colanic acid overproduction and a mucoid colony morphology. Bacteria are transformed by electroporation using a modification of published methods (25). Mid-exponential phase cells from an LB broth culture are washed once in an equal volume of 1 m*M* HEPES buffer, pH 7.0, once more in $\frac{1}{50}$ volume of HEPES buffer, and then resuspended in $\frac{1}{100}$ volume of 10% glycerol. For electroporation, 40 μl cells and 1 μl DNA (100–500 ng/μl) are placed in a chilled 0.2-cm electroporation cuvette from Bio-Rad (Richmond, CA). The pulse controller is set at 2.5 kV, 200 Ω, 25 μF. For recovery, either 1 ml of LB containing 0.2% D-glucose or 1 ml of SOC (2% Bacto-tryptone, 0.5% Bacto-yeast extract, 10 m*M* NaCl, 2.5 m*M* KCl, 10 m*M* MgCl$_2$, 10 m*M* MgSO, 20 m*M* glucose) is added, and the suspension is incubated at 37° C for 1 hr with shaking. The cells are then plated on LB medium containing the appropriate selective antibiotic.

Triparental Mating Experiments to Transfer pVK102-Based Recombinant Cosmids

pVK102 is Mob$^+$ and can be transferred when Tra functions are supplied in trans by pRK2013 (Mob$^+$ Tra$^+$ ColE1 Km^R). which is derived from RK2 (26, 27). pRK2013 is maintained in *E. coli* LE392. One milliliter each of overnight cultures of recipient (*E. coli* O9:K30:H12), donor (*E. coli* LE392[pWQ499, pWQ600, or pWQ602]), and helper (*E. coli* LE392[pRK2013]) are twice washed in 1 ml of fresh LB medium to remove antibiotics. Then 0.1 ml of each culture is mixed and plated onto LB agar plates. The plates are incubated at 37° C for 6 hr to allow conjugation, and then bacterial growth is scraped from the plate, diluted in 0.9% (w/v) saline, and spread on fresh plates of M9 minimal medium containing thiamine hydrochloride (1 μg/ml), niacinamide (10 μg/ml), and 0.2% D-glucose as a carbon source. Inclusion of 40 μg/ml each of L-histidine and L-tryptophan allows growth of the recipient while counterselecting against the donor, and 15 μg/ml tetracycline is used to select for recipients containing pVK102 plasmid derivatives.

Characterization of Cell Surface Polysaccharides in *Escherichia coli* K-12/K30 Hybrids

Genes involved in the biosynthesis of the K30 CPS (28, 29) and colanic acid (14) are both located near the *his* biosynthetic genes at 44 min on the *E. coli* chromosome. In an attempt to establish whether the K30 and colanic acid biosynthetic loci are allelic, hybrid K30/K12 strains were constructed. Because P1 transduction and lysate formation does not work in *E. coli* K30 strains (and perhaps in other strains with group I capsules), the approach used for hybrid construction involved moving the *his*-linked K30 biosynthesis genes from *E. coli* O9:K30:H12 into an *E. coli* K-12 His⁻ recipient. One strain, *E. coli* CWG13, was used in further experiments. This strain retained the chromosomal markers of the *E. coli* PA360 parent, with the exception of *hisG* and *serA*. Introduction of either pWQ499 (*rcsA⁺*) or pWQ600 (*rcsB⁺*, *rcsC⁺*) into *E. coli* CWG13 resulted in colonies with increased mucoid appearance. The polymer produced was characterized by HPLC and found to be the K30 CPS. In contrast, when the same plasmids were introduced into the parental strain, *E. coli* PA360, colanic acid production resulted. These observations suggest that the genetic loci for biosynthesis of colanic acid and the group IA K30 CPS are allelic, with the colanic acid *cps* gene locus being replaced by K30 *cps* in *E. coli* CWG13. Both are regulated by a common regulatory system comprising *rcs* gene products.

Experimental Procedures

Formation of Hfr Derivatives of Escherichia coli O9:K30:H12

To form an Hfr strain, homology in the *lac* region is used to direct the insertion of F'*ts*114*lac* into the chromosome of *E. coli* O9:K30:H12. This is a modification of the method of Chumley *et al.* (30), where homology in Tn*10* is used to direct the integration of the same F' plasmid. F'*ts*114*lac⁺zzf*-20::Tn*10* is transferred from *Salmonella typhimurium* TT628 (*pyrC7 rpsL1*) to *E. coli* CWG6 O9:K30:H12 (*trp lac rpsL*) by conjugation on LB plates for 4 hr as described above. The temperature is maintained at 30°C because of the temperature-sensitive nature of F'*ts*114*lac⁺zzf*-

20::Tn*10*. Merodiploids containing the F' plasmid are selected by dilution plating mating mixtures on M9 medium containing 0.2% D-glucose and 15 μg/ml tetracycline at 30°C. The strains are then tested for Lac$^+$ phenotype and K30 CPS. One isolate is taken, grown at 42°C for 6 hr, and then dilution plated on MacConkey lactose. The majority of colonies are Lac$^-$ due to the loss of F'*ts*114*lac*$^+$*zzf*-20:: Tn*10*, but small numbers of Lac$^+$ colonies result from integration of the F' into the chromosome to form an unstable Hfr. Four individual Hfr derivatives are then used as donors to transfer the *his–cps* region from *E. coli* CWG6 to *E. coli* PA360, as described below.

Construction of Escherichia coli K-12/K30 Hybrid

The recipient for these experiments is *E. coli* PA360 K-12 (*thi-1 leuB6 hisG1 serA1 argH1 lacY1 gal-6 malA1 xyl-7 mtl-2 rpsL9 tonA2 supE44 thr-1* λ^-). Conjugation is performed by plate mating at 37°C for 4 hr as described above. The growth is scraped into 0.9% (w/v) saline and serially diluted. Dilutions are plated onto M9 minimal medium supplemented with L-leucine, L-serine, and L-arginine (all at 40 μg/ml), thiamine hydrochloride (1 μg/ml), niacinamide (10 μg/ml), and 0.2% D-glucose, to select His$^+$ recombinants. The donor is counterselected by omission of tryptophan. When the other *E. coli* PA360 chromosomal markers are tested, most of the recombinants are found to be SerA$^+$, but the other parental markers were retained. Recipients from conjugation experiments are screened for the ability to produce K30 CPS, by the slide agglutination reaction with MAb4-15A (13) and by a sensitivity test using K30 CPS-specific bacteriophage coliphage K30 (18), as described above. Around 98% of the His$^+$ recombinants were K30$^+$.

Construction of Defined *rcsA* and *rcsB* Mutations in *Escherichia coli* O9:K30:H12

The use of multicopy plasmid constructs to study the effects of regulatory elements can potentially give misleading results, and direct extrapolation to the situation with a single chromosomal copy is not always straightforward. Therefore, to further understand the role of $rcsA_{K30}$ and $rcsB_{K30}$ in *E. coli* O9:K30:H12, the cloned genes were used to make precise chromosomal mutations. Plasmid pGP704 is a broad host range suicide delivery vector developed by Miller and Mekalanos (31). The plasmid requires the *pir* gene product for replication. In the absence of the *pir* gene, the plasmid can be rescued by recombination into chromosomal loci. The site of recombination is directed by homology of an internal gene fragment. Mutagenesis plasmids pWQ510 and pWQ650 were constructed using internal fragments from $rcsA_{K30}$ and $rcsB_{K30}$, respectively (Fig. 6). Because internal fragments of the genes are used to

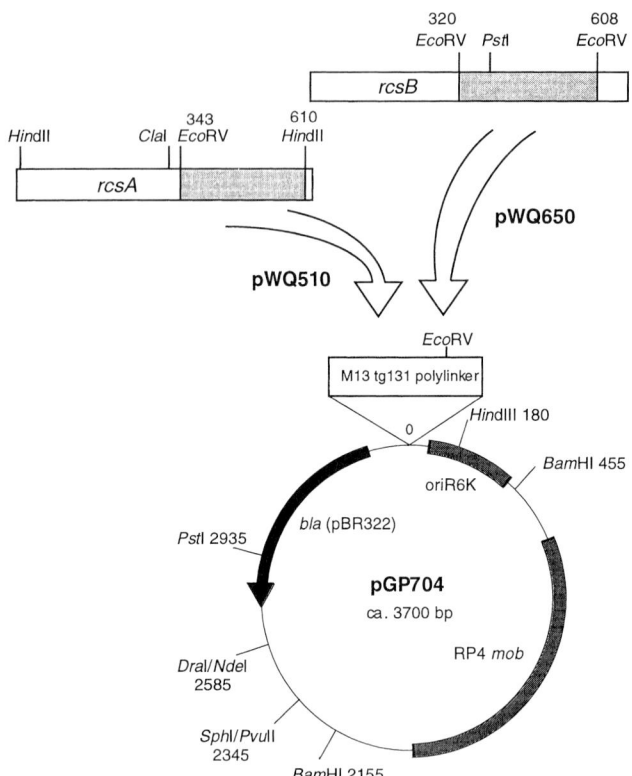

FIG. 6   Construction of plasmids pWQ510 and pWQ650 for the mutagenesis of *rcsA* and *rcsB* in *E. coli* O9:K30:H12. Plasmid pGP704 (*oriR6K mobRP4* Ap^R) (31) requires the *pir* gene product for replication, and this is supplied in the λ lysogenic host strain *E. coli* SY327 λ*pir* [K-12 F$^-$ λ*pir araD* Δ(*lac-pro*) *argE*$_{(Am)}$ *recA56 nalA* Rif^R]; this strain is used for making plasmid constructs. Internal fragments from each gene were cloned in pGP704, in this case, the 267-bp *Hin*dII–*Eco*RV fragment from *rcsA* or the 288-bp *Eco*RV fragment from *rcsB*. Both fragments were extracted from 1.2% agarose gels, purified using GeneClean (Bio 101, La Jolla, CA), and ligated into the *Eco*RV site of pGP704 to form pWQ510 (*rcsA*) and pWQ650 (*rcsB*), respectively. The constructs were used to transform *E. coli* SY327 λ*pir* to ampicillin resistance (100 mg/ml). [Data from Jayaratne *et al.* (21).]

direct recombination, disruption of the wild-type gene results from introduction of vector sequences into the appropriate open reading frame. The mutations in *E. coli* CWG120 (*rcsA*$_{K30}$) and CWG121 (*rcsB*$_{K30}$) were confirmed by Southern hybridization (Fig. 7).

Despite inactivation of *rcsA*$_{K30}$ or *rcsB*$_{K30}$, strains CWG120 and CWG121 still

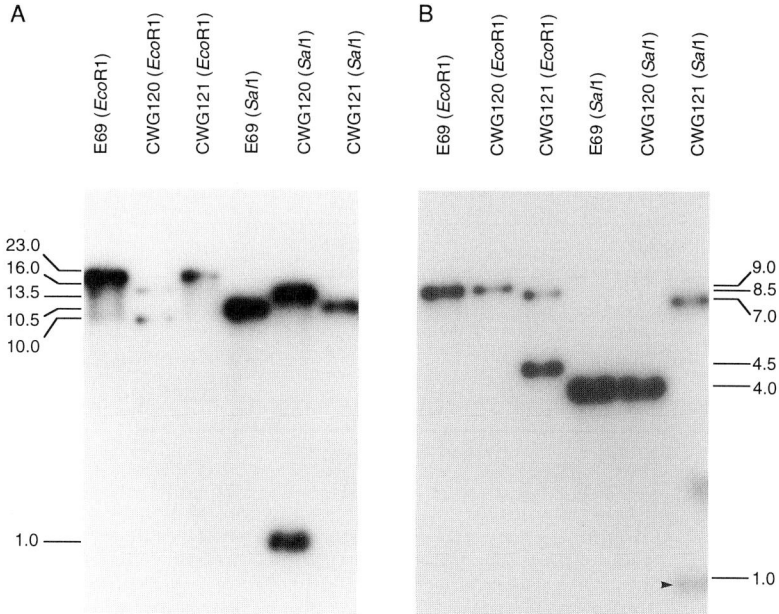

FIG. 7   Confirmation of the mutations in CWG120 (*rcsA*) and CWG121 (*rcsB*) by Southern hybridization. Stable Ap^R strains result from the integration of the pWQ510 and pWQ650 constructs into the chromosomal *rcsA* and *rcsB* genes. Chromosomal DNA from each strain was digested with *Eco*RI and *Sal*I, which cut within pGP704 but not in either of the target genes. Interruption of *rcsA* or *rcsB* is therefore indicated by the introduction of single *Eco*RI and *Sal*I sites into the inactivated gene. As expected, the *rcsA* probe hybridized to single 23-kb *Eco*RI and 10.5-kb *Sal*I fragments in both *E. coli* E69 and CWG121 (*rcsB*). Two fragments were detected in the *Eco*RI (16 and 10 kb) and *Sal*I (13.5 and 1 kb) digests of *E. coli* CWG120 (*rcsA*). The increase in total size of the fragments detected by hybridization corresponded to the size of pWQ510 (~4 kb). Similarly, the *rcsB* probe hybridized to single 9-kb *Eco*RI and 4-kb *Sal*I fragments in *E. coli* E69 and CWG120 (*rcsA*). In *E. coli* CWG121, two *Eco*RI (8.5 and 4.5 kb) and *Sal*I (1 and 7 kb) fragments were detected. The arrow in (B) indicates a faint 1 kbp fragment. [Data from Jayaratne *et al.* (21).]

produced K30 CPS, detectable by agglutination with the K30 CPS-specific MAb4-15A (13). The amount and distribution of the K30 CPS were estimated by a modification of the HPLC composition analysis (Table III). The CPS sample amounts were standardized and the uronic acid content determined by peak integration, after calibration of the HPLC using a D-glucuronic acid standard. These results indicate that the presence of Rcs regulatory components is not essential for K30 CPS production and formation of a K30 capsular structure at 37°C.

TABLE III   Influence of *rcs* Gene Products on Amount and Distribution of K30 Capsule Polysaccharide Synthesized by *Escherichia coli* Strains[a]

Strain	Cell-bound K30 CPS		Cell-free K30 CPS	
	Amount[b]	Increase (-fold)[c]	Amount[b]	Increase (-fold)[c]
E69	0.03	—	0.03	—
CWG121 (*rcsB*)	0.02	0	0.03	0
CWG121 (pWQ600: *rcsB+, rcsC+*)	0.24	8	1.47	49
CWG123 (*rcsC*)	0.34	11	0.63	21
CWG123 (pWQ602: *rcsC+*)	0.05	2	0.01	0

[a] Data from Jayaratne *et al.* (21).

[b] The amount of K30 CPS was determined by quantitation of glucuronic acid in individual samples. Glucuronic acid was chosen because it is not a component in other cell surface polymers (e.g., lipopolysaccharide O-antigens). Samples were extracted from bacteria as described above. However, viable cell counts were made by dilution plating to determine the number of cells used in each extract. Samples of known volume were hydrolyzed in trifluoroacetic acid and the amount of glucuronic acid was determined by peak integration of HPLC profiles. The amount of K30 CPS in each sample was then expressed as the amount of glucuronic acid (μg) per 10^7 bacteria. Figures are the average of triplicate samples.

[c] The increase is established by comparison to the values for the wild-type strain, E69.

Experimental Procedures

Mutagenesis of rcsA and rcsB in Escherichia coli O9:K30:H12

The strategy for the formation of mutagenesis delivery plasmids is shown in Fig. 6. The mutagenesis plasmids are transferred to *E. coli* O9:K30:H12 by conjugation in a triparental mating experiment. pGP704 is Mob$^+$ and can be transferred when Tra functions are supplied in trans by pRK2013, as described above for pVK102 derivatives. pRK2013 is maintained in *E. coli* LE392. One-tenth milliliter each of washed overnight cultures of recipient (*E. coli* O9:K30:H12), donor (*E. coli* SY327 λpir[pWQ510 or pWQ650]), and helper strains (*E. coli* LE392[pRK2013]) are used for mating, following the procedure described above. *Escherichia coli* E69 is a prototrophic strain. Therefore, mating mixtures are plated on M9 minimal medium containing 0.2% D-glucose as a carbon source, to select against the auxotrophic donor and helper strains, and 100 μg/ml ampicillin is included to select for integration of the plasmids into the chromosome. The mutations are confirmed by Southern hybridization experiments.

Southern Hybridization

Probes for *rcsA* and *rcsB* are generated by the polymerase chain reaction (PCR). The *rcsA* forward primer consists of nucleotides 1–18 (5'-ATGTCAACGATTATTATG), where nucleotide 1 is the first base of the initiation codon. The reverse primer is complementary to nucleotides 618–600 (5′-CATGTTGACAAAAATAC). For *rcsB*, the forward primer (5′-ATGAACAATATGAACGTAA) corresponds to nucleotides 1–19, and the reverse primer (5′-TTAGTCTTTATCTGCCGGAC) is complementary to nucleotides 651–632. DNA is amplified by using 24 cycles, each consisting of a 1.5-min denaturation step (94°C), a 1.5-min annealing step (54°C for *rcsA* and 40°C for *rcsB*), and a 2-min polymerization step (72°C). The reaction mixture contains 25 ng template DNA (plasmids pWQ499 or pWQ600), 100 ng oligonucleotide primers, 125 μM deoxynucleotide triphosphate mixture, and 2.5 U *Taq* polymerase in 100 μl *Taq* polymerase buffer. The amplified product is separated by agarose gel electrophoresis, excised from the gel, and purified using GeneClean (Bio 101, La Jolla, CA). To label the probe, 25 ng of DNA is denatured at 100°C in 5–20 μl distilled water and then cooled on ice. The probe is labeled using a random priming kit (GIBCO/BRL, Gaithersburg, MD) for 1.5–2 hr at 25°C, according to the manufacturer's instructions. The radiolabel is 2.5–5 μl [α-³²P]dATP (3000 Ci/mmol) in a total reaction volume of 50 μl. Southern transfers are performed using overnight alkaline transfer (0.4 M NaOH) and a Zeta-probe membrane (Bio-Rad). The membrane is soaked briefly in 2× SSC (20× SSC contains, per liter, 175.3 g NaCl, 88.2 g sodium citrate, adjusted to pH 7.0 using 10 M HCl), dried, and baked at 80°C for 2 hr. The hybridization solution contains 1 mM EDTA, pH 8.0/0.5 M NaH$_2$PO$_4$, pH 7.2/7% (w/v) sodium dodecyl sulfate (SDS). The radiolabeled probe is denatured by heating in a boiling water bath for 5 min and then added to the blot in hybridization solution; hybridization is carried out at 65°C for 18 hr. The membranes are washed twice in 1 mM EDTA, pH 8.0/40 mM NaH$_2$PO$_4$, pH 7.2/5% SDS, and then twice more in 1 mM EDTA, pH 8.0/40 mM NaH$_2$PO$_4$, pH 7.2/1% SDS. Each wash is performed at 65°C for 30 min. After washing, the blots are dried and exposed to Cronex 4 X-ray film.

Requirement of *rcs* Gene Products for High Level Expression of K30 Capsular Polysaccharide in *Escherichia coli*

In *E. coli* K-12, multicopy *rcsA* elevates expression of colanic acid in an *rcsB*-dependent fashion (32). To determine whether the same was true for K30 CPS expression, plasmids pWQ499 (*rcsA*$^+_{K30}$), pWQ600 (*rcsC*$^+_{K30}$, *rcsB*$^+_{K30}$), and pWQ603 (*rcsB*$^+_{K30}$) were introduced into *E. coli* strains E69, CWG120, and CWG121 (Table II). As with *E. coli* K12, multicopy *rcsA*$_{K30}$ caused increased mucoid phenotype only in the presence of a functional *rcsB*$_{K30}$ gene, and the requirement for *rcsA*$_{K30}$ was overcome in strains with multicopy *rcsB*$_{K30}$. In the presence of multicopy pWQ600 (*rcsB*$^+_{K30}$), the amount of cell-bound K30 CPS in *E. coli* CWG121 (*rcsB*$^-$) increased 8-fold, and the cell-free polymer was elevated 49-fold (Table III). In all of these experiments, the structure of the polysaccharide products was identified as K30 CPS by HPLC analysis.

In *E. coli* K-12, a mutation in *rcsC* (*rcsC137*) results in constitutive high level production of colanic acid (32). A collection of CPS overexpression mutants from *E. coli* K30 was screened for analogous defects, by complementation experiments. One mutant, *E. coli* CWG123, was complemented by pWQ602 (*rcsC⁺*) suggesting an *rcsC* defect (Table II). *Escherichia coli* CWG123 shows an 11-fold increase in cell-associated K30 CPS and a 21-fold increase in cell free CPS (Table III). Introduction of pWQ602 into *E. coli* CWG123 reduced the amounts of both forms of polymer to levels comparable with the wild-type strain. Reduction to wild-type levels of K30 polysaccharide synthesis also resulted from introducing an *rcsB* defect into *E. coli* CWG123, to form CWG131, using the mutagenesis plasmid pWQ650 as described above. The reduction in K30 synthesis in CWG131 could be overcome by multicopy *rcsB* (pWQ603) but not *rcsA* (pWQ504). These results show that the elevated production of K30 polysaccharide in the *rcsC* mutant is RcsB-dependent and confirm (i) the requirement for the *rcs* system in high level expression of K30 CPS and (ii) the central role for *rcsB* in K30 CPS expression.

Role of rcs Gene Products in Regulation of Capsular Polysaccharide Formation in Other Group K-Antigen Serotypes

To determine whether *rcs* gene products regulate all group I CPSs in *E. coli,* plasmid pWQ600 (*rcsB⁺, rcsC⁺*) is introduced into 22 additional strains by triparental mating experiments (Table IV). The compositions of polysaccharides produced in the presence or absence of pWQ600 are determined by the HPLC method described above (21). Group IB CPSs are distinguished from group IA CPSs by the presence of amino sugars in the group IB polymers (2). Amino sugars are not found in any group IA CPS.

In 10 strains with group IA CPSs, the introduction of pWQ600 increases the amount of polysaccharide produced, judging from colony morphology. In all cases, the only polymer produced in the presence or absence of pWQ600 is the authentic group IA CPS (Table IV). These results are consistent with the possibility that all of the *cps* gene clusters involved in expression of group IA CPSs are allelic and regulated in a similar fashion. However, it is also conceivable that some strains contain an inactive colanic acid biosynthesis gene cluster, and further detailed analysis of the appropriate chromosomal regions is required to resolve this possibility.

Escherichia coli strains with group IB K-antigens are more complicated, as they were found to have functional biosynthesis gene clusters for both the K-antigen and colanic acid. The chromosomal location has not been determined for any of the group IB K-antigen biosynthesis gene clusters, although they are assumed to map near *his* (2). At 37°C, no detectable colanic acid is produced by *E. coli* strains with

TABLE IV   Effect of Multicopy $rcsB_{K30}$ on *Escherichia coli* Group I Capsule Expression

K-antigen group	Serotypes	Polysaccharide[a] synthesized	
		Wild type (no plasmid)	Plus pWQ600
IA	K26, K27, K28, K30, K31, K34, K37, K39, K42, K43, K55	K-antigen	K-antigen
IB	K8 K9, K40, K41, K44, K45, K46, K47, K48, K49, K50, K87	K-antigen	K-antigen plus colanic acid
II	K1, K5	K-antigen	K-antigen plus colanic acid

[a] Plasmid pWQ600 was introduced into each strain by triparental mating procedures. The cell surface polysaccharides were then isolated and the composition determined by HPLC analysis. Data from Jayaratne *et al.* (21) and W. J. Keenleyside *et al.* (35).

group IB K-antigens unless multicopy *rcsB* is present, a situation which resembles *E. coli* K-12. In 12 strains with group IB CPSs, introduction of pWQ600 resulted in increased polymer formation, owing to the activation of colanic acid synthesis. However, the wild-type group IB CPS was also produced under these conditions (Table IV). It was not possible to determine whether the expression of the group IB CPS was also elevated in response to *rcsB,* owing to the large amount of colanic acid produced.

It has been known for some years that the *cps* genes for *E. coli* group I CPSs and the *kps* genes for group II K-antigens are not allelic (33). Also, Goebel (34) showed that a strain with a K1 (group II) CPS was able to produce colanic acid. Introduction of multicopy $rcsB_{K30}$ into strains with the well-characterized group II K1 and K5 CPSs resulted in a mucoid phenotype owing to activation of colanic acid synthesis. The polymer could be clearly distinguished from the respective K-antigens by HPLC analysis of acid-hydrolyzed extracellular polysaccharides. The polysaccharide purified from the *E. coli* K1 and K5 (pWQ600) recombinants had a composition and monosaccharide ratio corresponding to colanic acid. No colanic acid was detected in samples from the wild-type strains, in the absence of pWQ600. Although no group II CPS sugars were detected in samples from bacteria synthesizing colanic acid, this simply reflects the relative amounts of the two polymers produced. The amount of enzyme activity involved in either K1 or K5 CPS synthesis was not significantly affected by the simultaneous production of colanic acid and the K-antigen, and a group II capsular structure was clearly visible in electron micrographs of strains producing both types of polymers (35). From these results, it appears that strains with group II CPSs (at least in the case of the K1 and K5 prototypes) contain both *kps* (group II CPS synthesis) and *cps* (colanic acid synthesis) genetic loci. As in the case of *E. coli* K-12 and strains expressing group IB K-antigens, a negligible amount of colanic acid (undetectable by HPLC in the presence of the K-antigen) is produced unless the Rcs regulation is unbalanced.

Concluding Remarks

In *E. coli* K-12, colanic acid is synthesized at low temperature (below 30°C) or when grown on nitrogen-poor, carbon-rich media (4). Colanic acid production is also insignificant in *E. coli* strains with group IB and group II CPSs, when the bacteria are grown on rich media at 37°C. In contrast, the K30 and other group IA CPSs are produced and form a capsular structure at 37°C, despite the fact that these CPSs are regulated by the same system as colanic acid. The reason(s) for the lower amounts of colanic acid synthesis in *E. coli* K-12 at 37°C (4) is unclear, but it appears that the major difference between the colanic acid and K30 systems lies in the amount of "basal" polysaccharide synthesis (21). In *E. coli* O9:K30 and K-12 derivatives, high levels of CPS expression require functional Rcs proteins. Mutations in *E. coli* K-12 affecting *lon* and *rcsC* (*rcsC137*) increase transcription of two different *cps*::*lacZ* fusions by 10- to 46-fold and 116- to 226-fold, respectively. When colanic acid polymer formation is measured instead, this equates to 13- and 28-fold increases in synthesis in *lon* and *rcsC137* mutants, respectively (32). In *E. coli* K30, the $rcsC_{K30}$ mutation in CWG123 elevates cell-free K30 CPS by 21-fold. Whether the difference between the levels of synthesis in the basal and induced states in the two systems indicates an additional level of regulation, or merely reflects slightly altered interactions between the components of similar systems, has not been established. Differences in the surface organization and temperature of production of colanic acid and K30 CPS make direct comparison difficult. About 50% of the K30 CPS is surface attached to form a capsular structure in wild-type *E. coli* K30 (36), whereas colanic acid is a slime polysaccharide. However, any factor which increases CPS synthesis in *E. coli* K30 causes an altered distribution, resulting in more cell-free polymer. In addition, *rcsB* appears to interact with other cellular factors and is subject to complex regulation (20, 37), suggesting that colanic acid synthesis is potentially affected by a variety of factors which are in turn influenced by genetic background.

The key questions that remain to be addressed with these systems are severalfold. First, what is the nature of the *cps* promoter(s), and how does it interact with the regulatory elements? Second, what are the environmental cues that result in activation of the regulatory system *in vivo?* Third, what is the biological significance of high-level CPS expression? Finally, the retention of genetic loci for multiple cell surface polysaccharides suggests that each is important to the cell. The precise roles played by each polymer in the biology of *E. coli* await clarification.

Acknowledgments

Research was supported by a research operating grant to C.W. from the Medical Research Council of Canada. The authors thank Drs. J. Mekalanos, I. Ørskov, C. A. Schnaitman, V. Stout, and E. R. Vimr for generously supplying bacterial strains.

References

1. I. Ørskov, F. Ørskov, B. Jann, and K. Jann, *Bacteriol. Rev.* **41,** 667 (1977).
2. B. Jann and K. Jann, *Curr. Top. Microbiol. Immunol.* **150,** 19 (1990).
3. C. Whitfield, W. J. Keenleyside, and B. R. Clarke, *in* "*Escherichia coli* in Domestic Animals and Man" (C. L. Gyles, ed.), p. 437. CAB International, Wallingford, U.K. 1994.
4. A. Markovitz, *in* "Surface Carbohydrates of the Prokaryotic Cell" (I. W. Sutherland, ed.), p. 415. Academic Press, New York, 1977.
5. H.-M. Kuhn, U. Meier-Dieter, and H. Mayer, *FEMS Microbiol. Rev.* **54,** 195 (1988).
6. P. W. Taylor, *Microbiol. Rev.* **47,** 46 (1983).
7. M. A. Horwitz, *Rev. Infect. Dis.* **4,** 104 (1982).
8. S. Gottesman and V. Stout, *Mol. Microbiol.* **5,** 1599 (1991).
9. T. Ophir and D. L. Gutnick, *Appl. Environ. Microbiol.* **60,** 740 (1994).
10. C. Whitfield and M. A. Valvano, *Adv. Microsc. Physiol.* **35,** 135 (1993).
11. K. L. McCallum, D. H. Laakso, and C. Whitfield, *Can. J. Microbiol.* **35,** 994 (1989).
12. P. Jayaratne, D. Bronner, R. MacLachlan, C. Dodgson, N. Kido, and C. Whitfield, *J. Bacteriol.* **176,** 3126 (1994).
13. M. K. Homonylo, S. J. Wilmot, J. S. Lam, L. A. MacDonald, and C. Whitfield, *Can. J. Microbiol.* **34,** 1159 (1988).
14. P. Trisler and S. Gottesman, *J. Bacteriol.* **160,** 184 (1984).
15. V. Stout, A. Torres-Cabassa, M. R. Maurizi, D. Gutnick, and S. Gottesman, *J. Bacteriol.* **173,** 1738 (1991).
16. W. J. Keenleyside, P. Jayaratne, P. R. MacLachlan, and C. Whitfield, *J. Bacteriol.* **174,** 8 (1992).
17. J. H. Miller, "A Short Course in Bacterial Genetics: A Laboratory Manual and Handbook for *Escherichia coli* and Related Bacteria." Cold Spring Harbor Laboratory, Plainview, New York, 1992.
18. C. Whitfield and M. Lam, *FEMS Microbiol. Lett.* **37,** 351 (1986).
19. O. Westphal and K. Jann, *Methods Carbohydr. Chem.* **5,** 83 (1965).
20. V. Stout and S. Gottesman, *J. Bacteriol.* **172,** 659 (1990).
21. P. Jayaratne, W. J. Keenleyside, P. R. MacLachlan, C. Dodgson, and C. Whitfield, *J. Bacteriol.* **175,** 5384 (1993).

22. V. C. Knauf and E. W. Nester, *Plasmid* **8,** 45 (1982).
23. R. A. Hull, R. E. Gill, P. Hsu, B. H. Minshew, and S. Falkow, *Infect. Immun.* **33,** 933 (1981).
24. J. Sambrook, E. F. Fritsch, and T. Maniatis, "Molecular Cloning: A Laboratory Manual," 2nd Ed. Cold Spring Harbor Laboratory, Cold Spring Harbor, New York, 1989.
25. J. Binotto, P. R. MacLachlan, and P. R. Sanderson, *Can. J. Microbiol.* **37,** 474 (1991).
26. G. Ditta, D. Stanfield, D. Corbin, and D. R. Helinski, *Proc. Natl. Acad. Sci. U.S.A.* **77,** 7347 (1980).
27. D. H. Figurski and D. R. Helinski, *Proc. Natl. Acad. Sci. U.S.A.* **76,** 1648 (1979).
28. C. Whitfield, G. Schoenhals, and L. Graham, *J. Gen. Microbiol.* **135,** 2589 (1989).
29. D. H. Laakso, M. K. Homonylo, S. J. Wilmot, and C. Whitfield, *Can. J. Microbiol.* **34,** 987 (1988).
30. F. G. Chumley, R. Menzel, and J. R. Roth, *Genetics* **91,** 639 (1979).
31. V. L. Miller and J. J. Mekalanos, *J. Bacteriol.* **170,** 2575 (1988).
32. S. Gottesman, P. Trisler, and A. Torres-Cabassa, *J. Bacteriol.* **162,** 1111 (1985).
33. I. Ørskov, V. Sharma, and F. Ørskov, *Acta Pathol. Microbiol. Scand.* **84B,** 125 (1976).
34. W. F. Goebel, *Proc. Natl. Acad. Sci. U.S.A.* **49,** 464 (1963).
35. W. J. Keenleyside, D. Bronner, K. Jann, B. Jann, and C. Whitfield, *J. Bacteriol.* **175,** 6725 (1993).
36. P. R. MacLachlan, W. J. Keenleyside, C. Dodgson, and C. Whitfield, *J. Bacteriol.* **175,** 7515 (1993).
37. F. G. Gervais, P. Phoenix, and G. R. Drapeau, *J. Bacteriol.* **174,** 3964 (1992).

[17]   Genetic Mapping in *Bacillus subtilis*

Simon M. Cutting and Vasco Azevedo

Introduction

The simple spore-forming prokaryote *Bacillus subtilis* has for decades lent itself to genetic analysis. Second only to *Escherichia coli,* it has served as a model system in which to study gene regulation, biosynthetic pathways, development, etc. Methods have long existed for genetic transfer in this organism; indeed, genetic material can be exchanged by the classic procedures of transformation and transduction. In addition, DNA can be artificially introduced into *B. subtilis* by protoplast transformation or electroporation. The ease of genetic analysis has made *B. subtilis* the most-studied gram-positive organism. This chapter summarizes methods that exist for rapidly mapping genes. To map a gene two steps are usually taken. First, the approximate position of the gene is determined by long-range genetic mapping. Traditionally, the generalized transducing phage PBS1 has been used, but an alternative procedure is to use a yeast artificial chromosome (YAC) library of *B. subtilis* DNA fragments. The YAC library provides an extremely rapid and simple method for locating genes by direct physical mapping but is limited in that it is first necessary to have cloned DNA of the gene of interest. Having established an approximate position of the gene of interest, fine-structure mapping using DNA-mediated transformation is then the second step to obtain a precise map location.

This chapter presents a selection of the most frequently used mapping techniques. The provision of all *B. subtilis* genetic methodologies is beyond the scope of this chapter. The reader is referred to two books which provide voluminous information about *B. subtilis* methodology and biology: *Molecular Biological Methods for Bacillus,* edited by Harwood and Cutting (1), and *Bacillus subtilis and Other Gram-Positive Bacteria: Biochemistry, Physiology and Molecular Genetics,* edited by Sonenshein, Hoch, and Losick (2).

Long-Range Mapping of Genes Using A Yeast Artificial Chromosome Library

The approximate positioning of genes on the *B. subtilis* chromosome has been greatly simplified with the construction (3) of an ordered array of *B. subtilis* DNA cloned into yeast artificial chromosomes. The YAC library consists of a set of 59 overlapping clones that contain over 98% of the *B. subtilis* chromosome (see Fig. 1 and Table I). Each YAC clone contains a *B. subtilis* fragment of DNA ranging in size from 40 to 240 kb.

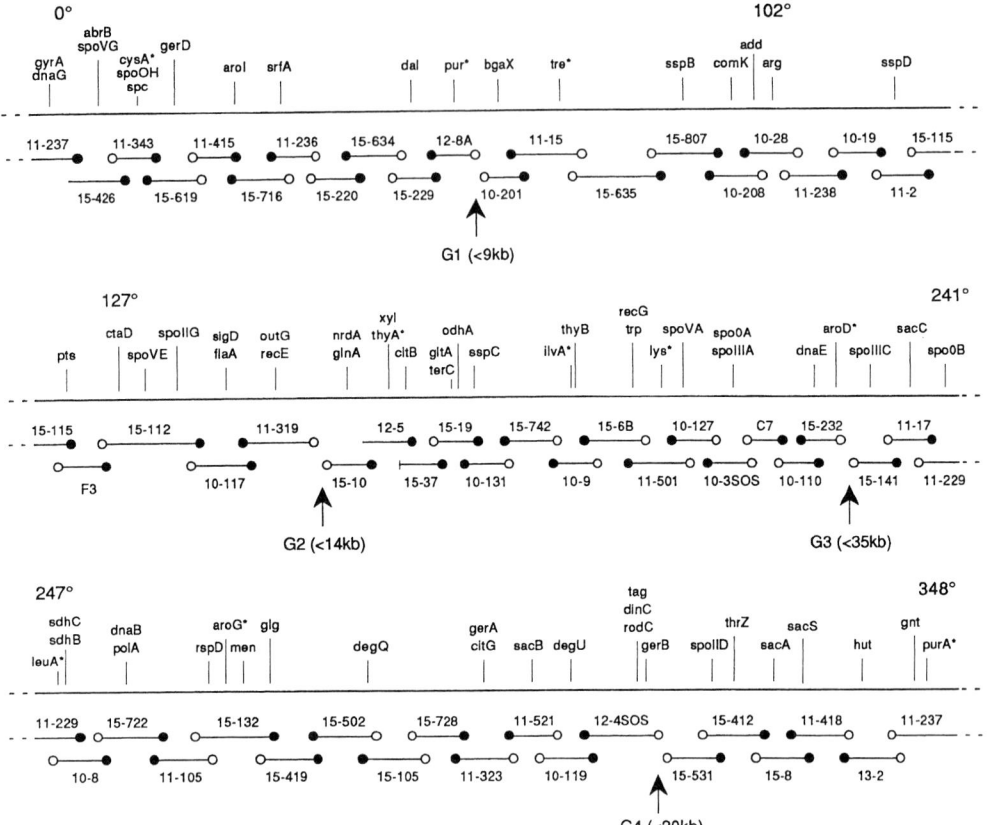

FIG. 1   Ordered *B. subtilis* DNA inserts carried in YAC clones. The *B. subtilis* DNA segments carried in individual YAC clones are represented by horizontal lines. They are arranged according to their relationships with the neighboring inserts and the *B. subtilis* genes they contain. YAC clone accession numbers are given under each line. *Bacillus subtilis* genes which hybridized with the corresponding YACs are indicated above the uninterrupted line symbolizing the chromosome. Genetic map positions (in degrees) are given for *dnaG, add, ctaD, spoOB, sdh,* and *gnt.* Asterisks (*) indicate markers which were identified by direct transformation of *B. subtilis* mutants. Filled and open circles represent left and right extremities of inserts, respectively. The vertical line at the left-hand end of insert 15-37 (close to *terC*) indicates that this end resulted from a coligation event, since it hybridizes with the nonadjacent 15–634 insert. The position of the four gaps in the collection is indicated by vertical arrows and the symbols G1 to G4. The numbers in parentheses give the maximum gap size.

The YAC collection can be used in two different ways to map genes. For genes that are already cloned or partially cloned, the DNA fragment can be used as a radioactive probe (or chemiluminescent probe) to screen a dot-blot filter containing DNA

TABLE I   Yeast Artificial Chromosome Clones Covering the
Bacillus subtilis Chromosome

Map position[a]	YAC clone	Map position[a]	YAC clone
358	11-237	184	15-742
5	15-426	190	10-9
10	11-343	196	15-6B
15	15-619	202	11-501
23	11-415	206	10-127
30	15-716	211	10-3S0S
35	11-236	216	C7
40	15-220	219	10-110
45	15-634	223	15-232
50	15-229	230	15-141
55	12-8A	236	11-17
62	10-201	242	11-229
68	11-15	246	10-8
78	15-635	252	15-722
86	15-807	259	11-105
94	10-208	267	15-132
99	10-28	274	15-419
104	11-238	282	15-502
110	10-19	289	15-105
116	11-2	294	15-728
120	15-115	302	11-323
123	F3	308	11-521
132	15-112	312	10-119
141	10-117	318	12-4SOS
150	11-319	327	15-531
154	15-10	334	15-412
165	12-5	340	15-8
169	15-37	346	11-418
173	15-19	353	13-2
178	10-131		

[a] Map position is indicative and corresponds to approximately the middle of the insert.

from all 59 YAC clones (see below). Physical mapping provides rapid positioning of a gene of interest and is particularly useful for multicopy genes which are ordinarily difficult to map. Okamoto *et al.* (4) have mapped 10 rRNA genes on the *B. subtilis* chromosome using the polymerase chain reaction (PCR) to amplify and map DNA fragments of related genes. The YAC library provides an extremely accurate tool for mapping genes for which cloned DNA sequences exist. A filter containing the complete YAC library can be obtained from Dr. S. D. Ehrlich at the address given in the Strains section at the end of the chapter.

For *B. subtilis* mutants for which no cloned DNA exists, the YAC library can also be used to map the allele by direct transformation of the mutant with DNA prepared from each of the 59 YAC clones (3 μg of yeast DNA should be sufficient for each transformation). In practice, DNA needs only be prepared from the YAC clones once and can be used thereafter with each new mutant to be tested. There are several drawbacks with this approach. First, only mutations with a selectable phenotype can be mapped. Thus, a null mutation in the *metA* operon can be mapped by direct selection for Met$^+$ on minimal agar plates, yet a suppressor allele that increases transcription of a particular gene could not be mapped by this procedure. So this method must be thought of as a crude, yet reliable, tool for positioning unknown genes on the chromosome. A second consideration is the time involved in performing 59 genetic crosses with the YAC DNA and mutant cells. Arguably, it is faster to map genes using PBS1-generalized transduction using the mapping kit (see below), where only 9 genetic crosses are needed followed by 22 prototrophic selections. Indeed, the YAC library should be thought of as an important mapping tool in *B. subtilis* laboratories but one to be used as a supplement to existing mapping techniques.

Rapid Physical Mapping of DNA Using A YAC Filter

The YAC filter consists of a dot-blot nylon filter containing an ordered arrangement of all 59 YAC clones. The filter can and should be reused by "stripping" the bound probe as described by Sambrook *et al.* (5). If the filter is to be reused it is important that the filter not be allowed to dry out at any point, and once used it should be sealed in Saran Wrap. The general procedure for using the YAC filter is to prehybridize the filter for a minimum of 2 hr at 42°C in hybridization solution [0.9 M NaCl; 48 mM sodium phosphate, pH 7.5; 5 mM EDTA; 0.1% (w/v) sodium dodecyl sulfate (SDS); 0.2% Denhardt's solution; 100 μg/ml salmon sperm DNA]. The radioactively labeled probe is then added and hybridization continued overnight at 42°C. (*Note:* It is important that the labeled probe does not contain any sequences homologous with pBR322.) The filter is washed two times (50% (v/v) formamide; 0.36 M NaCl; 20 mM sodium phosphate, pH 7.5; 2 mM EDTA; 0.1% (w/v) SDS) for 20 min each at 42°C. The filter is then wrapped in Saran Wrap and exposed to autoradiographic film.

Preparation of Total Yeast DNA from A YAC Library

The YAC library can also be supplied as viable yeast cells (*Saccharomyces cerevisiae* strain SX4-6A, *his3-532, ade2-1, ura3-1 322, trp1-289, inos$^-$ can^R α ρ^+ ω^+*) contained in large microtitre plate wells, with one YAC clone per well. The cells can be stored at 4°C, although it is best to repropagate the collection on receipt. Each yeast strain should be grown in 2 ml of yeast selective medium (see Reagents and Media section) for 24 hr at 30°C. The cells are then harvested, resuspended in 1 ml of yeast storage medium, and frozen at $-70°C$ in microcentrifuge tubes. DNA is prepared from individual YAC clones as follows.

1. Grow yeast cells for 48 hr at $30°C$ in an orbital shaker in selective media (10 ml).
2. Harvest cells by centrifugation (4000 g, 10 min, $4°C$) and wash pellet with a solution of 50 mM EDTA (pH 8.0).
3. Resuspend cell pellet in 2 ml freshly prepared zymolase solution (1.2 M sorbitol; 10 mM Tris-HCl, pH 7.5; 20 mM EDTA; 14 mM 2-mercaptoethanol; 200 μg/ml zymolase) and incubate at $37°C$ for 90 min with gentle shaking.
4. Harvest spheroplasts by centrifugation (2500 rpm, 15 min, room temperature).
5. Remove supernatant and resuspend in 0.5 ml of lysis solution [1% (w/v) lithium dodecyl sulfate; 100 mM EDTA; 10 mM Tris-HCl, pH 8.0; filter-sterilize using a 0.2-μm filter].
6. Vortex the suspension and check that it clarifies. Then extract two times with phenol–chloroform and once with chloroform.
7. Precipitate DNA and resuspend in 200 μl of TE buffer containing RNase A (20 μg/ml) and incubate at room temperature for 30 min.
8. Extract RNase A-treated DNA two times with phenol–chloroform and once with chloroform. Precipitate DNA and resuspend in 100 μl of TE buffer. This procedure should yield a total of 15 μg yeast DNA.

Long-Range Mapping of Genes Using PBS1-Generalized Transduction

Until relatively recently, the established method for positioning markers on the *B. subtilis* chromosome has been to use the generalized transducing phage, PBS1 (6). This large (>200 kb), pseudolysogenic phage is ideal for long-range mapping of genes because it is able to package up to 10% of the *B. subtilis* chromosome. This is over 10 times that transferred by DNA-mediated transformation. The first approach for mapping genes is to use PBS1-mediated generalized transduction to obtain an approximate position of a gene on the chromosome by use of two- or three-factor transduction crosses. After this, DNA-mediated transformation is used to obtain a more accurate position of the gene of interest. To facilitate the rapid mapping of genes a "kit" of nine auxotrophic strains has been created (7). These strains can be obtained from the *Bacillus* Genetic Stock Center (BGSC) and are listed in Table II. Each strain is isogenic to the "wild-type" *B. subtilis* strain 168, and each contains one or more auxotrophic alleles which, together, represent 22 selectable genetic markers that evenly cover the complete chromosome. An unknown mutation is mapped by first preparing a PBS1-generalized transducing lysate from a strain containing a mutation in the gene of interest, and this is used to infect each of the nine mapping kit strains; this is followed by selection of the Aux$^+$ phenotype (loss of the particular auxotrophic requirement) on selective media and screening for the mutant

TABLE II   *Bacillus subtilis* Mapping Kit

Strain	BGSC[a] strain no.	Genotype[b]
QB944	1A3	*cysA14* (11) *purA26* (348) *trpC2*
QB928	1A4	*aroI906* (26) *dal-1* (38) *purB33* (55) *trpC2*
QB934	1A5	*glyB133* (74) *metC3* (115) *tre-12* (62) *trpC2*
QB943	1A6	*ilvA1* (194) *pyrD1* (139) *thyA1* (168) *thyB1* (195) *trpC2*
QB922	1A7	*gltA292* (180) *trpC2*
QB935	1A8	*aroD120* (226) *lys-1* (210) *trpC2*
QB936	1A9	*ald-1* (280) *aroG932* (264) *leuA8* (247) *trpC2*
QB917	1A10	*hisA1* (298) *thr-5* (284) *trpC2*
QB123	1A11	*ctrA1* (324) *sacA321* (330) *trpC2*

[a] BGSC, *Bacillus* Genetic Stock Center.
[b] Numbers in parentheses are chromosomal map positions.

phenotype. This procedure should identify at least one reference marker that is linked to the mutation of interest and allow an approximate chromosomal position to be determined.

Two common problems encountered when using the mapping kit are as follows: (1) some of the auxotrophic alleles carried on the kit strains can revert to Aux⁺ spontaneously if they are not carefully maintained on minimal agar plates containing the correct supplements, and (2) some of the selection procedures required to identify Aux⁺ transductants require complex media preparation and, for reasons which are unclear, give low yields of transductants.

A more effective mapping procedure is to use the now extensive collection of "silent" chromosomal insertions of the transposon Tn917 (8). These strains, shown in Table III, are obtainable from the *Bacillus* Genetic Stock Center and are easier to use than the mapping kit strains because a transducing lysate is made from the Tn917 insertion strain and used to infect the mutant strain, followed by selection for MLSᴿ (resistance to erythromycin and lincomycin) which is encoded by the transposon. Antibiotic-resistant transductants are then screened for correction of the mutant phenotype. In this way, lysates of each Tn917 strain can be made and used repeatedly for mapping different mutations, with the need for only one selection procedure.

There are several problems encountered when performing PBS1-transduction experiments that should be mentioned here. First, no matter which of the above two strategies are used to map a mutation, it is necessary to produce a PBS1-generalized transducing lysate. As this step itself requires a source of PBS1 as a cleared lysate (referred to as the producer lysate), some confusion exists as to what exactly this is and how to obtain it. In practice, any PBS1-transducing lysate can be used as a producer lysate. PBS1 is normally obtained as a lysate kept over chloroform. A stock of

TABLE III   Silent Chromosomal Insertions
of Tn917[a]

Insertion	Map position (°)	BGSC strain no.
zaa-84 :: Tn917	0	1A627
zbj-82 :: Tn917	65	1A628
zca-82 :: Tn917	74	1A629
zce-83 :: Tn917	87	1A630
zdd-85 :: Tn917	121	1A631
zde-85 :: Tn917	126	1A632
zdi-82 :: Tn917	140	1A633
zei-82 :: Tn917	177	1A634
zej-82 :: Tn917	179	1A635
zfb-85 :: Tn917	183	1A636
zff-82 :: Tn917	200	1A637
zfg-83 :: Tn917	216	1A638
zgi-85 :: Tn917	245	1A639
zhb-83 :: Tn917	257	1A640
zhc-85 :: Tn917	260	1A641
zhi-83 :: Tn917	285	1A642
zib-82 :: Tn917	294	1A643
zii-83 :: Tn917	317	1A644
zif-85 :: Tn917	342	1A645

[a] From Ref. 8.

PBS1 bacteriophage as a cleared lysate can normally be obtained from investigators in the *Bacillus* community or from the author (S. Cutting). If the titer of the PBS1 lysate is in doubt, then simply use the lysate (0.1 ml) to infect a suitable standard spore-forming strain of *B. subtilis* (e.g., 168, PY79, or JH642). Next, prepare a PBS1-generalized transducing lysate (see below) and repeat the procedure using the newly acquired PBS1 lysate. Successive rounds of infection–lysis–infection will substantially increase the titer of phage, and if possible the stock of PBS1-producer lysate should be passaged every 6 months or so to maintain a high-titer working stock (although PBS1 lysates can remain viable for up to 10 years).

The last problem to be encountered is obtaining highly motile cells. PBS1 infects by adsorption to the flagellum, so highly motile cells are desirable in any transduction experiment. Some strains are difficult to make motile, whereas others appear weakly motile yet appear to be infected with PBS1 with little loss in efficiency. If poor cell motility is suspected as the problem in transduction experiments, then either an alternative mapping strategy must be used or, as a last resort, a motile derivative should be isolated (9).

Preparation of PBS1-Generalized Transducing Lysate

1. Inoculate 10-ml of brain–heart infusion broth (BHIB; see Reagents and Media section) (in a 250-ml culture flask) with a single, fresh colony of the appropriate strain. Incubate overnight at 30°C with slow orbital shaking.
2. The following morning, check the culture by microscopic examination for cell motility. If cells are motile, proceed to step 3; if not, remove the culture from the shaker, leave on the bench at ambient temperature for 15 min, and check again for motility.
3. Inoculate a test tube containing 10 ml of BHIB with 0.1 ml of motile culture. Incubate at 37°C on a roller drum shaker.
4. After 5 min inoculate the tube with 0.1 ml of a PBS1-producer lysate (see above) and return the tube to the roller drum shaker.
5. After a total of 8 to 9 hr of incubation, remove the tube from the shaker and leave on the bench at ambient temperature overnight.
6. In the morning, decant the contents of the tube into a centrifuge tube and spin hard to pellet any unlysed cells (8000 g, 15 min, room temperature).
7. Carefully decant the lysate supernatant into a sterile glass tube, filter-sterilize using a 0.45-μm filter, and store at 4°C over chloroform.

Generalized Transduction Procedure

1. Inoculate 10 ml of BHIB (in a 250-ml culture flask) with a fresh colony of the recipient strain. Incubate overnight at 30°C with slow orbital shaking.
2. The following morning, check cell motility. If cells are motile proceed to step 3; if not, remove from the shaker and leave on the bench for 15 min before checking again for motility.
3. Use 2 ml of motile culture to inoculate 10 ml of warm BHIB (in a 250-ml culture flask) and incubate at 37°C with medium orbital shaking.
4. After 3–4 hr, check for cell motility. If cells are motile, proceed to step 5; if not, remove from the shaker and leave on the bench for a further 15 min.
5. Add 1 ml each of the motile culture to two test tubes. To one tube add 0.1 ml of PBS1-transducing lysate prepared from the donor strain (see above). The other tube serves as an uninfected control.
6. Incubate both tubes for 30 min at 37°C in a roller drum shaker.
7. After 30 min the cells can be diluted in T base (see Reagents and Media section) and plated directly onto the appropriate selective agar plates. Remember that if selecting for prototrophy (e.g., with the mapping kit strains), the cells must be washed two times with T base (10 ml) before dilution and plating onto minimal agar plates to avoid carryover of the rich BHIB medium.

(*Note:* With a good donor lysate and motile recipient cells, expect to obtain only 1000–20,000 transductants/ml.)

Fine-Structure Mapping by DNA-Mediated Transformation

Having determined the approximate chromosomal position of a gene by one of the long-range mapping procedures described above, a gene can be mapped more precisely against other chromosomal loci (markers) in the region of interest by DNA-mediated transformation. *Bacillus subtilis* is able to become naturally competent, and this state is achieved simply by growing cells to the stationary phase of growth in a defined minimal medium (10) as outlined below. Cells are transformed with chromosomal DNA which, on entry into the cell, is linearized; only a single strand of DNA actually enters the cell, after which it is able to recombine with homologous DNA on the chromosome. Importantly, the average size of transforming chromosomal DNA is only about 30 kb; thus, two genetic markers must be separated by no more than this distance for the genetic linkage to be determined. Ordinarily, transformation is used first to determine the genetic distance or linkage between the gene of interest (*genX*) and a suitable reference marker by a "two-factor" genetic cross. Linkage between two genetic markers can be defined as a cotransformation frequency and is expressed as a percentage. Linkage is often defined as transformation map units by using the equation $(100 - \%$ transformation frequency$)/100$. For example, for two mutations that are 60% cotransformed, the distance between the two markers is 0.4 transformation map units. As implied by the name, map units are used by convention in assigning linkage relationships on genetic maps.

When mapping genes by transformation there are several important considerations. First, linkage between two markers is considered essentially linear only at distances less than 10 kb. Nonlinearity is due partly to random endonucleolytic fragmentation of chromosomal DNA on uptake by competent cells but also to the phenomenon termed congression. Congression is observed when a competent cell independently takes up two fragments of transforming DNA. Congression occurs at high, "saturating" concentrations of transforming DNA (1 to 5 μg/ml of competent cells) and at a frequency of 2–5%, which can produce spurious transformation linkages between unlinked genes (especially with weakly linked markers). To obtain accurate estimates of genetic distance, nonsaturating concentrations of transforming DNA must be used ($<$50 ng/ml of competent cells). Use of higher concentrations of transforming DNA may distort estimates of genetic distance.

A second consideration with fine-structure mapping by transformation concerns the nature of markers employed to determine genetic distance. When using a chromosomal marker that consists of an antibiotic-tagged insertion such as a transposon insertion, the resulting genetic distance obtained in such a cross will be exaggerated. Thus, linkage relationships can only be considered approximate in such situations.

Last, determining genetic distance by two-factor transformations will not order a genetic locus relative to other genetic markers. For example, if a gene of interest (*genX*) shows genetic linkage to two markers *leuA* and *pheA* which are ordered *leuA-pheA*, then the relative gene order could be (1) *genX-leuA-pheA*, (2) *leuA-genX-pheA*, or (3) *leuA-pheA-genX*. To determine the correct order of genes, a three-factor genetic transformation cross must be used if possible.

Preparation of Bacillus subtilis Chromosomal DNA

1. Inoculate 25 ml of Luria–Bertani (LB) medium (see Reagents and Media section) with a single, fresh colony. (*Note:* Small micropreparations of DNA can be prepared from 2–5 ml of culture, although those volumes produce less consistent DNA yields.) Incubate overnight at 30°C or for 6 hr at 37°C.
2. Harvest cells using a bench-top centrifuge (6000 *g*, 5 min; use a phenol-resistant polypropylene tube) and resuspend cell pellet in 4.5 ml of sterile lysis buffer (50 mM EDTA; 0.1 M NaCl; pH 7.5).
3. Add 4 mg of lysozyme powder, vortex briefly, and incubate at 37°C (static) for 10 min.
4. Add 0.3 ml of 20% (w/v) sarkosyl, gently invert the tube several times to clarify the cell suspension, and incubate on ice for 5–10 min, inverting occasionally.
5. Add 3–4 ml of phenol–chloroform and vortex vigorously for several minutes.
6. Centrifuge at high speed in a bench-top centrifuge (10,000 *g*, 10 min) to separate the organic and aqueous phases and carefully remove the upper, DNA-containing aqueous phase to a new screw-cap plastic centrifuge tube. (*Note:* To avoid contamination with phenol, do not try to remove all of the aqueous DNA solution. If contamination is suspected, repeat steps 5 and 6.)
7. To the extracted, aqueous phase add 0.1 volume of 3 M sodium acetate (pH 5.2) and approximately 2–3 volumes of ice-cold ethanol. Cap the tube and invert the tube repeatedly until the DNA precipitate has condensed to a small clump.
8. With a pipette tip or glass Pasteur pipette carefully remove the DNA clump to a microcentrifuge tube. Drain any excess ethanol, air-dry for 1 hr, then add 1 ml TE buffer (do not desiccate or lyophilize as this will prevent the solubilization of the DNA). Leave the DNA solution to dissolve overnight at 4°C. Determine the DNA concentration and store aliquots at −20°C.

Preparation of Competent Bacillus subtilis Cells

1. Using a fresh, single colony of the strain to be made competent, inoculate an LB agar plate (this is best made as a small patch of 2–3 inches in diameter on the plate). Incubate at 30°C overnight.

2. The following morning, use a wooden applicator stick to scrape the cell growth from the plate and resuspend it (by vortexing) in about 2 ml of freshly prepared SpC medium (see Reagents and Media section).
3. Use the cell suspension to inoculate 20 ml of prewarmed SpC medium in a 250-ml culture flask to give a starting OD_{600} of about 0.5 (with practice this can be done by eye).
4. Incubate the culture at 37°C with moderate orbital shaking. Take periodic OD_{600} readings to follow cell growth.
5. Just as or slightly before the culture reaches stationary phase (i.e., <5% increase in OD_{600} readings in a 15-min time span), use 2 ml of the culture to inoculate 20 ml of fresh, prewarmed, SpII medium in a 250-ml flask. Incubate at 37°C with moderate orbital shaking. (*Note:* Normally, 4–5 hr of growth is required to reach stationary phase in SpC medium. If problems are encountered, check that all nutritional supplements have been included in the minimal medium. It is best not to incorporate antibiotics, though.)
6. After 90 min of incubation pellet the cells (8000 *g*, 5 min, room temperature). Carefully decant the supernatant into a separate, sterile container and save it.
7. Gently resuspend the cell pellet in 1.6 ml of saved supernatant and add 0.4 ml of sterile 50% (v/v) glycerol.
8. The competent cells can be used at this point, or they can be frozen immediately at −70°C in small (0.5 ml) aliquots. (*Note:* Routinely, it is not necessary to use dry ice or liquid nitrogen to rapidly freeze competent cells. Indeed, well-prepared cells will remain viable for up to 5 years or longer.)

Transformation of Competent Cells

1. Quick-thaw a tube of frozen competent cells by immersion in a 37°C water bath.
2. For each transformation dispense 0.2 ml of competent cells into a test tube, add DNA, and incubate in a roller drum shaker for 30 min at 37°C. The transforming DNA must be in a volume of 100 μl or less. For temperature-sensitive strains (*ts*) or for strains containing *ts* plasmids, increase the time of incubation to 90–120 min.
3. Dilute the transformed cells in T base and plate onto appropriate selective media.

Selection Procedures

For all gene transfer experiments, there must, of course, be a method of selecting for recombinants. Normally, this requires direct selection for the marker being introduced. Typically, this might involve selection for conversion of an auxotrophic

TABLE IV   Selective Antibiotic Concentrations

Resistance	Antibiotic	Selective concentration	Solvent
Cm^R	Chloramphenicol	5 μg/ml	95% ethanol
MLS^R	Erythromycin[a]	1 μg/ml	95% ethanol
	Lincomycin	25 μg/ml	50% ethanol
Nm^R	Neomycin	5 μg/ml	Distilled water
Pm^R	Phleomycin[b,c]	0.1–0.5 μg/ml	Distilled water

[a] When selecting for MLS^R, both lincomycin and erythromycin are used.

[b] Pm^R is best selected using an agar-overlay technique. Plate cells onto plates containing no drug, incubate at 37°C for 2 hr, then add 4 ml of LB overlay agar (LB solidified with 0.7% agar) containing phleomycin to give the final concentration shown.

[c] Bleomycin is not recommended as an alternative to phleomycin as it increases the mutation rate. Phleomycin can be obtained from Sigma (St. Louis, MO).

marker to prototrophy or introduction of a drug-resistant marker. Remember that not every genetic marker uses a prototrophic or drug-resistant selection procedure; the two most common methods are described here.

For prototrophic selections, a selection must be made on a minimal agar plate that does not contain the nutritional supplement being selected for. Several media exist for prototrophic selections, but lactate glutamate minimal agar (LGMA) is commonly used because it has the added advantage that the sporulation phenotype of recombinants can be determined directly. Thus, Aux^+ recombinants that are Spo^+ develop a brown pigmentation that creates brown-opaque colonies on LGMA. Spo^- colonies, however, form white-translucent colonies. This allows the simultaneous mapping of an auxotrophic to a sporulation marker in one experiment.

Selection for drug resistance can be made directly on any rich medium, although Difco sporulation medium (DSM) is advantageous for mapping sporulation genes because the Spo phenotype of drug-resistant colonies can be determined directly. Table IV gives the required amounts of antibiotic that must be used in selective media. If necessary, kanamycin (at 2.5 μg/ml) can be used instead of neomycin. Generally, it is best to prolong the incubation time of transformed or transduced cells from 30 to 120 min when selecting for drug resistance.

Reagents and Media

Brain–Heart Infusion Broth (BHIB)

Difco (Detroit, MI) brain–heart infusion is prepared as recommended by the manufacturer but containing 0.5% (w/v) Difco Bacto-yeast extract. Sterilize by autoclaving.

Difco Sporulation Medium (DSM)

Bacto-nutrient broth	8 g
10% (w/v) KCl	10 ml (or 1 g KCl)
1.2% (w/v) $MgSO_4 \cdot 7H_2O$	10 ml (or 0.25 g $MgSO_4 \cdot 7H_2O$)
1 N NaOH	to pH 7.0 ($\sim$0.7 ml)

Dissolve in 1000 ml distilled water by stirring. KCl and $MgSO_4$ are presterilized solutions. Autoclave, allow to cool to 50°–60°C, and then add the following three sporulation supplements:

1 M $Ca(NO_3)_4$	1 ml
10 mM $MnCl_2$	1 ml
1 mM $FeSO_4$	1 ml

The three sporulation supplements are presterilized solutions. The $FeSO_4$ will form a red-ferrous precipitate following sterilization (which can be avoided by filter-sterilizing the iron solution, although the medium seems to work fine regardless). The $FeSO_4$ solution, however, must be remixed prior to adding to DSM. Store medium at room temperature.

Lactate Glutamate Minimal Agar (LGMA)

For 1000 ml:

Agar	15 g
$MgSO_4 \cdot 7H_2O$	10 g
Sporulation salts (see below)	900 ml

Mix and autoclave. After cooling to 60°C, add the following sterile solutions:

0.1 M $CaCl_2$	10 ml
5% (w/v) Glutamate	40 ml
L-Alanine (20 mg/ml)	5 ml
35% (v/v) Sodium lactate	8 ml
Growth supplements	see Table V

The sporulation salts mix is prepared from solution A and solution B (see below). Add 1 ml of solution A to about 100 ml of distilled water in a beaker. In a 1-liter measuring cylinder, add 10 ml of solution B and bring the volume to about 500 ml with distilled water. Now add the diluted solution A and bring the final volume in the cylinder to 1000 ml. Autoclave aliquots (90 ml, 180 ml, or multiples) and store at room temperature. Solutions A and B are added separately to water to avoid precipitation.

TABLE V   Prototrophy Growth Supplements[a]

Symbol	Compound	Stock solution (mg/ml)	Stock solution per liter medium	Final concentration in medium[b] (μg/ml)
DAL	D-Alanine	10	5	50
ARG	L-Arganine hydrochloride	10	5	50
ASG	L-Asparagine	10	5	50
ASP	L-Aspartic acid	50	10	500
CYS	L-Cysteine hydrochloride[c]	4	10	40
GLU	L-Glutamic acid[d]	50	10	500
GLN	L-Glutamine	10	5	50
GLY	L-Glycine	10	5	50
HIS	L-Histidine hydrochloride	10	5	50
ISO	L-Isoleucine	10	5	50
LEU	L-Leucine	10	5	50
LYS	L-Lysine hydrochloride	10	5	50
MET	L-Methionine	10	5	50
PHE	L-Phenylalanine[e]	10	5	50
PRO	L-Proline	10	5	50
SER	L-Serine	10	5	50
THR	L-Threonine	10	5	50
TRP	L-Tryptophan[f]	2	10	20
TYR	L-Tyrosine[g]	2	10	20
VAL	L-Valine	10	5	50
ADE	Adenine[h]	2	10	20
CYT	Cytidine	5	10	50
GUA	Guanine[i]	2	10	20
THY	Thymine[j]	5	10	50
URA	Uracil[j]	4	10	40
SKI	Shikimic acid	10	5	50

[a] All stock solutions can be autoclaved (15 min) unless indicated otherwise. Stocks are stored at $-20°$C (long-term) and 4°C (short-term).

[b] Use of greater concentrations may inhibit growth.

[c] Stock solution decomposes unless stored under mineral oil.

[d] Adjust to pH 7.0 with 5 M NaOH.

[e] Dissolve in 1 mM NaOH.

[f] Filter-sterilize.

[g] Dissolve in 10 mM NaOH.

[h] Adenine or adenosine; dissolve in 30 mM HCl.

[i] Guanine or guanosine; dissolve in 0.1 M KOH.

[j] Dissolve in 0.1 M KOH.

Solution A: $FeCl_3 \cdot 6H_2O$, 0.098 g; $MgCl_2 \cdot 6H_2O$, 0.830 g; $MnCl_2 \cdot 4H_2O$, 1.979 g. Dissolve in 100 ml of distilled water. Store at 4°C. Do not autoclave.
Solution B: NH_4Cl, 53.5 g; Na_2SO_4, 10.6 g; KH_2PO_4, 6.8 g; NH_4NO_3, 9.7 g.

Dissolve in 900 ml of distilled water. Adjust to pH 7.0 with 2 N NaOH and autoclave. Store at 4°C.

Luria–Bertani (LB) Medium/Agar

Per liter: Bacto-typtone, 10 g; Bacto-yeast extract, 5 g; NaCl, 10 g; 1 M NaOH, 1.0 ml. Sterilize by autoclaving. For plates, solidify with 1.5% (w/v) agar.

SpC Medium

Prepare fresh on the day of use from the following sterile solutions:

T base (see below)	20 ml
50% (w/v) Glucose	0.2 ml
1.2% (w/v) $MgSO_4 \cdot 7H_2O$	0.3 ml
10% (w/v) Bacto-yeast extract	0.4 ml
1% (w/v) Bacto-casamino acids	0.5 ml
Growth requirements	see Table V

SpII Medium

Prepare fresh on the day of use from the following sterile solutions:

T base	20 ml
50% (w/v) glucose	0.2 ml
1.2% (w/v) $MgSO_4 \cdot 7H_2O$	1.4 ml
10% (w/v) Bacto-yeast extract	0.2 ml
1% (w/v) Bacto-casamino acids	0.2 ml
0.1 M $CaCl_2$	0.1 ml
Growth requirements	see Table V

T Base

Per liter: $(NH_4)_2SO_4$, 2 g; $K_2HPO_4 \cdot 3H_2O$, 18.3 g; KH_2PO_4, 6 g; trisodium citrate dihydrate, 1 g. Sterilize by autoclaving. T base can be prepared as a 10× stock.

Yeast Selective Medium

Per liter: Yeast nitrogen base (containing no amino acids), 6.7 g; glucose, 20 g; tyrosine, 55 g; adenine, 55 g; Bacto-casamino acids, 14 g. Sterilize by autoclaving. For agar plates, solidify with 2% (w/v) Difco Bacto-agar.

Yeast Storage Medium

Per liter: Bacto-yeast extract, 10 g; Bacto-tryptone, 10 g; glucose, 20 g; glycerol, 300 ml. Sterilize by autoclaving.

Strains

Bacillus strains can be obtained from the *Bacillus* Genetic Stock Center, The Ohio State University, Department of Biochemistry, 484 West 12th Avenue, Columbus, OH 43210-1292. (Telephone: 614-292-5550; Fax: 614-292-1538). The YAC library can be obtained by written request from S. D. Ehrlich, Laboratoire de Genetique Microbienne, Institut National de la Recherche Agronomique, Domaine de Vilvert, 78352 Jouy en Josas Cedex, France.

Acknowledgments

We thank S. Dusko Ehrlich and Pascale Serror for permission to reproduce some of the YAC library methodology. Preparation of this chapter was supported by U.S. Public Health Services Grant GM49206 (S.M.C.) from the National Institutes of Health.

References

1. C. R. Harwood and S. M. Cutting (eds.), "Molecular Biological Methods for *Bacillus.*" Wiley, Chichester, 1990.
2. A. L. Sonenshein, J. A. Hoch, and R. Losick (eds.), "*Bacillus subtilis* and Other Gram-Positive Bacteria: Biochemistry, Physiology and Molecular Genetics." American Society for Microbiology, Washington, D.C., 1993.
3. V. Azevedo, E. Alvarez, E. Zumstein, G. Damiani, V. Sgaramella, S. D. Ehrlich, and P. Serror, *Proc. Natl. Acad. Sci. U.S.A.* **90,** 6047 (1993).
4. K. Okamoto, P. Serror, V. Azevedo, and B. Vold, *J. Bacteriol.* **175,** 4290 (1993).
5. J. Sambrook, E. F. Fritsch, and T. Maniatis, "Molecular Cloning: A Laboratory Manual." Cold Spring Harbor Laboratory, New York, 1989.
6. D. Karamata and J. D. Gross, *Mol. Gen. Genet.* **108,** 277 (1970).
7. R. A. Dedonder, J. A. Lepesant, J. Lepesant-Kejzlarova, A. Billault, M. Steinmetz, and F. Kunst, *Appl. Environ. Microbiol.* **33,** 989 (1977).
8. M. A. Vandeyar and S. A. Zahler, *J. Bacteriol.* **167,** 530 (1986).
9. S. M. Cutting and P. B. Vander-Horn, *in* "Molecular Biological Methods for *Bacillus*" (C. R. Harwood and S. M. Cutting, eds.), pp. 27–74. Wiley, Chichester, 1990.
10. D. Dubnau and R. Davidoff-Abelson, *J. Mol. Biol.* **56,** 209 (1971).

[18] Molecular and Biochemical Methods for Studying Chemotaxis in *Bacillus subtilis*

Mia Mae L. Rosario and George W. Ordal

Introduction

Chemotaxis in bacteria (for reviews, see Refs. 1 and 2 and references therein) is the oldest sensory–motor network in nature and antedates the evolutionary separation of the archaebacteria (archaea) from the true bacteria (eubacteria). Subsequently, the eukarya diverged from the archaea. The process continues to be used by archaea and eubacteria (called here "bacteria"). Through chemotaxis, bacteria migrate by a biased random process to high concentrations of attractants or low concentrations of repellent. Unstimulated, peritrichous (flagella projecting from entire surface) bacteria, like *Bacillus subtilis* and *Escherichia coli*, exhibit alternating smooth swims (runs) and tumbles and give the appearance of erratic swimming. The purpose of tumbling is to randomly reorient the bacteria for the succeeding run. When in a gradient of chemoeffector (attractant or repellent), the frequency of runs and tumbles is altered. For instance, when headed toward higher concentrations of attractant bacteria increase their tendency to run, and when headed toward lower concentrations they increase their tendency to tumble. This bias affords net migration into a more favorable environment.

In general, chemoeffectors bind to chemoreceptors (sensory receptors) and thereby initiate a cascade of events involving phosphoryl transfer to affect briefly the direction of rotation of the flagella. Simultaneously, events begin that involve methylesterification of the receptors to bring about adaptation, that is, a return to prestimulus behavior. Adding attractant to a suspension of *E. coli*, for instance, causes all the bacteria to run; soon, even with the attractant still present, the bacteria return to the prestimulus behavior. The necessity for adaptation can be readily seen from the fact that if attractant binding caused perpetual running, without tumbling, then the bacteria would have no way of biasing their behavior when headed in the favorable direction. (Owing to Brownian motion, bacteria always run in curves, rather than in straight lines, and hence will not head straight up an attractant gradient.)

Like many sensory processes in bacteria, most of which involve transcriptional activation, chemotaxis uses the two-component system (the autophosphorylating kinase and the response regulator) (Fig. 1). The heart of chemotaxis in both *B. subtilis* and *E. coli* is excitation mediated by an autophosphorylating kinase (CheA), which monitors the state of the receptors [the methyl-accepting chemotaxis proteins (MCPs)]. In *B. subtilis*, binding of attractant to the receptors stimulates CheA activity and leads to phosphorylation of CheY. CheY — P binds to the switch, consisting of FliM, FliG,

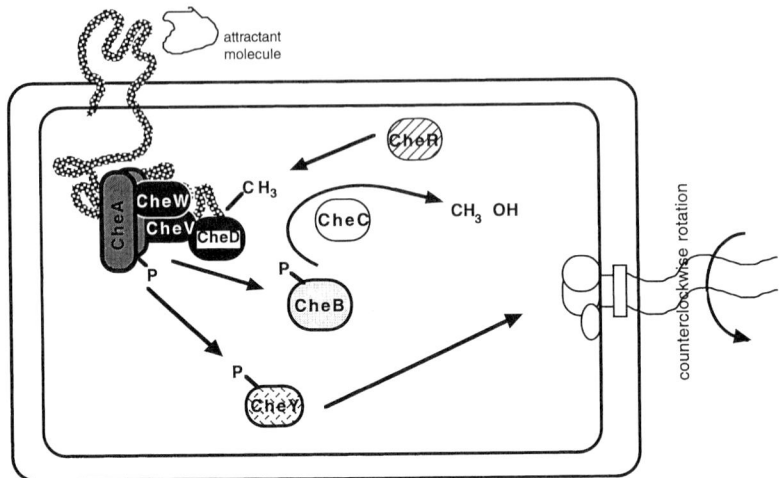

Fɪɢ. 1   Proposed model for the chemotactic signaling response in *B. subtilis.* See text.

and FliY, to cause counterclockwise (CCW) rotation of the flagella and hence running. The unique advance that chemotaxis has made over other systems is that it can return the generation of the response to background levels (adaptation) even with the signal still present. It achieves this goal by use of receptors that can be methylesterified on certain glutamate side chains. CheA — P also phosphorylates the CheB methylesterase, and CheB — P helps remove methyl groups from the MCPs, releasing it as methanol, so that adaptation can occur. In *B. subtilis,* removal of methyl groups from the receptors appears to be involved in adaptation to attractants.

Although chemotaxis in *B. subtilis* is similar to that in *E. coli* in that phosphorylation is involved in excitation and methylation in adaptation, it is different in several important respects. Through a study of these and other organisms, it should be possible to understand the probable nature of the ancestral mechanism and to examine eukaryotic cells for proteins that might have once served a chemotaxis function. The methods presented in this chapter will, it is hoped, serve well anyone interested in studying chemotaxis in *B. subtilis* and may well serve as a point of departure in the development of methods to study chemotaxis in other gram-positive bacteria.

The differences between *B. subtilis* and *E. coli* include the following:

In *B. subtilis* addition of attractants activates CheA. In *E. coli,* addition of repellents activates CheA (2).

In *B. subtilis,* methyl groups on the MCPs promote running; in *E. coli,* methyl groups on the MCPs promote tumbling (3).

In *B. subtilis,* attractant causes methyl groups to be transferred from the MCPs to an unidentified intermediate; in *E. coli,* repellent causes release of methyl groups as methanol directly from the MCPs (4). There are two unique proteins in *B. subtilis,* CheC and CheD, which appear to be involved in methyl transfer (5).

Uniquely in *B. subtilis,* there is a very effective methylation-independent adaptation system (6).

The switch, where CheY — P binds to control direction of flagellar rotation, includes FliM and FliG in both organisms but FliY (42 kDa) in *B. subtilis* and the homolog FliN (14 kDa) in *E. coli* (7).

In *B. subtilis* the chemotaxis genes are most located in a large operon that also includes genes encoding proteins of the basal body. This operon appears to be governed by the major vegetative sigma factor (σ^A). However, the receptor genes are governed by an alternate sigma factor (σ^D). In *E. coli* the chemotaxis genes are in several operons and are governed by the homologous alternate sigma factor (8, 9).

In *B. subtilis,* two partially redundant proteins, CheW and CheV, are believed to couple the binding of chemoeffector at the MCPs to changing CheA activity. CheV has two domains, one homologous to CheW and the other to CheY. *cheV* is governed by σ^D, and *cheW* is governed by σ^A. In *E. coli,* there is only CheW (10, 11).

Capillary Assay

The capillary assay (see Ref. 12 for examples) is a spatial gradient assay of chemotaxis and is generally taken to be the ultimate criterion by which a strain is judged to be wild type or mutant. Cells should be grown in minimal medium, rather than broth, for good taxis. Assays should be carried out at low density (A_{600} of 0.001). If the assay of a strong attractant is carried out at high titers of cells, the accumulation cells will deplete the oxygen and, by negative aerotaxis, impede further entry of cells into the capillary. Furthermore, by carrying out assays at fairly low titers of cells, the entire contents can be poured out on plates and colonies counted the next day. The "soft agar" used for pour plating needs to be quite high in agar to prevent colonies breaking through the surface and swarming over it.

1. At the end of the afternoon, streak out bacteria on a tryptose blood agar base (TBAB Difco, 33 g/liter) plate and incubate overnight at 30°C.
2. Using a 2-mm-diameter microbiological loop, scrape a loopful of cells and inoculate into 1 ml tryptone broth (TBr: 1% tryptone, 0.5% NaCl) and vortex. Add 40 μl of the suspension to 2 ml minimal medium, in an 18 × 150 mm test tube, and shake at 37°C for 4 hr. Minimal medium contains 50 mM phosphate buffer, (68.05 g/liter KH_2PO_4, 87.09 g K_2HPO_4; pH to 7.0 with KOH) pH 7, 1 mM $MgCl_2$, 1 mM $(NH4)_2SO_4$, 0.001 × sporulation salts (0.14 M $CaCl_2$, 10 mM

MnCl$_2$, 0.2 M MgCl$_2$), 50 μg/ml required amino acids, and 20 mM sorbitol. For assaying amino acid taxis, include 20 mM sorbitol in the growth medium; for assaying sugar taxis, use 20 mM of the sugar plus 0.1% glycerol.

3. Add 20 μl of a mixture of 0.5 M sodium lactate and 5% glycerol and continue shaking for 15 min. Omit this step if sugars are used as attractants.

4. During the growth period, clean the appropriate number of 1-μl Drummond capillaries by drawing up and expelling distilled water 10 times. (Unwashed capillaries have both toxins and attractants.) Seal one end in a flame. Distribute 1.5 ml attractant dissolved in chemotaxis buffer [CB (per 100 ml): 1 ml of 1 M phosphate buffer, pH 7, and 1.0 ml of che cocktail containing 10 mM EDTA, 4.2 mM CaCl$_2$, 5% (w/v) glycerol, 0.5 M sodium lactate, and 30 mM (NH$_4$)$_2$SO$_4$] in 15-ml polypropylene beakers. Flame and place open end of capillary tubes into the attractant solution. During the cooling of the air, the solution will enter the capillaries. Put small U-shaped glass rods on a glass plate on a slide warmer and put a glass coverslip over each (the "corral") (Fig. 2). Set the temperature to 37° C. Put several moist pieces of (artificial) sponge on the plate to retard evaporation from the bacterial suspensions.

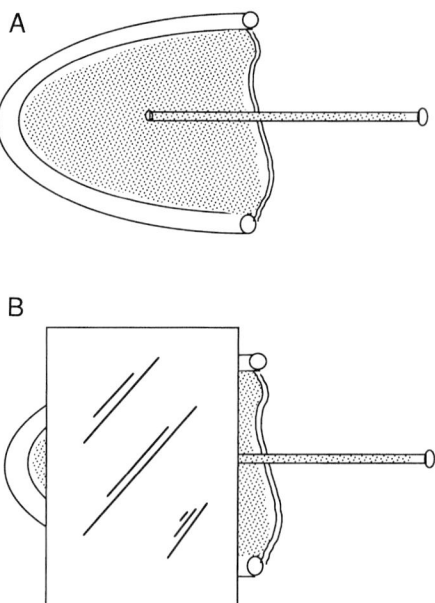

FIG. 2   The "corral" for capillary assays. (A) The glass U tube provides the boundary for the "pond" of bacteria. The open end of a capillary tube containing attractant is inserted into the pond, and bacteria swim into the tube. (B) Same setup as (A), but with the corral topped with a coverslip.

5. After incubation, filter the bacteria using a 0.45-μm nitrocellulose filter. Resuspend in 2 ml CB. Dilute with CB to an A_{600} of 0.001 (about 10–20 ml total volume) in a new 18 × 150 mm test tube. Avoid using a tube cleaned with detergent as residual detergent may harm the bacteria.

6. Pipette the bacterial suspension into a corral, pick up a capillary tube containing attractant with a pair of tweezers, dry the outside with a Kimwipe, and insert the open end into the corral. Incubate for 30 min. If there are several samples, it is recommended that a capillary tube be inserted into a corral every 30 sec.

7. Harvest the bacteria (again at 30-sec intervals) by rinsing the outside of the capillary with a stream of water, breaking off the sealed end, and inserting it into the medicine dropper that comes with the capillaries. Put a finger over the end, and expel the contents into a 13 × 100 mm test tube with 0.5 ml TBr. Plate out by adding 3–4 ml of molten TBr (1.5% agar) to the tube and pouring its contents onto solidified TBr plates containing 1.5% agar. Incubate overnight at 37°C. Count colonies the next day.

Each assay concentration should be performed at least in triplicate. Blanks where only CB is in the capillaries should be run as a negative control. The number of bacteria that enter such capillaries is sometimes very small, and, especially for some mutants, it is useful to carry out the experiment not at the usual A_{600} of 0.001 but at 0.02. Sometimes a culture shows vigorous motility and sometimes seems sluggish. To "control" for this, if S is the number of bacteria in the capillary containing attractant and N is the number in the capillary containing only CB, then the statistic, $(S - N)/N$, is a measure of chemotaxis. In this case, if the assay was carried out at the higher titer, N has to be divided by 50 (0.02/0.001). If the motility in all cultures is good, then, for convenience, the term $(S - N)$ may be used, often with the standard error.

Swarm Plate Assay

Swarm plates are the most convenient and fastest method for surveying colonies for the chemotaxis phenotype (see Ref. 13 for examples). Bacteria are inoculated at various positions on a semisolid agar plate. As they grow, they consume the attractant and migrate outward as a ring, "chasing" the attractant. Swarm plates are particularly useful in doing genetics, where, for example, transformants have to be scored for being wild type or mutant. The plates can contain minimal medium or tryptone broth. The latter develop in about 5 hr, and the former take about 8 hr. Sometimes, for unknown reasons, strains known to be mutant (e.g., *cheR* or *cheD* mutants) appear fairly normal on the tryptone broth swarm plates, rendering minimal swarm plates more useful. Moreover, the minimal plates allow subtle morphologies to be detected, which can be useful because mutants in different genes have noticeably different swarm phenotypes (Fig. 3) (13).

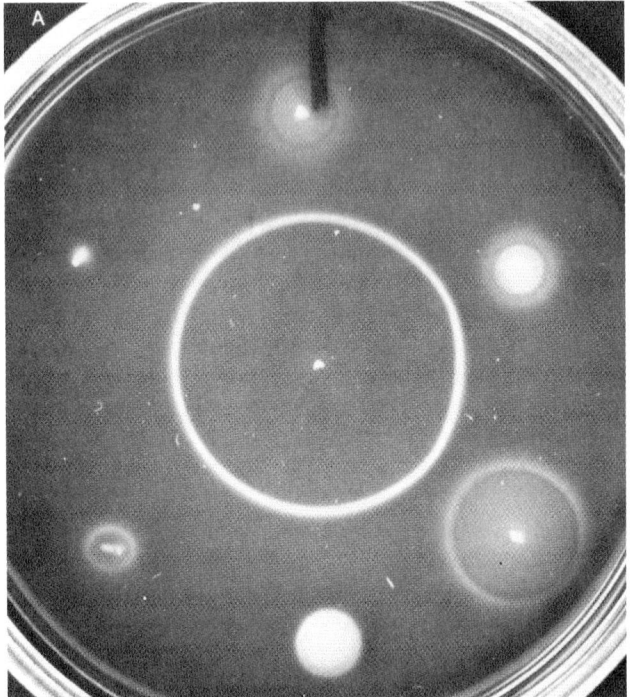

Fig. 3   Morphologies of mutant swarms. Experiments were performed as described by Ordal *et al.* (13). The plates contained 0.1 m*M* mannitol as the attractant and were incubated overnight in wet incubator at 37° C. The wild-type swarm is in the center of the plate. Mutant strains were stabbed around the center.

For broth swarm plates, autoclave 1% (w/v) tryptone, 0.5% (w/v) NaCl, and 0.30% (w/v) agar and pour into 100 × 15 mm petri plates. For minimal swarm plates, autoclave 0.25% agar and aseptically add the following presterilized components per 100 ml: 1 ml of 10× Spizizen's salts [10× Spizizen's salts contains 20 g (NH$_4$)$_2$SO$_4$, 140 g K$_2$H$_2$PO$_4$, 60 g KH$_2$PO$_4$, 10 g sodium citrate, and 2 g MgSO$_4$·7H$_2$O in 1 liter distilled water; dissolve in order given], 0.1 to 0.2 m*M* sugar or sugar alcohol (mannitol gives the sharpest and clearest swarm ring), 0.1 m*M* amino acid requirements, and 20 μl tryptone broth. In the absence of the last component, inoculated bacteria will not grow well. The following sugars can be used for swarm plates: mannitol, glucose, α-methylglucoside, β-methylglucoside, mannose, maltose, trehalose, fructose, sucrose, and gentiobiose. Unlike the case for *E. coli*, minimal amino acid swarm plates do not work very well in *B. subtilis*. However, proline taxis (0.42% agar, 1.3 m*M* proline, 0.7 m*M* sorbitol), glutamate taxis (0.42% agar, 1.3 m*M* glutamate,

1 mM sorbitol), and arginine taxis (0.42% agar, 1.3 mM arginine, 1 mM sorbitol) can be detected.

1. Streak the strain to be tested the night before use. It is important to obtain single colonies from the freshly streaked cells.
2. Using a sterile toothpick, stab into the colony to be tested and then into the swarm plate(s). Seven colonies can be conveniently scored on the same plate (six on the perimeter and one in the center). Alternatively, 50 μl of a culture can be inoculated. In some instances, in testing candidates for being mutant, many more samples (in our experience up to 18 or so) can be tested on the same plate. In this case, it helps to have a diagram under the swarm plate, showing where candidates should be inoculated.
3. The plates should be placed in a wet incubator at 37°C. If an incubator is not available, then plates can be placed inside a chamber to avoid becoming dried out. After 4 hr or so (later for minimal plates), briefly inspect the plates. The best way is by looking at them atop a cylinder having a light bulb in the bottom and an opaque plate toward the top so that all illumination comes from the side.

A variation of the swarm plate assay, often used in identifying new chemotaxis mutants, is called the miniswarm plate. Here, bacteria are included in the plates during pouring. In this case, about 50 bacteria per plate are usually poured in, and the agar has to be cooled to 60°C or lower (*B. subtilis* is much more resistant to higher temperatures than is *E. coli*). The individual cells grow into colonies, which appear, eventually, as small swarms (miniswarms). The agar concentration is ordinarily considerably higher for these. One recipe, for instance, that we developed calls for 0.55% agar, 0.1 mM potassium malate, 0.1 mM glycerol, and 0.3 mM mannitol.

Tethering Assays

Chemotaxis is traditionally evaluated using capillary assays and swarm plate assays, for they measure migration of a population of bacteria in a spatial gradient. However, to evaluate the tendency to swim smoothly or tumble and to observe the kinetics of excitation and adaptation on addition or removal of chemoeffectors, individual bacteria need to be scrutinized. The best method for doing this is to tether a bacterium by a flagellum to the ceiling of a laminar flow chamber and to observe the effect of stimuli on the rotational behavior (see Refs. 3, 5, and 6 for examples).

To carry out tethering experiments, a bacterial culture, grown on minimal medium, is passed several times through two fine needle syringes to shear the flagella. A small percentage of the bacterial culture would have only one flagellum. These bacteria can be tethered to a coverslip via an antiflagellar antibody (Fig. 4) leaving the cell body free to rotate. Bacteria with more than one flagellum will be multiply tethered to the

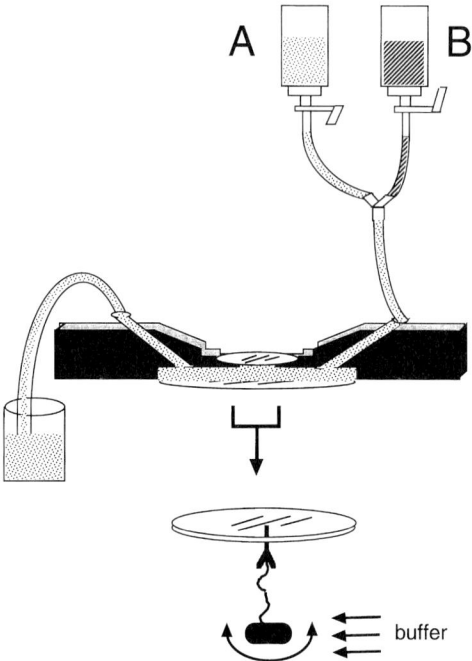

FIG. 4   Diagram of the laminar flow chamber. Bacteria with a single flagellum are tethered to a circular coverslip via an antiflagellar antibody. Chemotaxis buffer from reservoir A flows through the chamber. Bacteria rotate freely with respect to the immobilized flagellum. When reservoir A (containing CB) is closed and reservoir B opened, chemotaxis buffer with effector flows through the chamber. The rotation of the cell body is analyzed as described in text.

coverslip and will be unable to rotate. The rotation of the cell body, which rotates with respect to the flagellum, is observed under a light microscope connected to a camera that projects the image on the screen. If the bacterial flagella were sheared skillfully, one may be able to see up to 10 analyzable, rotating cell bodies in one field of view. Chemotaxis buffer is flowed from a reservoir through the chamber. To observe the response to attractant or repellents, the reservoir is then switched to a solution of chemotaxis buffer containing the appropriate chemoeffector. The reservoir is then switched back to chemotaxis buffer to evaluate the removal response. Behavior is then recorded on a videocassette.

A computer program has been developed to analyze the time-dependent behavior of a population of cells (16). Bacterial cells are observed one at a time. Each time a bacterium changes the direction of rotation, a particular key is pressed on the computer keyboard. Each keystroke records the time and direction of each change. The behavior is digitized by giving a clockwise rotation a value of 1 and a counterclock-

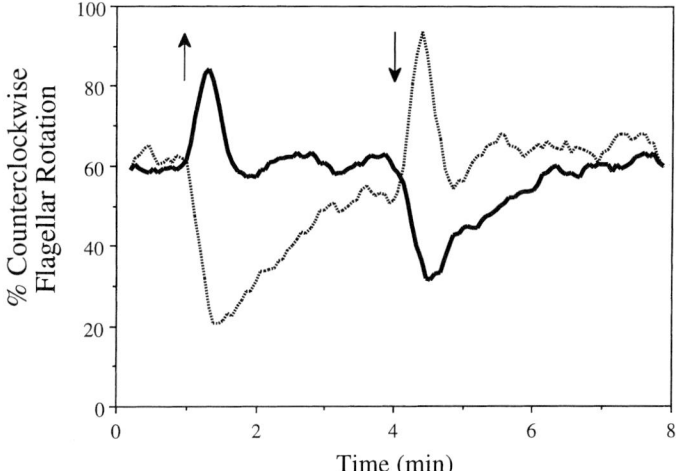

Fıg. 5   Tethering profile of wild-type *B. subtilis*. Experiments were performed as described by Kirsch *et al.* (6). Solid line represents the behavior in response to addition and removal of the attractant azetidine-2-carboxylic acid, a proline analog; dotted line represents the behavior in response to the repellent indole. Effector is added at 1 min and removed at 4 min (arrows). See text for detailed experimental procedure.

wise rotation a value of 0. After an entire set (15–30 bacteria) is entered, the program then averages the behavior of the set as 4-sec intervals. The values obtained are then smoothed (width of 5) and plotted using any graphing program (Fig. 5).

We usually observe the chemotactic response over a period of 8 min. Chemoeffector is added at 1 min and removed as 4 min. The behavior of wild-type *B. subtilis* (Fig. 5) has notable features. Unstimulated wild-type cells have a 60% smooth swimming (counterclockwise) bias. This rotation bias increases with positive stimuli (either addition of attractant or removal of repellent). Negative stimuli (either addition of repellent or removal of attractant) results in a decrease in the bias of counterclockwise rotation. For both stimuli, the smooth swimming bias goes back to prestimulus levels (adaptation). The adaptation period for positive stimuli is usually shorter than the adaptation period for negative stimuli. Interestingly, the population never reaches a 100% clockwise or counterclockwise bias.

Preparing Bacteria

1. Streak the strain on a TBAB plate (with appropriate antibiotic) and grow in a 30°C incubator overnight. It is important that the cells are freshly streaked for use the next day.

2. Resuspend a 2-mm loopful of bacteria in 1 ml TBr.
3. Add 10 μl of the culture to 1 ml minimal medium in a test tube.
4. Shake at 37°C for 4.5 hr.

Tethering Bacteria

To be sure that the motility is satisfactory, the freely swimming cells should be observed under the microscope before shearing the flagella.

1. Using a 3-ml syringe with a 26-gauge needle (we use Precision Glide needles by Becton Dickinson Parsippany, NJ), insert the needle into one end of a Teflon tubing 12–15 cm long (polyethylene 20, inner diameter 0.38 mm, outer diameter 1.09 mm). Be careful not to poke a hole or scrape bits of the tubing wall that could block the syringe needle.
2. Aspirate the bacterial culture into the syringe with tubing. Trim the tubing to about 2 to 2.5 cm in length. Insert a second syringe and needle into the other end of the tubing.
3. Shear the flagella by passing the culture back and forth between the two syringes. Do this for a minimum of 15 min. To obtain an optimum number of tethered, rotating cells, 30 min or more of shearing is best.
4. After shearing, separate the two syringes. Pull back slightly on the plunger so that the culture will not squirt out of the syringe.
5. Hold the syringe upright and push out the air. Place about 3 μl of antiflagellar antibody on a 12-mm round glass coverslip. Add a bead of sheared culture (about 3–5 μl) and mix (do not pipette up and down). Incubate for about 3–5 min.
6. Meanwhile, clean the flow cell (Fig. 4). Make sure the bottom 22-mm round coverslip is clean and intact. Remove all residual vacuum grease with ethanol. Connect all tubing and fill reservoirs. Eliminate air bubbles in the tubing by running buffer through it for a while.
7. After the 5-min incubation, put vacuum grease on the edges of the chamber where the top coverslip will lay. Fill the bottom of the chamber with CB. Using a pair of fine tweezers, carefully invert the coverslip and place it on the flow chamber. Gently press this in place to make a good seal.
8. Allow CB to flow through the chamber. Scan the slide for a field of view with good, rotating bacteria. Project onto the monitor. Several bacteria should be visible on the screen.
9. Start recording on a videocassette. Add and remove chemoeffectors via the stopcocks. Keep track of the counter number when events occur. Typically, buffers are switched at 1 min to CB plus chemoeffector and back to CB at 4 min. Bacterial behavior is recorded for a total of 8 min. Be sure that bacteria which are to be compared from different fields are treated identically.

Analyzing Bacteria

Boot up the computer program named BUGSPIN (developed by M. Kirsch) on any microcomputer or IBM-compatible computer. Analyze one bacterium at a time, pressing a key each time the bacterium changes its direction of rotation. Remember that the rotation of the cell body is due to the rotation of the flagella that is tethered to the coverslip. Clockwise rotation of the cell body reflects counterclockwise rotation of the flagellum and vice versa.

After analysis of a number of bacteria, the computer program will give a printout of the averaged behavior of the set at 4-sec intervals. These values are then smoothed with a width of 5 and plotted using any graph program.

Analysis of Methylation of Methyl-Accepting Chemotaxis Proteins

Like the MCPs of *E. coli,* the MCPs of *B. subtilis* are methylated at particular glutamate residues (14, 15). In fact, the methylating enzymes of either species can work on both the homologous and heterologous MCPs *in vitro* (16, 17), and *cheR* from *B. subtilis* can complement a defect in *E. coli cheR in vivo* (3). However, there are some interesting differences. First, the *B. subtilis* MCPs so far cloned are 72 kDa, larger than the *E. coli* MCPs, which are 58–60 kDa (15). Second, there is little net change in methylation on addition of attractant. For most attractants, there is, however, an immediate increase in methyl group turnover on the MCPs [methyl groups leaving the MCPs being replaced by new groups from S-adenosylmethionine (*S*-AdoMet)] when attractant is added. This turnover declines during the adaptation period (18). Third, for these attractants, methyl groups leaving the MCPs are not released directly as methanol, as in *E. coli.* The maximal release of radioactive methanol in a cold chase experiment in *B. subtilis* occurs on the third or fourth cycle of adding and removing the attractant, rather than on the first (4). Furthermore, formation of methanol occurs with both positive (addition of attractant and removal of repellent) and negative (removal of attractant and addition of repellent) stimuli (18). Fourth, the methyl groups return to the MCPs on removal of attractant (4). Fifth, there are two unique proteins, CheC and CheD, that seem to be involved in the methylation process. For most attractants, it is believed that methyl groups are transferred to a yet unidentified acceptor, perhaps CheD, during adaptation. From this, they can be released as methanol or return to the MCPs on removal of attractant (5).

Other attractants of *B. subtilis* are like repellents of *E. coli* in causing direct release of methyl groups from the MCPs to produce methanol. For these, there is no turnover of methyl groups on the MCPs. The major release of radioactive methanol occurs on the first cycle in a cold chase experiment. Only positive stimuli cause methanol formation; negative stimuli retard it (19).

Methylation Assays in Vivo

Basic Protocol

The methylation assay is usually the first experiment to be carried out for characterizing the MCPs of a particular mutant. Cells are grown to 180 Klett units (KU) (red filter), at which point cells are in early stationary phase and expression of the MCPs is maximal. They are then harvested, washed, and treated with lysozyme to create protoplasts. Unlike *E. coli, B. subtilis* cells do not readily lyse when boiled with sodium dodecyl sulfate (SDS); thus, to facilitate rapid lysis on addition of SDS, cells are first converted to protoplasts. L-[*methyl-^{3}H]Methionine is incubated with the protoplasts to label the MCPs, and the cells are frozen quickly. The cells are thawed on ice, centrifuged, and the proteins separated on an SDS–polyacrylamide gel. Labeled bands are visualized by fluorography (Fig. 6). Levels of methylation are analyzed relative to the wild-type strain.

1. Inoculate 1 ml of Luria broth (LBr: 1% tryptone, 0.5% NaCl, and 0.5% yeast extract) (with appropriate antibiotic) with a single colony. Incubate at 37°C overnight.
2. Inoculate 5 ml of LBr (with antibiotic) with 125 μl of overnight culture in a 125-ml sidearm flask.
3. Incubate with shaking at 37°C until the cell density reaches 180 KU (~4–5 hr). Make sure all cultures reach 180 KU at the same time (shake the more advanced cultures at room temperature for a while to allow the lagging cultures to catch up).
4. Divide the culture into two small (13 × 100 mm) test tubes.
5. Pellet the cells by centrifugation for 5 min at room temperature. Discard supernatant.
6. Resuspend the cell pellet in 3 ml CB with 100 μg/ml chloramphenicol (Cm) (250 μg/ml for Cm-resistant strains). Vortex gently.
7. Centrifuge for 5 min. Repeat the CB wash twice.

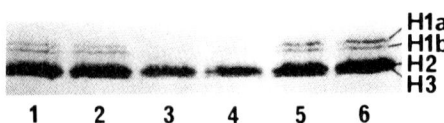

FIG. 6   Fluorogram of the effect of aspartate addition and removal on MCP methylation. Experiments were performed as described by Thoelke *et al.* (4). Cells were incubated in radioactive methionine. At 9 min, excess (10 *M*) nonradioactive methionine was added. At 11 min 30 sec 0.1 *M* aspartate was added, and at 12 min 30 sec it was removed. Samples were taken at following times: lanes 1 and 2, 11 min; lanes 3 and 4, 12 min; lanes 5 and 6, 14 min. Only the MCP region of the fluorogram is pictured.

8. Resuspend in 5 ml protoplast buffer (PB: 50 mM phosphate buffer, pH 7, 30 mM sodium lactate, 20 mM magnesium chloride, 0.1 mM EDTA, 20% sucrose) with Cm. Vortex gently.

9. Centrifuge for 10 min.

10. Pour off supernatant. Resuspend pellet with the residual PB in the tubes. Combine pellet suspensions.

11. In a 125-ml flask, dilute cells in 2.5 ml PB to an O.D.$_{600}$ = 1. Add 1 mg/ml lysozyme at 1 mg/ml.

12. Incubate cells for 20 min in 37°C bath, shaking slowly (just enough to swirl the solution around).

13. Check samples under a microscope to ensure that protoplasts have been made. Normal cells are rod-shaped and become spherical in the absence of the cell wall.

14. Add 75 μl of L-[*methyl-*^{3}H]methionine to each flask (30 μCi/ml cells).

15. After 5 min of incubation, freeze the flask in a dry ice/acetone bath. At this point, reactions can be stored at -70°C if desired.

16. Thaw flasks in a cold room in ice water.

17. Transfer the contents into labeled test tubes.

18. Spin down the protoplasts for 30 min at 4°C (in cold room). Discard the supernatant as radioactive waste.

19. Drain tubes upside down on paper towels in cold room. Remove excess moisture with Kimwipe or cotton swab.

20. Add 100 μl of 2× SDS solubilizer solution [2× SDS solubilizer is 0.12 M Tris base, 4% (w/v) SDS, 20% (v/v) glycerol, and 10% (v/v) 2-merceptoethanol, pH 6.8] and vortex. Transfer to microcentrifuge tubes.

21. Boil for 3 min.

22. Load 35 μl on a 10% SDS–polyacrylamide gel. Run at 25 mAmp. (*Note:* We use a 40:0.5 acrylamide–bisacrylamide solution in making our gels. We have found that such mixtures enhance the resolution of the MCPs. Different gel mixtures may be used by the researcher as desired.)

23. Stain with Coomassie blue and destain. Check to see that all lanes have approximately the same amount of protein.

24. Soak gel in about 300 ml DMSO I (dimethyl sulfoxide) for 1 hr or more.

25. Soak gel in DMSO II for 2.5 hr or more.

26. Soak gel in DMSO III for at least 1 hr but not more than 3 hr. {*Note:* DMSO can be used a number of times. After the first use, return DMSO to the bottle and label as DMSO I (for the first bottle to use for soaking) and DMSO II (for the second bottle to be used) and DMSO III [for the third bottle with 2,5-diphenyloxazole (PPO)]}. After a number of uses DMSO I has too much water. DMSO II then becomes DMSO I and a fresh bottle of DMSO becomes DMSO II.

27. Soak gel in water for 30 min to precipitate PPO.

28. Place gel on filter paper for support and desiccate on gel dryer.

29. Expose to X-ray film for 12 hr or more.

Determination of Turnover of Methyl Groups and Degree of Methylation of Methyl-Accepting Chemotaxis Proteins

The assays are used to determine the rate of turnover and the level of methylation of the MCPs. Cells are grown and prepared as described above. However, 10 μM L-[*methyl*-³H]methionine, which is not significantly consumed during the period of incubation, is added, and samples are harvested as a function of time. It takes 20 to 30 min for labeling of the MCPs to become maximal, when the "steady state" is reached, whereby the specific activity of methyl groups added to the MCPs equals that of methyl groups being removed (5). The turnover rate is obtained by quantitating the rate of labeling of the MCPs. The maximal amount of labeling reflects the number of methyl groups on the MCPs. The effect of chemoeffectors on the number of methyl groups associated with some or all MCPs can be determined by adding them to the steady-state-labeled cells and comparing the degree of labeling before and after.

Cold Chase Experiments to Determine Effects of Addition and Removal of Attractant on Turnover of Methyl Groups on Methyl-Accepting Chemotaxis Proteins

Nonradioactive methionine (10 μM) is added to the incubating protoplasts (after step 14, above) to dilute the specific activity of the methionine pool and the *S*-AdoMet pool. The specific activity of methyl groups on the MCPs changes slowly unless attractants like aspartate are added. Then, the increased turnover causes an immediate reduction of radioactivity associated with the MCPs (20). Time points are taken as desired. Because aliquots of 2.5 ml are taken at each time point, the starting volume of diluted cells (step 11, above) must be increased accordingly.

1. Follow the basic protocol, steps 1 to 7, except grow a 20-ml culture to 180 KU.
2. For the last CB wash, spin for 10 min.
3. Pour off supernatant. Resuspend pellet with the residual CB in tubes. Combine respective pellet suspensions. In this case protoplasts are not made because they would get crushed in step 10 below.
4. Dilute cells in 20 ml of CB at an A_{600} of 1 and incubate at 25° C.
5. Add L-[*methyl*-³H]methionine (10 μCi/ml).
6. After 9 min, add excess (10 μM) nonradioactive methionine.
7. At 11 min, take a 2.5-ml aliquot, place in a test tube, and freeze in a dry ice/acetone bath.
8. At 11 min 30 sec, add effector.
9. Freeze a 2.5-ml sample at 12 min.
10. At 12 min 30 sec, filter a 2.5-ml sample on a 0.45-μm nitrocellulose filter using a filter vacuum apparatus. Wash with 2 ml CB containing 10 μM nonradioactive methionine.

11. Using a pair of forceps, take the filter off the apparatus and place it in a small beaker (10 ml). Resuspend the filtered cells in 2.5 ml CB with 10 μM nonradioactive methionine. Transfer to a 13 × 100 mm test tube and freeze at 14 min.
12. Thaw samples at 4°C.
13. Add lysozyme to a final concentration of 1 mg/ml. Incubate for 30 min at 4°C.
14. Follow basic protocol, steps 18–28.
15. Expose to X-ray film for at least 3 days.

Assay for Methanol Production in Vivo (Flow Assay)

Cells are incubated with L-[*methyl-³H*]methionine allowing labeling of the MCPs by methyl transfer. The cells are then placed on a cellulose acetate filter. Chemotaxis buffer (with or without effector) containing cold methionine is flowed through the cells (Fig. 7), and the effluent is collected in fractions. Aliquots of each fraction are transferred to open microcentrifuge tubes which have been placed in scintillation vials containing scintillation fluid. The volatile product, [³H]methanol, is allowed to equilibrate between chemotaxis buffer and scintillation cocktail for 24 hr before counting. If desired, repeated cycles of addition and removal of effectors may be done in succession. Results of a representative experiment are given in Fig. 8.

1. Follow basic protocol, steps 1 to 7.
2. Dilute cells in CB to an A_{600} of 1.0. Only 3 ml will be used for methylation.
3. Add 200 μCi of L-[*methyl-³H*]methionine (specific activity of 80 Ci/mmol) to 3 ml of diluted cells. The resulting solution has approximately 1 μM L-[*methyl-³H*]methionine.

FIG. 7   Diagram of a flow assay setup. Radioactively labeled cells are placed on a syringe filter, and chemotaxis buffer with effector is passed through the filter. Fractions are collected and volatile counts are recorded. See text for detailed experimental procedure.

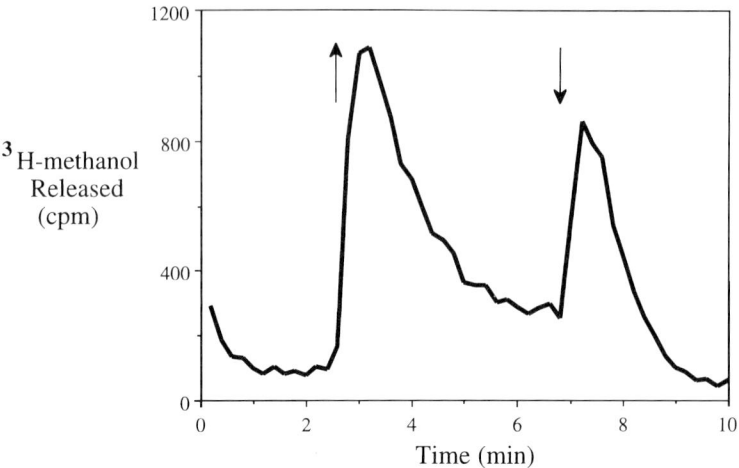

FIG. 8   Methanol production in a wild-type *B. subtilis* strain. Experiments were performed as described by Rosario *et al.* (11). Cells incubated with l-[*methyl*-³H]methionine were subjected to the addition (upward-pointing arrow) and removal (downward-pointing arrow) of 0.1 *M* aspartate in the presence of excess unlabeled methionine. Radioactive methanol release was determined as a function of time.

4. Incubate for 10 to 15 min at 37°C with slow shaking (100 rpm).
5. Meanwhile, prewash a syringe filter with CB (we use a 25-mm, 0.45-μm cellulose acetate membrane filter from Nalge Company, Rochester, NY) by flowing CB through with a pump (Fig. 7).
6. After incubation, aspirate the labeled cells into a 5-ml disposable syringe.
7. Transfer cells to the filter, taking care not to introduce air bubbles that can interfere with the flow of buffer through the filter membrane.
8. Connect all tubings and flow CB with Cm and unlabeled methionine (10 μM) through the filter using a peristaltic pump. Attach the filter to a continuous vortex with tape as shown in Fig. 7. Mild vortexing allows continuous flow through the filter without clogging by the cells. Wash the cells for 10 min to remove excess unincorporated L-[*methyl*-³H]methionine (flow rate 15 ml/min).
9. After 10 min, collect fractions every 12 sec for 2 min.
10. Quickly switch intake tube to a beaker containing CB plus Cm, 10 μM methionine, and effector.
11. Collect fractions for 4 min (0.2 min/fraction).
12. Switch intake tube back to buffer without effector. Collect fractions for another 4 min.
13. Switch intake tube to another effector if desired.
14. Cut the lids off 0.5-ml microcentrifuge tubes (the "ship"). Place 0.4 ml of each fraction into the tubes. Place each tube in 2.5 ml of aqueous scintillation fluid

(the "ocean") in scintillation vials (or just enough fluid to keep the ship upright and not get scintillation fluid in it).

15. Cap the vials and equilibrate at room temperature for 24 hr before counting.

Methylation Assays in Vitro

To evaluate the capability of MCPs to be methylated or ability of the cytoplasmic proteins to methylate or demethylate the MCPs, membrane and cytoplasmic extracts for different strains are first separated. Various membrane and extract combinations are mixed and incubated with *S*-adenosyl[*methyl*-³H]methionine. Glycerol is needed to stabilize CheB and is included in the reaction. The MCPs are separated by electrophoresis, and the gel is treated as in the basic methylation protocol. As a methyltransferase substrate, we use *cheR* mutant membranes so that the MCPs will be completely unmethylated at the beginning of the experiment. These membranes are also used to prepare radioactively labeled membranes as a substrate for the CheB methylesterase.

Preparation of Crude Soluble Extracts and Membrane Fractions

1. Inoculate 10 ml LBr (with antibiotic). Grow overnight.
2. Add 10 ml overnight culture to 1 liter LBr and incubate at 37°C for 10 hr. (Alternatively, inoculate the culture late at night so that it will be ready in the morning.)
3. Cool culture on ice for 10 min. All succeeding steps should be performed at 4°C.
4. Centrifuge at 5000 × g for 5 min. (Cells can be stored at −70°C at this point.)
5. Resuspend cells in cold 1 *M* KCl. Centrifuge 5 min at 5000 × g.
6. Repeat step 5 twice.
7. Resuspend cells in 500 ml of cold French press buffer (FPB) and centrifuge for 10 min at 10,000 × g. FPB contains 10 m*M* potassium phosphate, pH 7.0, 10 m*M* MgCl₂, 1.0 m*M* EDTA, 20 m*M* KCl, 0.1 m*M* phenylmethylsulfonyl fluoride (PMSF), 20% (v/v) glycerol, and 0.02% NaN₃.
8. Resuspend in 300 ml FPB.
9. Lyse cells using a French pressure cell or a cell disruptor.
10. Spin down cell debris at 16,000 rpm for 30 min.
11. Centrifuge at 45,000 rpm in an ultracentrifuge for 4 hr (or at 36,000 rpm overnight).
12. Remove supernatant and concentrate by ultrafiltration at 10–20 mg/ml protein. Divide samples into aliquots and freeze at −70°C ("soluble extract").
13. Resuspend membranes in 200 ml of 1 *M* KCl. Centrifuge as described in step 11.
14. Repeat step 13.
15. Repeat step 13 but use methyltransferase buffer (MTB: 0.1 *M* phosphate buffer, 1m*M* MgCl₂, 0.1 m*M* EDTA, 1 m*M* 2-mercaptoethanol, 0.02% azide, 0.1 m*M* PMSF; pH to 7.45) instead of 1 *M* KCl.

16. Resuspend membranes in MTB to a protein concentration of 10–15 mg/ml. Divide into aliquots and freeze at −70° C.

In Vitro Methylation Assay

1. For each reaction to be studied, set up a tube containing the following:

    20 μg membrane protein fraction
    200 μg soluble extract
    2 μl 0.1 M MgCl$_2$
    10 μl 100% glycerol
    30 μCi S-adenosyl[$methyl$-^{3}H]methionine (add this last)
    MTB to bring the total volume to 50 μl

2. Incubate reactions at 28° C for 1 hr.
3. Stop reaction by adding 16 μl 4× SDS solubilizer.
4. Analyze the samples on a standard SDS–polyacrylamide gel.
5. Treat gel as described in the basic protocol. Expose to film for 2–4 days.

Methylation of Membranes for Methylesterase Substrate

Membranes of a *cheR* mutant are radioactively methylated with extracts of an *E. coli* strain overexpressing CheR using the protocol above. The membranes are then reisolated by centrifuging at 45,000 rpm for 4 hr and stored at −70° C for use as a substrate for methylesterase assays.

Generation of Mutants

Mutagenesis Methods

Analysis of mutants is the cornerstone of understanding a biological process. Chemotaxis mutants can be obtained by a variety of techniques. Bacteria can be randomly mutagenized, and we have used ethyl methanesulfonate (EMS), insertion of the transposon Tn917, and insertion of the phage SPβc2 :: Tn917. The disadvantage of the two latter methods is that transcription of genes downstream of the disrupted gene may be impaired. However, it is possible to clone neighboring DNA if mutants are made using Tn917. The advantage of using the phage insertion is that specialized transducing phage can be obtained; however, the utility of these has declined as cloning has become easier.

Alternatively, individual genes can be mutagenized *in vitro* randomly or at particular sites. For the latter, the methods include using an oligonucleotide-directed mutagenesis system to make one or two point mutants, the polymerase chain reaction (PCR) to make extensive changes, and cassette mutagenesis, such as inserting a nonpolar chloramphenicol cassette. After the cells have been mutagenized or mutated DNA has been introduced into bacteria, the mutants have to be identified. Mutants can be obtained either by selection or by screening.

Ethyl Methanesulfonate Mutants

We have used EMS to obtain point mutants in chemotaxis genes (21). We suspect that we have obtained only single mutations because we never have had problems complementing them.

1. Grow bacteria in broth to 180 KU, wash twice, and suspend at 5×10^7 cells/ml in mutagenesis buffer (0.1 M potassium phosphate, pH 7, 2 mM potassium glutamate, and 0.2 mM EDTA.
2. Add 3% (w/v) EMS and shake to dissolve. Incubate at room temperature for 1 hr.
3. Wash cells with mutagenesis buffer and grow overnight in different tubes in TBr. To guarantee independent mutations, only one mutant of a particular phenotype should be saved from each overnight culture. A culture grown without shaking in TBr at 30° C is suitable for further work the following morning.

SPβc2dl2 :: Tn917 Insertion Mutants

The phage SPβc2:Tn*917*, which confers resistance to erythromycin and lincomycin, creates a turbid zone of lysis when spotted onto a law of SPβ⁻ bacteria and incubated overnight. The turbidity represents lysogens. If the strain used has a deletion of the attachment site for SPβ, then the phage will integrate in many locations throughout the chromosome. These bacteria will be resistant to erythromycin (Em) and lincomycin (Ln) and can be used as a source of chemotaxis/motility/flagellar mutants. The *B. subtilis* strain CU156 contains a deletion of the attachment site for SPβ. See Zahler (22) for more information about making these lysogens.

Tn917 Insertion Mutants

To obtain mutants having the transposon Tn*917* or Tn*917lacZ*, a plasmid which carries the transposon whose synthesis is temperature sensitive is used. In addition it should carry a selectable marker, like the Cm resistance gene. A Che⁺ bacterium carrying this plasmid is incubated overnight at the permissive temperature (30° C), in several tubes without shaking. Chloramphenicol (5 μg/ml) should be added during growth, along with Em (0.25 μg/ml) and Ln (6.25 μg/ml) to prevent loss of the transposon. One microliter of each culture is plated on LBr plates containing 1 μg/ml Em and 25 μg/ml Ln. Plates are incubated overnight at 50° C to allow segregation of the plasmid. (The higher concentration of Em and Ln should be used for carrying out selections, to avoid mutation to resistance; however, the bacteria have to be induced with at least 15 μg/ml Em before plating.) Bacteria are harvested from each plate and used as a source for mutants (by the miniswarm selection method) (23).

See Youngman (24) for more information about mutagenesis with these transposons and for subsequently cloning the genes into which the transposons were inserted.

Selection Methods

Swimming Down

It is a curious fact that smooth swimming bacteria swim downward, even against a concentration gradient of attractant (21). Mutagenized bacteria can be inoculated in amino acid attractant, such as asparagine, on top of a 5 to 20% glycerol or sucrose gradient in 1.5-ml Eppendorf tube. The attractant will diffuse downward, creating a gradient that will retard the downward migration of the wild type. However, smooth swimming bacteria will ignore the attractant and will swim down where, after 30 to 60 min, they can be harvested using a syringe. They can be plated and colonies tested individually on swarm plates.

Repeated Swarming

If mutagenized bacteria are inoculated in a swarm plate, they will migrate outward by chemotaxis (see section on Swarm Plate Assay above). Mutants, however, will remain at the site of inoculation or, at least, will migrate outward more slowly (some do move a lot although they do not form a ring). Bacteria from the site of inoculation are used to inoculate a new swarm, and this procedure is repeated many times. Eventually, the center of the swarm becomes substantially enriched for mutants, and, at that point, they can be streaked and individual colonies tested on swarm plates.

Selection against Movement

As described under the section on Swarm Plate Assay above (last paragraph on mini-swarm plates), the agar concentration can be 0.2 to 0.25% and 10^4 or 10^5 bacteria can be poured into the plate. After appropriate incubation, the plate can be carefully inspected for small colonies. The wild-type swarms will diffuse together, and mutants that are impaired in ability to migrate will be the only bacteria visible. Candidates are streaked and tested on normal swarm plates for their chemotaxis or motility phenotype (some will be nonmotile). Completely nonmotile cells are surprisingly rare from this selection, probably owing to the fact that a bacterium that cannot move will form a colony that is too small to see.

PBS1 Resistance

PBS1, the main generalized transducing phage used for *B. subtilis,* is a flagellotrophic phage. If bacteria are incubated with PBS1, diluted with 3 ml of soft agar (0.7% agar) in broth, and plated on a hard agar plate, clones of bacteria resistant to the phage will arise. These can be streaked to purify them and examined for motility phenotype. The bacteria should all lack flagella, as we have observed that bacteria with paralyzed flagella ("*mot*" mutants) still support PBS1 growth (23).

Screening Methods

Most of our mutants have been obtained by screening. The most convenient method is to use miniswarm plates (see Swarm Plate Assay section). After overnight incubation, colonies having an abnormal swarm phenotype can be streaked and tested as individual colonies on swarm plates.

Mapping

Three types of mapping are used for locating genes: transduction, transformation, and hybridization to a yeast artificial chromosome (YAC) library of *B. subtilis* DNA. Transduction has traditionally been used to locate genes on the chromosome; transformation has been used to carry out fine-structure mapping. However, with the development of a library of ordered fragments (25), the YAC method offers promise for being a rapid means of gross mapping of cloned genes.

Transduction

Two-factor crosses are used in a preliminary screen to locate mutations in the neighborhood of known genetic markers. To facilitate this, the *Bacillus* Genetic Stock Center (BGSC, Columbus, OH; 26) has two sets of mapping strains that cover the entire chromosome, one having largely nutritional markers and the other a set of Tn917 insertions. The Tn917 insertion confers $Em^R Ln^R$ on its host. Recombination frequencies can be used, in a rough way, to calculate physical distances from the algorithm $C = (1 - t) + \ln(t)$, where C is the cotransduction index of the markers and t is the fractional length of the transducing fragment separating the markers (27). PBS1 is a very large transducing phage, about 1/13.05 of the *B. subtilis* chromosome (27). After transduction, it may be helpful to carry out three-factor crosses, which order genetic markers. The methods presented are largely from a course taught by S. Zahler at Cornell University (28).

Growing PBS1
1. Streak the "donor" bacteria on a tryptose blood agar base (TBAB) plate and incubate at 30° or 37° C for about 18 hr.
2. Scrape a 2-mm microbiological loop across the bacteria and inoculate into 1 ml of Pennassay broth (PAB, Difco Antibiotic Medium 3) or MBr (LBr having 5 mM

MgSO$_4$). This gives moderate turbidity ($1-2 \times 10^8$ bacteria/ml). Use a 18 $\times$ 150 mm test tube and incubate at 37°C with shaking for 90 min.

3. Add 0.5 ml PBS1 grown on *Bacillus licheniformis* and continue shaking for 30 min.
4. Add 10 ml of PAB (or LBr) and continue shaking for 30 min.
5. Add 500 μg/ml chloramphenicol and continue shaking for 2 hr.
6. Incubate tube on its side without shaking for 18 to 24 hr.
7. Centrifuge the contents to pellet the bacteria. Take the supernatant (containing PBS1) and either filter sterilize through 0.45-μm syringe filters or add chloroform and vortex extensively. In the latter case, plate out 0.1 ml on a TBAB plate to ensure sterility. Add DNase.

Carrying out PBS1 Transduction

1. Streak "recipient" strain on a TBAB plate from a single colony and incubate at 30° or 37°C for 18 hr.
2. Inoculate the recipient strain in 0.5 to 2.5 ml MBr in an 18 $\times$ 150 mm test tube at 0.5 to 1 $\times$ 10^8/ml and incubate at 37°C with shaking for 4.5 hr. Examine the bacteria under a microscope to be sure they are motile.
3. Put 0.25 ml of the culture in a 13 $\times$ 100 mm test tube and add 0.25 ml or less PBS1 lysate obtained from growth of PBS1 on "donor" strain. As a control, add buffer instead of PBS1.
4. Incubate tubes with shaking for 15 to 20 min.
5. Add 2.5 ml SC (2.5 ml of 4 *M* NaCl plus 5 ml of 1 *M* sodium citrate in 100 ml sterile distilled water) and centrifuge for 10 min.
6. Vortex pellet and plate contents. Alternatively, resuspend in 0.25 ml SC and plate 0.1 ml on selection plates. (*Note:* If selection is for Em^{R}Ln^R, then 15 μg/ml Em should be present during the incubation with PBS1.)

Transformation

The fragments of DNA used in transformation are much smaller, averaging only about 1% of the chromosome. The algorithm mentioned above still applies; however, the physical distances involved are much smaller. Thus, transformation is used for fine-structure mapping, such as for ordering adjacent genes or markers within particular genes.

Isolation of Chromosomal DNA

1. Inoculate 20 ml TBr with the donor strain and grow overnight, preferably for less than 10 hr (longer periods make cells harder to lyse).
2. Add to 200 ml PAB and grow for 3 hr at 37°C.
3. Centrifuge at 10,000 $\times$ g for 10 min.

4. Start protease incubation (10 mg/ml; use 0.4 ml per culture). Incubate for 1 hr at 37°C.

5. Resuspend pellet in 4 ml of 0.1 *M* EDTA/0.15 *M* Tris, pH 8.0. Transfer to a large 16 × 125 screw-cap tube.

6. Add 4 mg of lysozyme and incubate without shaking for 30 min at 37°C.

7. Add 0.4 ml of the 10 mg/ml protease and incubate without shaking for 0.5 hr at 37°C.

8. Add 1.1 ml of 5% SDS (1% final concentration) and 0.2 ml of 4 *M* NaCl.

9. Mix on a rotator for 0.5 hr at room temperature.

10. Add an equal volume of chloroform–isoamyl alcohol (24 : 1 v/v), then mix using a rotator for 0.5 hr or more.

11. Centrifuge in a swinging-bucket rotor to separate phases. Transfer upper phase to a fresh tube using a Pasteur pipette whose end has been bent into a hook.

12. Repeat the chloroform–isoamyl alcohol extraction.

13. Repeat the extraction two more times with phenol; the extraction may be done overnight if desired.

14. Transfer upper layer to a large test tube (18 × 150 mm).

15. Layer with 2.5 volumes of ice-cold ethanol.

16. Spool out DNA using a hooked Pasteur pipette with a sealed end.

17. Rinse by dipping into three successive washes of ice-cold 70% (v/v) ethanol.

18. Air dry.

19. Break off the hook into 1 ml of TE: 10 m*M* Tris, 1 m*M* EDTA; pH 8.0 to dissolve.

Transformation Procedure

There are many different media and procedures for transformation. The one we most commonly use follows. Growth medium 1 (GM1) contains 1× Spizizen salts, 0.5% glucose, 0.02% casamino acids, and 20 m*M* MgCl$_2$. Growth medium 2 (GM2) contains 1× Spizizen salts, 0.5% (w/v) glucose, 0.01% (w/v) casamino acids, and 50 μg/ml required amino acids. The recipe for 10× Spizizen salts is given in the Swarm Plate Assay section.

1. Grow 1 ml overnight in minimal medium.

2. Spin down cells in a microcentrifuge (15 sec). Remove most of the supernatant, leaving 100 μl behind. Resuspend pellet.

3. Dilute 10 μl of suspension in 1 ml water. Measure A_{525}.

4. Add enough cells to 1 ml GM1 in a large test tube for the A_{525} to be between 0.2 and 0.3.

5. Shake at 37°C for 4–4.5 hr.

6. Add 100 μl of the culture to 800 μl of prewarmed GM2.

7. Shake at 37°C for 1.5 hr.

8. Add 100 μl of the donor DNA (diluted in TE).
9. Shake slowly at 37°C for 0.5 hr.
10. Plate the appropriate dilution of the transformation mixture on selection plates. Incubate at 37°C.

Inactivation of Chemotaxis Genes

Mutations of the genes are made on a cloned segment of DNA and propagated in *E. coli*. The mutation is then introduced into the chromosome of the desired strain by using an integrational plasmid or by double-crossover recombination. Integrational plasmids are used when the cloned DNA is only part of an operon, or for monocistronic genes. If the gene to be inactivated is part of an operon, it is better to insert a drug resistance marker within the gene to avoid polarity effects. Polar mutations occur when inactivation of upstream genes affect the expression of downstream genes.

pBGSC6 Integrational Vector

pBGSC6 (Fig. 9) is unable to replicate autonomously in *B. subtilis* but is capable of integration onto the chromosome. The cloned DNA segment is mutagenized *in vitro*, cloned into pBGSC6 (26), and propagated in *E. coli*. It is then transformed into *B. subtilis*. A single crossover event between homologous portions of the cloned DNA and the chromosome results in integration of the plasmid into the gene of interest. Because the plasmid carries a *cat* gene (encoding resistance to Cm), Cm^R colonies are selected for. Southern hybridization may be used to confirm the mutation.

pDB4 Antibiotic Resistance Cassette Vector

pDB4 (Fig. 9) contains the Cm resistance gene. Several restriction sites bounding the resistance gene can be used to clone into the gene to be inactivated. It is advisable to insert the cassette at the beginning of the gene. The plasmid containing the inactivated gene is linearized and transformed into *B. subtilis* with selection for Cm^R. Linearization of the plasmid is necessary to prevent integration of the plasmid on the chromosome. Cm^R transformants will arise only if a double-crossover event occurs involving both sides of the mutation. The efficiency of recombination is increased if there is at least 0.5 kb on either side of the gene cassette. Several antibiotic resistance cassette vectors (Km^R, Em^R) have been developed and are described by Perego (29).

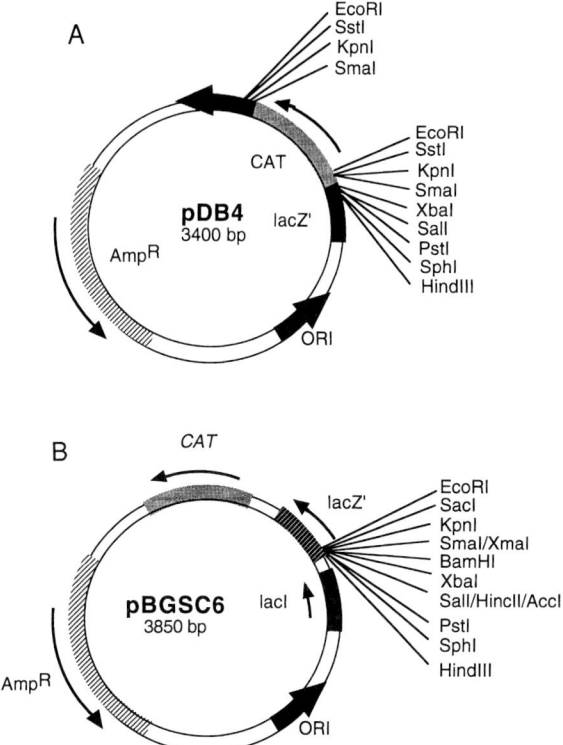

FIG. 9   Plasmids used to inactivate chemotaxis gene. (A) pBGSC6 from the *Bacillus* Genetic Stock Center (not to scale). A cloned segment of DNA is mutagenized *in vitro* and cloned into the multiple cloning site. The plasmid is then transformed into the wild-type strain and resistance to Cm selected. Transformants result from double crossover between homologous regions of the chromosome and the plasmid. (B) pDB4 (7) (not to scale). The restriction sites used to clone the *cat* gene are indicated. Directions of transcription are indicated by arrows.

Complementation of Chemotaxis

The polarity of a mutation can be checked by complementation. The wild-type gene is introduced into the mutant strain. Restoration of wild-type phenotype indicates nonpolarity of the mutation, that is, it does not affect expression of downstream genes. In the *che/fla* operon of 30 genes, we found this to be an important step in mutational analysis.

pEB112 is a shuttle vector that carries the Ap^R gene for selection in *E. coli* and the Km^R gene for selection in *B. subtilis* (Fig. 10) (30). The wild-type gene is cloned into

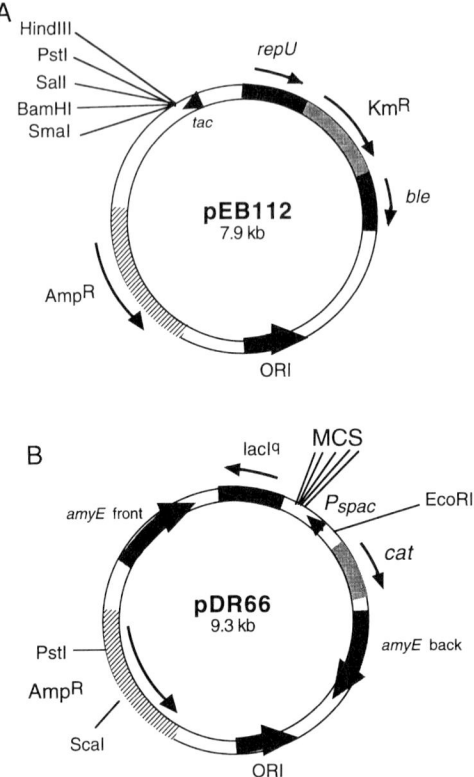

FIG. 10   Plasmids used to complement chemotactic mutants. Both contain the selectable marker *amp* and origin of replication for growth and maintenance in *E. coli.* (A) pEB112 (30) (not to scale) contains a selectable marker for *B. subtilis (kan).* Unique restriction sites are indicated. Direction of transcription of genes are indicated by an arrow. (B) pDR66 (31) (not drawn to scale) also contains a selectable marker for *B. subtilis (cat),* and regions of homology to the nonessential *amyE* locus (*amyE front* and *amyE back*). The multiple cloning site of pDR66 contains *Hind* III, *Xba* I, *Sal* I, and *Sph* I sites. The closely related pDR67 has *Hind* III, *Sma* I, *Xba* III, *Hpa* I, *Bgl* II, *Cla* I, and *Sph* I sites.

the plasmid and transformed into the mutant *B. subtilis* strain. Km^R transformants are tested, usually on swarm plates, for restoration of the wild-type phenotype. It is recommended that the complemented strain be "cured" to ensure that the plasmid did not integrate or recombine to repair the mutant gene. Curing is done by streaking out the strain several successive times on a non-drug-containing plate to lose the plasmid. The cured strain should be mutant.

To ensure that the wild-type gene is present in single copy, it can be integrated into another part of the chromosome, the *amyE* locus. The DNA is cloned into the multiple cloning site of pDR66/67 (Fig. 10) (31), which is bound by the front and back portions of the *amyE* gene. The plasmid is then linearized using a restriction site outside the *amy–cat* region and transformed into the *B. subtilis* mutant, selecting for Cm^R. The transformants would have had a double-crossover recombination event occur between the *amyE* locus and the homologous region of DNA in *amy(front)* and *amy(back)*. This single-copy integrant is stable. To confirm integration in the *amyE* locus, a starch hydrolysis test is performed on the transformants.

References

1. R. B. Bourret, J. F. Hess, and M. I. Simon, *Annu. Rev. Biochem.* **60,** 401 (1991).
2. G. W. Ordal, L. Marquez-Magana, and M. J. Chamberlin, *in* "*Bacillus subtilis* and Other Gram-Positive Bacteria: Biochemistry, Physiology, and Molecular Genetics" (A. L. Sonenshein, J. Hoch, and R. Losick, ed.), p. 765. American Society for Microbiology, Washington, D.C., 1993.
3. M. L. Kirsch, A. R. Zuberi, D. Henner, P. D. Peters, M. A. Yazdi, and G. W. Ordal, *J. Biol. Chem.* **268,** 25350 (1993).
4. M. S. Thoelke, J. R. Kirby, and G. W. Ordal, *Biochemistry* **28,** 5585 (1989).
5. M. M. L. Rosario, J. R. Kirby, D. A. Bochar, and G. W. Ordal, *Biochemistry,* **34,** 3823 (1995).
6. M. L. Kirsch, P. D. Peters, D. W. Hanlon, J. R. Kirby, and G. W. Ordal, *J. Biol. Chem.* **268,** 18610 (1993).
7. D. S. Bischoff and G. W. Ordal, *Mol. Microbiol.* **6,** 2715 (1992).
8. A. R. Zuberi, C. Ying, M. R. Weinreich, and G. W. Ordal, *J. Bacteriol.* **173,** 1870 (1990).
9. K. Kutsukake and T. Iino, *J. Bacteriol.* **176,** 3598 (1994).
10. K. L. Fredrick and J. D. Helmann, *J. Bacteriol.* **176,** 2727 (1994).
11. M. M. L. Rosario, K. L. Fredrick, G. W. Ordal, and J. D. Helmann, *J. Bacteriol.* **176,** 2736 (1994).
12. G. W. Ordal and K. J. Gibson, *J. Bacteriol.* **129,** 151 (1977).
13. G. W. Ordal, H. M. Parker, and J. R. Kirby, *J. Bacteriol.* **164,** 802 (1985).
14. J. A. Ahlgren and G. W. Ordal, *Biochem. J.* **213,** 759 (1983).
15. D. W. Hanlon and G. W. Ordal, *J. Biol. Chem.* **269,** 14038 (1994).
16. A. Burgess-Cassler and G. W. Ordal, *J. Biol. Chem.* **257,** 12835 (1982).
17. D. O. Nettleton and G. W. Ordal, *J. Bacteriol.* **171,** 120 (1989).
18. M. S. Thoelke, J. M. Casper, and G. W. Ordal, *J. Biol. Chem.* **265,** 1928 (1990).
19. C. Kristich, J. R. Kirby, and G. W. Ordal, unpublished data (1994).
20. M. S. Thoelke, H. M. Parker, E. A. Ordal, and G. W. Ordal, *Biochemistry* **27,** 8453 (1988).
21. G. W. Ordal, D. O. Nettleton, and J. A. Hoch, *J. Bacteriol.* **154,** 1088 (1983).
22. S. A. Zahler, *in* "The Molecular Biology of the Bacilli" (D. Dubnau, ed.), p. 269. Academic Press, New York, 1982.
23. A. R. Zuberi, C. Ying, H. M. Parker, and G. W. Ordal, *J. Bacteriol.* **172,** 6841 (1990).

24. P. Youngman, *in* "Plasmids: A Practical Approach" (K. Hardy, ed.), p. 79. IRL Press, Oxford and Washington, D.C., 1987.

25. V. Azevedo, E. Alvarez, E. Zumstein, G. Damaini, V. Sgaramella, S. D. Erlich, and P. Serror, *Proc. Natl. Acad. Sci. U.S.A.* **90,** 6047 (1993).

26. *Bacillus* Genetic Stock Center, Department of Biochemistry, Ohio State University, Columbus, OH.

27. D. J. Henner and J. A. Hoch, *Microbiol. Rev.* **44,** 57 (1980).

28. S. A. Zahler, personal communication.

29. M. Perego, *in* "*Bacillus subtilis* and Other Gram-Positive Bacteria: Biochemistry, Physiology, and Molecular Genetics" (A. L. Sonenshein, ed.), p. 615. American Society for Microbiology, Washington, D.C., 1993.

30. H. Leonhardt and J. C. Alonso, *J. Gen. Microbiol.* **134,** 605 (1988).

31. K. Ireton, D. Z. Rudner, K. J. Siranosian, and A. D. Grossman, *Genes Dev.* **7,** 283 (1993).

[19] Cloning of *Thermomonospora fusca* Cellulase Genes in *Escherichia coli* and *Streptomyces lividans*

David B. Wilson

Introduction

Actinomycetes play a major role in the recycling of plant cell walls in nature (1). *Thermomonospora fusca* is a moderately thermophilic actinomycete that is a major degrader of cellulose and xylan in heated organic material such as compost heaps, manure piles, and rotting hay; however, it does not appear to degrade lignin (2). There have been a number of studies of *T. fusca* crude cellulase because of its relatively high activity, thermostability, and broad pH optimum (3–5). In addition, the crude cellulase has been fractionated and six different cellulases (E1–E6) purified and characterized (6, 7). Three of the enzymes, namely, E1, E2, and E5, are endocellulases, E3 and E6 appear to be cellobiohydrolases, and E4 is an unusual enzyme with exocellulase activity but it can degrade carboxymethylcellulose (CMC) and has weak endocellulase activity (7). The structural genes encoding five of the cellulases have been cloned into *E. coli* and sequenced (8–11). In addition, all five genes have been cloned into *Streptomyces lividans,* where they are expressed from native promoters and are secreted into the culture supernatant (12, 13). The *T. fusca* cellulase genes all are expressed at a higher level in *S. lividans* than in *E. coli,* and the one gene that was tested (E5) was not expressed in *Bacillus subtilis, Streptococcus faecalis,* or *Saccharomyces cerevisiae* (14). It was expressed in all the gram-negative bacteria tested: *Erwinia amylovora, Erwinia herbicola,* and *Enterobacter agglomerans.* However, expression was stable only in *E. coli;* all of the other strains quickly lost the ability to express cellulase activity although they retained the antibiotic resistance carried by the plasmid.

The catalytic domains of the five sequenced cellulases belong to three different cellulase families (13, 15). The endocellulase E1 and the unusual enzyme E4 both belong to family E, whereas the endocellulase E2 and the exocellulase E3 both belong to family B. The endocellulase E5 belongs to family A. All five cellulase genes contain a cellulose-binding domain and a peptide hinge. The cellulose-binding domains show homology with one another and with other bacterial cellulose-binding domains (13).

All five genes contain the identical 14-base inverted repeat sequence TGGGA-GCGCTCCCA which is located upstream of the translational initiation codon in every gene. In the two genes where the promoter has been identified (E3 and E5), the repeat sequence is downstream from the promoter. The same sequence has been found in every actinomycete cellulase gene that has been sequenced (13). This sequence is the binding site for a *T. fusca* protein that appears to function in induction of cellulase synthesis by cellobiose (16). There are no other common sequences in the 5' upstream regions of these genes, so that the promoters are different for each gene.

In this chapter we describe the methods used to clone and express *T. fusca* cellulase genes in *E. coli* and *S. lividans.*

Procedures

DNA Isolation

An essential first step in cloning genes from any organism is the isolation of high molecular weight DNA. A number of procedures have been tried on *T. fusca,* and it is necessary to take strict precautions against DNA degradation. The nuclease inhibitor diethyl pyrocarbonate (DEPC) must be fresh; the procedures should be done quickly and in the cold. The following procedure yields about 1 mg of pure chromosomal DNA from a culture of *T. fusca* strain YX that has been grown for 18 hr at 55°C in 200 ml of Hagerdahl medium containing 0.2% cellobiose (17). The cell pellet is resuspended in 50 ml of 50 mM Tris-HCl, pH 8.0, 25 mM EDTA, and 1.4 ml of DEPC and 6.5 ml of 10% (w/v) sodium dodecyl sulfate (SDS) are added and mixed well. The suspension is treated in a French press at 5000 psi and placed directly into 50 ml of phenol–$CHCl_3$–isoamyl alcohol (PCI) (25:24:1 v/v). This is mixed well and allowed to stand 5 min. After centrifugation the upper phase is transferred to a clean tube, and the lower phase is reextracted with 30 ml of 50 mM Tris-HCl, pH 8.0, 25 mM EDTA. After centrifugation the second aqueous phase is added to the first and 75 ml of $CHCl_3$ added, and the sample is mixed well and centrifuged. The aqueous (upper) phase is transferred to a large flask. NaCl or sodium acetate, pH 5.2, is added to 0.3 M. Cold ethanol, 180 ml, is added gradually with stirring. The DNA is spooled onto a glass hook and washed by dipping the hook successively for 30 sec in 20 ml of 70, 80, and 90% (v/v) ethanol. The DNA on the hook is allowed to dry briefly and then resuspended in 1.5 ml of 10 mM Tris-HCl, pH 7.5, containing 1 mM EDTA (TE). A sample of the DNA is analyzed on a 0.6% agarose gel to check its concentration and size. If necessary, the DNA may be extracted several times more with $CHCl_3$, precipitated with either NaCl or LiCl and cold ethanol, and spooled again on a glass hook.

Partial Digestion with Sau3A

The next step is to isolate DNA fragments and ligate them into an appropriate vector. Partial digestion with *Sau*3A works well to obtain fragments of *T. fusca* DNA, and this step can be followed by sucrose gradient centrifugation of the digest to isolate inserts of the appropriate size. The following conditions have been worked out on a small scale using agarose gel electrophoresis to monitor the extent of digestion by *Sau*3A. The extent of digestion will vary depending on the purity of the chromosomal DNA. *Thermomonospora* YX DNA (80 μg in 300 μl), 500 μl water, 90 μl of concentrated restriction enzyme buffer (200 mM Tris-HCl, pH 7.4, 300 mM NaCl, 100 mM MgCl$_2$), and 38 μl restriction enzyme buffer containing 3.3 units of *Sau*3A are combined and incubated at 37°C for 30 min. *Sau*3A is then inactivated by heating the reaction mixture to 65°C for 20 min.

Sucrose Gradient Centrifugation

A linear 10–40% sucrose gradient in 20 mM Tris-HCl, pH 8.0, 10 mM EDTA, and 50 mM NaCl is formed in a Beckman SW41 rotor tube. The DNA partial digest is heated to 65°C for 1 min and layered on top of the gradient. After centrifugation at 23,000 rpm for 16 hr at 25°C, 0.3-ml fractions are collected from the bottom of the tube. Small samples (2 μl) of each fraction are spotted on 0.8% agarose gels containing 0.5 μg/ml ethidium bromide, electrophoresed, and examined on a UV transilluminator to locate fractions containing DNA. The size distribution of fragments in the fractions is analyzed by agarose gel electrophoresis. The 11 fractions containing the larger DNA fragments (2.9 ml total volume) are pooled. NaCl (4 M) is added to a final concentration of 0.3 M, and 2.5 volumes of 95% ethanol is added. The DNA is precipitated at -20°C overnight and then centrifuged at 8000 g for 20 min. The dried pellet is dissolved in 50 μl of TE.

Screening Transformants

When cloning into an *E. coli* plasmid such as pUC19, fragments of 2 to 6 kb are used, whereas when cloning into a vector such as λ 2001, larger fragments from 5 to 14 kb are used. In most of our experiments we have used *E. coli* as the host for identifying the desired clones. *Escherichia coli* has no endogenous cellulase activity, so that transformed colonies or plaques can be screened for cellulase activity using the CMC overlay method of Teather and Wood (18). The best procedure that we have found is to pick colonies or plaques onto duplicate plates and then overlay one set of plates after they have grown up using 8 ml per plate of a melted overlay that contains, per 100 ml, 0.7 g agarose, 50 ml of 0.1% CMC in water, and 10 mM buffer; either potassium phosphate, pH 6.5, or sodium acetate, pH 5.5, works well. After the overlay has solidified the plates are incubated at the desired temperature for a period of 4 to 24 hr. Stain the plates with 0.1% Congo red for 30 min and destain with 1 M NaCl.

The brand of CMC is critical in the assay as the sensitivity of the assay varies greatly depending on the lot of CMC that is used. We have used 4H1F from Hercules Incorporated (Wilmington, DE) or 4HI from the Aqualon Company (Wilmington, DE); however, low viscosity CMC from Sigma (St. Louis, MO), which we use for our liquid cellulase assay, did not work well. To get the strongest response, the overlay should be boiling when it is poured on the replica plate that is being screened. This method is very sensitive so that it is possible to identify clones even when the level of expression is low. We have been able to detect colonies that produce as little as 0.008 units/mg protein of CMCase activity by this method. However, the exact value may vary by a factor of 5 depending on the specific enzyme being assayed.

One problem that can arise in the assay is that some bacteria bind the Congo red dye, creating a clear zone around the colony. Colonies that produce cellulase have a yellow ring around the colony that is distinguishable from the clear ring produced by dye binding. Thus, it is a good idea to have observed a positive colony in the assay before attempting to screen transformants. The sensitivity of the assay can be varied by changing the time and temperature at which the plates are incubated after they have been overlayed. We normally incubate at 50°C for either 1 hr, 4 hr, or overnight depending on the level of activity that is produced. However, the correct time and temperature depend on the properties of the expressed enzyme. Endocellulases show high activity on CMC, but a few exocellulases have low activity on CMC. We were able to clone the *T. fusca* E4 gene which codes for an exocellulase using the overlay method (11). Furthermore, a plasmid carrying the E3 gene which we isolated by screening transformants with an oligonucleotide probe, designed to code for an amino acid sequence determined from the purified enzyme, does give a signal in the overlay assay even though we had not detected the gene in our screening of *T. fusca* libraries with the CMC overlay assay.

Another powerful general method for screening transformants is to use an overlay containing a fluorescent substrate such as 4-methylumbelliferyl β-D-cellobiopyranoside (MUCB) or -cellotrioside (MUCT), where the positive colonies fluoresce. The pH of the overlay can be adjusted to optimize the activity, and the concentration of the compound can be varied. Because of the cost of the compounds, concentrations between 0.01 and 0.1 mM are generally used. Not all cellulases are active against these substrates, but many are. Furthermore the substrates do not distinguish between exocellulases and endocellulases, as some enzymes from each class are active whereas other enzymes from both classes are inactive. Of the six *T. fusca* cellulases that were tested, E1 and E6 had high activity on MUCB, E2 had low activity, and E3, E4, and E5 had no activity on MUCB. E5 and E6 had high activity on MUCT, E2 and E3 had very low activity, and E1 and E4 had no activity on MUCT (7). As discussed earlier, E1, E2, and E5 are endocellulases, E3 and E6 are exocellulases, and E4 has properties of both but behaves more like an exocellulase. A similar but less sensitive assay for cellulase-positive transformants is to use an overlay containing *p*-nitrophenyl-β-D-cellobioside or -cellotrioside, where positive colonies are yellow owing to the release of *p*-nitrophenol.

In addition to activity assays that will identify any clone that contains a cellulase gene that is active on the screening substrate, it is possible to use screens that identify a specific gene. These require purification to homogeneity of the cellulase that is being studied. One procedure is to make a specific antibody to the pure cellulase. Both monoclonal and polyclonal antibodies have been used successfully (19, 20) to screen for cellulase genes. The major problem with this approach is nonspecific labeling, which can cause a high background that obscures positive colonies. The extent of the problem depends on the actual antiserum or antibody preparation that is used, and nonspecific labeling usually does not occur with monoclonal antibodies. This approach requires expression of the gene being cloned by the transformants.

The most powerful screen is one that does not require expression of the gene. One screen with this property utilizes a synthetic oligonucleotide that could encode a peptide segment present in the cellulase. This requires the sequencing of a segment or several segments of the protein to be cloned. This is usually the N-terminal sequence, but if that is blocked, the protein can be cleaved specifically either by cyanogen bromide or an enzyme like trypsin. Then peptides are isolated and their N-teminal sequences determined. A degenerate oligonucleotide probe is designed and synthesized that will code for a section of the sequence (21). Because of the degeneracy of the genetic code, the design process is very important because it is desirable to minimize the number of different oligonucleotides in the probe. The optimal size is around 17 nucleotides. If the codon usage of the organism from which the gene is being cloned is known, that information can be used to minimize the degeneracy.

We have succeeded in cloning two *T. fusca* cellulase genes by screening a plasmid library in *S. lividans* with the CMC overlay assay (11, 12). Because it is much more difficult to transform *S. lividans* than *E. coli* and the efficiency of transformation is lower in *S. lividans,* this approach is used only when a gene cannot be cloned in *E. coli.* The library is constructed from a partial *Sst*I digest of *T. fusca* DNA that is purified on a sucrose gradient as described earlier. The purified DNA is ligated with *Sst*I-digested pIJ702 DNA (22) and used to transform *S. lividans* TK24 using the media and procedures described in Ref. 22. One modification is at a key step in the transformation procedure and increases the yield of spheroplasts. After 15 min of incubation of the washed cells with lysozyme, cells are drawn up and down in a pipette (triturated) three times and then examined in a microscope. If they are completely converted to spheroplasts, proceed directly to the next step, if not incubate for 10 min and repeat the trituration. Approximately 1 μg of plasmid DNA is required for each transformation.

The level of expression of a plasmid in *S. lividans* appears to vary between different transformants, so we always screen the transformants with the CMC overlay assay, even when we are transforming with pure plasmid DNA, in order to pick up a transformant with the highest level of expression. It is also important to select a transformant that grows well, as the growth rate of different transformants can vary greatly.

Streaking for single-colony isolates does not work well for *S. lividans*. Instead, the desired colonies are picked into 50 μl of TSB in a sterile tube and manipulated with a toothpick to break up the mycelia. This solution is spread on a fresh plate, and then the spreader is used to make a single swipe on a second plate. One of the plates should have well-separated colonies.

Single colonies are picked to duplicate plates and grown at 30°C for 2 days. One plate is tested for cellulase-positive colonies by the CMC overlay assay. The desired colonies are picked from the duplicate plate as above, then the solution is used to inoculate a 10-ml Tryptone soya broth liquid culture containing 5 μg/ml thiostrepton (22) in a 125-ml flask containing a spring. Grow the culture (2–6 days) until dense. Add 0.5 ml of 50% glycerol to 1 ml of cells and freeze at −70°C. Use the cells from 1 ml of culture to isolate plasmid DNA to check for the correct plasmid construction. Regrow the culture several times to check the stability of the strain. A generous supply of stock cultures should be kept frozen at −70°C in 15% glycerol. Not all transformants will produce spores, and our plasmids do not seem to be stable in spores. It is not advisable to propagate cultures by successive rounds of mass culture (22).

To prepare plasmid DNA from *S. lividans* cells, spin down 2 ml of a dense culture of cells in a 2.2-ml microcentrifuge tube. Resuspend the cells in 200 μl of 50 m*M* glucose plus 25 m*M* Tris, pH 8, 10 m*M* EDTA, and 2 mg/ml lysozyme. Incubate at 37°C for 15 min. Triturate. Now follow the Insta-Prep (5 Prime–3 Prime, Boulder, CO) procedure: add 300 μl of phenol–chloroform–isoamyl alcohol (PCI) mix and pour into a prepared Insta-Prep tube. Extract with an additional 200 μl PCI followed by extraction with 300 μl of chloroform. Recover the plasmid DNA as recommended. An alternative plasmid preparation method is given by Irwin *et al.* (7).

It is possible to use an *E. coli–S. lividans* shuttle plasmid such as pSH1.7 to move genes into *S. lividans,* which simplifies the process (23). The plasmid contains a number of unique restriction sites, and the ligation mixture between the insert and the shuttle plasmid can be transformed into *E. coli* to allow identification of a plasmid with the proper restriction map. Plasmid DNA from an *E. coli* transformant can then be used to transform *S. lividans.*

Plasmid stability can be a problem in *S. lividans,* especially when the *E. coli– S. lividans* shuttle plasmid pSH1.7 is used. In several cases we have found that the stability of a gene cloned into pSH1.7 can be increased by deleting the *E. coli* plasmid DNA from the construction and transforming directly into *S. lividans.* Therefore, it is best to make a construction in the shuttle vector that allows the removal of the *E. coli* plasmid DNA by digestion with a single restriction enzyme. This makes it possible to produce large amounts of plasmid DNA in *E. coli,* digest out the *E. coli* DNA, ligate the plasmid shut, and then transform into *S. lividans.*

For large-scale production of cellulases, use 250 ml of a well-grown inoculum per 10 liters of medium. Growth of an *S. lividans* transformant is at 30°C in TSB or NMMP medium usually for 48 hr (22). Growth in NMMP is poorer than in TSB medium but appears to give less proteolysis of the secreted cellulases. The cellulases are secreted into the culture supernatant so that they can be separated from most *S. lividans* proteins by centrifugation to remove the mycelia. The cellulase can be recovered from the culture supernatant by absorption to phenyl-Sepharose. Phenylmethylsulfonyl fluoride is added to the culture supernatant to give a final concentration of 0.1 mM to inhibit proteolysis, and all procedures are carried out at 4°C. First the supernatant is passed through a 0.45-μm Durapore filter (Millipore, Bedford, MA). Then solid ammonium sulfate is added to the culture supernatant from a 10-liter culture to give a final concentration of 1.0 M, and the supernatant is loaded on a 150-ml column of phenyl-Sepharose CL-4B (Sigma) at a flow rate of up to 300 ml/hr.

The column is washed with 1 volume of a buffer containing 0.6 M $(NH_4)_2SO_4$, 10 mM NaCl, 5 mM potassium phosphate, pH 6, and then with 2 volumes of the buffer diluted 1:1 with 5 mM potassium phosphate, pH 6. Bound proteins are eluted with 5 mM potassium phosphate, pH 6, and then with water. Fractions should be assayed for activity and then electrophoresed on an SDS–polyacrylamide gel to test sample purity. The best fractions are combined to give partially purified enzyme.

Acknowledgments

This work was supported by Grant DE-FG02-84 ER 13233 from the U.S. Department of Energy.

References

1. A. J. McCarthy, *FEMS Microbiol. Rev.* **46**, 145 (1987).
2. A. J. McCarthy, A. S. Ball, and S. L. Bachmann, *in* "Biology of Actinomycetes '88," p. 283. Japan Scientific Societies Press, Tokyo, 1989.
3. F. J. Stutzenberger, *Appl. Microbiol.* **24**, 83 (1972).
4. B.-G. Hagerdal, J. D. Ferchak, and E. K. Pye, *Microbiology* **36**, 606 (1976).
5. A. R. Moreira, J. A. Phillips, and A. E. Humphrey, *Biotechnol. Bioeng.* **23**, 1325 (1981).
6. R. E. Calza, D. C. Irwin, and D. B. Wilson, *Biochemistry* **24**, 7797 (1985).
7. D. C. Irwin, M. Spezio, L. P. Walker, and D. B. Wilson, *Biotechnol. Bioeng.* **42**, 1002 (1993).
8. A. Collmer and D. B. Wilson, *Bio/Technology* **1**, 594 (1983).
9. Y.-J. Ho and D. B. Wilson, *Gene* **71**, 331 (1988).
10. G. Lao, G. S. Ghangas, E. D. Jung, and D. B. Wilson, *J. Bacteriol.* **173**, 3397 (1991).
11. E. D. Jung, G. Lao, D. C. Irwin, B. K. Barr, A. Benjamin, and D. B. Wilson, *Appl. Environ. Microbiol.* **59**, 3032 (1993).
12. G. S. Ghangas and D. B. Wilson, *Appl. Environ. Microbiol.* **54**, 2521 (1988).
13. D. B. Wilson, *Crit. Rev. Biotechnol.* **12**, 45 (1992).

14. G. S. Ghangas and D. B. Wilson, *Appl. Environ. Microbiol.* **53,** 1470 (1987).

15. N. R. Gilkes, B. Henrissat, D. G. Kilburn, R. C. Miller, Sr., and R. A. J. Warren, *Microbiol. Rev.* **35,** 303 (1991).

16. E. Lin and D. B. Wilson, *J. Bacteriol.* **170,** 3843 (1988).

17. B. G. R. Hagerdal, J. P. Ferchak, and E. K. Pye, *Appl. Environ. Microbiol.* **36,** 606 (1978).

18. R. M. Teather and P. J. Wood, *Appl. Environ. Microbiol.* **43,** 777 (1982).

19. N. R. Gilkes, D. G. Kilburn, M. L. Langsford, R. C. Miller, Jr., W. W. Wakurchuk, R. A. J. Warren, D. J. Whittle, and W. K. R. Wong, *J. Gen. Microbiol.* **130,** 1377 (1984).

20. P. Hu, S. K. Kahrs, T. Chase, Jr., and D. E. Eveleigh, *J. Ind. Microbiol.* **10,** 103 (1992).

21. J. Sambrook, E. F. Fritsch, and T. Maniatis, "Molecular Cloning: A Laboratory Manual," 2nd Ed., p. 11.3. Cold Spring Harbor Laboratory, Cold Spring Harbor, New York, 1989.

22. D. A. Hopwood, M. J. Bibb, K. F. Chater, T. Kiersen, C. J. Bruton, H. M. Kieser, D. J. Lydiate, C. P. Smith, J. M. Ward, and H. Schrempf, "Genetic manipulation of streptomyces". A laboratory manual. The John Innes Foundation, Norwich, England, 1985.

23. S. E. Jensen, B. K. Leskiw, J. L. Doran, A. K. Petrich, and D. W. S. Westlake, *in* "Genetics and Molecular Biology of Industrial Microorganisms" (C. L. Hershberger, S. W. Queener, and G. Hegeman, eds.), p. 239. American Society for Microbiology, Washington, D.C., 1989.

[20] Computational Analyses Aiding Identification and Characterization of Proteins, Genes, and Operons

Milton H. Saier, Jr., and Jonathan Reizer

Introduction

A new discipline of biological research has emerged termed "computational biology." As laboratories gear up for megabase genome sequencing of both prokaryotic and eukaryotic organisms (1), and as the volume of DNA sequence information continues to increase in an exponential fashion, the need for computational analyses increases superexponentially. The human element implies that genome sequences will be riddled with errors, errors that can be recognized when the appropriate tools are applied. We have come to accept the fact that the human brain lacks many of the capacities of a computer. On the other hand, although the computer is an indispensible helpmate for searching and comparing sequences, biochemists and molecular biologists must guard against the blind acceptance of any sophisticated algorithmic output; given the choice, one should think like a biologist and not like a computerized statistician (2).

In this chapter our goal is to expose the reader to the utility of computer techniques, approached from the rationale of the molecular biologist, for the correction of sequencing and assignment errors and for gleaning maximal information from DNA sequence data. The compendium of references provided will allow excursion into the use of currently available programs in much greater depth than described here. We begin with presentation of sections outlining computer-aided approaches that facilitate proper identification and characterization of operon, gene, and protein structures. We then exemplify the use of these approaches, drawing on a large body of representative published data concerned primarily with operons including genes encoding proteins of the phosphoenolpyruvate:sugar phosphotransferase system (PTS) and other transport systems.

Principles and Techniques of Computer-Aided Operon and Gene Analyses

Particularly in prokaryotes, structural genes concerned with a particular function are usually found within an operon or a set of coordinately regulated operons called a regulon. Regulons are frequently self-contained. They include not only the structural genes required for function, but also regulatory genes concerned with transcriptional control and, more rarely, genes concerned with translational or posttranslational control (3). Recognition of regulatory as compared with structural genes can be facilitated by the search for specific motifs (4). These signature sequences are invaluable for the recognition of zinc finger motifs, leucine zipper motifs, and ATP-binding motifs, all of which occur in distinct types of transcription factors. Of particular utility in the identification of DNA-binding proteins in prokaryotes (and to a lesser extent in eukaryotes) are motifs for formation of two α helices connected by a β turn (5). The helix–turn–helix motif occurs in numerous prokaryotic DNA-binding proteins as well as in eukaryotic homeodomain proteins. As is usually the case, however, the nonjudicious application of motif recognition programs can lead to judgmental errors (5).

As mentioned above, operons almost always include genes concerned with a common metabolic function. For most sugar catabolic operons these functions include sugar transport and phosphorylation, conversion of the phosphorylated carbohydrate to a common metabolic intermediate, and transcriptional regulation. The mannitol catabolic operon, for example, includes three genes, *mtlADR* (6). The *A* gene encodes a mannitol-specific permease/kinase, the Enzyme IICBA protein of the PTS (7) that transports and phosphorylates mannitol; the *D* gene encodes the mannitol-phosphate dehydrogenase that converts the hexitol phosphate product of the PTS reaction to fructose 6-phosphate; and the *R* gene encodes the *mtl* operon-specific transcriptional regulator. The glucitol operon includes six genes, *gutABDMRQ* (8, 9). *gutA* and *B* encode the glucitol permease/kinase, the Enzyme II complex of the PTS equivalent to the *mtlA* gene product of the mannitol operon; *gutD* encodes the glucitol-6-phosphate dehydrogenase that converts glucitol 6-phosphate to fructose 6-phosphate; and the remaining 3 genes (*gutM, R,* and *Q*) all seem to encode proteins that are concerned with transcriptional regulation (9; M. Yamada and M. H. Saier, Jr., unpublished results). This principle of functional operon containment is not restricted to catabolic operons; it is also applicable to biosynthetic and amphibolic operons. A few operons are known to coordinate distantly related functions [i.e., DNA replication, transcription, and translation (10)].

The identification of specific genes relies on the use of programs that allow recognition of peptide coding regions, transcriptional terminators, repeated sequences, and other consensus patterns. Identification of open reading frames is facilitated by the use of programs that analyze sequence composition (11). For example, CODON-PREFERENCE is a frame-specific gene finder that recognizes protein coding sequences by virtue of the similarity of their codon usage to a codon frequency table or by the bias of their composition (usually GC composition) in the third position of each codon. The TESTCODE program identifies protein coding sequences by plotting a measure of the nonrandomness of the composition at every third base. In contrast to CODONPREFERENCE, the TESTCODE program does not require a codon frequency table. FRAMES reveals open reading frames for the six reading frames of a DNA sequence. FRAMES can superimpose a pattern of rare codon choices if provided with a codon frequency table. COMPOSITION determines the composition of a sequence. For nucleotide sequences, COMPOSITION also determines dinucleotide and trinucleotide content.

Identification of a correct prokaryotic start codon for an open reading frame (ORF) is facilitated by identification of a Shine–Dalgarno (ribosome binding) sequence preceding the gene (12). However, a recognizable Shine–Dalgarno sequence is not always present for lowly expressed genes. Promoter and terminator sequences serve to define the beginning and end of a transcription unit. Programs for promoter recognition sequences (13) and terminator recognition sequences (14) search for a consensus-like sequence similar to other well-defined transcriptional signals. Repetitive extragenic palindromes (REP) or palindromic units (PU) serve to define intergenic regions, as such sequences virtually never occur within coding regions (15). Finally, when two genes overlap, even by a single nucleotide, it is probable that the genes are translationally coupled. That is to say, the ribosome need not dissociate from and rebind to the RNA in order to translate the two genes.

Principles and Techniques of Computer-Aided Protein Analyses

Programs concerned with protein analysis are available for (i) plotting average degrees of similarity as a function of residue position for a set of homologous proteins; (ii) locating signature sequences or motifs that are common to all members of a particular family but absent from other proteins; (iii) deriving probable primordial (ancestral) sequences from current sequences; (iv) predicting membrane protein topologies; (v) averaging hydropathy and amphipathicity plots; (vi) detecting internal duplications, deletions, and insertions within coevolving proteins; (vii) estimating regions of relative flexibility; (viii) estimating the probability that a sequence will lie within a particular secondary structural element (i.e., α helix, β strand, or β turn); (ix) averaging secondary structure and flexibility predictions for a group of homologous proteins, thereby increasing reliability; and (x) averaging comparison scores, similarity scores, or percent identities for estimation of relative degrees of evolutionary divergence for coevolving homologous proteins or protein domains or for approximating the relative ages of independently evolving protein families. The reader is referred to Volume 183 of *Methods in Enzymology* for detailed consideration of specific computer programs concerned with the analysis of protein and nucleic acid sequences (2).

The principles underlying the study of protein evolution and many of the computer-aided methods for quantifying the relatedness of any two proteins have been described in detail by Doolittle (2, 16–18). A primary goal of the protein evolutionist is to reconstruct past events leading to the structures of current proteins, from which the relationships of present-day proteins that share a common ancestry can be defined (19, 20). A single sequence can evolve into different sequences either by speciation and divergence (such genes are called orthologous) or by gene duplication within a single organism followed by divergence (such genes are called paralogous). The latter event can give rise to proteins of different function or to isoforms of proteins having the same function. Because the rates of change of many proteins are very slow, ancient events can be inferred by examining protein families. The sequence of a slowly evolving prokaryotic enzyme may, for example, exhibit as much as 40–50% identity with the corresponding enzyme from a eukaryote, even though eukaryotes diverged from prokaryotes over 1.5 billion years ago. High degrees of sequence similarity found in evolutionarily distant organisms, or low degrees of sequence similarity found in closely related organisms, may be due to horizontal transmission of genetic information (21, 22), even between prokaryotes and eukaryotes (23), and this information can probably be passed between the two kingdoms in both directions (see Ref. 24 and discussion therein). Homologous genes in a single organism that are acquired laterally (horizontally) are said to be xenologous. It is important to realize that different proteins evolve at very different rates, and that different degrees of sequence similarity observed for two proteins will in general reflect both the time and the rate of divergence of the two structural genes (25–27).

The first task in defining the ancestry of a group of proteins is always to identify all sequenced members of a particular family by screening available databases. Families or clusters of closely related proteins are first identified and then interrelated if possible. Signature sequences of established families can be used to identify newly sequenced members of these families (4). When several families of proteins are shown to share a common ancestry, we say that they comprise a superfamily. Because they share a common origin, the members of a superfamily are said to be homologous. Proteins are either homologous or not homologous; there are no degrees of homology although there are ranges of similarity or identity (16).

A binary comparison score [expressed in standard deviations (SDs)] can be calculated from the comparison of two amino acid sequences by using any of several programs [ALIGN (28), LOS ALAMOS (29), and RDF2 (30)]. The degree of similarity between the two sequences is compared with a large number of random shuffles of the two sequences (thus eliminating discrepancies arising from unusual amino acid compositions) to establish significance. When two sequences give a comparison score of 3 SDs or less, there is little or no evidence for homology. However, if they give a comparison score of 6 SDs, the probability (P) that the degree of similarity exhibited by these two sequences arose by chance is about 10^{-9}. This suggests that such sequences probably arose from a common ancestor by divergent evolution, but this degree of sequence similarity could conceivably have arisen by a convergent evolutionary process, particularly if the sequences compared are short. Thus, homology is not established. On the other hand, when a comparison score is high (i.e., ≥ 9 SDs; $P \leq 10^{-19}$), the degree of similarity is considered to be too great to have arisen either by chance or by a convergent evolutionary process, so the two sequences are considered to be homologous (16). Methods for assessing the statistical significance and reliability of particular molecular sequence features have been presented (see, e.g., Refs. 11, and 31–33). It should be noted that the term "convergence" has been used loosely by different investigators to refer to the independent evolution of a similar sequence (sequence convergence), a similar specificity or catalytic function (functional convergence), or a similar topology or structural scaffold (structural convergence) (34).

Several questions can be asked to further substantiate the conclusion of common descent. (i) Do the two proteins (or protein domains) share a common function? (ii) Are the sequences compared derived from comparable portions of the proteins or protein domains? (iii) Are the two proteins or protein domains of about the same lengths? (iv) Do they have similar topologies or three-dimensional structures? If the answers to these questions are yes, one gains additional confidence in the conclusions derived solely from the computer-based statistical analyses described above. Simultaneous convergence of sequence, structure, and function is considered to be an exceptionally improbable evolutionary event.

Once homology has been established for a group of proteins or protein domains, the sequences are optimally aligned, yielding a single multiple alignment. Relative evolutionary distances and phylogenetic positions can be determined using available computer programs. Results obtained with two distinct programs, based on different assumptions, namely, the TREE and PAPA programs (35, 36), can be used to evaluate the reliability of the results obtained for any one program (37, 38). The construction of phylogenetic trees allows one to see the relatedness of members of a group of homologous proteins at a glance.

Application of Computer-Aided Approaches to the Identification of Genes Encoding Phosphotransferase System Proteins

The phosphoenolpyruvate:sugar phosphotransferase system (PTS) in any bacterium consists of at least two energy-coupling proteins, Enzyme I and HPr, as well as three proteins or protein domains, designated IIA, IIB, and IIC, which serve as the sugar-specific PTS permease/kinase. The IIA and IIB proteins or protein domains serve as phosphoryl transfer units, whereas the IIC domain is the sugar recognition element and transmembrane transport protein. The IIA, IIB, and IIC domains/proteins comprise the Enzyme II complex (Fig. 1) (7, 39).

Table I lists newly sequenced genes encoding PTS proteins, and Fig. 2 shows the structures of well-characterized bacterial operons encoding energy-coupling PTS proteins. In *Escherichia coli* the tricistronic *pts* operon (*ptsHIcrr*) encodes three energy-coupling proteins, HPr, Enzyme I, and the glucose-specific IIA (IIA^{glc}) protein. The last protein is a central regulatory protein in enteric bacteria (40). In *B. subtilis* where the IIA^{glc} protein apparently does not play a central regulatory role, IIA^{glc} is the

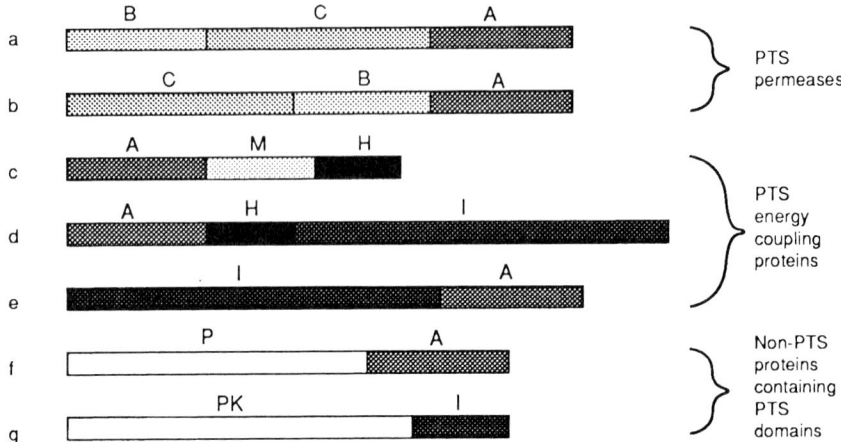

FIG. 1   Prototypes of currently sequenced PTS proteins showing domain orders. PTS-related domains include the following: I, Enzyme I; H, HPr; A, first permease-specific phosphorylation site domain; B, second permease-specific phosphorylation site domain; C, transmembrane permease domain. M (modulator), P (permease), and PK (pyruvate kinase) represent non-PTS-related protein domains. Examples are as follows: (a) the β-glucoside (Bgl) Enzyme II of *E. coli;* (b) the mannitol (Mtl) Enzyme II of *E. coli;* (c) the diphosphoryltransfer protein (DTP) of enteric bacteria; (d) the multiphosphoryltransfer protein (MTP) of *Rhodobacter capsulatus* [L.-F. Wu, J. M. Tomich, and M. H. Saier, Jr., *J. Mol. Biol.* **213,** 687 (1990)]; (e) the putative energy-coupling protein, Enzyme I, in the pyruvate–formate-lyase region of the *E. coli* genome [F. R. Blattner, V. Burland, G. Plunkett III, H. J. Sofia, and D. L. Daniels, *Nucleic Acids Res.* **21,** 5408 (1993)]; (f) the lactose : H^+ symport permease of *Streptococcus thermophilus;* and (g) the pyruvate kinase of *Bacillus stearothermophilus* which contains 110 residues that are homologous to the active site domain of Enzyme I. In all cases examined, interdomain splicing can occur, giving rise to two or more proteins in which the homologous or functionally equivalent domains are associated with distinct polypeptide chains. Tandem domain duplications (IIB^{fru} of *E. coli* and *R. capsulatus*), domain deletions (IIA^{scr} and IIA^{tre} of *E. coli* and IIA^{scr} of *B. subtilis*) have also been documented. Additionally, a IID domain is found in a minor family of PTS permeases [e.g., the mannose enzyme II complex in *E. coli;* H. Imaishi, M. Gomada, S. Inouye, and A. Nakazawa, *J. Bacteriol.* **175,** 1550 (1993); B. Erni, B. Zanolari, and H. P. Kocher, *J. Biol. Chem.* **262,** 5238 (1987)].

C-terminal part of a single polypeptide chain that comprises the entire Enzyme II^{glc} complex (41), and the bicistronic *pts* operon includes only the structural genes encoding HPr and Enzyme I. In *Mycoplasma capricolum* the *ptsH* gene encoding HPr comprises a monocistronic operon, whereas Enzyme I and IIA^{glc} comprise a distinct transcription unit that is bicistronic (42).

Rhodobacter capsulatus and *E. coli* contain PTS energy-coupling proteins in fructose (*fru*) catabolic operons (Fig. 2). In the former of these two organisms, IIA^{fru} is

TABLE I   Newly Sequenced Genes Encoding Proteins of the Bacterial Phosphotransferase System in *Escherichia coli*[a]

Gene designation	Encoded protein[b]	Postulated function	Map position (centisome)	Identifer (SWISSPROT)	Ref.
frvA	IIA^{frv}	Fructose-like permease	88.1	P32155	c, d
frvB(C)	IIBC^{frv}	Fructose-like permease	88.1	P32154	c, d
pflC	IIC^{pfl}	Fructose-like permease[e]	89.2	P32672	f
pflB	IIB^{pfl}	Fructose-like permease[e]	89.2	P32673	f
pflQ	IIB^{pfl}	Fructose-like permease[e]	89.3	P32676	f
glvB	IIB^{glv}	Glucoside-like permease	83.2	P31451	c, g
glvC	IIC^{glv}	Glucoside permease	83.2	P31452	c, g
fruB	DTP	Fructose permease	48.7	—	h
rpoP	IIA^{ntr}	Regulation	72.1	P31222	i, j
rpoR	NPr	Regulation	72.1	P33996	j
cmtA	IICB^{cmt}	Mannitol permease	66.3	P32059	k
cmtB	IIA^{cmt}	Mannitol permease	66.3	P32058	k
pflI	Enzyme I^{pfl}	?	89.1	P32670	f
ptsJ	PtsJ	Transcriptional regulation	54.6	—	l

[a] All of the listed genes except for the *ptsJ, rpo,* and *cmt* genes were sequenced as a result of the *E. coli* genome sequencing project. Genetic designations are in accordance with earlier publications cited above, as well as proposed functions.

[b] PTS protein nomenclature is in accordance with accepted designations (7; 39). Designations of the encoded proteins are based on biochemical analyses (*fruB; rpoP; rpoR; cmtA; cmtBC*) or homology studies (all other proteins). Precise map positions were provided by Ken Rudd (personal communication).

[c] J. Reizer, V. Michotey, A. Reizer, and M. H. Saier, Jr., *Protein Sci.* **3**, 440 (1994).

[d] G. Plunkett III, V. Burland, D. L. Daniels, and F. R. Blattner, *Nucleic Acids Res.* **21**, 3391 (1993).

[e] A detailed description of these genes is provided in J. Reizer, A. Reizer, and M. H. Saier, Jr., *Microbiol.* **141** (1995), in press.

[f] F. R. Blattner, V. Burland, G. Plunkett III, H. J. Sofia, and D. L. Daniels, *Nucleic Acids Res.* **21**, 5408 (1993).

[g] V. Burland, G. Plunkett III, D. L. Daniels, and F. R. Blattner, *Genomics* **16**, 551 (1993).

[h] J. Reizer, A. Reizer, H. L. Kornberg and M. H. Saier, Jr., *FEMS Microbiol. Lett.* **118**, 159–162 (1994).

[i] H. Imaishi, M. Gomada, S. Inouye, and A. Nakazawa, *J. Bacteriol.* **175**, 1550–1551 (1993).

[j] D. H. A. Jones, F. C. H. Franklin, and C. M. Thomas, *Microbiology* **140**, 1035–1043 (1994).

[k] G. A. Sprenger, *Biochim. Biophys. Acta* **1158**, 103 (1993).

[l] Titgemeyer *et al.,* unpublished.

linked to HPr and Enzyme I domains, but in the latter organisms only IIA^{fru} and a fructose-specific HPr, FPr, are present in the polypeptide chain, linked to one another by a modulator domain.

New energy-coupling PTS proteins have been found in *E. coli* as a result of operon and genome sequencing projects (Fig. 2). In the 89.1 centisome region of the *E. coli* genome, a gene encoding an Enzyme I homolog (Enzyme I^{Ani}) linked to an Enzyme IIA^{fru} homolog is found. In contrast to the usual Enzyme I, the enzyme appears to be cryptic under all conditions tested (J. Reizer and M. H. Saier, Jr., unpublished results). Other genes encoding PTS proteins, one homologous to IIC^{fru} and two homologous to IIB^{fru}, are found in an adjacent operon at 89.2 to 89.3 centisomes. In the

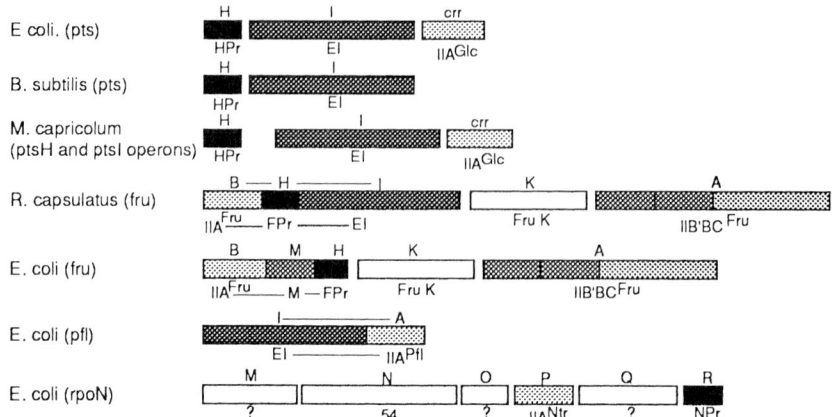

FIG. 2 Operons encoding PTS energy-coupling proteins in various bacteria. Sources and operon designations are indicated at left. Gene designations are indicated above the corresponding gene. The encoded protein or protein domain is indicated below the gene. For fused energy-coupling domains resembling HPr, "FPr" indicates the fructose-inducible HPr-like domain rather than the intact protein. The *E. coli* Enzyme I–IIA^{pfl} encodes a protein linking an Enzyme I to a IIA^{fru}-like domain [F. R. Blattner, V. Burland, G. Plunkett III, H. J. Sofia, and D. L. Daniels, *Nucleic Acids Res.* **21**, 5408 (1993)]. The *E. coli rpoN* operon includes two PTS proteins, a IIA^{fru}-like protein designated IIA^{ntr} [J. Reizer, A. Reizer, G. R. Jacobson, and M. H. Saier, Jr., *Protein Sci.* **1**, 722 (1992)] and an HPr-like protein, NPr, which may serve in coordination of carbon and nitrogen assimilation [D. H. A. Jones, F. C. H. Franklin, and C. M. Thomas, *Microbiology* **140**, 1035–1043 (1994)].

72.1 centisome region of the *E. coli* chromosome an operon encoding RpoN, the nitrogen-specific RNA polymerase σ factor, includes two downstream genes, one encoding a IIA^{fru}-like protein designated IIA^{ntr} and a second encoding an HPr-like protein, designated NPr (Fig. 2). In all examples involving multidomain, multifunctional proteins, the gene structures were elucidated through the use of computer techniques (43–45; J. Reizer and M. H. Saier, Jr., unpublished results). Further, recognition of the genes encoding PTS proteins within the *rpoN* operon led to the identification of novel regulatory functions for these proteins. It is worth noting that in all recognized cases, operon structure reflects the functional complexity of the encoded proteins.

Two novel PTS protein-containing operons, presumably cryptic, have been characterized using computer techniques (Table I; Figs. 3 and 4) (45). The glucoside-like (*glv*) operon (Fig. 3) encodes an Enzyme IIC, an Enzyme IIB, and a putative phosphoglucosidase. The *glvB* and *glvC* genes overlap by a single base pair, suggesting translational coupling. Computer-aided approaches provided evidence for a *glvC* gene structure that is different from and more probable than the one originally proposed (45). The fructose-like (*frv*) operon (Fig. 4) includes two genes encoding IIA^{frv}

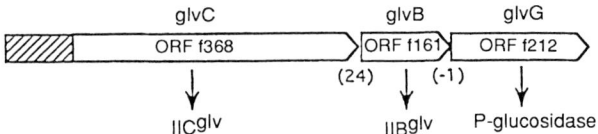

FIG. 3   Genetic map of the *glv* locus. The genetic map shows the proposed names of the genes that encode IIC^{glv}, IIB^{glv}, and a putative phosphoglucosidase. The number of base pairs in the two intergenic regions is given in parentheses; a minus sign denotes an overlapping region. The hatched region in ORFf368 denotes the previously included 5' region in this ORF which, therefore, was termed ORFf455 (see Ref. 45 for details).

and IIBC^{frv} proteins, as well as two genes encoding proteins of unknown function (*frvX* and *frvR*). FrvX and FrvR lack sequence similarity to any protein in current databases, but FrvR is believed to be a transcription factor on the basis of the identification of two N-terminal helix–turn–helix motifs (45). A similar computer-aided approach led to the discovery of the *fruB* gene that encodes the diphosphoryltransfer protein (DTP) of *E. coli* (46).

Elegant studies (47) resulted in the sequencing of the cellobiose-catabolic (*cel*) operon of *E. coli*. It was suggested that the cellobiose-specific PTS permease of *E. coli* consists of an Enzyme II (II^{cel}; CelB) which does not exhibit sequence similarity with any of the sequenced PTS permeases and an Enzyme IIA (IIA^{cel}; CelC) which is homologous to the lactose Enzyme IIA (IIA^{lac}) of *Staphylococcus aureus* (48). Functionally important characteristics of the cellobiose permease were not identified, and it was noted that the *cel* operon contained a transcribed and translated open reading frame (CelA) of unknown function (47).

The deduced amino acid sequences of the proteins encoded within the *cel* operon were subjected to computer analyses (49). The analyses revealed that the CelA, CelB, and CelC proteins all exhibit sufficient sequence similarity with distinct domains of the lactose permease of *S. aureus* to establish homology. Moreover, the sum of molecular weights was found to be equal to that of a typical PTS permease. Evidently, the three proteins comprise the cellobiose permease. Sequence comparisons with well-characterized PTS permeases allowed prediction of the specific functions of each of the three proteins in the transmembrane group translocation of cellobiose.

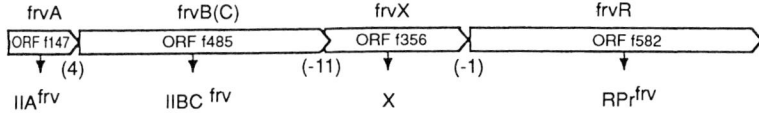

FIG. 4   Genetic map of the *frv* locus. The genetic map shows the proposed gene assignment of the DNA regions that encode IIA^{frv}, IIBC^{frv}, a functionally unidentified protein (X), and RPr^{frv}. The number of base pairs in the three intergenic regions is given in parentheses (a minus sign denotes an overlapping region).

Characteristic features as well as presumed phosphorylation sites within the permease were also identified.

Application of Computer-Aided Approaches to Other Classesof Transport Proteins

Our laboratory has been involved in computer analyses of transport protein sequences in addition to those comprising the PTS (see Ref. 50 for a comprehensive review). Among the most significant findings were (a) definition of protein domain and interdomain linker structures (7, 41, 51); (b) functional predictions regarding specific sequence motifs (52–54); (c) identification of relative domain conservation and postulation of the functional significance (55–58); (d) establishment of independent evolutionary pathways for several distinct families of transport proteins (50, 59); (e) provision of evidence for transport protein specificity, on the basis of regional sequence conservation (58, 60); and (f) evaluation of phylogenetic relationships among established members of a single protein family.

Future Perspectives

The discipline of computational biology relies on a spectrum of computer techniques that are ever increasing in applicability, reliability, and refinement. These techniques allow postulation of protein structure and function merely on the basis of a DNA sequence. Eventually, these tools will allow predictions not only at the molecular level but also at the organismal level, providing applications to diagnosis and molecular characterization of human disease. They are expected to allow the scientific community to fight diseases at their sources using gene therapy techniques (61). The trend in biology during the twentieth century, from organism to molecule, will thus be reversed in the twenty-first century with the integration of molecular details into comprehensive informational maps allowing precise understanding of genome organization, of the principles of genome evolution, and of organismal functions. The role of the computational biologist will become increasingly important until he, too, will be largely replaced by computerized systems that integrate the information content of entire genomes to allow formulation of molecular processes as well as of organismal structure and function.

Acknowledgments

We thank Mary Beth Hiller for providing expert assistance in the preparation of the manuscript. The work was supported by U.S. Public Health Services Grants 5RO1AI 21702 and 2RO1AI 14176 from the National Institute of Allergy and Infectious Diseases.

Note added in proof: Updated designations of PTS genes that are included in Table I can be found in M. H. Saier, Jr., and J. Reizer. *Molec. Microbiol.* **13**, 755–764 (1994); B. S. Powell, D. L. Court, T. Inada, Y. Nakamura, V. Michotey, X. Cui, A. Reizer, M. H. Saier, Jr., and J. Reizer. *J. Biol. Chem.* **270**, 4822–4839 (1995); and J. Reizer, A. Reizer, and M. H. Saier, Jr. *Microbiol.* **141** (1995), in press.

References

1. R. Wilson, R. Ainscough, K. Anderson, C. Baynes, M. Berks, J. Bonfield, J. Burton, M. Connell, T. Copsey, J. Cooper, *et al., Nature (London)* **368,** 32 (1994).
2. R. F. Doolittle (ed.), "Methods in Enzymology," Vol. 183. Academic Press, New York, 1990.
3. L. Lindahl and A. Hinnebusch, *Curr. Biol.* **2,** 720 (1992).
4. A. Bairoch, *Nucleic Acids Res.* **20**, S2013 (1992).
5. I. B. Dodd and J. B. Eagen, *Nucleic Acids Res.* **18,** 5019 (1990).
6. R. M. Figge, T. M. Ramseier, and M. H. Saier, Jr., *J. Bacteriol.* **176,** 840 (1994).
7. M. H. Saier, Jr., and J. Reizer, *J. Bacteriol.* **174,** 1433 (1992).
8. M. Yamada and M. H. Saier, Jr., *J. Biol. Chem.* **262,** 5455 (1987).
9. M. Yamada and M. H. Saier, Jr., *J. Mol. Biol.* **203,** 569 (1988).
10. R. H. Doi, *in* "Modern Microbial Genetics" (U. N. Streips and R. E. Yasbin, eds.), p. 15. Wiley, New York, 1991.
11. J. Devereux, P. Haeberli, and N. O. Smithies, *Nucleic Acids Res.* **12,** 387 (1984).
12. J. Shine and L. Dalgarno, *Proc. Natl. Acad. Sci. U.S.A.* **71,** 1342 (1974).
13. M. C. O'Neill, *Nucleic Acids Res.* **19,** 313 (1991).
14. V. Brendel and E. V. Trifonov, *Nucleic Acids Res.* **12,** 4411 (1984).
15. E. Gilson, W. Saurin, D. Perrin, S. Bachellier, and M. Hofnung, *Nucleic Acids Res.* **19,** 1375 (1991).
16. R. F. Doolittle, "Of Urfs and Orfs: A Primer on How to Analyze Derived Amino Acid Sequences." University Science Books, Mill Valley, California, 1986.
17. R. F. Doolittle, *in* "Prediction of Protein Structure and the Principles of Protein Conformation" (G. D. Fasman, ed.), p. 599. Plenum, New York, 1989.
18. R. F. Doolittle, *Protein Sci.* **1,** 191 (1992).
19. L.-F. Wu, A. Reizer, J. Reizer, B. Cai, J. M. Tomich, and M. H. Saier, Jr., *J. Bacteriol.* **173,** 3117 (1991).
20. L.-F. Wu and M. H. Saier, Jr., *Res. Microbiol.* **142,** 943 (1991).
21. E. A. Groisman, M. H. Saier, Jr., and H. Ochman, *EMBO J.* **11,** 1309 (1992).
22. P. Reeves, *Trends Genet.* **9,** 17 (1993).
23. J. A. Heinemann and G. F. Sprague, Jr., *Nature (London)* **340,** 205 (1989).
24. P. Bork and R. F. Doolittle, *Proc. Natl. Acad. Sci. U.S.A.* **89,** 8990 (1992).
25. M. O. Dayhoff, R. M. Schwartz, and B. C. Orcutt, *in* "Atlas of Protein Sequence and Structure" (M. O. Dayhoff, ed.), Vol. 5, Suppl. 3, p. 345. National Biomedical Research Foundation, Silver Spring, Maryland, 1978.
26. M. Nei, "Molecular Evolutionary Genetics." Columbia Univ. Press, New York, 1987.
27. W.-H. Li and D. Graur, "Fundamentals of Molecular Evolution." Sinauer, Sunderland, Massachusetts, 1991.

28. M. O. Dayhoff, W. C. Barker, and L. T. Hunt, *in* "Methods in Enzymology" (C. H. W. Hirs and S. N. Timasheff, eds.), Vol. 91, p. 524. Academic Press, New York, 1983.

29. M. Kanehisa, *Nucleic Acids Res.* **10**, 183 (1982).

30. W. R. Pearson and D. J. Lipman, *Proc. Natl. Acad. Sci. U.S.A.* **85**, 2444 (1988).

31. J. Felsenstein, *Ann. Rev. Genet.* **22**, 521 (1988).

32. S. F. Altschul, W. Gish, W. Miller, E. W. Myers, and D. J. Lipman, *J. Mol. Biol.* **215**, 403 (1990).

33. S. Karlin and S. F. Altschul, *Proc. Natl. Acad. Sci. U.S.A.* **87**, 2264 (1990).

34. P. Bork, C. Sander, and A. Valencia, *Protein Sci.* **2**, 31 (1993).

35. D.-F. Feng and R. F. Doolittle, *in* "Methods in Enzymology" (R. F. Doolittle, ed.), Vol. 183, p. 375. Academic Press, San Diego, 1990.

36. R. F. Doolittle and D.-F. Feng, *in* "Methods in Enzymology" (R. F. Doolittle, ed.), Vol. 183, p. 659. Academic Press, San Diego, 1990.

37. C. C. Trandinh, G. M. Pao, and M. H. Saier, Jr., *FASEB J.* **6**, 3410 (1992).

38. M. Van Rosmalen and M. H. Saier, Jr., *Res. Microbiol.* **144**, 507 (1993).

39. P. Postma, J. Lengeler, and G. R. Jacobson, *Microbiol. Rev.* **57**, 543 (1993).

40. M. H. Saier, Jr., *Microbiol. Rev.* **53**, 109 (1989).

41. S. L. Sutrina, P. Reddy, M. H. Saier, Jr., and J. Reizer, *J. Biol. Chem.* **265**, 18581 (1990).

42. P.-P. Zhu, J. Reizer, A. Reizer, and A. Peterkofsky, *J. Biol. Chem.* **268**, 26531 (1993).

43. L.-F. Wu, J. M. Tomich, and M. H. Saier, Jr., *J. Mol. Biol.* **213**, 687 (1990).

44. F. R. Blattner, V. Burland, G. Plunkett III, H. J. Sofia, and D. L. Daniels, *Nucleic Acids Res.* **21**, 5408 (1993).

45. J. Reizer, V. Michotey, A. Reizer, and M. H. Saier, Jr., *Protein Sci.* **3**, 440 (1994).

46. J. Reizer, A. Reizer, H. L. Kornberg, and M. H. Saier, Jr., *FEMS Microbiol. Lett.* **118**, 159–162 (1994).

47. L. L. Parker and B. G. Hall, *Genetics* **124**, 455 (1990).

48. F. Breidt, Jr., W. Hengstenberg, U. Finkeldei, and G. C. Stewart, *J. Biol. Chem.* **262**, 16444 (1987).

49. J. Reizer, A. Reizer, and M. H. Saier, Jr., *Res. Microbiol.* **141**, 1061 (1990).

50. M. H. Saier, Jr., *Microbiol. Rev.* **58**, 71 (1994).

51. C. F. Higgins, *Annu. Rev. Cell Biol.* **8**, 67 (1992).

52. J. Reizer, A. Reizer, and M. H. Saier, Jr., *Biochim. Biophys. Acta* **1197**, 133–166 (1994).

53. J. Reizer, S. Buskirk, A. Bairoch, A. Reizer, and M. H. Saier, Jr., *Protein Sci.* **3**, 853 (1994).

54. I. Yamato, M. Kotani, O. Yumiko, and Y. Anraku, *J. Biol. Chem.* **269**, 5720 (1994).

55. J. Kuan and M. H. Saier, Jr., *Crit. Rev. Biochem. Mol. Biol.* **28**, 209 (1993).

56. M. D. Marger and M. H. Saier, Jr., *Trends Biochem. Sci.* **18**, 13 (1993).

57. J. Reizer, A. Reizer, and M. H. Saier, Jr., *Crit. Rev. Biochem. Mol. Biol.* **28**, 235 (1993).

58. R. Tam and M. H. Saier, Jr., *Microbiol. Rev.* **57**, 320 (1993).

59. M. H. Saier, Jr., *BioEssays* **16**, 23 (1994).

60. J. Reizer, A. Reizer, and M. H. Saier, Jr., *Res. Microbiol.* **141**, 1069 (1990).

61. M. A. Findeis, *Technol. Rev.* **97**, 46 (1994).

Plasmids and Phages: Replication, Transcription, Assembly

[21] Techniques to Isolate Specialized Templates for *in Vitro* Transcription Assays

Susan Garges, Hyon E. Choy, and Sangryeol Ryu

Introduction

In the study of transcription and transcription factors, the value of *in vitro* experiments is obvious. *In vitro* transcription assays are frequently employed to understand regulation at the level of transcription initiation. For all *in vitro* transcription work, regardless of the system, an important aspect is the quality and state of the DNA. In this chapter, we explain how specialized templates for transcription can be prepared. The specialized templates are necessary for certain types of experiments, and possible reasons for using such templates are discussed in each section.

A One-Promoter Transcription System: The Minicircle

Current understanding of transcription initiation involves multiple stages including binding of RNA polymerase to the promoter (closed promoter complex formation), isomerization of the promoter, and transcription idling during which RNA oligomers are made and aborted by the initial transcribing complex. In addition, clearance of the RNA polymerase from the promoter can be considered a step in initiation (1). Any of the steps could be factor-affected. The measurement of closed complex formation and subsequent isomerization routinely uses an abortive initiation assay (2). Assessment of the idling and the promoter clearance steps requires a transcription assay that uses a DNA template with a promoter(s) that allows one to analyze both abortive and full-length transcripts directly and simultaneously without the possibility of interference by the products of other promoters, or problems with the other promoters competing for RNA polymerase. Because the topological conformation of DNA can greatly influence the interactions between DNA and transcriptional regulatory proteins (3–5), it is relevant if the unitary promoter DNA template is supercoiled. Choy and Adhya have developed a plasmid vector that can generate supercoiled DNA templates, "minicircles," convenient for *in vitro* study of transcription at individual promoters under physiological conditions (6). The minicircles are generated *in vivo* by exploiting the mechanism of site-specific integrative recombination of bacteriophage λ.

Construction of Plasmids

For *in vivo* synthesis of supercoiled minicircles, we have constructed a plasmid, pSA508, which contains a DNA segment with a multiple cloning site (MCS) and a transcription termination sequence between the phage λ attachment site, *attP'OP*, and the corresponding bacterial site *attB'OB* (6) (Fig. 1). pSA508 was derived from pIBI24 (International Biotechnologies, New Haven, CT) by inserting the DNA elements between the *Eco*RI and *Hind*III sites in the following order: the 46-bp bacterial attachment site *attB'OB* (from -19 to $+26$; Ref. 7), a 49-bp MCS, the 54-bp ρ-independent transcription terminator of the *rpoC* gene of *E. coli* (8), and the 408-bp *attP'OP* derived from pPH54 (from $+247$ to -160; Ref. 9).

Preparation of Minicircle DNA in Vivo by Bacteriophage λ Integrative Recombination

We have used the following protocol to isolate minicircles. Plasmid pSA508 or a derivative is introduced by transformation at 32°C into *E. coli* strain SA1751 {F$^-$ *relA rpsL lac trp* (λ *int*$^+$ *xis439 cI857* [*cro-chlA*] ΔH1)}. In this strain, expression of the *int* gene from the cryptic prophage is under control of the temperature-sensitive *cI857* repressor. To prepare the DNA minicircles, SA1751 containing the parental plasmid is grown to late logarithmic phase (A_{600} around 0.7) at 32°C to accumulate substrate plasmid, shifted to 42°C for 10 min to induce synthesis of λ Int (Integrase) protein from the cryptic λ prophage by inactivating the thermolabile repressor, and then cooled to 32°C. The culture is returned to the lower temperature because the integrative recombination reaction is thermosensitive (10). After 30 min at the lower temperature, the cells are harvested. The 10-min heat shock (42°C) and the 30-min incubation at 32°C should result in sufficient expression of Int protein and maximal recombination between the *attB'OB* and *attP'OP* elements catalyzed by Int protein. Total plasmid DNA can be isolated by any conventional DNA isolation protocol [we use an alkaline lysis protocol (11)].

Figure 2 shows products of recombination reaction for the parental plasmid, pSA508. As expected, there was no recombination product before heat induction of Int synthesis (lane 2). The supercoiled pSA508 is 3.3 kb and migrates as a monomer at the position marked S1 and as a dimer at the position marked S2 (Fig. 2). Lanes 3 and 4 show the two recombination product circles of 2.9 kb (arrow b) and 0.4 kb (minicircle; arrow a) as expected. However, the two lanes show, in addition, multiple ladders of discrete sizes. The ladders comprising bands of ascending sizes and decreasing intensities, visible between arrows a and b, contain oligomeric forms of the minicircles. Another ladder present above the larger product circles at arrow b contains the larger circular product with increasing numbers of minicircle units. The production of the multimerized circles is unavoidable because the Int-mediated recombination between *att* elements present in the substrate plasmids and product circles are very efficient *in vivo* (12). The use of a RecA mutant host for generation of minicircles does not reduce oligomerization of the products.

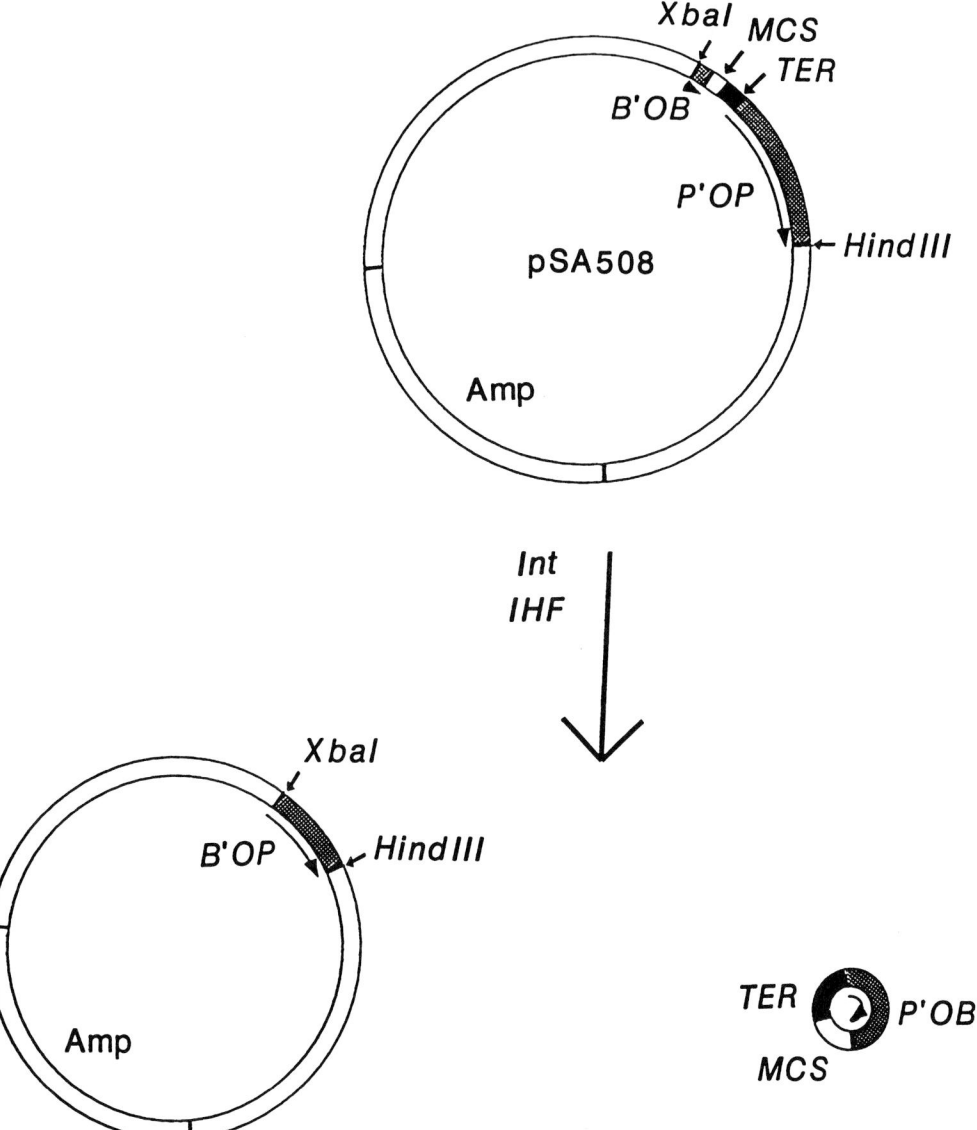

FIG. 1   Arrangement of functional elements of the unitary promoter vector pSA508 and generation of a DNA minicircle as a result of Int/IHF-mediated recombination. Details of the
recombination reaction are explained in the text. The shaded areas (*B'OB, P'OP, B'OP,* and
P'OB) indicate the sequences involved in λ phage integrative recombination before and after
the reaction. *TER* is the transcription terminator; *MCS,* multiple cloning site; *Amp,* ampicillin
resistance gene; Int, integrase; IHF, integration host factor.

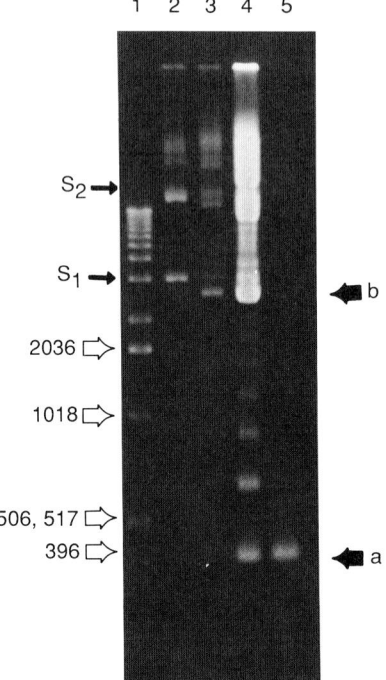

FIG. 2   DNA profiles of plasmid pSA508 before and after Int/IHF-mediated *in vivo* recombination displayed on a 1.5% agarose gel. Lane 1 contains a 1-kb ladder (BRL, Gaithersburg, MD); lane 2, plasmid DNA before recombination; lanes 3 and 4, DNA profile of recombination products [lane 4 was loaded with 5-fold more DNA than lane 3 to make the minicircle product (arrow a) clearly visible]; lane 5, monomer DNA minicircle after purification. Arrows denote the following: a, monomer minicircles; b, larger product circles; S1, monomer substrate; S2, dimer substrate. Numbers at left are sizes of marker DNA fragments (in bp).

The monomer minicircles are separated from other proteins in a 1% agarose gel in TAE (40 mM Tris–acetate, 1 mM EDTA, pH 8.0) and extracted by electroelution in dialysis tubing (11). Briefly, the agarose block containing minicircles is placed in $\frac{1}{4}$-inch dialysis tubing. The agarose block should not be too big, so that it can move easily inside of the tubing. The dialysis tubing is completely filled with TAE buffer and placed in a horizontal electrophoresis chamber. TAE buffer is added to the chamber to barely cover the dialysis tubing. Electrophoresis is carried out at an appropriate voltage (5 V/cm between the two electrodes) for 2 hr. Just before harvesting the eluent, the current is reversed and run at 100 V for 30 sec. NaCl is added to the eluent to a final concentration of 0.75 M to process the eluent through a Qiagen tip (tip 20)

(Chatsworth, CA). The minicircles are precipitated with 2-propanol and redissolved in an appropriate volume of 10 mM Tris–acetate (pH 7.8), 0.1 mM EDTA. Note that we have found in our laboratory that protocols using phenol fail to yield DNA of sufficient quality for transcription assays.

In Vitro Transcription

The transcription can be done in a single round by quenching unbound RNA polymerase in the reaction with heparin or in a multiple-round fashion as described below. The protocol described for minicircles can be used on the other templates described in the chapter. Every measure should be taken to prevent contamination by RNase.

The basic reaction mix contains the following in a 50-μl volume: 20 mM Tris–acetate (pH 8.0), 10 mM magnesium acetate, 100 mM potassium glutamate, 1 mM dithiothreitol (DTT), 1 mM ATP, 0.2 mM each of GTP and CTP, 10 μM UTP, 10 μCi [α-^{32}P]UTP (800 Ci/mmol), 2 nM DNA template, 20 nM *E. coli* RNA polymerase holoenzyme, and 5% (v/v) glycerol (6, 13). Note that the optimal concentration of magnesium and potassium salts should be titrated for each promoter. For example, 3 mM magnesium acetate and 200 mM potassium glutamate are the optimal concentrations for the *lac* promoter, whereas 10 mM magnesium acetate and 100 mM potassium glutamate are optimal for the *gal* promoter. All components except nucleotides are incubated at 37°C for 10 min. Transcriptions are started by the addition of nucleotides and terminated after 10 min by addition of 50 μl formamide loading buffer [80% (v/v) formamide, 1× TBE (89 mM Tris borate, 89 mM boric acid, 2 mM EDTA), 0.05% (w/v) bromphenol blue, 0.05% (w/v) xylene cyanol]. RNA is resolved directly by electrophoresis on an 8.0% (w/v) polyacrylamide/8 M urea sequencing gel. Transcript sizes between 80 and 150 nucleotides resolve well in such a gel. We have found that phenol extraction and ethanol precipitation, followed by dissolving in water and loading buffer, may be necessary when analyzing transcripts longer than 200 nucleotides or if the background of the gel is high. To examine abortive initiation products, a 25% polyacrylamide/8 M urea gel is used. Gels are usually dried, and the transcripts are visualized by autoradiography or measured with a β-scanner for quantitation. For obtaining an autoradiogram of a 25% gel, the gel is exposed to X-ray film directly as a wet gel, as it is difficult to transfer a 25% gel to filter paper. The number of [^{32}P]UMPs incorporated into the transcript should be considered when comparing the different length transcripts.

Figure 3 shows transcripts from the promoter-free minicircles derived from the vector, pSA508, and transcripts initiated from *E. coli gal* promoters in 8 and 25% polyacrylamide gels. Transcription reactions were carried out in the presence of various concentrations of cAMP and excess cAMP receptor protein (CRP; 100 nM) to demonstrate that the cAMP–CRP complex modulates the *gal* promoters, P1 and P2, inversely. Figure 3 shows clearly that the vector minicircles (pSA508) are transcriptionally sterile; no full-length and very few short transcripts were seen in 8% or 25%

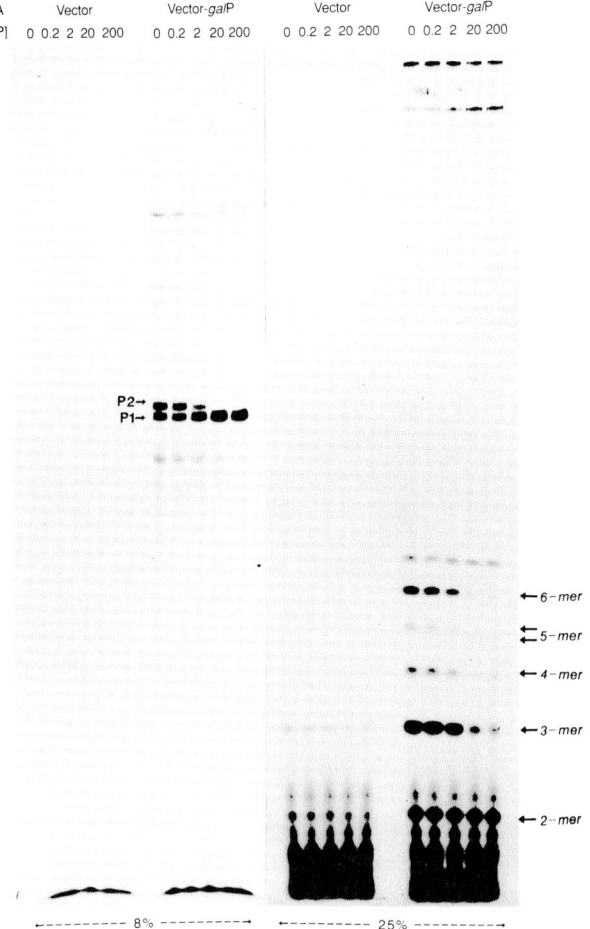

FIG. 3   Effect of cAMP concentrations on transcription using the vector minicircle (from pSA508) or *gal* minicircle (Vector–*galP*). The *E. coli gal* operon is driven by two independent promoters, *P1* and *P2,* separated by 5 bp (6). In this reaction, cAMP receptor protein (CRP) was present in excess (100 n*M*). Transcription products were analyzed on 8% (left) and 25% (right) urea/polyacrylamide gels.

polyacrylamide gels. The DNA minicircles containing the *gal* promoter segment, on the other hand, yielded two distinct full-length transcripts in the absence of cAMP: a 120-nucleotide RNA from the *galP1* promoter and a 125-nucleotide RNA from the *galP2* promoter are visible when the 8% gel is used. The *gal*-specific RNA bands constituted more than 90% of the total transcripts. The 54-bp stretch of DNA containing the stem–loop structure of the *rpoC* terminator was extremely effective in terminating transcription initiated from either *gal* promoter.

Electrophoresis of the transcription products on a 25% polyacrylamide gel showed the presence of large amounts of oligomeric transcripts from the *gal* promoters (Fig. 3). The di-, tri-, and hexamers, as well as the minor tetramers, were inversely proportional to the concentration of cAMP, suggesting that they originated from *galP2*. It appears that more than 90% of the transcription initiated from *galP2* was aborted.

Advantages of using the minicircle as template in *in vitro* transcription studies are not limited to studying promoter clearance. The minicircles can allow one to study *in vitro* promoters that are intrinsically weak and supercoiled DNA-dependent. Use of plasmid DNA is often unsuccessful with weak promoters that fail to compete for RNA polymerase in the presence of other promoters present in plasmid. In such cases, the minicircles are the best way to study the promoters *in vitro*.

Catenated Templates

Catenanes can be used to test whether one transcription factor can act in trans when the factor binding site and the promoter can be cloned on each circle of DNA separately (13, 14). Catenated DNA can be prepared *in vitro* using the site-specific recombination system either from *E. coli* bacteriophage λ (15) or from the transposon Tn*3* resolvase (14). We describe here the method using the λ recombination system which has been used in our laboratory to test whether the *E. coli* transcription activator CRP could act in trans on the *lac* promoter (13). The vector is derived from pBluescript (Stratagene, La Jolla, CA) by cloning the transcription terminator from the *rpoC* gene with a multiple cloning site upstream between the λ phage attachment site *attP′OP* and the corresponding bacterial site *attB′OB*. The *lac* promoter is cloned between *attP′OP* and *attB′OB*, and the CRP binding site is cloned outside (Fig. 4A). Catenanes are made *in vitro* by incubating the substrate DNA in a reaction mixture containing 44.2 mM Tris-HCl (pH 7.5), 4.2 mM EDTA, 67.5 mM KCl, 6.25 mM spermidine, 5 μM DNA, 25 μM integrative host factor (IHF), and 130 μM Int for 50 min at 25°C as described by Mizuuchi *et al.* (15). After one phenol extraction, catenanes are separated by gel electrophoresis and recovered by electroelution. We use a preparative 1.2% agarose gel in TAE run at 2 V/cm overnight. Species of catenanes differing in linking number (16) migrate as a diffuse band, faster than the substrate DNA (Fig. 4B). The catenanes are purified from the agarose as described above.

When catenated DNA molecules are used as templates for *in vitro* transcription, conditions are the same as described above for minicircles. Note that when whole

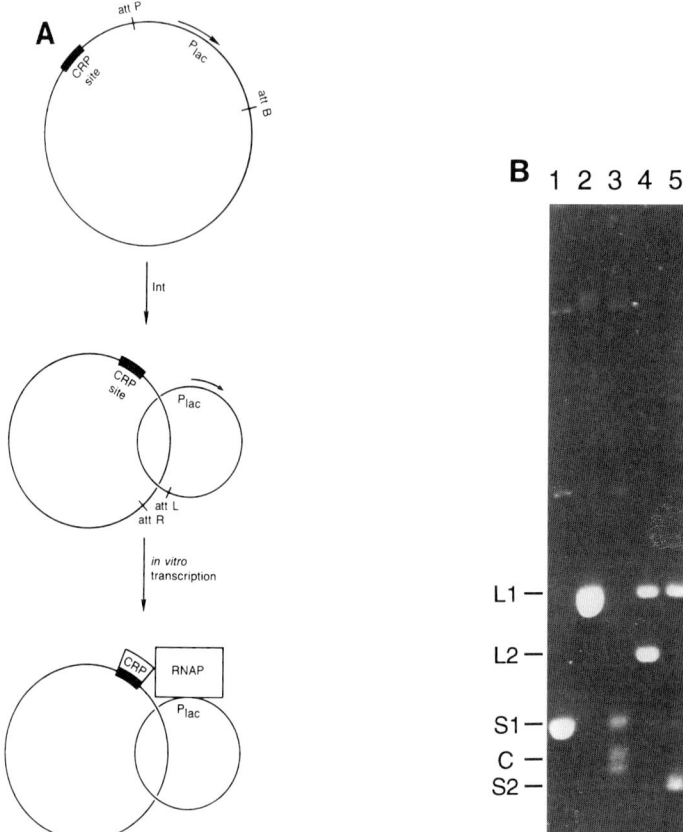

FIG. 4   (A) Scheme to make catenated DNA circles using the recombination system of bacteriophage λ. *attP* is *P′OP*, *attB* is *B′OB*, *attL* is *P′OB*, and *attR* is *B′OP*. (B) Products of the recombination. The samples were run on a 1.2% agarose gel as described in the text. L1 denotes linear substrate DNA, cut with *Hin*dIII; L2, linear large-circle DNA formed after recombination, cut with *Hin*dIII; S1, supercoiled substrate DNA; S2, supercoiled large-circle DNA formed after recombination; C, catenanes. Lane 1 contains supercoiled substrate DNA; lane 2, linear substrate DNA cut with *Hin*dIII; lane 3, catenanes made after recombination; lane 4, DNA cut at the *Hin*dIII site which is located in the large circle, after the recombination reaction (this shows that recombination was about 70% efficient; the 500-bp supercoiled small circle which ran out of the gel is not shown); lane 5, DNA cut with *Pst*I located in the small circle after recombination reaction (the 500-bp linear small-circle DNA which ran out of the gel is also not shown).

plasmids are used, the *rep* RNAs (106 and 108 bases long), transcribed from the plasmid origin, serve as useful internal controls (13).

Nicked and Gapped Templates

There may be some situations where it would be useful to introduce specific structural changes in a DNA template. The introduction of a nick or a single-stranded gap in a template has been used to evaluate interaction between transcription factors (17) and between a transcription factor and RNA polymerase (13).

DNA with Single-Stranded Nicks

The principle of preparing DNA with single-stranded nicks is shown in Fig. 5A using the *lac* promoter as an example. A plasmid is constructed by creating a blunt-end

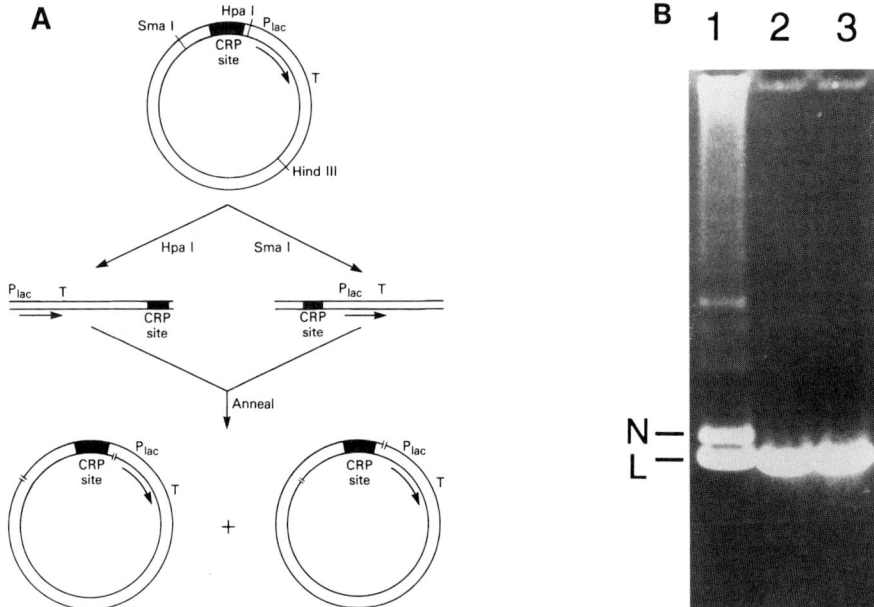

Fig. 5   (A) Schematic representation of making DNA templates with single-stranded nicks at positions shown. (B) The 1.2% agarose gel showing nicked DNA and its parental linear DNA. N denotes nicked circular DNA; L, linear DNA. Lane 1 contains DNA with single-stranded nicks produced as described in the text (the remaining mixture of linear DNAs and high molecular weight aggregates are shown); lane 2, linear DNA cut with *Sma*I; lane 3, linear DNA cut with *Hind*III.

restriction site in a specific location by site-directed mutagenesis. The single-stranded nicked DNA is prepared by denaturing and annealing a mixture of equal amounts of two different portions of plasmid, one cut at the specific location and the other cut elsewhere in the plasmid. About 15 nM of each DNA is mixed in TES buffer (10 mM Tris-HCl, pH 8.0, 1 mM EDTA, 100 mM NaCl). The mixture is boiled for 3 min and allowed to cool to room temperature slowly over 2 hr. (This can be done easily by boiling a sample in a heating block and then turning it off.) Theoretically, the resulting DNA should contain an equal amount of the two different starting linear DNAs and single-stranded nicked DNA, but the mixture contains high molecular weight DNA aggregates, probably because of entanglement of DNA during the annealing step (Fig. 5B). We have found that the formation of aggregates depends on the sequence of the plasmid and on the restriction sites used. To reduce aggregation, it is important to determine empirically the highest DNA concentration to make the most single-stranded nicked DNA for each different plasmid, by doing a small-scale annealing experiment with various amounts of DNA. The conformation of the single-stranded nicked DNA formed is a relaxed circular DNA with two single-strand nicks which moves more slowly than linear DNA in an agarose gel (Fig. 5B). The recovery yield of a desired form of DNA is usually less than 10%.

Homogeneous Nicked DNA

The single-stranded nicked DNAs obtained by the above methods are mixtures containing nicks in either strand. Homogeneous nicked DNA, that is, DNA containing a nick in one specific strand, can be made as described by Hochschild and Ptashne (17). Circular single-stranded DNA can be obtained by using f1 helper phage if the plasmid contains the f1 origin of replication. Either one of the two strands from the parental plasmid can be produced, depending on the orientation of the f1 origin sequence. This single-stranded DNA is mixed with DNA that has been linearized by cutting at a site where the nick is desired (molar ratio of 3 : 1 of single-stranded DNA to double-stranded DNA), denatured, and annealed. The homogeneous nicked DNA migrates more slowly than the linear DNA and is purified from an agarose gel as described above.

DNA with Single-Stranded Gaps

DNA with single-stranded gaps can be made by the same approaches as described for DNA with the single-stranded nicks (see above) and has been used successfully in our laboratory (13). A blunt-end restriction site is introduced in the area of interest, and corresponding plasmids with extra bases in the center of the site are made by site-directed mutagenesis. The restriction site is cut with the restriction enzyme, and the partner plasmids with extra bases are cut with a second enzyme elsewhere. Equal amounts of the cut plasmid pairs are mixed, denatured, and renatured as described above. The gapped DNA circles migrate more slowly than linear DNA in an agarose gel and can be purified as described above.

Acknowledgments

We thank Drs. Ding Jin and Darren Sledjeski for critical reading of the manuscript and Dr. Sankar Adhya for discussions and support.

References

1. M. J. Chamberlin, *Annu. Rev. Biochem.* **43,** 771 (1974).
2. W. R. McClure, *Annu. Rev. Genet.* **54,** 171 (1985).
3. H. Kramer, M. Amouyal, A. Nordheim, A. Nordheim, and B. Muller-Hill, *EMBO J.* **7,** 547 (1988).
4. A. L. Meicklejohn and J. D. Gralla, *J. Mol. Biol.* **207,** 661 (1989).
5. S. Hahn, W. Hendrickson, and R. Schleif, *J. Mol. Biol.* **188,** 355 (1986).
6. H. Choy and S. Adhya, *Proc. Natl. Acad. Sci. U.S.A.* **90,** 472 (1993).
7. M. Mizuuchi and K. Mizuuchi, *Nucleic Acids Res.* **13,** 1193 (1985).
8. C. Squires, A. Krainer, G. Barry, W.-F. Shen, and C. L. Squires, *Nucleic Acids Res.* **9,** 16827 (1981).
9. D. L. Hsu, W. Ross, and A. Landy, *Nature (London)* **285,** 85 (1980).
10. G. Guaneros and H. Echols, *Virology* **52,** 30 (1973).
11. F. Ausubel, R. Brent, R. Kingston, D. Moore, J. Seidman, J. Smith, and K. Struhl, "Current Protocols in Molecular Biology." Greene and Wiley, New York, 1989.
12. H. Echols, *J. Mol. Biol.* **47,** 575 (1970).
13. S. Ryu, S. Garges, and S. Adhya, *Proc. Natl. Acad. Sci. U.S.A.* **91,** 8582 (1994).
14. A. Wedel, D. S. Weiss, D. Popham, P. Droge, and S. Kustu, *Science* **248,** 486 (1990).
15. K. Mizuuchi, M. Gellert, R. A. Weisberg, and H. A. Nash, *J. Mol. Biol.* **141,** 485 (1980).
16. J. Bliska and N. R. Cozzarelli, *J. Mol. Biol.* **194,** 205 (1987).
17. A. Hochschild and M. Ptashne, *Cell (Cambridge, Mass.)* **44,** 681 (1986).

[22] Protein–Protein Interactions of DNA-Binding Proteins: Studies on Replication Initiator Protein, RepA, of Plasmid P1

Gauranga Mukhopadhyay, Justin A. Dibbens, and
Dhruba K. Chattoraj

Introduction

Site-specific DNA-binding proteins often engage in protein–protein interactions which are crucial to the function of the protein. Such interactions can occur between identical or functionally related proteins and often influence the DNA binding process itself. Protein–protein interactions are likely to occur when multiple DNA binding sites for the same protein or functionally related proteins are present in the region of interest: origins and terminators of replication (1, 2), centromere-like sites for plasmid partition (3), promoters and enhancers of transcription (1, 4), and sites of recombination (1). DNA-binding proteins also interact with proteins that do not interact with DNA themselves (5). Because protein–protein interactions can alter the affinity of DNA–protein interactions (cooperative DNA binding), initial evidence for the presence of protein–protein interactions may be obtained from measurements of DNA–protein affinity. In this chapter we describe such approaches to the study of protein–protein interactions involving the replication initiator protein, RepA, of plasmid P1.

The M_r of monomeric RepA is 32,000, and the protein exists in solution in monomer–dimer equilibrium, with a K_D of about 2 μM at physiological salt concentrations at which most DNA-binding experiments are performed (6). RepA purified from overproducing strains is largely (>99%) inactive for DNA binding. The equilibrium constant mentioned above refers to the dimerization of inactive monomers. A third species, active monomers, can bind DNA specifically. The presence of a huge excess of inactive species, that can bind DNA nonspecifically, complicates *in vitro* studies of RepA binding considerably. Significantly increased specific binding can be obtained by activating the inactive species with *Escherichia coli* chaperones, DnaJ and DnaK, in the presence of ATP (6). An increase in active fraction reduces nonspecific binding indirectly, and the background can be further reduced by adding nonspecific DNA [calf thymus DNA, or poly(dI·dC)]. An appropriate excess, determined empirically, of unlabeled nonspecific DNA over labeled specific (probe) DNA titrates most of the inactive protein and thereby reduces interference in the study of specific binding to the probe. A second source of complexity may come from partially active

Methods in Molecular Genetics, Volume 6

species that may not bind DNA specifically at low concentrations but may do so at somewhat higher DNA concentrations. These nonidealities are common to DNA-binding proteins, but the sheer magnitude of the proportion of inactive species has made activation of RepA binding a discipline in its own right (6).

RepA has 14 binding sites on plasmid P1 (7). The sites are each 19 bp long and nearly identical in sequence. For this reason they have been called iterons. They are present in two clusters. One cluster of five iterons is present in the origin. The requirement for multiple iterons most likely reflects the formation of a higher order structure for initiation (8). The structure may require cooperative DNA–protein and protein–protein interactions and participation of host factors such as the host initiator protein, DnaA, for which there are also multiple sites present at the plasmid origin (8). The cluster of nine iterons is involved in control of replication. This involves pairing with the cluster of five iterons via interactions between RepA molecules bound to the two clusters. This reaction we call DNA pairing. Alternate names include DNA coupling, handcuffing, and formation of sandwich structures. We have studied interactions between adjacent RepA–iteron complexes as may occur in the origin, or between complexes on separate regions of plasmid DNA that control DNA replication (action at a distance). In the latter case, the complexes can be present at sites hundreds of base pairs (bp) away on the same molecule (in cis) or on separate DNA molecules (in trans). The cis interactions considered here include the heterologous interaction between RepA–iteron and DnaA–DNA complexes.

Analysis of Cooperative DNA Binding

When a protein binds to DNA that has multiple sites, binding to one site can change the affinity of binding to one or more of the remaining sites. The affinity can either increase (positive cooperativity) or decrease (negative cooperativity). Evidence for cooperativity strongly suggests direct protein–protein contact. In principle, cooperativity could be mediated by the intervening DNA separating the sites without protein contacts by as yet unknown mechanisms. Similarly, absence of cooperativity does not constitute a proof for lack of protein–protein contact. Here, we discuss measurement of cooperativity for the simplest case of RepA binding to two iterons in cis.

Preparation of DNA Probe

For determination of the affinity of DNA–protein interactions, the concentrations of DNA and protein need to be known. The DNA concentration is determined by UV absorbance. However, the UV absorbance spectrum of DNA fragments, when gel purified, may deviate from that expected of native double-stranded DNA, indicating the presence of significant impurities. To circumvent the problem of UV-absorbing or other impurities that might interfere with protein-binding studies, we separate the fragments from the vector by chromatography on Mono Q HR 5/5 anion-exchange column [Pharmacia, Piscataway, NJ, FPLC (fast protein liquid chromatography) system].

About 400 μg Cs-banded plasmid DNA is digested to completion with restriction enzymes in a volume of 300 μl, and the protein is removed with a phenol–chloroform–isoamyl alcohol (1 : 0.94 : 0.04 v/v) mixture, hereafter called phenol–chloroform. To the aqueous layer, ammonium acetate is added to 2.5 M and the DNA precipitated with 3 volumes of ethanol. The precipitate is washed with 2 volumes of ice-cold 70% (v/v) ethanol (without dislodging the pellet from the tube wall) and finally resuspended in 100 μl TE (10 mM Tris-HCl, pH 8, 1 mM EDTA). The entire sample is loaded onto a 1-ml Mono Q column, and the fragments are eluted using a NaCl gradient. The fractions carrying the fragments are detected by the UV monitor of the FPLC system and are confirmed by polyacrylamide gel electrophoresis (PAGE). The salt concentrations at which the different size fragments elute can be initially determined by using commercial DNA (we have used MspI-digested pBR322 DNA). In this way the collection of fractions and their assay by PAGE can be minimized considerably. The peak and two flanking fractions are pooled (total volume 1.5 ml) and the DNA allowed to precipitate with 3 volumes of ethanol overnight at $-20°$C in a 15-ml Corex tube and centrifuged in a JS 13.1 rotor (Beckman Instruments, Fullerton, CA) at 10,000 rpm (16,000 g) for 20 min at 4°C. The pellet is gently layered with 6 ml ice-cold 70% (v/v) ethanol and the tube is spun again for 5 min as above. The supernatant is removed by aspiration. The pellet is dried under vacuum for 5 min, resuspended in 100 μl TE, and the UV absorbance spectrum measured using a microcuvette (Pharmacia). Recovery of the fragment is routinely about 20%. The fragment is stored at $-20°$C.

Labeling DNA 3' Ends with Klenow Polymerase

Usually 100 ng of the fragment, enough for several experiments, is labeled with [α-^{32}P]deoxynucleoside triphosphate [2 μl (20 μCi) of 3000 Ci/mmol dNTP in a total reaction volume of 50 μl] using reagents from a Multiprime labeling kit (Amersham, Arlington Heights, IL). The reaction is terminated by adding 2 μl of a 0.5 M EDTA solution, and two aliquots (1 and 2 μl) are removed for trichloroacetic acid (TCA) precipitation before the sample is subjected to any further manipulations. (Similar aliquots are also removed from a control tube that has identical reagents but no DNA fragment. These are used to determine the level of unincorporated radioactivity in DNA samples after TCA precipitation, as described below.) The remainder of the reaction mixture (49 μl) is treated with phenol–chloroform once, and the aqueous layer is passed twice through Sephadex G-25 spin columns, preequilibrated in TE (5 Prime–3 Prime, Boulder, CO), to remove all small molecular weight impurities including traces of phenol–chloroform. Two aliquots (1 and 2 μl) are removed from the final eluent for TCA precipitation.

Precipitation with Trichloroacetic Acid

Aliquots for TCA precipitation are transferred directly to 50-μl drops of water previously distributed to 10-ml disposable glass tubes and stored on ice. When all aliquots have been transferred, the tubes are then processed together. One milliliter of an ice-cold solution of 20% (w/v) TCA and 20 mM sodium pyrophosphate (hereafter called TCA solution) is added to each tube. The mixtures are incubated for at least 15 min on ice and filtered under gravity through GF/C filters (Whatman, Clifton, NJ) placed on a 10-place filtration manifold (Hoefer, San Francisco, CA). It is important that the filters be presoaked briefly in the TCA solution and be wet when the samples are applied to the filter. Each tube is rinsed with 3 ml TCA solution, which is then passed through the filter. The process is repeated once more. The filter is then washed twice with 5 ml TCA (total amount of TCA applied to the filter is 17 ml). Finally, each filter is washed two times with 5 ml of cold ethanol, dried under a heating lamp, and counted using liquid scintillation solution.

The washing steps should reduce the counts per minute (cpm) to background levels in control filters without DNA ($\leq$100 cpm), whereas the filters containing DNA should have at least 10^5 cpm. The efficacy of washing can also be made evident by comparison of the radioactivity of the pairs of filters with 1- and 2-μl aliquots. To determine whether the washing has been excessive and has removed some DNA in addition to free radioactivity from the filters, an aliquot of the Sephadex G-25-purified DNA is applied to a dry filter, dried, and counted directly, and a second aliquot is precipitated with TCA as before and counted. Comparison of the radioactivity of the two filters determines if there has been any significant loss of DNA during washing, and the amount of loss, if any, can then be included in calculation of DNA concentration. The first two filters containing aliquots of the labeling mixture give the relationship of DNA concentration and cpm (specific activity), and this value is used to calculate DNA concentration from cpm in the final Sephadex G-25 eluent.

DNA Probe Preparation Directly from Plasmid DNA

A simpler way to make DNA probe that bypasses the FPLC fragment purification steps is to cleave a known amount of plasmid DNA (Cs-banded and concentration determined by UV absorbance) at one end of the region of interest. The linearized DNA is then end-labeled in the same reaction tube without purification, by supplementing the restriction digestion mixture with nucleotides (radiolabeled and unlabeled) and Klenow polymerase. The reaction is terminated by adding EDTA. Aliquots are removed for TCA precipitation, and the reaction volume is recorded for determination of specific activity. The DNA is purified by phenol–chloroform and Sephadex G-25 columns, then cleaved again to separate the region of interest from the vector. The two fragments can then be separated in a gel and their relative specific activities determined by autoradiography. The fragment is gel-purified and counted, and the DNA concentration is determined from the previously determined specific activity of the linearized plasmid DNA. A correction factor of 2 or that actually determined from the autoradiograph is introduced because only one end of the fragment is radioactive. The limitation of this simplified method is that the fragment can be irreversibly contaminated with unknown impurities from the gel matrix.

Determination of Active Protein Fractions

Like DNA concentration, protein concentration can also be determined by UV absorbance at 280 nm, and this requires a highly purified protein preparation. This by itself is not enough, however, as the extinction coefficient of proteins, unlike DNA, varies significantly from protein to protein depending on the precise residue composition. To convert absorbance units to molar concentration, an aliquot of the protein solution of known absorbance is subjected to amino acid analysis. Once the molar extinction coefficient is known, absorbance measurements suffice to determine concentrations of future preparations of the protein. The future preparations need not be highly pure: concentrations in these cases are determined not by absorbance but by comparing band intensities in sodium dodecyl sulfate (SDS)-PAGE using the highly pure preparation as standard. These procedures determine the concentration of total protein, not that of the species active in DNA binding. Accurate affinity measurements require knowledge of the active fraction. The value is simple to determine with a DNA probe of known concentration (see Preparation of DNA probe) and known binding stoichiometry (moles of protein bound per mole of DNA sites at saturation of binding). Determination of the latter value requires preparation of radioactive protein of known specific activity and is technically difficult to achieve accurately (9, 10). The value is often assumed from indirect evidence. Conclusions regarding the nature of cooperativity and several other characteristics of the binding reaction do not depend on the assumed value.

The active fraction is determined by titrating a fixed concentration of protein with varying concentrations of DNA (Fig. 1A). The binding is followed by band-shift assays (see Experimental Methods). In the example shown in Fig. 1A, each 20-μl reaction mix was composed of 10 μl RepA solution, 1 μl probe DNA (10 pM, enough to be able to follow binding after overnight exposure), and varying amounts of cold DNA (0.5 to 12 nM) in 9 μl. Stock solutions of the components were diluted in 1 $\times$ binding buffer before mixing. In the case of RepA, the results were similar when unlabeled fragment DNA was replaced with equimolar amounts of supercoiled plasmid DNA. Therefore, we used supercoiled DNA to spare the more expensive fragment DNA. The liberty to use plasmid DNA directly encouraged us to extend the titration curves up to 160 nM in site concentration. (This is equivalent to about 5 μg

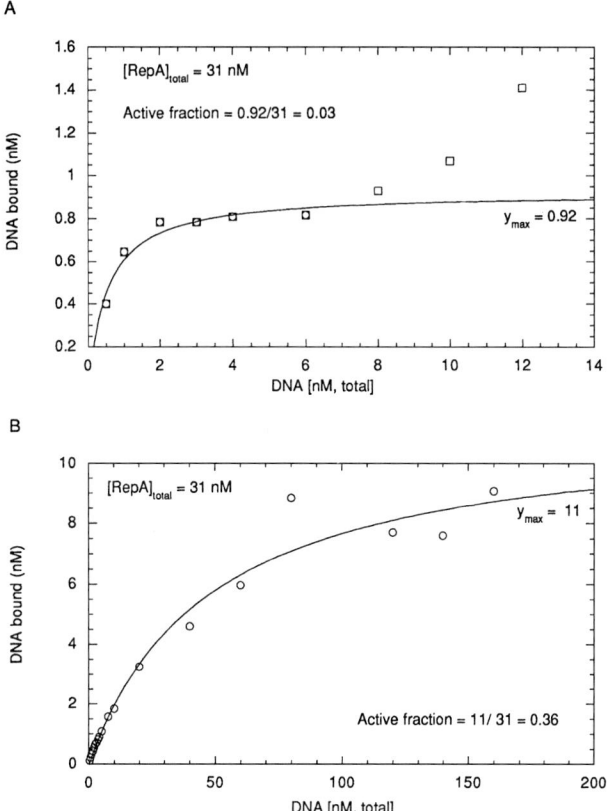

FIG. 1   Titration of RepA with DNA to determine the fraction of RepA active in DNA bind-ing. (A) Note that the titration curve is biphasic. From the initial points, the maximal concen-tration of DNA present as RepA–DNA complexes (y_{max}) is calculated to be 0.92 nM by fitting the points to the equation $y = y_{max}x/(y_{1/2} + x)$ (6). Assuming 1:1 binding stoichiometry (one RepA monomer–one iteron), the concentration of bound RepA is also taken as 0.92 nM out of a total RepA concentration of 31 nM. The active fraction (equaling bound/total) of 0.03 means that only 3% of the protein is active in DNA binding. Note that at the end of the titration curve, the concentration of bound DNA starts to rise again and can reach a new plateau at about 11 nM, equivalent to an active fraction of 0.36 (B). The reason for this biphasic titration curve is not clear, but a simple explanation could be that the proteins which show binding at high DNA concentrations are partially active in DNA binding. The second phase of the titration curve is not relevant, as in most experiments only picomole concentrations of specific DNA have been used.

of 2.7-kb plasmid DNA per 20 μl of reaction mixture. In fact, instead of using 5 μg of plasmid DNA with one cloned iteron, we used 1 μg of plasmid DNA carrying five cloned iterons to avoid any problem of overloading the gel.)

The active fraction of the DNA probe was not determined separately, as it is usually close to 1. The exact value can be determined from a binding curve (Fig. 2A), where a fixed concentration of DNA probe is titrated with increasing protein concentrations. The active fraction is the ratio of the probe bound at infinite protein concentration (y_{max}) and the input probe. y_{max} is determined by curve-fitting using the equation given in the legend to Fig. 1. Note that the active fractions do not approach 1 in the examples of Fig. 2.

Calculation of Cooperativity Parameter, ω

The change in affinity of a protein (P) for its site (O) arising from the influence of a second protein bound to a second site is formally represented by the cooperativity parameter, ω. This is the ratio of the dissociation constants for the binding of the protein to its site, in the absence and the presence of the second bound protein. Binding of the two proteins to a DNA fragment with two sites ($O_I O_{II}$) is described by the following overall and four partial reactions, each with its own dissociation constant:

$$P_I + P_{II} + O_I O_{II} \stackrel{K_D}{\rightleftharpoons} P_I P_{II}$$

$$P_I + O_I O_{II} \stackrel{K_{D1}}{\rightleftharpoons} P_I O_{II}$$

$$P_I + O_I P_{II} \stackrel{K_{D2}}{\rightleftharpoons} P_I P_{II}$$

$$P_{II} + O_I O_{II} \stackrel{K_{D3}}{\rightleftharpoons} O_I P_{II}$$

$$P_{II} + P_I O_{II} \stackrel{K_{D4}}{\rightleftharpoons} P_I P_{II}$$

where $P_I O_{II}$, $O_I P_{II}$, and $P_I P_{II}$ are DNA–protein complexes with sites I, II, and both occupied, respectively. The value of ω for site I is K_{D1}/K_{D2} and for site II, K_{D3}/K_{D4}. The analysis is the same when the two proteins are identical.

The affinities are derived from binding isotherms (Fig. 2). Traditionally, cooperativity is assessed by transforming binding isotherms to Scatchard or Hill plots. However, when the same protein is involved in binding to the two sites, it is possible to derive the K_D values from the binding isotherms directly. First K_{D1} is determined from a binding isotherm, where the DNA carries site I only, by fitting the data to the equation

$$f_{bound} = 1/(K_{D1}/[P] + 1)$$

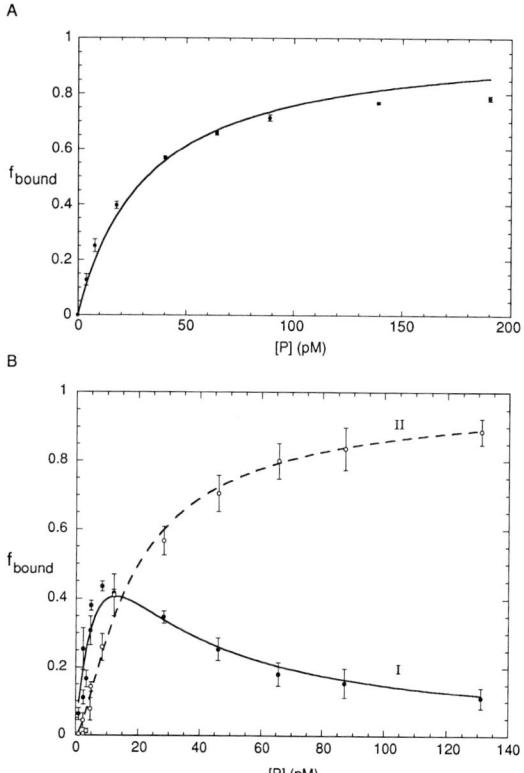

FIG. 2   Binding isotherms for RepA–DNA interactions where the fragments contain (A) one iteron or (B) two iterons. The fraction of DNA complexed with protein(s) (f_{bound}) was determined by the band-shift assay. The concentration of free RepA, [P], equals total active RepA − the bound RepA. The bound protein concentration was calculated from f_{bound} assuming 1 : 1 binding stoichiometry. The data points were fitted to equations as described in the text, using the Kalidagraph software. (B) A value of ω greater than one was calculated to be 1.9, indicating that RepA–iteron interactions under the conditions of our experiments are positively cooperative, but weakly so.

where [P] represents free protein concentration (Fig. 2A). Constant K_{D2} is determined from the binding isotherms with two sites by fitting the data for the first and second bound species to the equations

$$f_I = 1/(1/\{A[P]\} + [P]/\{AK_D\} + 1)$$

$$f_{II} = 1/(K_D/[P]^2 + AK_D/[P] + 1)$$

where $A = 1/K_{D1} + 1/K_{D3}$, and $K_D = K_{D1}K_{D4} = K_{D2}K_{D3}$ (Fig. 2B). Substitution of K_{D1} derived from the binding isotherm for a single site (Fig. 2A) allows calculation of the remaining dissociation constants, and hence the cooperativity parameters.

The extent of cooperativity can be also assessed qualitatively from the maximum fraction (f_{max}) of the first bound species (Fig. 2B). A f_{max} value less than 0.5 is indicative of positive cooperativity, whereas a value greater than this indicates negative cooperativity. The magnitude of deviation from 0.5 is a measure of the extent of cooperativity. In fact, ω can be calculated directly from f_{max} in a system where the same protein binds to two identical sites using the equation $\omega = (1/f_{max} - 1)^2$ (11). This method has the advantage that knowledge of the probe concentration and the active fraction of the protein is not required.

Experimental Methods

Gel Electrophoretic Mobility Shift Assay

Gel electrophoresis has been widely used to detect the formation of DNA–protein complexes because of the ease of performance and because of the ability to resolve different molecular species. The gel electrophoretic mobility shift assay, simply called the band-shift assay, has been described in this series earlier (12). Here, we only illustrate how the assay has revealed many of the complex properties of RepA discussed earlier: low active fraction, high nonspecific binding of the inactive species, chaperone-mediated activation of the inactive species and consequent reduction of nonspecific binding, and apparent increase of active fraction possibly because of binding of partially active species at high iteron concentration (Fig. 1B). The results of Fig. 3 indicate that detection of DNA pairs by the band-shift assay may not be straightforward. The assay may not record action at a distance if the interactions are not strong enough to survive gel electrophoretic conditions, as may be the case with RepA and GalR (13), although successes have been reported in other systems: EBNA1 (14), LacI (15), and AraC (16). The method, however, is excellent for detecting cooperative interactions between adjacent sites (17) and for heterologous interactions (8).

Ligation Kinetics

For detection of DNA pairing, a gentler alternative to the band-shift assay is to use DNA ligase in the binding reaction (18). The basis of the method is that, by pairing DNA, a protein can increase the local concentration of DNA ends and, thereby, can increase the rate of intermolecular ligation, if ligase is included in the reaction. Usually, a ladder of multimers is seen because of the reversibility of the pairing reaction. If it were irreversible, a dimeric length dead-end product would be expected. It is likely that the ladder is formed because the protein–DNA or protein–protein contacts are made and broken spontaneously during the time course of the ligation experiments. The method is quantitative enough that mutants modestly reduced in the pairing activity can be identified as such (18).

The fragment length is an important consideration for the success of the assay. For RepA, we have varied lengths between 220 and 400 bp, with the iterons centrally located in the fragment. This configuration may be important for two reasons: in case the pairing is antiparallel, asymmetric location of the sites will locate the ends away from one another in paired complexes which could reduce, rather than increase, the rate of ligation. Second, protein binding too close to the end could cause steric hindrance to the ligation reaction. For this reason, we have allowed at least 100 bp of flanking DNA. If the flanking DNA is too long, then the enhancement of ligation caused by pairing will be lost owing to increased flexibility of the DNA arms [the same reason DNA cyclization shows length dependence (19)]. A fragment size around 300 bp is also convenient for resolution of the multimers in an agarose gel. Use of blunt-ended DNA fragments reduces the rate of ligation, so that sampling can be done every 5 to 10 min.

When comparing pairing efficiencies between wild-type and mutant proteins, it is important that the fraction of protein-bound DNA used is similar in all cases. We have performed our ligation reactions under conditions where the RepA-bound fraction of probe DNA was around 50%. When the DNA concentration is low, the frequency of bimolecular association reactions can be greatly increased by addition of molecular crowding agents such as polyvinyl alcohol (up to 4%, w/v), polyethylene glycol (up to 15%, w/v), or spermidine (up to 10 mM). To determine the extent of ligation stimulation caused by nonspecific binding, a similar length fragment but without the specific binding site should be used as a control. In the case of RepA, nonspecific binding is reduced by the activation of the protein with chaperone proteins. In systems where nonspecific binding is not reduced by chaperones, covalently closed circular molecules can be used to reduce nonspecific DNA binding to the probe without reducing ligase activity significantly. Some loss of ligase activity is unavoidable because of nonspecific DNA binding activity of ligase. Reduction of nonspecific binding by increased salt concentration is not recommended, as ligase activity itself is salt sensitive. Moreover, the protein–protein interactions required for the pairing reaction may also be sensitive to salt concentrations. Finally, although the technique has been used primarily to determine interactions in trans, the method can also be used to determine interactions between appropriately placed sites in cis (20). In that case, molecular crowding agents are not used, and the circular products of ligation are separated from linear fragments in gels containing ethidium bromide or in two-dimensional gels (the linear fragments run in a diagonal and the circles above the diagonal).

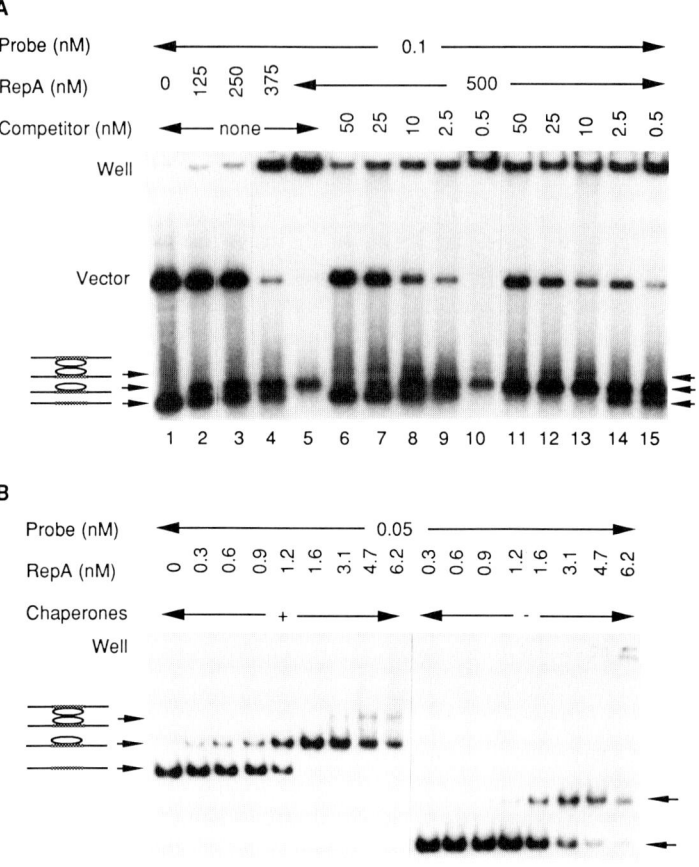

FIG. 3   Detection of bimolecular and higher order complexes of RepA and iteron DNA.
(A) Agarose gel electrophoresis. Binding reactions were performed at room temperature for
10 min in 20-μl volumes in a buffer containing 20 mM Tris-HCl, pH 8, 40 mM KCl, 60 mM
NaCl, 10 mM MgCl$_2$, 1 mM DTT, 0.1 mM EDTA, and 7.5 μg/ml BSA. Before loading onto
the gel, the mixture was supplemented with 4 μl of 30% glycerol (v/v) and 0.3% (w/v) bromphenol blue. Lane 1 shows the vector (1655 bp) and a 230-bp fragment (bottom arrow) carrying a single iteron (stippled). In the presence of RepA (ellipse), a slower migrating species is
seen that represents RepA–iteron complexes (middle arrow, lanes 2–5). As the RepA concentration is increased, most of the DNA does not enter the gel owing to the formation of RepA–
DNA aggregates which include the vector DNA. The aggregates, seen at the position of the
gel well, form because of high nonspecific DNA-binding and DNA-pairing activities of RepA.
On addition of excess of unlabeled competitor DNA (linear fragments of pUC19 DNA), most
of the nonspecific binding now involves the unlabeled DNA, allowing more of the probe DNA
to enter the gel (lane 6). If the unlabeled DNA also contains iterons (lanes 6–10), then binding
to probe DNA decreases as expected (cf. lanes 6 and 11). More interestingly, a second retarded

Ligation Method

For the ligation experiment, 50–100 pM of fragment is mixed with RepA in T buffer [50 mM Tris-HCl (pH 7.4), 50 mM NaCl, 50 mM KCl, 10 mM MgCl$_2$, 0.1 mM EDTA, 30 μg/ml bovine serum albumin (BSA)] supplemented with 5 mM ATP, 20 mM dithiothreitol (DTT), 4% polyvinyl alcohol (PVA) at 30°C for 10 min to allow equilibration of the binding reaction. RepA is activated with 2-fold molar excess of DnaJ and 5-fold molar excess of DnaK in the supplemented T buffer at room temperature for 40 min prior to use in the binding reaction. A 5-μl aliquot is withdrawn and loaded onto a 1% agarose gel to monitor the fraction of bound DNA. To the remainder, 4 U of T4 DNA ligase is added. Aliquots of 10 μl are withdrawn every 10 min and mixed with an equal volume of a 2× stop solution (20 mM EDTA, 1% SDS, and 0.02% bromphenol blue). The entire sample is loaded onto a 1% running

---

species can now be seen, presumably representing DNA pairs (top arrow). These structures were most likely trapped in the aggregates and, therefore, were not seen without competitor DNA (lanes 2–5). The DNA pairs were seen at a reduced frequency in the presence of nonspecific competitor DNA (lanes 11–13). The higher frequency in the presence of iteron-containing competitor DNA is expected owing to the increase in total (unlabeled and labeled) concentration of RepA–iteron complexes. The unlabeled complexes contribute to the frequency when their partners are labeled. We further note that even when the unlabeled DNA was present in 100-fold molar excess (lane 8), specific complexes with probe DNA were still present. This is most likely because the active fraction of RepA increased when the concentration of iteron DNA increased (Fig. 1B). In other words, RepA binding to unlabeled DNA did not necessarily happen at the expense of binding to probe DNA. The gel was 1.5% in agarose and was run at 6 V/cm for 1 hr at room temperature. The gel buffer contained 25 mM Tris base plus 0.19 M glycine and 1 mM EDTA (pH 8.5). After the run, the gel was soaked in paper towels for 5 min, transferred to Bio-Rad gel dryer paper, and dried under vacuum in a gel dryer for 30 min without heat and then for 90 min at 60°C. (B) Polyacrylamide gel electrophoresis showing reduction of nonspecific binding in the presence of chaperone proteins DnaJ and DnaK, and ATP. A purified 340-bp fragment carrying a single iteron was used in the experiments. Note that in the presence of chaperones, saturation of binding (complete conversion of probe DNA to RepA–DNA complexes) was achieved without significant nonspecific binding (lane 6). At higher RepA concentrations, nonspecific binding is apparent due to smearing of RepA–DNA complexes (lanes 7–9). In the absence of chaperones, the fraction of active RepA being less, a higher protein concentration is required to achieve saturation of binding, and at this concentration nonspecific binding becomes significant (lane 17). It should be noted that nonspecifically bound DNAs do not migrate as a discrete species in the gel. Comparison of lanes 6 and 7 indicates that nonspecific binding contributes to the formation of DNA pairs or to their stability during electrophoresis. The binding reactions were performed in a 20-μl volume in T buffer supplemented with 5 mM ATP, 20 mM DTT, and 6% glycerol. The binding reaction was allowed to equilibrate for 10 min after which the samples were loaded onto a running 5% polyacrylamide gel at 10 V/cm. The gel was transferred to Bio-Rad gel dryer paper and dried under vacuum at 80°C for 30 min.

agarose gel in TBE [89 m*M* Tris, 89 m*M* Boric acid, 2 m*M* EDTA, pH 8.3] buffer at 10 V/cm. The gel is dried for quantification as described for the band-shift assay (Fig. 3).

DNase I Footprinting

One of the most powerful means of demonstrating action at a distance is by DNase I footprinting. Like the ligation assay, the method is gentle, and the probing is done under solution conditions with minimal perturbation of the equilibrium. Another advantage of the method is that the sites can be present in supercoiled DNA, to mimic the situation *in vivo*. The method has been used to demonstrate interactions between sites in cis separated by anywhere from 5 to about 20 turns of B-form DNA. Usually the affinity of one of the sites for the protein is measured alone and then in the presence of a second site in cis. If the affinity is altered owing to the presence of the second site, then the binding is cooperative by definition, and direct contact between the proteins bound to the two sites is suggested.

The effect of cooperativity can be made more apparent if one of the sites used is of lower affinity, so that the weaker site hardly binds when present alone. For the two proteins to contact one another, the binding sites need to be on the same face of DNA. This is accomplished by adjusting the center-to-center distance between the sites to an integral number of turns of DNA helix. For B-form DNA, a helical turn is taken as 10.5 bp, but the exact number depends on the base composition of the intervening region. It is best, therefore, to try a set of substrates where the separation between the sites varies continuously over a full helical turn in steps of 1 bp. Cooperative binding by protein–protein contact is strongly suggested if the center-to-center distance in the substrates that show maximal binding differs from those with minimal binding by about half a helical turn. In the latter case, the sites are on the opposite sides of the helix, and, therefore, the intervening DNA has to twist by half a turn, an energetically costly step, or the proteins themselves must be contorted in order for the contacting protein surfaces to meet one another.

There is a second diagnostic of loop formation. The distortion of the DNA helix due to bending of intervening DNA may change the DNase I sensitivity to the region, although the region is not directly contacted by proteins. A periodic enhancement and diminution of cleavage separated by roughly 5 bp has been observed in several systems, and the appearance of this pattern of cleavage is, therefore, taken as diagnostic of loop formation (21, 22). This characteristic of DNA looping is a helpful indicator in systems where loop formation may not lead to cooperative binding. The technique is also equally applicable to demonstrate protein–protein interactions between heterologous proteins (8). We have shown that the binding of DnaA and RepA to the P1 origin to sites separated by more than 100 bp is reciprocally cooperative: at subsaturating concentrations, specific binding of both proteins increases when present together. In this case, although cooperative binding is indicated from affinity considerations, the reactivity of the intervening DNA was not altered. Binding of RepA and DnaA to supercoiled DNA is described below.

Method of Supercoiled Footprinting with DNase I

Supercoiled DNA (600 ng) is reacted with 800 ng of DnaA and 1 μg of RepA in a volume of 100 μl at 37°C in T buffer supplemented with 5 mM ATP and 0.5 mM DTT. The ATP is a special requirement for DnaA to bind DNA. After 10 min, 10 μl of a solution containing 0.02 units of DNase I (Promega, Madison, WI; stock diluted a few minutes before use in T buffer supplemented with 50 mM CaCl$_2$) is added to the mixture and incubated for 1.5 min at 37°C. The reaction is terminated by adding 10 μl of a solution containing 200 mM EDTA and 5% SDS. The samples are heated for 3 min at 80°C and treated with proteinase K (final concentration 50 μg/ml) at 37°C for 1 hr. The DNA is extracted with phenol–chloroform, and the supernatant is passed through two Sephadex G-50 spin columns saturated with water. The DNA is subsequently analyzed by primer extension.

Primer Extension Analysis

Two synthetic oligonucleotide primers homologous to the top and bottom strands and preferably 20 to 50 bases away from the region containing the binding sites were used. The oligonucleotides are 5′-end-labeled with [γ-^{32}P]ATP and purified by passing through a 1-ml Sephadex G-50 column equilibrated with water. The DNase I-treated sample is hybridized with the radiolabeled primers in the presence of extension buffer (50 mM Tris-HCl, pH 7.4, 10 mM MgSO$_4$, and 0.2 mM DTT) at 45°C for 5 min. The sample is cooled in ice, and 0.1 volume of a dNTP solution (final concentration 0.5 mM) containing 3 units of Klenow polymerase is added. The mixture is incubated at 50°C for 10 min, the reaction quenched by adding an equal volume of a solution of 1 M Tris-HCl, pH 7.4, 30 mM EDTA, and 20 μg glycogen/ml, and the DNA precipitated with ethanol. The DNA is dissolved in 20 μl of water, and an aliquot of desired radioactivity is lyophilized, resuspended in a dye solution [98% (v/v) deionized formamide, 10 mM EDTA, 1% (w/v) xylene cyanol, 1% (w/v) bromphenol blue], boiled for 5 min, cooled on ice, and loaded on a 6% denaturing polyacrylamide gel. For a size standard, a similar primer extension is done using the same plasmid and the primers, but the DNA is not reacted with DNase I and dideoxynucleotides are included together with dNTPs. If need be, the band intensities can be improved by amplified primer extension in a DNA thermal cycler using a double-stranded DNA cycle sequencing kit (Life Technologies, Gaithersburg, MD) (23).

DNase I footprinting is ideal for determining interactions in cis when the binding sites are separated by few to about 200 bp. The enzyme itself being a DNA-binding protein can destabilize weak DNA–protein complexes and, therefore, may not record authentic DNA-binding reactions in some cases. The cutting efficiency of DNase I is quite sequence dependent and is determined by the width of the minor groove of DNA (24). There can be regions of DNA where the enzyme may not cut even when the DNA-binding protein is absent. For this reason, the two strands of DNA may show very different sensitivity to the enzyme, and probing both the strands is a good idea. It is also imperative to keep the average number of cleavages in the probe DNA to less than 1. This is determined by comparing the intensities of full-length fragments after primer extension from samples with and without treatment with DNase I. The intensity in the former should be at least 37% of the latter if the average number of cleavages per probe molecule is no greater than 1 (from the Poisson distribution).

Electron Microscopy

As it is possible to visualize proteins on DNA, electron microscopy (EM) has been used to obtain direct evidence for action at a distance in many systems, including RepA. The DNA loops are easily visualized if the sites are a few hundred base pairs away, although interactions between sites separated by only five turns of helix have been visualized (25). The technique is also ideal for determining both cis and trans interactions. The success of the techniques depends on how well the complexes are preserved during air-drying. The DNA should stick to the EM grids strongly enough so that it does not distort or aggregate during air-drying. Similarly, the protein should be fixed with chemical cross-linking agents so that it does not flatten out during air-drying, a process which can significantly compromise recognition of the protein on the contour of DNA. In the case of RepA, we have used alcian blue to make carbon-coated EM grids sticky to DNA, and this treatment has preserved at least 7-kb-long DNA fragments in extended conformations (Fig. 4) (26). The integrity of proteins has been preserved by cross-linking with glutaraldehyde and to a lesser extent with uranyl acetate.

Restriction digested and phenol–chloroform extracted plasmid DNA is used without gel purification. One fragment carries the iterons, and the second fragment representing the vector serves as a control for looping arising from nonspecific binding. For the preparation of EM grids, the alcian blue stock solution (0.2%; see below) is diluted between 100- to 200-fold with glass distilled water in a microcentrifuge tube. One-half milliliter of the solution is dispensed with a 1-ml plastic pipette tip to form a drop on the surface of a plastic petri dish. Grids are then floated on the drop with the carbon-coated side touching the solution. Care should be taken so that the un-coated side of the grid stays dry. As alcian blue-coated grids are not stored, only enough grids for a single day (usually about 10) are coated at a time, and they are individually transferred to the same drop. After 5 min of floating, the grids are transferred horizontally to the surface of a 0.5-ml drop of water, so that the carbon-coated side touches the water. After 10 min the grids are transferred to a piece of filter paper with the carbon side facing up and are used within 1 hr.

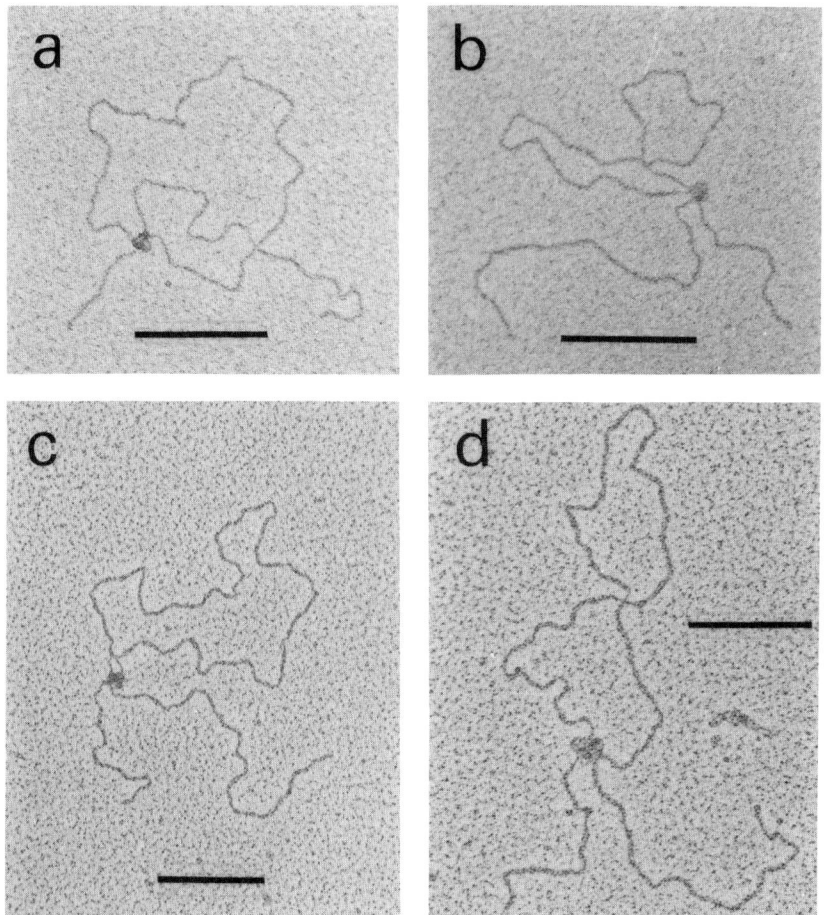

FIG. 4   Electron micrographs of RepA–DNA complexes showing RepA-mediated DNA loops. Glutaraldehyde-fixed complexes were deposited on alcian blue-coated carbon films. The contrast comes from uranyl acetate staining and tungsten shadowing. The bars represent 0.5 kb.

DNA–protein complexes are made in 20 μl in a buffer as described in Fig. 3A, but without DTT and BSA. The DNA concentration needs to be higher in these experiments, about 1 nM for 4-kb fragments. It is convenient to monitor binding using the band-shift assay by mixing a tracer amount of radioactive DNA into the binding mixture. The RepA concentration has to be carefully titrated to ensure minimal non-specific binding. The binding reaction is allowed to equilibrate for 10 min, after which 2 μl of a 1% solution of glutaraldehyde is added; the mixture is incubated for 10 min at room temperature and subsequently diluted 50-fold by adding 1 ml of the buffer solution used in the binding reaction. Ten microliters of the diluted solution is applied with a plastic pipette tip to the coated surface of the grid, kept in a horizontal position with a clamping forceps, so that the coated surface faces up and the back of the grid does not contact the tabletop. After 1 min, the excess sample is removed from the grid surface by touching the edge of the grid with a 25-μl glass capillary pipette, and the grid is rinsed with 10 μl of water and floated on a 100-μl drop of water (one grid per drop). After 10 min the grid is removed from water with a clamped forceps, and the sample side is rinsed with 30 μl of a 0.5% uranyl acetate solution two times, removing the excess liquid with a capillary tube as above. The grid is rotary shadowed with tungsten (W) vapor by heating a 0.025-inch-diameter wire for 2 min at 35 A.

Care is taken throughout the sample preparation steps to ensure that the back side of the grid stays dry, as it never gets washed, and use of filter papers to remove liquid from the grid surface is avoided as they can be a source of impurities. The dilution after glutaraldehyde fixation and the washing steps after sample application on the grid are not essential but improve the background significantly. The glutaraldehyde step has the disadvantage that the cross-linker can bridge two proteins that normally do not interact significantly. For RepA, use of glutaraldehyde is not obligatory. After 1 min of applying the DNA–protein complexes (10 μl) to the grid, the excess sample is not removed but uranyl acetate solution added directly. The grid is dried after rinsing it with uranyl acetate a second time as described above.

Contrast

The primary source of contrast comes from shadowing. Tungsten is desired over more conventionally used materials such as Pt–Pd or Pt, because of the smaller grain size of the metal vapor after deposition on the grid surface and consequent higher resolution. One disadvantage of W is that the shadowed grids are best when examined on the same day. Stored grids lose contrast, possibly because of oxidation of W. If storage is required, it is best to store the grids under vacuum in the shadow chamber. The W filament can be reused a few times, but the time of heating needs to be reduced with continued use (e.g., from 2.25 to 1.75 min between the first and the fourth use). It is better to shadow the sample conservatively, because it can be repeated later if the first shadowing is too light. Uranyl acetate adds to the contrast significantly and, in fact, can be made the only source of contrast, albeit poor. It is helpful if the discrimination between protein and DNA needs to be highlighted (27).

Alcian Blue Stock Solution

Make a 0.2% (w/v) stock solution of alcian blue 8GS (Electron Microscopy Sciences, Fort Washington, PA 19034, Cat. No. 10350) in 3% (v/v) acetic acid. Most of the dye should dissolve readily. Spin in a microcentrifuge for 5 min. Save the supernatant (OD_{630} of ~9; measure after dilution with water) which can be used for months when kept in the dark at room temperature. DNA-like filaments can be seen sometimes with alcian blue coating of the grids without any added DNA. Usually with fresh stock solutions and at higher dilutions, the filaments are not a problem. Too dilute a solution will not allow DNA to stick to the grid, in which case the molecules may collapse completely into dots.

Magnetic Separation

Streptavidin-coated paramagnetic beads provide a simple, gentle, and rapid separation of biotinylated nucleic acids and nucleoprotein complexes and are convenient to study DNA pairing. The technique takes advantage of the high affinity biotin–streptavidin interaction and magnetic separation. One of the interacting DNAs is first tagged with biotin and then coupled to the beads via a streptavidin–biotin linkage. Proteins and target DNAs are added and the complexes formed with the DNA bound to beads (trans pairs) are concentrated on the wall of the container with a magnet. The supernatant is removed while the complexes are held on the wall by the magnetic field. Once removed from the magnetic field, the complexes are easily resuspended and assayed for the presence of target DNA. We have used this technique to study RepA–DNA interactions using supercoiled DNA.

Preparation of Biotinylated Supercoiled DNA

About 5 μg plasmid DNA carrying iterons is linearized with a restriction enzyme that cleaves far away from the iterons and which leaves a 5' overhang. After restriction, the protein is removed with phenol–chloroform. To the aqueous layer, ammonium acetate is added to 2.5 M and the DNA precipitated with 3 volumes of ethanol. The precipitate is washed with 70% (v/v) ethanol and finally resuspended in 5 μl TE (10 mM Tris-HCl, pH 8, 1 mM EDTA). The end-filling is done with 40 μM of biotin-14-dATP or biotin-14-dCTP (GIBCO/BRL, Gaithersburg, MD), 0.5 mM each of the other three dNTPs, and 5 U Klenow (New England Biolabs, Beverly, MA) in a volume of 50 μl. The buffer is 10 mM Tris-HCl, pH 8, 5 mM MgCl$_2$, 100 mM NaCl, and 1 mM 2-mercaptoethanol. After 1 hr at 30°C, the biotinylated nucleotides are chased with 0.5 mM of all four dNTPs for 2 min at room temperature and the reaction terminated by adding EDTA to 20 mM. The mixture is extracted with phenol–chloroform once and the aqueous layer passed through Sephadex G-25 spin columns equilibrated in TE. An aliquot is ligated in the presence of 1 μg/ml ethidium bromide and 3 U ligase (GIBCO/BRL) in a volume of 50 μl and at a DNA concentration of no more than 4 μg/ml to promote formation of circles rather than linear polymers. The buffer is 50 mM Tris-HCl, pH 7.6, 10 mM MgCl$_2$, 1 mM ATP, 1 mM DTT, and 5% (w/v) polyethylene glycol 8000 (GIBCO/BRL).

After 16 hr at 18°C, the mixture is extracted twice with phenol–chloroform and the aqueous phase extracted several times with 1-butanol to reduce the volume to 10 μl. These extractions also remove ethidium bromide from the DNA and make it negatively supercoiled. [Intercalation of ethidium bromide unwinds DNA, and the unwinding is made permanent by ligase joining of ends. On removal of the intercalating drug, the DNA can wind again to the B-form, and the positive twists are compensated by negative supercoils owing to the invariance of linking number (28).] The labeled DNA is mixed with 1 μg of the starting supercoiled DNA (not labeled with biotin) and run in an agarose minigel. The excess of unlabeled DNA serves to locate the position of supercoiled DNA when the gel is stained with ethidium bromide after the run. The band corresponding to the supercoiled DNA is excised from the gel and is transferred to a Poly-Prep column (Bio-Rad Laboratories, Richmond, CA), frozen in dry ice, thawed at room temperature, and spun at 1000 rpm in a Sorvall GLC-2 table-top centrifuge (about 120 g at the tube bottom) for 10 min. The flow through is extracted with phenol–chloroform and the DNA precipitated with ethanol in the presence of ammonium acetate as before.

Preparation of ^{3}H-Labeled Supercoiled DNA

Supercoiled plasmid DNA is ^{3}H-labeled *in vitro* in a reaction mixture containing 50 mM Tris-HCl, pH 7.5, 10 mM EDTA, 5 mM 2-mercaptoethanol, 5 μCi *S*-adenosyl-L-[*methyl*-^{3}H]methionine (stock 80 Ci/mmol, Amersham, Arlington Heights, IL), 10 μg plasmid DNA, 10 units of *Hpa*II methylase (New England Biolabs) in a volume of 100 μl. After 1 hr at 37°C, methylation is stopped by addition of SDS to 0.5% and the mixture extracted once with phenol–chloroform. From the aqueous phase, the DNA is precipitated with ethanol and dissolved in 50 μl TE, and the solution is passed through Sephadex G-25 spin columns two times. The specific activity is 80,000 counts/min (cpm)/μg DNA, as determined by TCA precipitation.

Immobilization of Biotinylated DNA to Dynabeads

One hundred microliters of Dynabeads (M280-streptavidin; Dynal Inc., Great Neck, NY), enough for 10 reactions, is washed two times by resuspending in 200 μl of 10 mM Tris-HCl, pH 7.5, 1 mM EDTA, and 3 M NaCl using an MPC-E magnetic particle concentrator (Dynal) and finally in 94 μl of the same solution and 6 μl of biotinylated DNA. The mixture is incubated at 45°C for 45 min. The beads are resuspended once in 100 μl of the same solution, but with 2 M NaCl, and then two times in 100 μl of T buffer supplemented with 5 mM ATP and 4% PVA.

Intermolecular Interactions

Binding reactions are set up in 50 μl of T buffer supplemented with 5 mM ATP, 20 mM DTT and 4% PVA, containing 10 μl of immobilized biotinylated DNA, ^{3}H-labeled supercoiled DNA (about 10,000 cpm per reaction), and the proteins. The binding reactions are allowed to equilibrate for 10 min at room temperature and then transferred to an MPC-E concentrator. After 30 sec, the bulk of the supernatant is removed gently with a pipette and the remaining drops by aspiration with a capillary tip. The tubes are not washed as this dissociates the complexes significantly. The beads are resuspended in 50 μl of TE containing 0.5% SDS solution by gentle tapping and separated from the solution using an MPC-E concentrator. Ten microliters of the supernatant is mixed with 10 ml of Econofluor (National Diagnostics), and the radioactivity is determined. In the absence of proteins, this procedure retains less than 2% of the input counts. The method is used to study both RepA–RepA and RepA–DnaA interactions in trans.

Epilogue

In this chapter, we have described some of the simpler techniques to demonstrate protein–protein interactions of DNA-binding proteins *in vitro*. Emphasis has been given to studying the interactions quantitatively, distinguishing specific from nonspecific interactions, and using supercoiled DNA. The inference about protein–protein contacts is mostly indirect when studied by the techniques described here. More direct evidence for interactions can be demonstrated using modified enzyme-linked immunosorbent assay (ELISA) tests (29). Further confirmation can be supplied by characterization of mutant proteins. Correlation of the strengths of pairing activities of mutant proteins with their *in vivo* phenotypes will not only help identify the domains involved, but, more importantly, will establish to what extent the interactions are significant *in vivo*.

Acknowledgments

The authors are indebted to Sam Wilson, Mark Lewis, Steve Widen, Bill Beard, Barbara Funnell, Manolo Espinosa, Roger McMacken, and Deepak Bastia for contributing to our education and to Michael Yarmolinsky for critical reading of the manuscript.

References

1. H. Echols, *J. Biol. Chem.* **265**, 14697 (1990).
2. S. Natarajan, S. Kaul, A. Miron, and D. Bastia, *Cell (Cambridge, Mass.)* **72**, 113 (1993).
3. B. E. Funnell, *J. Biol. Chem.* **266**, 14328 (1991).
4. W. Su, S. Porter, S. Kustu, and H. Echols, *Proc. Natl. Acad. Sci. U.S.A.* **87**, 5504 (1990).
5. M. Dodson, H. Echols, S. Wickner, C. Alfano, K. Mensa-Wilmot, B. Gomes, J. Lebowitz, J. D. Roberts, and R. McMacken, *Proc. Natl. Acad. Sci. U.S.A.* **83**, 7638 (1986).
6. S. DasGupta, G. Mukhopadhyay, P. P. Papp, M. S. Lewis, and D. K. Chattoraj, *J. Mol. Biol.* **232**, 23 (1993).
7. P. P. Papp, D. K. Chattoraj, and T. D. Schneider, *J. Mol. Biol.* **233**, 219 (1993).
8. G. Mukhopadhyay, K. M. Carr, J. M. Kaguni, and D. K. Chattoraj, *EMBO J.* **12**, 4547 (1993).

9. C.-C. Yang and H. A. Nash, *Cell (Cambridge, Mass.)* **57,** 869 (1989).

10. G. J. Schneider and E. P. Geiduschek, *J. Biol. Chem.* **265,** 10198 (1990).

11. S. Y. Tsai, M.-J. Tsai, and B. W. O'Malley, *Cell (Cambridge, Mass.)* **57,** 443 (1989).

12. L. P. Freedman and I. Alroy, *in* "Methods in Molecular Genetics" (K. W. Adolph, ed.), Vol. 1, p. 280. Academic Press, San Diego, 1993.

13. M. Brenowitz, E. Jamison, A. Majumdar, and S. Adhya, *Biochemistry* **29,** 3374 (1990).

14. G. Milman and E. S. Hwang, *J. Virol.* **61,** 465 (1987).

15. R. Fickert and B. Müller-Hill, *J. Mol. Biol.* **226,** 59 (1992).

16. J. H. Carra and R. F. Schleif, *Nucleic Acids Res.* **21,** 435 (1993).

17. C. Mao, N. G. Carlson, and J. W. Little, *J. Mol. Biol.* **235,** 532 (1994).

18. G. Mukhopadhyay, S. Sozhamannan, and D. K. Chattoraj, *EMBO J.* **13,** 2089 (1994).

19. D. Shore, J. Langowski, and R. L. Baldwin, *Proc. Natl. Acad. Sci. U.S.A.* **78,** 4833 (1981).

20. S. Mukherjee, H. Erickson, and D. Bastia, *Proc. Natl. Acad. Sci. U.S.A.* **85,** 6287 (1988).

21. A. Hochschild, *in* "DNA Topology and Its Biological Effects" (N. R. Cozzarelli and J. C. Wang, eds.), p. 107. Cold Spring Harbor Laboratory, Cold Spring Harbor, New York, 1990.

22. J. Plumbridge and A. Kolb, *Mol. Microbiol.* **10,** 973 (1993).

23. S. Sozhamannan and D. K. Chattoraj, *J. Bacteriol.* **175,** 3546 (1993).

24. M. Serrano, M. Salas, and J. M. Hermoso, *Science* **248,** 1012 (1990).

25. J. Griffith, A. Hochschild, and M. Ptashne, *Nature (London)* **322,** 750 (1986).

26. D. K. Chattoraj, R. J. Mason, and S. H. Wickner, *Cell (Cambridge, Mass.)* **52,** 551 (1988).

27. B. F. Funnell, T. A. Baker, and A. Kornberg, *J. Biol. Chem.* **262,** 10327 (1987).

28. R. Bowater, F. Aboul-ela, J. McClellan, A. I. H. Murchie, and D. M. J. Lilley, *in* "Methods in Molecular Genetics" (K. W. Adolph, ed.), Vol. 1, p. 241. Academic Press, San Diego, 1993.

29. J. Marszalek and J. M. Kaguni, *J. Biol. Chem.* **269,** 4883 (1994).

[23]   Transcriptional Regulators: Protein–DNA Complexes and Regulatory Mechanisms

Fernando Rojo and Margarita Salas

Introduction

Processes such as DNA replication, transcription, and recombination involve the concerted action of diverse DNA-binding proteins acting on the DNA. Knowledge of the structure and working mechanisms of these nucleoprotein complexes has greatly improved since the 1980s, owing mainly to the appearance of a number of techniques that are relatively simple and give detailed information about the protein–DNA complexes. In this chapter we describe a number of these techniques that have been applied in the past years to study the switch from early to late transcription in *Bacillus subtilis* phage ϕ29. We show the results obtained and describe how the techniques allowed us to elucidate the mechanism used by the transcriptional regulatory protein p4 to activate transcription from the viral late promoter.

Methods

Band-Shift Assays

Reactions for band-shift assays (1, 2) should be performed with DNA fragments preferably smaller than 350–400 bp. A double-stranded DNA fragment as small as a 15–20 bp can be used if the protein of interest still binds to it, but fragments larger than 350 bp give poor results. The DNA should be labeled at one or at both ends; about 2000–4000 counts/min (cpm) of labeled DNA per sample is used. Reactions are performed in a total volume of 20 μl, in 25 mM Tris-HCl (pH 7.5), 10 mM MgCl$_2$. About 2 μg of poly[d(I·C)] and a variable amount of salt (sodium or potassium chloride from 50 to 200 mM, or ammonium sulfate from 20 to 90 mM) are also included to minimize both nonspecific binding of the protein to the DNA and possible aggregation of purified proteins. We add about 0.5–1 μg of purified protein p4 and/or 0.25 μg of purified *B. subtilis* σ^A-RNA polymerase. The mixture is incubated for 20 min at 37°C or at room temperature and then transferred to ice. Samples are loaded onto a 4% polyacrylamide gel (80:1 acrylamide/bisacrylamide; about 15 × 15 × 0.1 cm; the polyacrylamide concentration should be higher for DNA fragments shorter than 100 bp) after the addition of 4 μl of 30% (v/v) glycerol. Marker dyes are usually not included in the samples because they are found to reduce severely protein p4 binding to DNA, though this effect is not normally important for most other proteins. A solution of marker dyes can be run in parallel to monitor the electrophoresis. To minimize dissociation of the protein from the DNA, the gel should be run at 4°C and with a running buffer of low ionic strength. We routinely use 120 mM Tris–acetate (pH 7.5), 2 mM $MgCl_2$, although 45 mM Tris–borate, 1 mM EDTA (0.5× TBE) also gives good results. The electrophoresis is run at 15 to 30 mA for 2.5–3 hr. There is no need to fix the gels, which are directly dried and exposed for autoradiography. This technique has been exhaustively reviewed (3).

DNase I Footprinting

Binding reactions for DNase I footprinting (4) are very similar to those used for band-shift assays, with two restrictions: the DNA fragments should be labeled at only one of the ends (either at the 5' end with polynucleotide kinase, or at the 3' end by Klenow filling), and salt should be kept low as it impairs DNase I cutting. The size of the DNA fragment is not critical, but the binding site for the protein should be located at a distance of approximately 50 to 200 nucleotides from the labeled end. About 15,000 cpm of the labeled DNA is used per sample. Binding reactions are carried out in a final volume of 20 μl, in 25 mM Tris-HCl (pH 7.5), 10 mM $MgCl_2$, in the presence of 2 μg of poly[d(I·C)] as nonspecific competitor DNA. We add 0.5–1 μg of protein p4 and/or 0.25–0.5 μg of B. subtilis σ^A-RNA polymerase. After incubation for 15 min at 37°C, 50 ng of DNase I is added (the amount of DNase I to be added should be determined empirically, so as to obtain an average of only one cut per DNA molecule). Digestion is allowed to occur for 2 min and stopped with 1 μl of 0.5 M EDTA. DNA is precipitated and analyzed in denaturing polyacrylamide gels. A size standard should be run in parallel to locate the protected region on the DNA fragment used.

We routinely use either chemical sequencing reactions for purines and pyrimidines of the same DNA fragment (5) or a dideoxy sequencing reaction (6) for a DNA of known sequence. It should be noted that DNase I cuts the DNA between the sugar and the phosphate, so that the nick has an OH group at the 3' end and a phosphate at the 5' end. This is not the case with the fragments obtained by chemical sequencing, where the chemical treatment leads to a loss of the sugar and the base, leaving phosphate groups at both 3' and 5' ends. Therefore, a DNA molecule of a given size obtained by chemical sequencing migrates 1 nucleotide ahead of the corresponding DNA molecule obtained by DNase I digestion if the DNA is labeled at the 3' end; the difference increases to 1.5 nucleotides if the DNA fragment is labeled at the 5' end.

Hydroxyl-Radical Footprinting

Hydroxyl-radical footprinting of the binding of a protein to DNA (7) gives much better resolution than DNase I footprinting, though it is technically more demanding.

Binding reactions are exactly as those for DNase I footprinting, except that glycerol should be omitted from the reaction because it quenches the hydroxyl radicals. This is a problem as purified proteins are usually stored at $-20°C$ in the presence of 50% (v/v) glycerol. We have observed that the reaction tolerates up to 0.5–1% (v/v) glycerol if the amount of the reagents generating the hydroxyl radicals is increased 10–20 times with respect to that originally described. If the purified protein is concentrated enough and is diluted in reaction buffer without glycerol, the final glycerol concentration can be kept under 1%. Fresh solutions of 40 mM sodium ascorbate, 12 mM ammonium iron(II) sulfate hexahydrate, and 5% hydrogen peroxide should be prepared. The iron(II) solution is mixed with an equal amount of 24 mM EDTA. A mixture is prepared containing equal volumes of the iron(II)/EDTA, sodium ascorbate, and H_2O_2 solutions, which is used immediately after mixing. Three microliters of this solution is added to 20 μl of binding mixture (prepared as for DNase I footprints), and the samples are incubated for 4 min at room temperature. The reaction is stopped with 2 μl of 100 mM thiourea and 2 μl of 0.5 M EDTA. The DNA is precipitated with sodium acetate (0.3 M final concentration) and 3 volumes of ethanol. The samples are then subjected to denaturing polyacrylamide gel electrophoresis side by side with size standards prepared as for DNase I footprints. Because hydroxyl radical attack on the DNA leads to the loss of the complete nucleoside, the migration of a DNA molecule of a given size is the same as that of the equivalent DNA molecule obtained by chemical sequencing (see above).

The purity of the DNA fragment used for the assay is critical for the quality of the results; otherwise, the bands tend to be fuzzy. We routinely purify the labeled DNA fragment by electrophoresis through a 4% nondenaturing polyacrylamide gel. The gel is briefly exposed without drying or fixing, and the band of interest is excised from the gel, chopped into small pieces, and the DNA recovered by overnight diffusion in 0.5 M ammonium acetate, 0.1% (w/v) sodium dodecyl sulfate (SDS), 1 mM EDTA, followed by ethanol precipitation.

Methylation Interference Assays

Treatment of DNA with dimethyl sulfate leads to methylation of guanine residues at the N-7 position (accessible through the major groove) and, with reduced efficiency, modification of adenine residues at the N-3 position (protruding through the minor groove). Methylation interference experiments (8, 9) allow one to identify those adenine or guanine residues that, when methylated, interfere with protein binding. End-labeled DNA is partially methylated with 50 mM dimethyl sulfate in 50 mM sodium cacodylate buffer (pH 8) containing 25 mM $MgCl_2$ and 0.1 mM EDTA. The reaction is carried out for 10 min at room temperature and stopped with 0.2 M 2-mercaptoethanol. The DNA is precipitated twice with ethanol, washed with 80% (v/v) ethanol, and dried. Methylated DNA (about 50,000–100,000 cpm) is then incubated with the DNA-binding protein in the presence of 2.5 μg of poly[d(I·C)], in a total volume of

20 μl. Free and protein-bound DNA are resolved in a band-shift gel as described for band-shift assays, and the gel is exposed without drying for about 1 hr to locate each band. Bands are excised and eluted from the gel (either by electroelution or by overnight diffusion of the crushed polyacrylamide in 0.5 M ammonium acetate, 0.1% SDS, 1 mM EDTA), ethanol precipitated, and dried. Because dimethyl sulfate does not cleave the DNA backbone directly, it is necessary to treat methylated DNA at high temperature under basic conditions, which leads to a loss of the modified base and cleavage of the DNA backbone. This is achieved by resuspending DNA in 100 μl of 1 M piperidine and incubating it for 30 min at 90°C. The reaction is stopped by addition of 1 volume of ethanol, and the DNA is concentrated under vacuum until it is dry, resuspended in formamide loading buffer, and analyzed by denaturing polyacrylamide gel electrophoresis side by side with a DNA size standard. The DNA fragments migrate exactly as those obtained by chemical sequencing.

Hydroxyl-Radical Interference: Missing Nucleoside Assay

The hydroxyl-radical interference method allows one to identify those nucleosides that are important for the binding of a protein to DNA (10). Although the initial attack of the hydroxyl radical is at the DNA sugar–phosphate backbone, the reaction leads to the loss of the complete nucleoside, so that the final product is a DNA molecule with a gap. The assay identifies which gaps allow or impair protein binding. A 200- to 400-nucleotide end-labeled DNA fragment is treated with hydroxyl radicals as indicated above, to obtain DNA fragments which statistically have no more than one gap per molecule. The DNA (about 100,000 cpm) is precipitated and resuspended in 20 μl of a solution containing 25 mM Tris-HCl (pH 7.5), 10 mM MgCl$_2$, and 2.5 μg poly[d(I·C)]. The purified protein is then added, and the mixture is incubated for 15 min at room temperature and 10 min at 4°C. Bound and unbound DNA are separated in a nondenaturing 4% polyacrylamide gel as indicated for the band-shift assays, identified and purified as indicated for methylation interference assays, and analyzed in denaturing polyacrylamide gels, side by side with a DNA size standard.

Circular Permutation Assays

Circular permutation assays (11) allow one to determine if a DNA region is curved and whether a protein bends the DNA on binding to it, and they permit one to locate the bending site and to estimate the degree of bending. The method is based on the empirical observation that the migration of a DNA fragment with a DNA curvature in a polyacrylamide gel is less than that of a noncurved DNA fragment of the same size, the effect being more pronounced when the curvature is located at the center of the DNA fragment than when it is close to an end. Therefore, if a sequence of interest is cloned as a tandem repeat in a plasmid (see, e.g., Refs. 12 or 13 for two different strategies to achieve this), the unique target sites for restriction enzymes (which are duplicated because the fragment is cloned as a tandem repeat) can be used to obtain a series of DNA fragments of identical size but with the sequence of interest at different positions relative to the fragment ends. This method allows the study of DNA fragments 150–300 bp long, as the differences in mobility owing to curvatures are less apparent in larger DNA fragments. An alternative possibility is to clone the sequence under study in a vector, such as pBEND2 (14), that is specifically designed for this purpose. This plasmid allows one to clone the DNA fragment as a single copy at a site that is flanked by two identical DNA segments 120 bp long, containing restriction sites for several enzymes. Therefore, the total size of the DNA fragment as excised from the vector is that of the cloned insert plus about 120 bp of extra DNA. Although the method is simpler, the DNA fragments to be cloned should not exceed 200 bp. Once a collection of DNA fragments of identical size with the sequence under study located at different distances from the fragment ends has been obtained, the DNA should be end-labeled and mobilities analyzed in polyacrylamide gels. We use 4% polyacrylamide gels, with the ratio of acrylamide to bisacrylamide being 40:1 for the study of naked DNA fragments and 80:1 when protein-induced curvatures are analyzed. The degree of bending can be estimated using the empirical equation $\mu_m = [\cos(\alpha/2)]\mu_e$, where α is the bending angle and μ_m and μ_e indicate the migration of the DNA fragments (or protein–DNA complexes) of minimal and maximal mobility, respectively (15).

In Vitro Run-off Transcription Assays

Run-off transcription reactions are performed in a final volume of 25 μl and contain 25 mM Tris-HCl (pH 7.5), 10 mM MgCl$_2$, 30–90 mM ammonium sulfate (alternatively 50–200 mM potassium chloride), 200 μM each ATP, CTP, and GTP, 80 μM [α-^{32}P]UTP (2 μCi), 2–10 nM DNA template, and 20–40 nM RNA polymerase. An adequate amount of the regulatory protein under study should be included also. The samples are incubated for 10 min at 37°C, and the reaction is stopped by the addition of 50 μl of a solution containing 2% SDS, 10 μg tRNA, and 100 mM EDTA. Nonincorporated nucleoside triphosphates (NTPs) can be eliminated through a Sephadex G-50 spun column. The RNA is precipitated with ethanol and transcripts analyzed by denaturing polyacrylamide gel electrophoresis. The optimum amount of salt added in the transcription reaction should be adjusted for each promoter. Many promoters are salt sensitive, probably because the RNA polymerase has a weak binding affinity for them. Nevertheless, it is not advisable to lower the salt concentration too much since many purified proteins aggregate under low salt conditions.

The reaction performed as described above leads to the production of multiple rounds of transcription for each template. If a single round of transcription is desired for specific purposes, the reaction mixtures should be incubated for 10 min at 37° C in the absence of NTPs to allow RNA polymerase to bind to the promoter. If the precise transcription start site is known, the first two initiating NTPs or the adequate dinucleotide can also be added at this step (40 μM final concentration) to force a precise initiation point. Transcription is then started by the addition of 3 μl of a preheated solution containing heparin and the four NTPs. The final concentration of heparin depends on the sensitivity of each promoter but can range from 10 to 100 μg/ml. Incubation is continued for 10 min, after which it is stopped as indicated above. Heparin acts as a strong competitor that inhibits binding of RNA polymerase to the promoter so that after its addition transcription starts only at those template molecules at which the RNA polymerase is already bound as a stable open complex. Because formation of the complex is not irreversible for all promoters (i.e., it can go back to closed complex and to dissociation of the RNA polymerase from the promoter; see Ref. 16 for a discussion), heparin can compete with the open complex at some promoters. If this is the case, the complexes can be stabilized by the addition of the first two initiating nucleotides, as formation of the first phosphodiester bond significantly stabilizes the open complexes (16). DNA supercoiling can also stabilize open complexes at some promoters (17). An alternative for sensitive promoters is to use poly[d(I·C)] as competitor, which is milder than heparin.

Analysis of RNAs by Primer Extension

This method has the advantage over the run-off transcription assay that allows the analysis of the transcripts produced from supercoiled templates or from long linear templates that would give rise to run-off transcripts of excessive length. The transcription reaction is performed exactly as described for the run-off assay, but radioactive UTP is not used and the concentration of nonlabeled UTP is increased to 200 μM. The same concentration of DNA template is used (2–10 nM). The reaction is stopped by the addition of 1 μl of 0.5 M EDTA, 1 μg of carrier tRNA, 3 μl of 3 M potassium acetate, and 70 μl of ethanol. RNAs are precipitated and resuspended in a solution containing 40 mM Tris-HCl (pH 7.5), 20 mM MgCl$_2$, 50 mM NaCl, 10 units of RNasin ribonuclease inhibitor, and 2–10 pmol (a large excess) of an oligonucleotide designed to hybridize 50–150 nucleotides downstream from the transcription start site. The mixture is heated for 5 min at 70° C and then allowed to cool slowly to 20° C. The solution is put on ice for 10 min, and the RNA is precipitated by the addition of potassium acetate (0.3 M final concentration) and 3 volumes of ethanol. The RNA is resuspended in 5 μl of water, and the primer is extended in a solution containing 50 mM Tris-HCl (pH 7.5), 40 mM KCl, 7 mM magnesium acetate, 2 mM dithiothreitol, 200 μM each dGTP, dCTP, and dTTP, 100 μM [α-^{32}P]dATP (400 Ci/mmol; 2 μCi), 10 units of RNasin, and 5 units of AMV reverse transcriptase, in a total volume of 10 μl. The reaction mixture is incubated at 42° C for 60 min, and then stopped by the addition of 0.5 μl of 0.5 M EDTA and 30 μl of 10 mM Tris-HCl (pH 7.5), 1 mM EDTA. The sample is filtered through a 1-ml Sephadex G-50 spun column to eliminate the nonincorporated labeled nucleotide, and the eluted cDNA is precipitated by the addition of 4 μl of 3 M potassium acetate and 200 μl of ethanol. The cDNA is analyzed by electrophoresis in 6% urea–polyacrylamide gels, side by side with an adequate size standard. Note that the use of [α-^{32}P]dATP can be circumvented if the oligonucleotide used as primer is labeled at the 5' end with polynucleotide kinase and [γ-^{32}P]ATP.

Potassium Permanganate Footprinting

Potassium permanganate reacts much more rapidly with single-stranded DNA than with double-stranded DNA (18), being rather selective for thymidines. Therefore, it is useful for probing DNA melting at the transcription initiation region of a promoter on formation of an open complex (19). An end-labeled DNA fragment containing the promoter under study is incubated at 37°C for 20 min in a solution containing 25 mM Tris-HCl (pH 7.5), 10 mM MgCl$_2$, 2 μg of poly[d(I·C)], 0.5 μg of RNA polymerase (except in the control sample), in a total volume of 25 μl. The first two initiating NTPs (40 μM each) and a regulatory protein can be included. Then KMnO$_4$ is added up to 4 mM and incubation continued for 30 sec at 37°C. The reaction is stopped by the addition of 5 μl of a solution containing 1 M 2-mercaptoethanol and 1.5 M sodium acetate. The DNA is precipitated, cleaved with piperidine as described above for methylated DNA (5), and analyzed by electrophoresis in 6% urea–polyacrylamide gels. The melted region should appear as a block of hypersensitive thymidines in both strands, though some of them can be protected by the RNA polymerase.

Preparation of Plasmids with Different Superhelical Densities

DNA supercoiling is known to affect the efficiency of several promoters (20). This effect can be studied by cloning the promoter of interest in a plasmid and obtaining a collection of plasmid samples with different supercoiling densities. This can be obtained by treating the supercoiled plasmid with topoisomerase I in the presence of ethidium bromide concentrations ranging from 0 to 20 μM (21). The superhelical densities of the samples obtained can be calculated from the linking number difference between the center of the topoisomer distribution of a given sample and the center of the topoisomer distribution corresponding to relaxed DNA (22). It is known that changes in temperature or ionic strength alter the average rotation angle between adjacent base pairs in the DNA helix which, in a closed-circular DNA, leads to a change in the superhelical density (23). Therefore, a DNA relaxed under the conditions optimal for topoisomerase I acquires certain superhelical density at the ionic conditions used in the transcription assays, which are frequently higher. This should be taken into consideration when calculating the superhelical densities of the different plasmid preparations. Because topoisomerase I can work efficiently under a wide range of ionic conditions, the simplest solution is to relax the plasmid with topoisomerase I under the same ionic strength conditions used in the transcription assays. The plasmids obtained can be used as templates for *in vitro* transcription reactions as indicated above, and transcripts detected by primer extension.

Kinetic Measurements

The process leading to the formation of a transcription initiation complex can be described by the following simple scheme (24–28), where R and P represent the RNA polymerase and the promoter, respectively:

$$R + P \underset{k_{-1}}{\overset{k_1}{\rightleftharpoons}} RPC \xrightarrow{k_2} RPo \xrightarrow{k_3} ITC \xrightarrow{k_4} \text{elongation complex}$$

The initial binding of RNA polymerase to form a transcriptionally inactive closed complex (RPc) is a reversible process. This equilibrium is represented by the affinity constant K_B ($K_B = k_1/k_{-1}$). A subsequent isomerization of the closed complex leads to the formation of an open complex (RPo), in which the binding of the RNA polymerase to the promoter is much tighter and there is melting of the $+1$ region. This step is essentially irreversible for most promoters, although there are exceptions. Incorporation of the first NTPs modifies the complex to the so-called initial transcribing complex (ITC), in which there is a continuous production of short abortive transcripts but the RNA polymerase does not escape from the promoter. Eventual promoter clearance allows the formation of an elongation complex.

The strength of a promoter is determined by the efficiency of the above steps. Those promoters that require a transcriptional activator are limited in at least one of the steps, the role of the activator being to accelerate the inefficient step. The τ plot analysis (25, 29) has been extensively used to determine the kinetic parameters of promoters. This method is based on the observation that, under pseudo-first-order conditions (i.e., a molar excess of RNA polymerase), a transcription reaction that is initiated by addition of the RNA polymerase shows a lag time before the reaction reaches steady state. The lag time is related to the time required for free enzyme and free promoter to combine and isomerize to start transcription. Plotting the observed lag times versus the inverse of the enzyme concentration leads to a rate equation in which the slope represents the inverse of the promoter strength ($K_B k_f$, where k_f is the forward rate constant, see below), and the time axis intercept is the inverse of k_f. The meaning of k_f depends on the transcription product analyzed. For promoters in which formation of the open complex is an irreversible process, and which produce a significant amount of short abortive transcripts, the products most frequently analyzed are these abortive transcripts. In this case, $k_f = k_2$ and describes only the efficiency of open complex formation. When the promoter forms low amounts of abortive transcripts, or if the open complexes are unstable and are in equilibrium with free enzyme, the reaction product to be analyzed should be a run-off transcript; in this case $k_f = k_2k_3k_4$ and represents both open complex formation and the rest of the steps leading to the formation of an elongation complex.

As open complexes formed by the $\phi29$ promoters analyzed are unstable (16), we perform the τ plots with run-off transcription assays. These should be carried out as indicated in the above section, with the following modifications. The assays are initiated by adding different amounts of RNA polymerase (14–176 nM final concentration) to preheated reaction mixtures, the concentration of DNA template being 1–2 nM. Samples are taken at different times after the addition of RNA polymerase and the amount of radioactive run-off transcript produced from the promoter determined. The lag observed (τ_{obs}) before a steady-state rate of transcription is achieved is estimated for each RNA polymerase concentration. Kinetic constants are deduced from the equation $\tau_{obs} = 1/k_f + 1/k_fK_B[R]$ (29).

Experimental Results and Interpretation

We have used the methods described above to study the transcriptional regulatory protein p4 of *Bacillus subtilis* phage $\phi29$. The genes of this lytic phage are transcribed in two stages, named early and late (reviewed in Ref. 30). Protein p4, the product of the early viral gene 4, directs the switch from early to late transcription in phage $\phi29$, acting as an activator of the promoter for the viral late genes, named A3, and simultaneously repressing the early A2b promoter (see Fig. 1). We focus first on

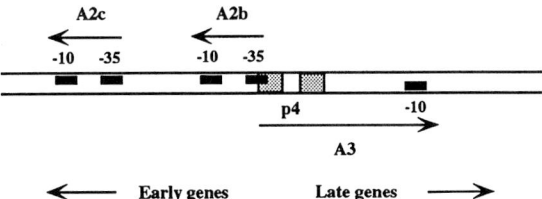

FIG. 1 Location of the promoters for early and late genes in the region of the $\phi29$ late A3 promoter. Black boxes denote the -10 and -35 regions having consensus sequences for *B. subtilis* σ^A-RNA polymerase, and open boxes are protein p4-binding sequences [I. Barthelemy and M. Salas, (32); B. Nuez, F. Rojo, I. Barthelemy, and M. Salas, (38)]. Note that the protein p4 binding site overlaps the -35 box of the early PA2b promoter, and that the late PA3 promoter lacks a -35 consensus box. The direction for early and late transcription is indicated.

the characterization of the complex formed by protein p4 with its target site on the viral genome; its role in transcription is discussed afterward.

Characterization of Complex Formed by Protein p4 with Its Target Site

Purification of protein p4 and the setup of an *in vitro* transcription assay (31) allowed the characterization of this protein as a transcriptional activator of the late A3 promoter. Gel retardation assays (Fig. 2) performed with a labeled DNA fragment containing the A3 promoter showed that protein p4 could bind to this DNA fragment under specific conditions, that is, in the presence of a large excess of nonspecific DNA. DNase I footprints allowed us to locate more precisely the binding site for protein p4, which was shown to span positions -58 to -104 relative to the transcription start site of the A3 promoter (32). As shown in Fig. 3A, binding of protein p4 to its target site protects some regions from DNase I cutting, with certain positions becoming hypersensitive to the cleaving agent. These positions are spaced about one helix turn from one another and give important information about the geometry of the complex.

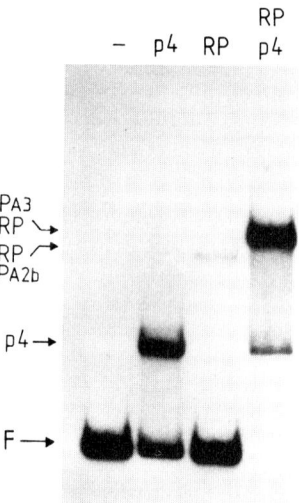

FIG. 2   Binding of protein p4 and RNA polymerase at the early A2b and late A3 promoters: gel retardation assay. Proteins included in each binding reaction are shown at top; RP indicates σ^A-RNA polymerase. In the absence of protein p4, RNA polymerase binds to the early PA2b promoter, although the binding is not stable. As will become clear in Fig. 3, binding of protein p4 to its target site displaces RNA polymerase from the early promoter and directs it to the late PA3 promoter. Note that binding of protein p4 and RNA polymerase to the late A3 promoter is cooperative, suggesting that they interact on binding to the promoter. F denotes free DNA.

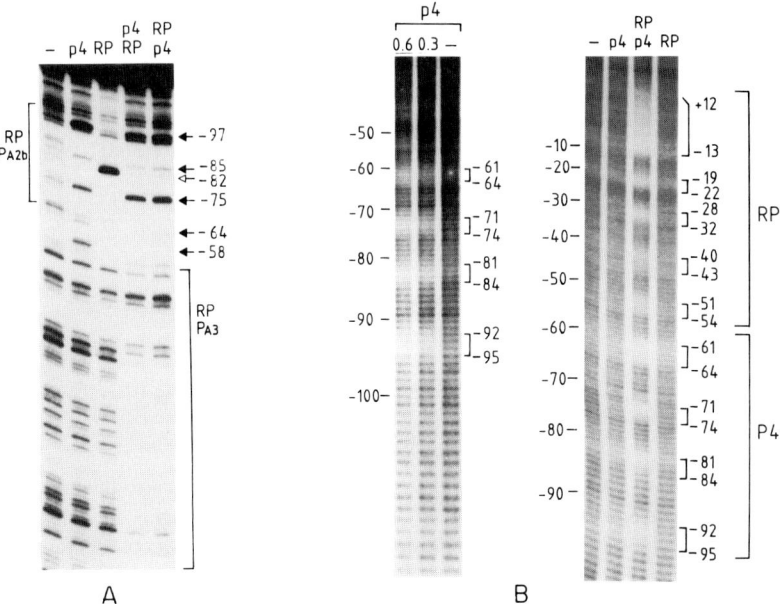

FIG. 3   Binding of protein p4 and RNA polymerase at the early A2b and late A3 promoters: footprinting assays. (A) DNase I footprinting. Proteins present in each binding reaction are indicated at top; RP stands for σ^A-RNA polymerase. In lane p4/RP, protein p4 was added 15 min before σ^A-RNA polymerase and the mixture incubated 15 min more; in lane RP/p4, the order of addition was opposite, which shows that protein p4 displaces σ^A-RNA polymerase from PA2b, driving it to the late A3 promoter. Positions hypersensitive to DNase I digestion on binding of protein p4 to its recognition site (black arrowheads) or σ^A-RNA polymerase to PA2b (position -82 relative to PA3 or -41 relative to PA2b, white arrowhead) are indicated, as well as the region protected by the binding of σ^A-RNA polymerase to PA2b and PA3. The footprint corresponds to the early strand (that coding for the early genes, which is the top strand on the sequence shown in Fig. 6C). (B) Hydroxyl-radical footprinting. Positions of the DNA backbone protected from the attack of hydroxyl radicals by protein p4 (left) or both protein p4 and σ^A-RNA polymerase (right) are indicated at the right of each autoradiogram. Numbering refers to the PA3 transcription start site; the footprint shown corresponds to the late strand (that coding for late genes, corresponding to the bottom strand of the sequence shown in Fig. 6C). A spatial representation of the protected regions can be seen in Fig. 5. Both figures have been previously published [F. Rojo and M. Salas, (37)].

DNase I cleavage rates vary greatly as a function of sequence, which is related to its manner of binding to DNA and its mechanism of action. DNase I interacts mainly with the phosphate backbone of the DNA, with binding affinity being very sensitive to the ionic strength. After binding, the DNA is bent about 20° away from the enzyme, so that the minor groove is widened to facilitate cutting (33). This is thought

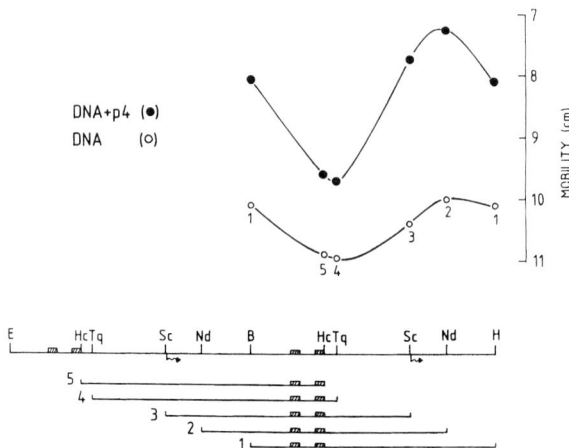

FIG. 4   Circular permutation assay. DNA fragments in which the protein p4 binding site (the putative bent site) had been circularly permuted were obtained by cloning as a tandem repeat a DNA fragment containing the p4 binding site; fragments of identical size but in which the p4 binding site is at different distances from the fragment ends were generated taking advantage of the fact that the single restriction sites on the cloned fragment are duplicated on the tandem repeat. The fragments were labeled at either end and their electrophoretic mobilities determined in 4% polyacrylamide gels in the absence or in the presence of protein p4. The mobility of a fragment is maximal when the bend center is close to one end of the molecule, whereas it is minimal when the bend center is at the middle of the fragment. The angle of curvature for the p4 binding site, estimated as indicated in the text, is about 45° in the absence of protein p4 and about 80°–85° in its presence [Rojo *et al.* (35); F. Rojo and M. Salas, (37); M. Mencía, M. Salas, and F. Rojo, (41)]. The figure has been modified from that previously published [Rojo *et al.* (35)].

to be the reason why DNase I cuts the DNA preferentially at positions being particularly flexible or already bent in a precise direction (33, 34), as bending of the DNA widens the minor groove in the outermost face of the bend and would facilitate cutting. With this mechanism in mind, it was proposed that the DNase I hypersensitive sites induced by protein p4 appeared as a consequence of a p4-induced increase in DNA curvature (32). This was later confirmed by circular permutation assays, which indicated that protein p4 binding to DNA increased the curvature of its binding site from about 45° to 85°, without changing the bend center (Fig. 4) (35). The presence of sites hypersensitive to DNase I inside of the binding site also suggested that protein p4 may not cover the complete DNA helix, but would rather bind to just one side of it. Because hypersensitive positions should lie in the DNA face being most accessible to the enzyme and with the minor groove widened (the outermost face of the bend),

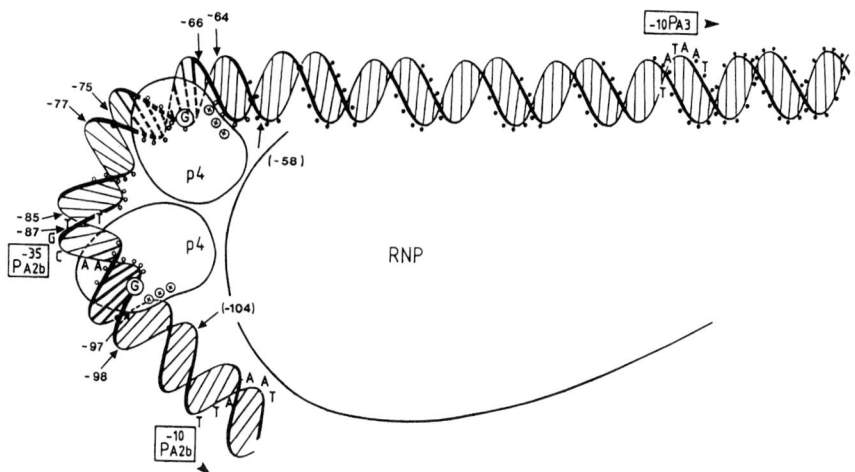

FIG. 5   Spatial model representing protein p4 and RNA polymerase bound at the phage ϕ29 late PA3 promoter. Thickened base pair lines correspond to the inverted repeat containing protein p4 target sequences. The -10 region for PA3 as well as the -10 and -35 regions for the divergently transcribed early PA2b promoter are indicated in rectangles. The positions that become hypersensitive to DNase I cleavage on protein p4 binding to DNA (arrows), guanine residues whose methylation interferes with protein p4 binding (enclosed in circles), and the positions that become protected from hydroxyl-radical attack by protein p4 (open circles) or by RNA polymerase bound at PA3 (filled circles) are also shown [Rojo *et al.* (17); Barthelemy and Salas (32); F. Rojo and M. Salas, (37)]. The DNase I hypersensitive sites represented in parentheses disappear when RNA polymerase binds to PA3. Plus signs ($+$) within protein p4 represent the basic carboxyl end of the protein, involved in maintaining the p4-induced DNA bending through electrostatic interactions with the DNA backbone [Rojo *et al.* (35); M. Mencía, M. Salas, and F. Rojo, (41)]. RNP indicates σ^{A}-RNA polymerase. Positions are numbered relative to the PA3 transcription start site as described by Rojo *et al.* (17). Reproduced from a previously published figure [M. Mencía, M. Salas, and F. Rojo, (41)].

a model was proposed showing which positions should lie in the inner and outer faces of the induced bend (see Fig. 5).

   Obtaining mutants of protein p4 lacking basic amino acids at the carboxyl end gave additional information. The mutants induced less bending than the wild type, as judged by circular permutation assays, and had lost the ability to induce the hypersensitive sites at both ends of the binding site, although they could still induce those located at the center of the binding site (35). This allowed us to propose a model in which binding of two protein p4 monomers (we currently think that it is two dimers, see below) to the inverted recognition sequences, and a subsequent interaction between them, would bend the DNA between the inverted repeats. Bending at both sides of the binding site would arise as a consequence of a nonspecific electrostatic interaction between the positively charged carboxyl end of the protein and the negatively charged DNA backbone.

The model was confirmed and extended by hydroxyl-radical footprinting. The sensitivity to hydroxyl radicals of the binding site in the absence of protein p4 is not even, but rather shows alternate regions of higher and lower susceptibility (32) (Fig. 3B). This is characteristic of DNA regions having a sequence-directed curvature (36) and reflects the curvature observed by circular permutation assays. The regions with lower sensitivity are located in the inner side of the curvature, where the narrower minor groove is less susceptible to the cleaving agent. Binding of protein p4 to DNA protects from hydroxyl radicals short stretches of nucleotides spaced about one helix turn, which coincide with those having a lower sensitivity to the cleaving agent in the absence of the protein (37). This indicates that protein p4 is binding to only one side of the DNA helix (otherwise the protection pattern would have been continuous) and that it binds to the inner side of the sequence-directed bend. As stated above, circular permutation assays indicated that protein p4 increases the sequence-directed bending considerably on binding, but they did not indicate the direction of bending. Both the hydroxyl-radical and DNase I footprints indicate that p4 binds to the inner side of the static curvature and that the increase in bending angle does not change direction but just increases the preexisting curvature. This makes sense, as the DNA should wrap around and accommodate protein p4 at the expense of the free energy obtained by protein binding. Binding stability would therefore be greater if the DNA is distorted in a direction of maximum flexibility, in this case the direction to which it naturally tends to bend in the absence of p4.

The protein p4 binding site at the late A3 promoter contains an 8-bp inverted repeat that includes protein p4 recognition targets. This was deduced from several approaches. First, data supporting this idea came from methylation interference assays, which indicated that guanine residues whose methylation interfered with protein p4 binding were those located in the inverted sequences (32). Methylation of any other guanine residue did not impair p4 binding (Fig. 6B). Hydroxyl-radical interference assays showed that the nucleosides whose removal was detrimental for protein p4 binding lay precisely in the above-mentioned inverted sequences, which confirmed their involvement in p4 recognition (Fig. 6A) (38). The target site was shown to be 5′-CTTTTT–15-bp spacer–AAAATG-3′. The symmetry of the binding site, together with the small size of the protein and the relatively long distance separating the inverted sequences (2.5 helix turns), suggested that protein p4 probably binds DNA as a dimer or as a tetramer. Protein p4 is a dimer in solution, as judged by gel filtration (F. Rojo, M. Mencía, and M. Salas, unpublished). Gel retardation assays performed with mixtures of wild-type and deletion mutants of protein p4 suggest that two p4 dimers bind to DNA (M. Mencía, F. Rojo, and M. Salas, unpublished).

Besides indicating the location of p4 target sequences, methylation interference

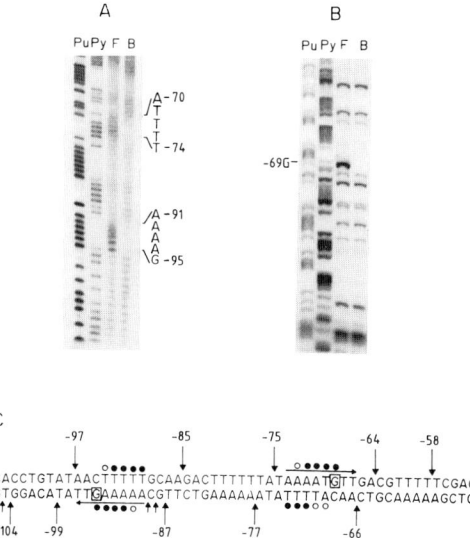

FIG. 6   Identification of protein p4 recognition sequences. (A) Hydroxyl-radical interference assay. An end-labeled DNA fragment containing the protein p4 binding site was treated with hydroxyl radicals to get, on average, one missing nucleoside per molecule. The DNA was then incubated with protein p4, and protein-bound (lane B) and free (lane F) DNA were resolved in a nondenaturing 4% polyacrylamide gel, eluted from the gel, and analyzed in a 6% denaturing polyacrylamide gel. Nucleosides whose absence interferes with protein p4 binding are indicated at the side. Lanes Pu and Py contain Maxam–Gilbert purine and pyrimidine sequencing reactions carried as size markers. Numbers indicate positions relative to the PA3 transcription start site. The results shown correspond to a DNA fragment labeled at the late strand. Nucleosides from the early strand critical for p4 binding are indicated in (C). Reproduced from a previously published figure [Nuez et al. (38)]. (B) Methylation interference assay. A 198-bp DNA fragment containing the A3 promoter was labeled at one end and partially methylated with dimethyl sulfate. The DNA fragment was incubated with protein p4; protein-bound and free DNA were resolved in nondenaturing band-shift gels and, after elution from the gel, treated with piperidine and analyzed in a 6% denaturing polyacrylamide gel. Free DNA (lane F) is enriched in DNA fragments in which guanine residue −69 has been methylated, whereas protein-bound DNA (lane B) contains only DNA not methylated at this residue. Chemical-cleavage purine (lane Pu) and pyrimidine (lane Py) sequencing reactions of the same DNA fragment were used as size standards. The result shown corresponds to the early strand, from a figure previously published [Rojo et al. (35)]; when the assay was performed with a DNA fragment labeled at the late strand, the G residue whose methylation interfered with protein p4 binding was that at position −95 [Barthelemy and Salas (32)]. (C) Protein p4 binding site. The inverted sequences are indicated by arrows. Nucleosides critical for protein p4 binding are denoted with filled circles, whereas open circles correspond to nucleosides whose absence diminishes but does not totally impair protein p4 binding. Guanine residues whose methylation interferes with protein p4 binding are enclosed in rectangles, and DNase I hypersensitive sites are indicated by vertical arrows. Reproduced from a figure previously published [Nuez et al. (38)].

assays also give useful information about whether the protein interacts with the bases through the major or minor groove. In double-stranded DNA, dimethyl sulfate methylates guanines at the N-7 position (accessible through the major groove) and, with a lower efficiency, adenines at the N-3 position (accessible through the minor groove) (8, 9). Protein p4 recognition sequences contain guanines and adenines, but only methylated guanines interfere with binding. This suggests that recognition of the target site takes place through the major groove (32).

Role of Protein p4 in Transcription Regulation

Gel retardation assays with a DNA fragment containing both the late A3 promoter and the adjacent but divergently transcribed early A2b promoter show that RNA polymerase binds weakly to the fragment in the absence of protein p4 (Fig. 2). DNase I footprintings indicate that in the absence of protein p4, RNA polymerase binds to the early A2b promoter; no binding is observed at the late A3 promoter (Fig. 3A). The fact that the binding is detectable by DNase I footprinting, but not by gel retardation assays, suggests that the A2b–RNA polymerase complex is weak and dissociates during electrophoresis in the band-shift assay, a process that lasts about 3 hr. DNase I footprints take no more than 3–4 min and are more appropriate in this case. Indeed, binding efficiency of RNA polymerase to the A2b promoter varies greatly from assay to assay, in agreement with an unstable binding. That the binding efficiency is low, despite the fact that the A2b promoter is very strong, suggests that this promoter is not optimized for RNA polymerase binding, but rather for other steps of the initiation process. In the presence of protein p4, gel retardation assays indicate that binding of the RNA polymerase is very efficient, but they do not show clearly to which promoter (A2b or A3) the polymerase is binding (Fig. 2). DNase I footprints show that protein p4 displaces RNA polymerase from the early A2b promoter, facilitating its binding to the late A3 promoter (Fig. 3A). Protein p4 is therefore repressing the early A2b promoter while activating the late A3 promoter. The two assays also suggest how protein p4 activates the late A3 promoter: because RNA polymerase does not bind to A3 in the absence of p4, the role of p4 must be to stabilize the RNA polymerase at the promoter. Measurement of the kinetic parameters of transcription initiation of the A3 promoter confirms this hypothesis. The τ plots indicate that protein p4 increases the affinity constant K_B by a factor of 8, whereas the rest of the steps leading to the formation of an elongation complex (open complex formation and promoter clearance) are accelerated 1.5-fold (39). Therefore, protein p4 increases the total promoter strength about 12-fold, mainly by facilitating the formation of the closed complex.

Hydroxyl-radical footprinting of the complex formed by p4 and RNA polymerase bound at the late A3 promoter shows a continuous array of protected and sensitive positions, spaced about one helix turn, with a large protected region around the transcription start site (37; Fig. 3B). The protection pattern of RNA polymerase indicates that it binds to only one face of the DNA helix, except in the region between around −10 and +15, where the protection is complete. This is similar to the pattern described for *Escherichia coli* RNA polymerase bound to the T7 A1 promoter (40). A spatial representation of the protected and sensitive positions shows that protein p4 and RNA polymerase bind to the same face of the DNA helix, and one just next to the other (Fig. 5). This suggests that the proteins could contact one another, an aspect that was confirmed by a variety of observations. First, it is clear from gel retardation assays that binding of protein p4 and RNA polymerase to the A3 promoter is a cooperative event (Fig. 2). This was confirmed using modified promoters with deletions or mutations at protein p4 recognition sequences (38). Second, protein p4 can form complexes with RNA polymerase, detectable by band-shift assays, in the presence of DNA fragments containing exclusively either the binding site for protein p4 or the binding site for RNA polymerase (39). Third, site-directed mutants of protein p4 were obtained that could bind to DNA as did the wild-type protein but did not stabilize the binding of RNA polymerase to the promoter, did not form complexes with RNA polymerase as judged by the band-shift assay described above, and were unable to activate transcription in run-off assays (41). On the basis of these results, it was concluded that Arg-120 of protein p4 is a likely candidate to interact with RNA polymerase, stabilizing it at the promoter and thereby activating transcription.

Conclusion

In summary, the techniques described above have allowed us to gain a detailed knowledge of both the physical structure of the complex formed by protein p4 and RNA polymerase at the ϕ29 late A3 promoter and the molecular mechanism underlying the activation of the promoter. The methods have been successfully used for several other regulated promoters and are a useful means for the characterization of any protein–DNA complex of interest.

Acknowledgments

This investigation has been aided by Research Grant 5R01 GM27242-15 from the National Institutes of Health, by Grant PB90-0091 from Dirección General de Investigación Científica y Técnica, and by an Institutional Grant from Fundación Ramón Areces. We thank I. Barthelemy, M. Mencía, B. Nuez, and M. Serrano for contributions.

References

1. M. G. Fried and D. M. Crothers, *Nucleic Acids Res.* **9,** 6505 (1981).
2. M. M. Garner and A. Revzin, *Nucleic Acids Res.* **9,** 3047 (1981).
3. D. Lane, P. Prentki, and M. Chandler, *Microbiol. Rev.* **56,** 509 (1992).

4. D. J. Galas and A. Schmitz, *Nucleic Acids Res.* **5,** 3157 (1978).
5. A. M. Maxam and W. Gilbert, *in* "Methods in Enzymology" (L. Grossman and K. Moldave, eds.), Vol. 65, p. 499. Academic Press, New York, 1980.
6. F. Sanger, S. Niklen, and A. R. Coulson, *Proc. Natl. Acad. Sci. U.S.A.* **74,** 5463 (1977).
7. T. D. Tullius and B. D. Dombrosky, *Proc. Natl. Acad. Sci. U.S.A.* **83,** 5469 (1986).
8. U. Siebenlist and W. Gilbert, *Proc. Natl. Acad. Sci. U.S.A.* **77,** 122 (1980).
9. U. Siebenlist, R. B. Simpson, and W. Gilbert, *Cell (Cambridge, Mass.)* **20,** 269 (1980).
10. J. J. Hayes and T. D. Tullius, *Biochemistry* **28,** 9521 (1989).
11. H.-W. Wu and D. M. Crothers, *Nature (London)* **308,** 509 (1984).
12. T. T. Stenzel, P. Patel, and D. Bastia, *Cell (Cambridge, Mass.)* **49,** 709 (1987).
13. F. Rojo and J. C. Alonso, *J. Mol. Biol.* **238,** 159 (1994).
14. J. Kim, C. Zwieb, C. Wu, and S. Adhya, *Gene* **85,** 15 (1989).
15. J. F. Thompson and A. Landy, *Nucleic Acids Res.* **16,** 9687 (1988).
16. R. L. Gourse, *Nucleic Acids Res.* **16,** 9789 (1988).
17. F. Rojo, B. Nuez, M. Mencía, and M. Salas, *Nucleic Acids Res.* **21,** 935 (1993).
18. H. Hayatsu and T. Ukita, *Biochem. Biophys. Res. Commun.* **29,** 556 (1967).
19. T. V. O'Halloran, B. Frantz, M. K. Shin, D. M. Ralston, and J. G. Wright, *Cell (Cambridge, Mass.)* **56,** 119 (1989).
20. G. J. Pruss and K. Drlika, *Cell* (Cambridge, Mass.) **56,** 521 (1989).
21. A. L. Meiklejohn and J. D. Gralla, *J. Mol. Biol.* **207,** 661 (1989).
22. J. C. Wang, L. J. Peck, and K. Becherer, *Cold Spring Harbor Symp. Quant. Biol.* **47,** 85 (1982).
23. W. Keller, *Proc. Natl. Acad. Sci. U.S.A.* **72,** 4876 (1975).
24. M. J. Chamberlin, *Annu. Rev. Biochem.* **43,** 721 (1974).
25. W. R. McClure, *Proc. Natl. Acad. Sci. U.S.A.* **77,** 5634 (1980).
26. A. J. Carpousis and J. D. Gralla, *J. Mol. Biol.* **183,** 165 (1985).
27. D. C. Straney and D. M. Crothers, *J. Mol. Biol.* **193,** 267 (1987).
28. B. Krummel and M. J. Chamberlin, *Biochemistry* **28,** 7829 (1989).
29. D. K. Hawley and W. R. McClure, *Proc. Natl. Acad. Sci. U.S.A.* **77,** 6381 (1980).
30. M. Salas and F. Rojo, *in* "*Bacillus subtilis* and Other Gram-Positive Bacteria: Biochemistry, Physiology, and Molecular Genetics" (A. L. Sonenshein, J. A. Hoch, and R. Losick, eds.), p. 843. American Society for Microbiology, Washington, D.C., 1993.
31. I. Barthelemy, J. M. Lázaro, E. Méndez, R. P. Mellado, and M. Salas, *Nucleic Acids Res.* **15,** 7781 (1987).
32. I. Barthelemy and M. Salas, *J. Mol. Biol.* **208,** 225 (1989).
33. D. Suck, A. Lahm, and C. Oefner, *Nature (London)* **332,** 464 (1988).
34. M. E. Hogan, M. W. Roberson, and R. H. Austin, *Proc. Natl. Acad. Sci. U.S.A.* **86,** 9273 (1989).
35. F. Rojo, A. Zaballos, and M. Salas, *J. Mol. Biol.* **211,** 713 (1990).
36. A. M. Burkhoff and T. D. Tullius, *Cell (Cambridge, Mass.)* **48,** 935 (1987).
37. F. Rojo and M. Salas, *EMBO J.* **10,** 3429 (1991).
38. B. Nuez, F. Rojo, I. Barthelemy, and M. Salas, *Nucleic Acids Res.* **19,** 2337 (1991).
39. B. Nuez, F. Rojo, and M. Salas, *Proc. Natl. Acad. Sci. U.S.A.* **89,** 11401 (1992).
40. P. Schickor, W. Metzger, W. Werel, H. Lederer, and H. Heumann, *EMBO J.* **9,** 2215 (1990).
41. M. Mencía, M. Salas, and F. Rojo, *J. Mol. Biol.* **233,** 695 (1993).

[24] Use of Coliphage λ and Other Bacteriophages for Molecular Genetic Analysis of *Erwinia* and Related Gram-Negative Bacteria

Vincent Mulholland and George P. C. Salmond

Introduction

Since the earliest developments in recombinant DNA techniques, bacteriophages have been important molecular biology "reagents." Phages (and their derivatives phagemids and cosmids) have been used as cloning vectors for sequencing and for expression systems. The *Escherichia coli* phage λ in particular has been exploited as one of the workhorses of prokaryotic molecular biology. However, the utility of λ as a cloning vector or as a vector for delivery of cosmids or transposons into a bacterium has been restricted by its narrow host range. In this chapter we discuss techniques which allow the exploitation of coliphage λ in bacteria other than *E. coli* K12, and focus particularly on the enteric phytopathogen *Erwinia*.

It is arguable that an indirect consequence of advances in recombinant DNA technology (as applied to bacteria other than *E. coli*) has been a steady shift away from any attempts to develop classical genetic tools in novel organisms. One of the classical tools of bacterial genetics is generalized transduction, mediated by either virulent or temperate phages. However, as amply demonstrated by the power of their use in *E. coli* genetics, generalized transducing phages offer quick and facile routes to strain construction, localized mutagenesis, and linkage analysis. When used in conjunction with a transposon mutagenesis system, generalized transduction can be especially useful by combining the technical ease of gene transfer with the selective power of antibiotic markers. In this review we intend to show how some phages can be used in a series of genetic manipulations in gram-negative bacteria. The use of such techniques is dependent on either isolation of novel phages for specific tasks, for example, transduction, or on the extension of the host range of currently available phages such as λ. When all of the tools are simultaneously applicable to an organism of interest, the organism then becomes as genetically tractable as *E. coli*. We use the development of a series of phage-related tools in *Erwinia* to highlight this point.

Part I: Phage λ-Based Gene Transfer Systems

Phage λ-based systems require the introduction of a gene encoding the λ receptor (LamB) into the bacterium and depend on both expression of this protein and correct processing to the outer membrane. To allow this system to operate efficiently, it is desirable that λ is unable to replicate and form plaques. However, the important ingredient in this system is the efficient injection of the DNA contained in the phage head.

Construction of λ Sensitive Strains

Conjugal Transfer of lamB⁺ Plasmids

The transfer of *lamB⁺*-containing broad-host-range plasmids (Table I) into gram-negative bacteria is usually accomplished by conjugal transfer or electroporation. Conjugal transfer requires the presence of a mobilization (Mob) site (commonly from RP4) containing an origin of transfer which is nicked, permitting the transfer of a

TABLE I   Bacteria, Plasmids, and Phages

Name	Characteristics	Ref.
Escherichia coli strain		
DH1	*recA hsdR17 gyrA supE endA*	a
HB101	*recA13 hsdR20 ara14 proA2 lacY1 galK2 rps120 xyl5 leu1 supE44*	b
LE392	*hsdR514 supE44 supF58 lacY1 galK2 galT22 metB1 trpR55*	c
Plasmids		
pHC79	Apᴿ Tcᴿ *cos⁺*	d
pHCP2	Apᴿ *lamB⁺*	e
pSF6	Spᴿ Smᴿ Mob⁺ *cos⁺*	f
pTroy⁹	Tcᴿ *lamB⁺*	g
RK2013	Knᴿ Tra⁺ (RSF1010 derivative)	h
λ derivatives		
λ467	$\lambda b_{221} O_{am29} P_{am3} rex$:: Tn5	c
λ840	$\lambda gt7$ -*his* $cI_{857} P_{am80} nin5 zzz$:: TN*10* Δ4 HH104	i
λTn*blaM*	λNM627 -[Δ(*slrλ1-2*) $cI_{857} slrλ4^C nin5 slrλ5^C S_{am7}$]:: Tn*blaM* (Spʳ)	j
λTn*phoA*	$\lambda b_{221} cI_{857} P_{am3} rex$:: Tn*phoA*	k

[a] D. Hanahan, *J. Mol. Biol.* **155**, 557 (1983).

[b] H. W. Boyer and D. Roulland-Dussoix, *J. Mol. Biol.* **41**, 459 (1969).

[c] F. J. de Bruijn and J. R. Lupski, *Gene* **27**, 131 (1984).

[d] B. Hohn and J. Collins, *Gene* **11**, 291 (1980).

[e] J. M. Clement, D. Perrin, and J. Hedgpeth, *Mol. Gen. Genet.* **185**, 302 (1982).

[f] G. Selvaraj, Y. C. Fong, and V. N. Iyer, *Gene* **32**, 235 (1984).

[g] G. E. de Vries, C. K. Raymond, and R. A. Ludwig, *Proc. Natl. Acad. Sci. U.S.A.* **81**, 6080 (1984).

[h] G. Ditta, S. Stanfield, D. Corbin, and D. R. Helinski, *Proc. Natl. Acad. Sci. U.S.A.* **77**, 7347 (1980).

[i] J. C. Way, M. A. Davis, D. Morisato, D. E. Roberts, and N. Kleckner, *Gene* **32**, 369 (1984).

[j] M. Tadayyon and J. K. Broome-Smith, *Gene* **111**, 21 (1992).

[k] C. Gutierrez, J. Barondess, C. Manoil, and J. Beckwith, *J. Mol. Biol.* **195**, 289 (1987).

single strand of the plasmid via a sex pilus. Sex pilus formation is usually mediated by the presence of a transfer-proficient plasmid conjugated into the strain containing the mobilizable plasmid during a triparental mating. A $lamB^+$ plasmid which is not self-transmissible is used for both simplicity of construction and for containment of the recombinant plasmid (in some cases this is of vital importance). The RP4 plasmid transfers to the strain carrying the mobilizable $lamB^+$ plasmid, which can then transfer to the recipient bacterium.

Mating is usually carried out by spreading the bacteria onto a nitrocellulose filter placed on a nonselective agar medium. RP4 has an inflexible sex pilus (unlike the flexible F plasmid sex pilus) which is fragile and sensitive to the shear forces present in liquid matings. Selection for the recipient after conjugation can be carried out in a variety of ways, including the use of minimal media or antibiotic selection depending on the phenotypic characteristics of the recipient strain (Tables II and III). Antibiotic selection is appropriate if the recipient strain expresses a resistance not found in the

TABLE II   Media and Buffers

Medium	Constituents (per liter)[a]
LB	10 g Bacto-tryptone, 5 g Bacto-yeast/extract, 5 g NaCl
2× YT	16 g Bacto-tryptone, 10 g Bacto-yeast extract, 5 g NaCl
Nutrient agar	28 g Oxoid nutrient broth agar
DDA	20 g Bacto-tryptone, 8 g NaCl, 9 g Bacto-agar for plates (or 2.5 g Bacto-agar for soft agar), (10 ml of 1 M MgSO$_4$)
Phage buffer	10 mM Tris, 10 mM MgSO$_4$, 0.01% (w/v) gelatin
50× Phosphate	350 g K$_2$HPO$_4$, 100 g KH$_2$PO$_4$
Minimal medium	(20 ml of 50× phosphate), (10 ml of 10%, w/v, ammonium sulfate), (0.41 ml of 1 M MgSO$_4$), (10 ml of 20%, w/v, carbon source)

[a] Items in parentheses are added as sterile solutions after autoclaving.

TABLE III   Antibiotics[a]

Antibiotic	Abbreviation	Stock solution (final concentration)
Ampicillin (sodium salt)	Ap	5 mg/ml (50 μg/ml)
Kanamycin sulfate	Km	5 mg/ml (50 μg/ml)
Nalidixic acid	Nx	5 mg/ml (50 μg/ml)
Spectinomycin	Sp	5 mg/ml (50 μg/ml)
Streptomycin sulfate	Sm	10 mg/ml (100 μg/ml)
Tetracycline	Tc	1 mg/ml (10 μg/ml)

[a] Tetracycline is dissolved in 50% (v/v) ethanol and stored at $-20°$C. Nalidixic acid is dissolved in 30 mM NaOH. All other antibiotics are dissolved in sterile water. The antibiotic stock solutions are filter sterilized and stored at 4°C (except Tc). Some antibiotics, for example, Ap, Kn, and Sp, should be stored for no longer than 1 month at 4°C.

E. coli strains used in the mating. A spontaneous antibiotic-resistant recipient may be isolated by plating out the bacterium at high cell density onto an agar plate containing either nalidixic acid (Nx) or streptomycin (Sm) (Table III). Such a mutant, resistant to Nx or Sm, may then be used as a selectable recipient in the triparental mating. Alternatively, selection on minimal media may be used to counterselect growth of the auxotrophic donor bacteria. Clearly, this is not possible if auxotrophic mutants of the recipient are required, for example, to assay if a phage is a generalized transducer (see below). An alternative, used in the case of isolation of *Erwinia* spp. mutants of various types, is the use of minimal medium containing either raffinose (1) or sucrose (2), neither of which can be used by *E. coli* as a carbon source.

Electroporation

The technique of electroporation of bacterial cells involves the application of a high voltage charge, across a narrow gap, through a high resistance sample of bacteria mixed with DNA (3). This technique is the most efficient transformation system yet devised. The equipment used in the process commonly relies on the discharge of a capacitor through a cuvette, with two opposing faces of metal, in which the cell suspension is placed. The actual mechanism of ingress of the DNA molecule is unknown, but it is likely to occur during transient permeabilization of specific locations in the bacterial membrane. A high ion concentration in either the bacterial suspension or the DNA can cause arcing across the electrodes which can damage the apparatus. As the electroporation of bacteria requires very high field strengths, a pulse controller should always be used to limit the possibility of arcing, achieved by placing a resistor in series with the sample which will limit the current if arcing occurs. In addition to presenting a mechanical danger (such as shattering the cuvette), arcing can cause the formation of aerosols which may present a significant biological hazard if the sample used is pathogenic. Therefore, the apparatus should be placed in a level of containment appropriate to the organism being used. Should arcing occur (and providing the cuvette remains intact), the cuvette should be left unopened for at least 30 min to allow the aerosol to settle, or disposed of without opening.

A number of intrinsic and extrinsic parameters govern the efficiency of DNA transfer into a bacterium. These include field strength, buffer composition, temperature, pH, DNA size, membrane/cell wall composition, restriction–modification barrier, growth phase, and outgrowth medium. As a starting point, one can use the standard conditions for *E. coli* (4), which may then be altered to increase transformation frequency.

Phage λ-Mediated Mutagenesis and Gene Transfer

The exploitation of λ-based methods requires the transfer, and constitutive expression of the *E. coli lamB* gene (encoding the λ phage receptor), to gram-negative

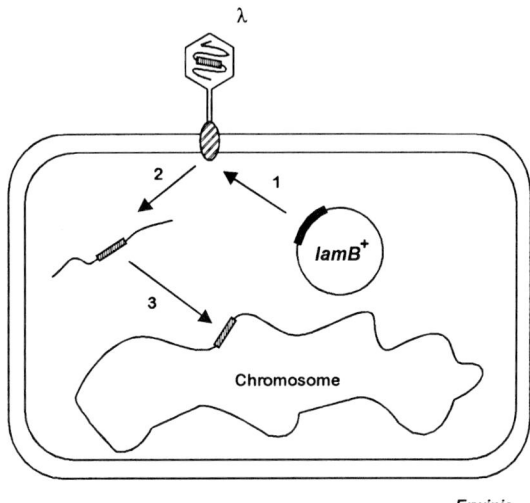

Fig. 1 Phage λ-mediated transposon mutagenesis of *Erwinia*. In step 1, the *lamB*⁺ plasmid, after introduction into the bacterium, synthesizes the λ receptor protein (LamB), which is then targeted to the outer membrane. In step 2, The λ phage binds to the LamB receptor and injects its DNA, carrying a transposon, into the bacterium. In step 3, λ is unable to replicate in *Erwinia*, enabling the positive selection of *Erwinia* in which transposition, from the λ DNA, has occurred.

bacteria that are not naturally sensitive to this phage (Fig. 1). The λ particle may then be able to adsorb to the outer membrane and inject its DNA into the cell. Two applications of λ-mediated gene transfer are considered: the delivery of transposons and the transfer of cosmids into non-*E. coli* hosts. A number of bacteria have been shown to be suitable for this technique including *Klebsiella oxytoca* (formerly *K. pneumoniae*), *Salmonella typhimurium* (5), *Vibrio cholerae*, numerous *Erwinia* spp. (1, 2, 6, 7), *Agrobacterium tumefaciens*, *Pseudomonas aeruginosa*, *Rhizobium meliloti*, *Azorhizobium sesbaniae* (8), and *Xenorhabdus bovienii* (9).

Plasmid-based suicide vectors for the delivery of transposons, although more promiscuous in their host range than *lamB*⁺-dependent techniques, may suffer from several drawbacks including cointegrate formation and stable maintenance of the plasmid in the recipient. The *sacB* gene from *Bacillus subtilis* (which mediates the synthesis of levansucrose in the presence of sucrose, lethal to most gram-negative bacteria) has been employed for the positive selection of loss of the delivery plasmid. However, such a system developed for *Xenorhabdus nematophilus* produced fewer than 20 mutants per conjugation (10).

The usefulness of λ-mediated transposon mutagenesis has been amply demonstrated in *Erwinia* spp. (1, 2, 7). Examples include isolation of auxotrophs, enzyme secretion mutants, global regulation mutants, growth rate mutants, and reduced virulence mutants of *Erwinia carotovora* subsp. *atroseptica* using Tn*5* mutagenesis (11, 12) and Tn*10* and Tn*phoA* mutagenesis of *Erwinia carotovora* subsp. *carotovora* (13). The use of Tn*phoA* mutagenesis allows the isolation of fusions to both exported and secreted proteins. Alkaline phosphatase, as part of a hybrid protein, is only catalytically active if exported to the periplasm. Alkaline phosphatase activity can be assayed by the breakdown of the chromogenic substrate 5-bromo-4-chloro-3-indolyl phosphate (XP) (14). This means that colonies which are blue on XP-containing plates have *phoA* fusions in genes encoding secreted, exported, or transmembrane proteins. In addition to mutagenesis, *lamB⁺ E. carotovora* subsp. *carotovora* has been shown to be capable of accepting cosmids by direct λ-mediated transfer, which has permitted the screening of a library of clones for complementation (15).

One a transposon-induced mutation is isolated, there are several techniques for further exploitation of the transposon for genetic analysis. If λ840 is used for mutagenesis, the Tn*10* mutants isolated can be used to obtain deletion derivatives by eduction around the insertion site (16). This technique, using a combination of fusaric acid and chlortetracycline which is toxic to cells carrying Tn*10,* has been used extensively over many years, and some improvements should be noted. The medium should be autoclaved for a little longer than is necessary for sterility (20 min, 121°C); serial dilutions of the culture to be cured of Tn*10* insertion should be plated onto the semiselective medium; and fast growing colonies should be picked and purified as soon as possible.

A series of λTn*phoA* derivatives has been constructed (17) which allows the switching of fusion type (and antibiotic resistance carried) by homologous recombination. An existing Tn*phoA* fusion may be exchanged to allow the production of translational or transcriptional *lacZ* fusions, or the construction of a variety of Tn*phoA* derivatives. Some of these are transposition-defective derivatives which are therefore stable after initial insertion. Some of the constructs cause polar mutations, whereas others contain an outward promoter to allow the transcription of downstream genes providing information on possible polar effects caused by the original insertion of Tn*phoA*. The use of this family of transposons, together with the construction of λ-sensitive strains of a variety of gram-negative bacteria, allows detailed genetic analysis of Tn*phoA* insertion mutants in many different bacteria.

Protocols for λ-Based Gene Transfer

Triparental Mating

1. Equal volumes of overnight cultures of the strains are mixed: (i) donor strain, for example, *E. coli* DH1 (pTroy9), (ii) mobilizing strain, for example, *E. coli* HB101 (RK2013), and (iii) recipient strain. Spread onto a sterile nitrocellulose filter on a nonselective nutrient agar plate and incubate for 6–16 hr at 30°C.

2. The nitrocellulose filter is then removed and placed in a universal bottle containing 2 ml LB (Table II). Resuspend the bacteria by vortexing for 30 sec.
3. Serial dilutions of the bacterial suspension are plated out onto the appropriate selective medium (see above).
4. Transconjugants are purified on the same selective medium.

Electroporation

1. A 10-ml culture is grown overnight in $2 \times$ YT (Table II) with antibiotic selection at $30° C$ with shaking. If high efficiency cells are required, a log phase culture (A_{600} of 0.5) should be grown from a 1 : 100 dilution of an overnight culture of the bacterium.
2. The culture is centrifuged (4500 rpm, 10 min, $4° C$).
3. The cells are washed twice with an equal volume of water, then twice with $\frac{1}{10}$ volume of 10% (v/v) glycerol.
4. The cells are then resuspended, in $\frac{1}{50}$ the starting volume, in water. If long-term storage is required, resuspend in 10% (v/v) glycerol and freeze at $-70° C$.
5. Plasmid DNA (5–50 ng) is added to 40 μl of cells (avoiding any bubbles) in a prechilled electroporation cuvette (0.2-cm electrode, Bio-Rad, Richmond, CA) and incubated on ice for 1–10 min.
6. Electroporation is carried out using a Bio-Rad Gene Pulser with pulse controller (capacitance 25 μF, voltage 2.5 kV, pulse controller resistance 200 Ω). Theoretically, a time constant of 5 msec is predicted; in practice the time constant is usually about 4.8 msec.
7. Immediately after pulsing, 1 ml of $2 \times$ YT medium is added to the cuvette and then incubated for 1 hr at $30° C$ to allow recovery, before being plated onto selective medium.

Escherichia coli cells can be kept for at least 6 months at $-70° C$ (see step 4) without significant loss of electroporation efficiency.

High-Titer λ Lysates

Note: The chloroform used in the isolation steps and for storage of phages should be treated with sodium hydrogen carbonate before use: add 1 g $NaHCO_3$ per 20 ml $CHCl_3$, shake well, and leave for 1 hr; decant the supernatant into a foil-covered glass bottle. Phage λ derivatives (containing conditional mutations) are prepared on the suppressing strain *E. coli* LE392.

1. An overnight culture of *E. coli* LE392 is grown in LB plus 10 mM $MgSO_4$.
2. Two λ plaques [10^5–10^6 plaque forming units (pfu)] are resuspended in 1 ml phage buffer (Table II), and 10, 50, and 100 μl is added to separate 200-μl aliquots of the overnight culture.
3. Following the addition of 3 ml soft DDA ($45° C$; Table II), the mixture is over-

layed on a fresh, wet DDA plate (40 ml DDA per 9 cm petri dish). Set up plates late in the day and incubate plates overnight at 37°C.

4. The next day inspect plates (by comparison with a phage-free control lawn) and harvest the plate which shows confluent lysis. The top agar is removed with a glass spreader, the plate is washed with 3 ml of phage buffer, and the washing pooled with the top agar. One-half milliliter chloroform is added and vortexed for 15 min to break up the agar overlay. The agar is pelleted in a centrifuge (4500 rpm, 20 min at 4°C), and the supernatant is decanted.

5. Lysates are stored over a few drops of chloroform at 4°C. This method yields 1–5 × 10^{10} pfu/ml for λ 467 or λ Tn*phoA*.

6. Lysates are titered by spotting 10 μl of 10^{-2}, 10^{-4}, 10^{-6}, and 10^{-8} dilutions onto a freshly poured lawn of *E. coli* LE392 and incubating at 37°C overnight. The plaque-forming units per milliliter can then be estimated by counting the number of plaques at an appropriate dilution.

Phage λ-mediated Transposon Mutagenesis of Erwinia

1. An *Erwinia* derivative carrying a *lamB*$^+$ plasmid (pHCP2 or pTroy9) is used and must be grown in the presence of the appropriate antibiotic (Table III) to select for maintenance of the plasmid [ampicillin (Ap) or tetracycline (Tc), respectively].

2. An overnight culture is diluted 1:20 into 10 ml of LB plus 10 mM MgSO$_4$ and antibiotic. Incubate at 30°C, with shaking, to an A_{550} of 0.8 (~2.5 hr). Alternatively, a 10-ml fresh overnight culture can be used directly.

3. The cells are pelleted (4500 rpm, 5 min), and the pellet is resuspended in 1 ml of LB containing 10 mM MgSO$_4$ and antibiotic. [If the strain under investigation produces large amounts of extracellular polysaccharides, the LamB receptor may be occluded. It may be possible to increase transduction efficiency by washing the culture in 1 M NaCl for between 1 and 20 min with gentle shaking (7)].

4. One hundred microliters is removed as a control for spontaneous antibiotic resistance. Then 1–5 × 10^8 pfu of λ carrying a transposon is added and incubated at 30°C for 30 min, after which 9 ml LB is added and shaken at 30°C for 1 hr.

5. The cells are pelleted (4500 rpm, 5 min) and resuspended in 1 ml LB. Aliquots (100 μl) are spread on nutrient agar plates containing the appropriate antibiotics and incubated at 30°C for 36 hr.

The procedure yields approximately 200 kanamycin-resistant (Km^R) colonies following infection of *E. carotovora* subsp. *carotovera* SCRI193 (pHCP2) [≡HC131] with 10^7 pfu of λ467. There are a variety of permutations of methods for this type of mutagenesis, and, to some extent, the outcome (efficiency of transpositions, spectrum of mutants, and gene fusions isolated) is a reflection of the particular conditions used.

Preparation of λ Lysate Carrying a Cosmid Library

1. A 25-ml culture of *E. coli* carrying the cosmid library to be packaged, for example, *E. coli* DH1, is grown to mid log phase in LB with antibiotic selection in

a 200-ml flask (37°C, 300 rpm). As fast growth is essential, good aeration is required, and antibiotics which can slow the growth of resistant strains (e.g., Tc) should be avoided where possible.

2. MgSO$_4$ is added to 10 mM, along with 250 μl λ *cI857 lysate stock* ($\sim 10^{10}$ pfu), and the culture is incubated at 30°C, 275 rpm, for 20–30 min.

3. The cells are heat-shocked (42°C, 20 min), then incubated (38°C, 275 rpm) until about 30 min after the first signs of lysis are detected (3–4 hr in total).

4. The sample is then transferred to a universal bottle with 1 ml chloroform (treated with NaHCO$_3$; see above) and vortexed for 15 min at 4°C. After centrifugation (4500 rpm, 10 min, 4°C), the supernatant is decanted into a fresh tube and MgSO$_4$ added to 10 mM. The lysate is stored over chloroform, at 4°C.

The titer of lysates prepared by this method tends to be slightly lower and more variable than for lysates prepared by the plate method. However, the main advantage of superinfection is that large volumes of the lysate can be easily obtained.

Complementation by λ-Mediated Cosmid Transduction

1. A mutant *Erwinia* carrying a *lamB*$^+$ plasmid (see above) is used as the recipient. A 5-ml overnight culture is grown at 30°C in LB plus 10 mM MgSO$_4$ plus antibiotic selection for the plasmid.

2. The cells are pelleted (4500 rpm, 5 min) and resuspended in 0.5 ml LB plus 10 mM MgSO$_4$. One hundred microliters is removed and spread onto an appropriate antibiotic-containing medium [e.g., if the cosmid library is based on pHC79, use Tc; if based on pSF6, use spectinomycin (Sp)] as a control for spontaneous antibiotic resistance.

3. One hundred microliters of the cosmid lysate is added to the remaining cells and incubated at 30°C for 30 min. Then 10 ml LB is added and incubated with shaking at 30°C for 2 hr.

4. The cells are pelleted in a microcentrifuge for 1 min, then resuspended in 100 μl of LB and plated out onto medium with the appropriate antibiotics. Incubate at 30°C for 48 hr and pick colonies to test for complementation.

Part II: Generalized Transducing Phages

Transduction is one of the three major routes of genetic exchange in bacterial systems, transformation and conjugation being the others. Transduction can be divided into two broad categories, specialized and generalized. Specialized transduction occurs when a prophage, integrated into the bacterial chromosome, excises aberrantly so that the released phage DNA carries part of the region surrounding the attachment site at one end, for example, the *gal* or *bio* genes either side of the *attB* site of λ integration in *E. coli*. Uses for this method of transduction are limited as there are size constraints on the amount of DNA packaged and only the region adjacent to the site of integration is transferred.

More useful tools for genetic manipulation and analysis are the generalized transducing phages which are able to package any region of the chromosome (although the frequency of packaging/transduction may vary from region to region). Parts of the bacterial chromosomal DNA are packaged into phage heads instead of the normal substrate, the phage genomic DNA. The process by which this occurs seems to vary among different phages. However, a detailed consideration of the mechanisms of formation of transducing particles is unnecessary in this consideration of generalized transduction as a tool for molecular manipulation and analysis. The manifold uses of generalized transduction include recombination mapping, strain construction, localized mutagenesis, and plasmid transfer. However, it is in combination with transposons that the full potential of generalized transduction is apparent. An obvious example of this is the analysis of transposon-induced mutants using a generalized transducing phage that can be used to transfer the transposons into a genetically clean background, allowing the cotransduction of the phenotypes to be assayed. Conventional molecular techniques, such as Southern hybridization, can identify the number of transposons present in a mutant, but generalized transducing phages allow the separation of multiple insertions to isolate the mutation of interest.

Isolation of Phages

Phages have been isolated from a wide variety of sources, including water (ponds and lakes, rivers, runoff from agricultural land), soil, excreta, foodstuffs (including uncooked meat products), and sewage. The choice of source for isolation may be governed by the distribution of hosts and reservoirs of the bacterium under investigation, but this is not true of the isolation of *Erwinia* spp. phages. One of the richest environmental sources for phages is sewage, and such samples have been used to isolate *Erwinia carotovora* bacteriophages in our laboratory (18, 19). The phages isolated include many which infect the environmentally rare *E. carotovora* subsp. *atroseptica*. These observations suggest that most of the novel phages have replicated naturally in other, more common, non-*Erwinia* hosts, implying a degree of promiscuity in phage host range.

The number of phages isolated can be increased by the use of a strain carrying a *lamB*$^+$-containing plasmid in the isolation procedure. A *lamB*$^+$-containing *Serratia* has been used in our laboratory to isolate phages from sewage, and 7 of the 12 phages isolated have been shown to be *lamB*$^+$-dependent (20).

Some phages are known to require cofactors for adsorption, so to isolate the maximum possible number of different phages the cofactors should be added to all media and buffers used in the enrichment process. Cofactors include $CaCl_2$ (10 mM), $MgSO_4$ (10 mM), casamino acids (0.2%, w/v), and tryptophan (0.2%, w/v). After isolation, one can then assay the requirements of the individual phage (e.g., the requirements may be for Mg^{2+} and casamino acids so, for that isolate, Ca^{2+} and tryptophan may be omitted).

The phages isolated should be assayed for the ability to transduce into various mutants, for example, auxotrophs isolated by transposon mutagenesis. Additionally, the ability to transduce plasmids between strains may also be assayed. To assay transduction of nutritional markers, a lysate prepared on the progenitor strain is used to transduce auxotrophic mutants to prototrophy, selecting on minimal medium. To establish the generalized nature of a potential transducing phage, it is vital to transduce a variety of markers.

Because the conditions required for successful transduction cannot be predicted, the method for transducing both *E. carotovora* subsp. *carotovora* (with ϕKP) and *E. carotovora* subsp. *atroseptica* (ϕM1) (18) is given as a starting point. This protocol usually works with ϕKP transductions into *E. carotovora* subsp. *carotovora*. However, if there are problems with efficiencies of transduction or killing of transductants caused by phage carryover, modifications outlined below may help to alleviate the difficulties. The standard transduction method (see below) may be modified to increase the frequency of transduction by (i) altering the temperature of incubation in step 4, (ii) varying the multiplicity of infection (MOI) ratio, or (iii) adding divalent cations (Mg^{2+}, Ca^{2+}, etc.).

Exploitation of Generalized Transducing Phages

Transposon Trapping

Transposon trapping allows the isolation of a transposon in the vicinity of a gene of interest, to permit transfer of the linked gene in strain construction or for localized mutagenesis. A wild-type isolate is mutated using λ-mediated transposon mutagenesis (see above). The resulting pool of insertion mutants is harvested and used to prepare a heterogeneous culture which, in turn, is used to produce a transducing phage lysate. The heterogeneous lysate is then used to transduce a strain carrying the mutation of interest, and transductants are selected on antibiotic-containing plates. The transductants are screened (or selected) for the presence of the wild-type gene, that is, examined for cotransduction of the transposon and the wild-type gene into the mutant strain. Strains which have a presumptive linked gene–transposon are then analyzed in detail for cotransduction of the two phenotypes to confirm linkage. Lysates prepared on the strains may be used in strain construction or localized mutagenesis.

Localized Mutagenesis

A region which is linked to a transposon can be specifically mutated using chemical mutagenesis of a transducing lysate. This allows the isolation of subtle mutants (e.g., carrying temperature-sensitive mutations), which may be moved between strains (by generalized transduction) using the transposon tag as a marker (21). Initially, a test mutagenesis is performed to assay phage inactivation by the mutagen for a particular phage lysate, that is, the inactivation of plaque-forming activity is used as an indirect monitor of mutagenicity.

Trapping of Bla⁺ fusions

Tn*blaM* encodes constitutive spectinomycin resistance and will give rise to ampicillin-resistant mutants (Bla⁺) when fused in-frame to an exported, secreted, or transmembrane protein. When bacteria containing random Tn*blaM* insertions are patch replicated onto agar plates containing 10 μg/ml Ap, three classes are found. Class one insertions produce no growth (or a few isolated colonies), class two gives rise to many (>20) isolated colonies and class three gives confluent growth. The third class are the true Bla⁺ fusions. The first two classes arise from the Tn*blaM* inserting in nonfusion sites (out of frame, reversed with respect to the direction of transcription, within a noncoding region, or in a gene whose product is not targeted to the membrane or exported). The major difference between class one and class two is predicted to be the rate of secondary transposition events, giving rise to true Bla⁺ fusions which are positively selected as secondary transpositions of the original Tn*blaM* insertion. The class two colonies (termed "high hoppers") were used to isolate Bla⁺ fusion mutants of *E. carotovora* subsp. *carotovora* without the need for further λ-mediated transposon mutagenesis (22). Selection on increasing amounts of Ap, up to at least 1 mg/ml, allowed the isolation of a variety of fusions, which could then be transduced into the progenitor strain using ϕKP (an *E. carotovora* subsp. *carotovora* generalized transducing phage). The value of this method is that strains which give rise to only a few mutants after λ-mediated transposon mutagenesis (because of inefficient infection by λ) may be tested for the presence of high hoppers of Tn*blaM*. This approach would circumvent the need to increase the efficiency of mutagenesis.

Use of Coliphages in Gram-Negative Bacteria

An additional consideration in the use of phages in some gram-negative bacteria is the possibility of employing *E. coli* phages, such as P1, T4GT7, and Mu. P1 is able to package plasmids carrying an *inc* region cloned from P1, and deliver them to myxobacteria, although P1 and the vectors themselves are unable to replicate (23). This system is used in delivery of transposons, marker exchange, and insertional mutagenesis in myxobacteria, for example, in the identification of developmentally regulated genes, after mutagenesis using P1 :: Tn*5lac* (24). In addition, P1 has been shown to be capable of infecting *Klebsiella aerogenes, K. oxytoca, Erwinia amylovora, Citrobacter,* and *Enterobacter* species, and P1-mediated transduction between *Klebsiella* and *E. coli* has been used to create hybrid strains (25). Phage T4 must be engineered with a series of mutations (e.g., T4GT7) before it is useful as a transducing phage (26). T4 receptors have been found in some strains of both *E. carotovora* subsp. *carotovora* and *E. carotovora* subsp. *atroseptica,* and lytic growth was observed in some sensitive strains (27). T4GT7 was used to transfer plasmids from *E. coli* to both *Erwinia carotovora* subspecies. Phage Mu has also been shown to infect a variety of *Erwinia* spp., and this observation has allowed the use of Mu derivatives both as mutagenic agents and for the isolation of fusions in the study of gene expression in *Erwinia chrysanthemi* (28).

A filamentous phage-based transposon delivery system has been developed in *Erwinia amylovora* (29). A mutant fd phage was used, in which a gene required for phage replication had been replaced by a fragment of DNA containing Tn5. The fd::Tn5 phage was propagated in bacteria carrying a cloned complementing gene. The *E. amylovora* derivative used in the mutagenesis carried an F plasmid required to produce the F sex pilus needed for infection of fd, which is a male-specific phage.

The use of coliphages in gram-negative bacteria other than *E. coli* is far more extensive than the examples given above might suggest. However, a comprehensive review of all the systems which have been developed to take advantage of this group of phages is beyond the limits of this chapter.

Protocols for Generalized Transduction

Isolation of Phages

1. Samples (500 ml) of untreated sewage, activated sludge, and effluent are collected from appropriate points of a domestic sewage treatment plant.
2. The samples are pooled, and the particulate matter is pelleted in a centrifuge (250-ml aliquots, 9000 rpm, 4°C for 15 min).
3. The supernatant is decanted and 5 ml chloroform added. The sample is then centrifuged (9000 rpm, 4°C for 15 min) and the supernatant transferred to a sterile 2-liter Erlenmeyer flask containing the constituents required for LB broth (Table II).
4. Five milliliters of an overnight culture of the chosen bacteria is added to each flask (and antibiotics if required) and incubated with shaking (250 rpm) at 15°, 25°, and 30°C for 24–48 hr. The range of temperatures which allows the replication of individual phage isolates can be narrow, so the optimum temperature must be determined empirically after isolation. The potential cofactors (Ca^{2+}, Mg^{2+}, etc.) should also be added (see above).
5. The bacteria are pelleted (9000 rpm, 4°C for 15 min), and the supernatant is stored over 5 ml chloroform (treated with $NaHCO_3$; see above).
6. The supernatant is serially diluted in phage buffer. Samples (100 μl) of the diluted supernatants are added to 300-μl aliquots of an overnight bacterial culture and incubated at the temperature used in the isolation of the phage for 15 min.
7. The bacterium/phage mixture is then added to 3.5 ml of LB containing 0.7% agar at 45°C (also containing Mg^{2+}, Ca^{2+}, tryptophan, and casamino acids) and poured as an overlay onto a nutrient agar plate containing the appropriate antibiotic. The plate is then incubated at the temperature used in the phage isolation.

Phages isolated from an enrichment can be used to select phage-resistant mutants (by spotting a phage onto a lawn of bacteria and selecting colonies which grow within the plaque). The original lysate (from step 5) can then be used to infect the phage-resistant mutants (as in steps 6 and 7). In this way it is possible to isolate other phages from the enrichment which are far less abundant than the phages which are "dominant" postenrichment. This technique is very efficient and increases the number of phages isolated.

Generalized Transduction of Erwinia

1. A 10-ml overnight culture is set up in LB plus 10 mM MgSO$_4$ with shaking at 30°C, for each transduction.
2. The overnight cultures are centrifuged (4500 rpm, 10 min) and resuspended in 1 ml LB plus 10 mM MgSO$_4$. An aliquot of 100 μl is removed and plated onto selective medium as a control for spontaneous antibiotic resistance and/or reversion.
3. Phage lysate to an MOI of 1.0 is added (add 10^9 phage assuming 10^9 bacteria in a 10-ml overnight culture). It is prudent to perform three transductions per experiment, adding 10-fold more and 10-fold less than the "ideal" number of phage.
4. The cell/phage mixtures are incubated at 25°C for 15–20 min.
5. The transductions are centrifuged (4500 rpm, 10 min) and washed twice with 10 ml LB.
6. The pellet is resuspended in 10 ml LB and incubated at 30°C with shaking for 40 min.
7. The cells are pelleted and resuspended in 1 ml LB.
8. One hundred microliters is plated onto an appropriate selective medium. The remaining cells are pelleted and spread onto another plate.
9. Incubate at 30°C for 48–72 hr.

Modified Transduction Method (Used with ϕKP)

1. One-half milliliter phage lysate is diluted in 4.5 ml phage buffer and transferred to a 35-mm glass petri dish.
2. A short-wavelength UV lamp is used to irradiate the phage lysate (at a fluence of 8 μW/cm^2 $\times$ 100). The lid is removed to begin irradiation, and 250-μl samples are removed at 0, 20, 40, 60, 80 sec, etc. The aliquots are stored on ice until use.
3. Whole-plate titers of the irradiated lysates are performed (10 μl, serially diluted), and transductions are carried out using 200 μl of irradiated lysate at each time point.
4. The optimum transduction frequency versus survival of plaque forming units is determined.

Hydroxylamine Mutagenesis

1. One milliliter of a high titer (transposon trapped) transducing lysate is mixed with 2 ml phosphate EDTA buffer [6 ml of 1 M K$_2$HPO$_4$ is added to 43.9 ml of 1 M

KH$_2$PO$_4$ (pH 6.0), mixed with an equal volume of 10 mM EDTA, and autoclaved], 3 ml water, and 4 ml hydroxylamine solution (560 μl of 4 M NaOH added to 350 mg hydroxylamine and made up to 5 ml with water) and incubated at 37° without shaking.

2. A similar reaction is set up, substituting the hydroxylamine solution with water, which is incubated in parallel with the sample containing the mutagen as a control.

3. Samples (100 μl) are removed every 6 hr, for up to 48 hr, and titered to assay phage inactivation. A graph, plotting phage titer against time, is constructed for the mutagenized phage and for the control phage not exposed to hydroxylamine. The time that gives a 1000-fold reduction in phage viability is used for the full-scale mutagenesis.

4. A mutagenesis is performed (as in step 1), using the optimum incubation time found in step 3, and transferred to a 50-ml Oakridge centrifuge tube.

5. The mutagenized phage lysate is pelleted (2.5 hr, 17,000 rpm at 23°C; Beckman J2-21 centrifuge, JA-17 rotor), the supernatant is carefully discarded, and the tube is blotted dry with tissue paper.

6. The pellet is resuspended in 1 ml of phage buffer overnight at 4°C and then used to transduce the wild-type bacterium (to antibiotic resistance, selecting for the transposon). The transductants are then screened for mutant phenotypes.

Final Comments

In this survey of the techniques available to exploit phages, we have given several examples of the strategies and methods used for the isolation of bacteriophages for novel hosts and for screening for transducing phages. We have also described methods for transposon delivery, gene fusion construction, and direct cosmid complementation in *Erwinia,* by exploiting the extended host range conferred by the transfer of the λ receptor gene (*lamB*). Many of these methods are already applicable to other gram-negative bacteria, particularly enteric bacteria. However, in principle, most of the methods should be suitable for a wider range of hosts than the literature currently suggests. As our studies with *Erwinia* demonstrate, success with such techniques will allow the researcher to open up the molecular genetic analysis of any new host in an effective and powerful way.

Acknowledgments

The authors thank Drs. Fiona Ellard, Jay Hinton, Nick Housby, and Ian Toth for contributions to the development of some of the methods reported in this review. We thank Dr. Susan Wharam for helpful discussions. This work was supported by National Environmental Research Council award GR3/8003 and Agriculture and Food Research Council award PG88/P01449 to G.P.C.S.

References

1. G. P. C. Salmond, J. C. D. Hinton, D. R. Gill, and M. C. M. Pérombelon, *Mol. Gen. Genet.* **203,** 524 (1986).
2. F. M. Ellard, A. Cabello, and G. P. C. Salmond, *Mol. Gen. Genet.* **218,** 491 (1989).
3. D. Hanahan, J. Jesse, and F. R. Bloom, *in* "Methods in Enzymology" (J. H. Miller, ed.), Vol. 204, p. 63. Academic Press, San Diego, 1991.
4. W. J. Dower, J. F. Miller, and C. W. Ragsdale, *Nucleic Acids Res.* **16,** 6127 (1988).
5. G. E. de Vries, C. K. Raymond, and R. A. Ludwig, *Proc. Natl. Acad. Sci. U.S.A.* **81,** 6080 (1984).
6. E. T. Palva, A. Harkki, H. Karkku, H. Lang, and M. Pirhonen, *Microb. Pathog.* **3,** 227 (1987).
7. E. M. Steinberger and S. V. Beer, *Mol. Plant–Microbe Interact.* **1,** 135 (1988).
8. R. A. Ludwig, *Proc. Natl. Acad. Sci. U.S.A.* **84,** 3334 (1987).
9. M. S. Francis, A. F. Parker, R. Morona, and C. J. Thomas, *Appl. Environ. Microbiol.* **59,** 3050 (1993).
10. J. M. Xu, M. E. Olsen, M. L. Kahn, and R. E. Hurlbert, *Appl. Environ. Microbiol.* **57,** 1173 (1991).
11. J. C. D. Hinton, J. M. Sidebotham, L. J. Hyman, M. C. M. Pérombelon, and G. P. C. Salmond, *Mol. Gen. Genet.* **217,** 141 (1989).
12. V. Mulholland, J. C. D. Hinton, J. M. Sidebotham, I. K. Toth, L. J. Hyman, M. C. M. Pérombelon, P. J. Reeves, and G. P. C. Salmond, *Mol. Microbiol.* **9,** 343 (1993).
13. J. C. D. Hinton and G. P. C. Salmond, *Mol. Microbiol.* **1,** 381 (1987).
14. C. Manoil and J. R. Beckwith, *Proc. Natl. Acad. Sci. U.S.A.* **82,** 8129 (1985).
15. P. J. Reeves, D. Whitcombe, S. Wharam, M. Gibson, G. Allison, N. Bunce, R. Barallon, P. Douglas, V. Mulholland, S. Stevens, D. Walker, and G. P. C. Salmond, *Mol. Microbiol.* **8,** 443 (1993).
16. B. R. Bochner, H.-C. Huang, G. L. Schieven, and B. N. Ames, *J. Bacteriol.* **143,** 926 (1980).
17. M. R. Wilmes-Riesenberg and B. L. Wanner, *J. Bacteriol.* **174,** 4558 (1992).
18. I. K. Toth, Ph.D. Thesis, Univ. of Warwick, UK (1991).
19. I. K. Toth, M. C. M. Pérombelon, and G. P. C. Salmond, *J. Gen. Microbiol.* **139,** 2705 (1993).
20. N. Thompson and G. P. C. Salmond, unpublished observations (1991).
21. J. S. Hong and B. N. Ames, *Proc. Natl. Acad. Sci. U.S.A.* **68,** 3158 (1971).
22. G. P. C. Salmond, unpublished observations (1991).
23. L. J. Shimkets, R. E. Gill, and D. Kaiser, *Proc. Natl. Acad. Sci. U.S.A.* **80,** 1406 (1983).
24. L. Kroos and D. Kaiser, *Proc. Natl. Acad. Sci. U.S.A.* **81,** 5816 (1984).
25. R. B. Goldberg, R. A. Bender, and S. L. Streicher, *J. Bacteriol.* **118,** 810 (1974).
26. G. G. Wilson, K. K. Y. Young, G. J. Edlin, and W. Konigsberg, *Nature (London)* **280,** 80 (1979).
27. M. Pirhonen and E. T. Palva, *Mol. Gen. Genet.* **214,** 170 (1988).
28. J. Ji, N. Hugouvieux-Cotte-Pattat, and J. Robert-Baudouy, *J. Gen. Microbiol.* **133,** 793 (1987).
29. P. Bellemann and K. Geider, *J. Gen. Microbiol.* **138,** 931 (1992).

[25]   Isolation, Purification, and Function of Assembly Intermediates and Subviral Particles of Bacteriophages PRD1 and φ6

Dennis H. Bamford, Päivi M. Ojala, Mikko Frilander, Laura Walin, and Jaana K. H. Bamford

Introduction

Cellular functions are associated with macromolecular assemblages. For example, replication, transcription, protein synthesis, energy production, and protein secretion take place in complex structures, where proteins, nucleic acids, and membranes actively interact. The cellular concentration of proteins is extremely high, some 100 mg/ml, and it is possible that there does not exist a single separate enzyme, but rather the cellular functions are strictly coordinated and structurally organized to complex functional units that intimately interact with one another. It is of crucial importance to obtain information on biological macromolecule complexes in order to understand the critical interactions leading to the assembly of the subunits and the function of the final structure. In many instances, it has been shown that the assembly information is built within the primary protein structure. The proteins "know" how to fold, how to interact with other proteins and other biological molecules, as well as how to carry out their specific functions. The information is not usually regulated by gene expression, but the components exist simultaneously in the cell and build assembly pathways to form the final structure.

It has been relatively difficult to obtain complete cellular assemblages in highly pure, intact form and in large quantities for detailed studies. Viruses have offered a model system for such studies: The basic rules for assembling a virus inside the cell must be governed by the same rules as the assembly of cellular complexes. It has been possible to obtain large quantities of pure viruses and their subassemblies and assembly intermediates for biochemical and biophysical studies. The principal means used for production of different virus particles are (i) naturally producing particles from infected cells, (ii) producing particles from infections with mutant viruses, (iii) disrupting the virions by chemical or physical means, and (iv) using expression systems based on recombinant DNA technology. In this chapter, we describe the techniques used in our laboratory for obtaining large quantities of relatively pure viral and subviral particles for biochemical and structural studies. The model systems in use are two bacterial viruses, PRD1 and φ6, which both have a lipid membrane as their structural component. Thus, the particles in both cases include nucleic acid, protein, and lipid constituents.

In this chapter, all protein concentration measurements are carried out by the Coomassie blue method of Bradford (1) using bovine serum albumin (BSA) as a standard. The sodium dodecyl sulfate–polyacrylamide gel electrophoresis (SDS-PAGE) system is a modified Laemmli system and is described in detail by Olkkonen and Bamford (2). To separate the PRD1 and the φ6 structural proteins, we routinely use a 16% acrylamide concentration. For the preparation of reproducible sucrose gradients, we use a tilted tube rotation device (BioComp Fredericton, New Brunswick, Canada) (3), where the sucrose gradient is formed from the light and heavy components by rotation of the tube. When the centrifugation procedures do not indicate the rotor, we refer to a microcentrifuge (Eppendorf or equivalent). The LB growth medium used in all phage propagations is described elsewhere (4).

Bacteriophage PRD1

The double-stranded DNA (dsDNA) bacteriophage PRD1 infects a variety of gram-negative bacteria harboring a broad-host-range IncP-type plasmid. The common hosts for PRD1 are *Escherichia coli, Salmonella typhimurium,* and *Pseudomonas aeruginosa.* The highest titers are obtained with *S. typhimurium.* The phage receptor is encoded by a conjugative plasmid (such as RP4 or RP1) which can easily be transferred to a desired host using appropriate antibiotic resistance selections (5). The plasmid is only needed for the receptor synthesis, as plasmid-free cells can produce viruses if, for example, DNA containing the terminal proteins (see below) is electroporated into the cells (6). In a liquid culture, virus production takes about 1 hr and several hundred infective virus particles are liberated to the medium on host cell lysis (7). Approximately 20% of the produced particles are empty, containing no DNA.

The virion is about 65 nm in diameter and is surrounded by a very stable protein coat composed of a major capsid protein P3 and a minor protein P5 (8). Both capsid proteins exist as trimers. The protein shell encloses the viral membrane, which is composed of approximately half protein and half lipid (9). A schematic picture of PRD1 is presented in Fig. 1. There are some 20 membrane-associated proteins that can be divided into integral membrane proteins and those peripherally associated with the lipid bilayer (10). The dsDNA genome resides inside the lipid membrane. The DNA is a 15-kb linear double-stranded molecule with a covalently linked protein at both 5' ends (11). The terminal protein is used as a primer in the initiation of DNA replication (12–14). Because of its structure, PRD1 is being used as a model for understanding the membrane structure and its translocation into the virus particle. In addition, the replication system yields information on the protein-primed replication.

For all these studies, large quantities of pure viral particles and their subassemblies are needed. Below we describe new methods to obtain ultrapure virus preparations, particles lacking DNA, pure protein shells composed of the major protein P3 alone, and trimeric P3 capsomers. In addition, methods are described for purification of the

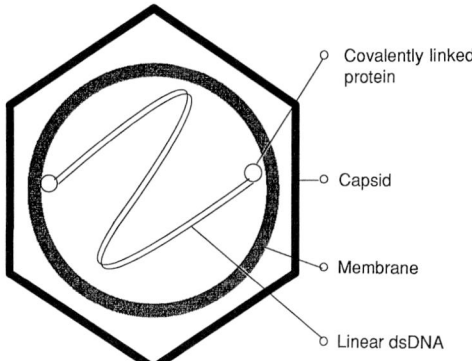

FIG. 1   Schematic presentation of phage PRD1 particle. The outermost protein capsid is composed of two proteins (P3, P5) which comprise some 80% of the protein mass of the virion. Underneath the capsid resides the viral membrane with which the rest of the virion proteins are associated. The membrane vesicle encloses the linear dsDNA molecule with covalently linked terminal proteins.

viral DNA with and without the genome terminal protein. For a more detailed description of the PRD1 system, see Refs. 15 and 16.

Purification of PRD1 Particles with Rate Zonal Centrifugation and Ion-Exchange Chromatography

Wild-type PRD1 bacteriophages are propagated in a liquid culture of the host bacterium *S. typhimurium* DS88 (17). Virus particles are liberated to the growth medium on host cell lysis and are concentrated from the lysate with polyethylene glycol (PEG) 6000 and NaCl precipitation. Further purification is done as a combination of rate zonal sucrose gradient and differential centrifugation (18). Better purification results can be obtained if the differential centrifugation step is replaced with an anion-exchange chromatographic step (29).

1. PRD1 stock can be obtained as follows. Plate appropriate dilutions of the phage with top agar on LB plates using the overnight grown *Salmonella typhimurium* DS88. After overnight incubation at 37°C, take the semiconfluent plates (500–1000 plaques/plate), collect the top agar, and add 5 ml of LB per collected plate. Incubate the suspension with aeration at 37°C for 3 hr. Remove the cell debris and the top agar by centrifugation (Sorvall SS-34 rotor, 10,000 rpm, 10 min, 4°C). Determine the number of viable phages in the supernatant by plating serial dilutions with *S. typhimurium* DS88 cells on LB plates. Incubate the plates overnight

at 37°C, count the plaques, and determine the titer. The PRD1 stock can be stored at 4°C for weeks without loss of titer.

2. Set up a 0.2-liter culture of the host bacterium *S. typhimurium* DS88 in LB and grow overnight with aeration at 37°C. Next morning, dilute the culture to obtain about 2.5×10^8 cells/ml (e.g., make 1 liter) and grow as above. Follow the increase in cell density with spectrophotometer (A_{540}) or with Klett colorimeter. When the cell density is about 1×10^9 cells/ml, infect the culture with PRD1 at a multiplicity of infection (MOI) of 5–10. In about 2 hr, the culture becomes clear on lysis of the cells.

3. Remove the cell debris from the lysate with low speed centrifugation (Sorvall GSA or GS-3 rotor, 8000 rpm, 20 min, 4°C). Add solid PEG 6000 to a final concentration of 10% (w/v) and NaCl to 0.5 *M*. Dissolve the PEG by magnetic stirring at 4°C. Centrifuge the virus precipitate (Sorvall GSA or GS-3, 8000 rpm, 20 min, 4°C). Remove the supernatant carefully, wash the residual PEG from the walls of the bottles, and resuspend the virus pellet overnight on ice into 12 ml of 20 m*M* potassium phosphate, pH 6.0.

4. Separate the aggregates from the virus preparation with low speed centrifugation (Sorvall SS-34 rotor, 8000 rpm, 10 min, 4°C). Prepare 5–20% sucrose gradients in 20 m*M* potassium phosphate, pH 6.0, and load 2 ml of the concentrated phage preparation on top of each gradient and centrifuge (Sorvall AH-629 rotor, 24,000 rpm, 60 min, 15°C).

5. Two light-scattering bands can be observed. The upper band contains DNA-less empty particles and the lower one infectious viruses. Collect the light-scattering virus band from the gradients. The material can be concentrated with differential centrifugation (Sorvall T-865 rotor, 33,000 rpm, 2 hr 30 min, 5°C). Remove the supernatant and resuspend the virus material on ice into 2 ml of 20 m*M* potassium phosphate, pH 6.0.

6. For anion-exchange chromatography purification of the virus obtained from the sucrose gradient, equilibrate the MemSep1000 (Millipore, Bedford, MA) DEAE cartridge with 5 bed volumes (7 ml) of 20 m*M* potassium phosphate, pH 6.0, 5 bed volumes of 1 *M* NaCl in 20 m*M* potassium phosphate, and 10 bed volumes of 20 m*M* potassium phosphate at a flow rate of 1.5 ml/min.

7. Pump the collected sucrose band to the cartridge by a peristaltic pump at a flow rate of 5 ml/min. Wash with 20 bed volumes of the phosphate buffer and elute the bound virus material with a linear NaCl gradient (0–1 *M* NaCl). Use a flow rate of 3.5 ml/min in the washing and elution. Collect 1.5-ml fractions. The virus elutes at about 330 m*M* NaCl.

Analyze the protein concentration of the fractions. Assay the infectivity of the peak fraction to ensure that the biological activity of the virus is preserved over the purification procedure (the specific infectivity should be approximately 1×10^{10} pfu/μg protein). The virus preparation can be stored on ice and will retain infectivity for about 1 week. PRD1 particles, purified as described above, contain less than 0.5%

protein impurities (as measured by SDS-PAGE). The amount of UV-absorbing impurities is reduced to one-half and the residual fluorescent impurities are reduced to about one-sixth of that found in the sucrose gradient-purified material. In addition, particles are uniform in size in electron micrographs (19).

Isolation of DNA-Less Empty Virus Particles and Membrane-Free Protein Shell

To be able to accurately assign the viral proteins to their actual positions in the virion, techniques for controlled removal of specific proteins are needed. In the case of membrane-containing viruses, methods for the virus membrane extraction must also be developed. This is most easily done with different detergent treatments. Detergents that differ in properties such as critical micellar concentration and ionic nature remove different membrane-associated proteins from a particle. The order of protein removal is dependent on the nature of the bonds that the protein makes with the rest of the particle. In the case of PRD1, it has been possible to separate integral and peripheral membrane proteins from one another (10). In the next section, the preparation of pure major coat protein P3-containing shells from DNA-less packaging-deficient *sus1* particles (20) is described. PRD1 *sus1* is an amber mutant in gene *IX*. It can be propagated in *Salmonella typhimurium* PSA (pLM2) as described in the PRD1 purification protocol (step 1). The PRD1 *sus1* has to be titered both with the suppressor host PSA and with the nonsuppressor host DS88. Typically, the suppressor host yields a titer at least 10^3 times higher than that of the nonsuppressor.

1. Make a *S. typhimurium* DS88 culture in LB and grow it as in the case of the wild-type virus. Infect the culture with *sus1* stock (MOI of around 10) obtained using the suppressor cells. After cell lysis, concentrate and purify the particles as described for wild-type virus (rate zonal centrifugation and ion-exchange chromatography). Use 20 mM Tris-HCl, pH 7.4, as a buffer. Only a single light-scattering zone can be seen in the gradients.
2. Determine the protein concentration of the purified concentrated *sus1* virus suspension and adjust to 2 mg/ml with the same buffer.
3. Add 20% (w/v) SDS to obtain a final concentration of 1%. Incubate at room temperature for 15 min.
4. Prepare 5–20% (w/v) sucrose gradients (in 20 mM Tris-HCl, pH 7.4) and add 2 ml of SDS–virus suspension on top of each gradient tube and centrifuge (Sorvall AH-629 rotor, 24,000 rpm, 1 hr 45 min, 15°C).
5. Collect the light-scattering zone which now represents the empty P3 shells. Concentrate the material either with differential centrifugation or ion-exchange chromatography (see above).
6. Determine the protein concentration of the material and analyze protein composition with SDS-PAGE.

Isolation and Purification of Major Capsid Protein Trimers (Capsomers)

The interactions between the capsomers as well as between the capsid protein and the membrane components can be disrupted by guanidinium hydrochloride (GuHCl) treatment. The GuHCl treatment releases the major capsid protein P3 and the minor capsid protein P5 as trimeric capsomers from the rest of the virus material that forms an aggregate. The trimers can then be further purified by standard biochemical methods.

1. Add 6 M GuHCl to 10 mg purified PRD1 in 20 mM Tris-HCl, pH 7.4 (0.5 – 1 mg/ml) to obtain 2.5 M final concentration and incubate at 23°C for 15 min.
2. Dilute with 20 mM Tris-HCl, pH 7.4, buffer to reduce the GuHCl concentration to 1.5 M. Remove the membrane and the DNA aggregate by differential centrifugation through a 20% (w/v) sucrose cushion in the same buffer (Sorvall T-865 rotor, 35,000 rpm, 3 hr, 15°C).
3. After centrifugation, remove the layer that contains the major and minor capsomers from the top of the cushion and dialyze it overnight against 20 mM Tris-HCl, pH 8.5, 50 mM NaCl at 4°C.
4. Filter the dialyzed fraction through a 0.22-μm filter and pump it (1 ml/min) into a Mono Q HR5/5 (Pharmacia, Piscataway, NJ) FPLC (fast protein liquid chromatography) column equilibrated with the same buffer. Wash the column with the same buffer until A_{280} reaches baseline values. Elute with a salt gradient (0 – 0.5 M NaCl in the same buffer, 1 ml/min) for 30 min. The P3 capsomers will elute as a sharp peak at approximately 0.3 M NaCl.
5. Measure the protein concentration of the collected peak fraction and analyze the protein content by SDS-PAGE in 16% gels.

We typically obtain about 3 mg P3 capsomers from 10 mg virus (which can be obtained from about 1.3 liters of culture). The theoretical amount (about 5.5 mg) is based on the approximations that the virus contains 15% DNA, 15% lipid, and 70% protein, of which 81% is in the coat proteins and 19% in the membrane. On SDS-PAGE, two isomorphs of P3 can be seen: a major one at about 45 kDa and a minor one at about 40 kDa (the molecular mass calculated from the sequence is 43.1 kDa). The minor capsomers (P5) also bind and elute from the Mono Q column, but owing to the large excess of P3, the P5-containing fraction is contaminated with P3.

Isolation of Genomic DNA from PRD1 Particles

The linear dsDNA genome can be isolated from purified wild-type virus particles either with or without the covalently linked terminal protein (P8). As the presence of the genome terminal protein is a prerequisite for the initiation of DNA replication, the DNA must be isolated for this purpose with a functional P8. On the other hand, it is essential to remove the terminal protein from the DNA used for cloning experiments because it is hydrophobic and affects the behavior of the DNA during agarose gel electrophoresis and in organic solvent extractions. For this reason, the DNA containing the terminal proteins is not used in physical studies. Rather harsh treatments have to be used to liberate the DNA from the stable capsid and from the hydrophobic membrane components. Two methods are described here: disruption of the virus with guanidinium hydrochloride (GuHCl) followed by a cesium chloride (CsCl) density gradient centrifugation (originally described for phage ϕ29; see Ref. 21) and disruption with SDS or SDS and protease treatment followed by phenol extractions (6).

Guanidinium Hydrochloride–Cesium Chloride Method to Isolate Genomic DNA with Covalently Linked Terminal Protein

1. Take 0.5 ml of purified virus (1–3 mg/ml) and add $\frac{1}{10}$ volume of 10× TES buffer (500 mM Tris-HCl, pH 7.8, 100 mM EDTA, 1 M NaCl). The final concentration of EDTA should be at least 10 mM.
2. Add an equal volume of 8 M GuHCl (in 1× TES buffer) and incubate on ice for 1 hr.
3. Add 1.710 g CsCl to the sample and bring the volume to 1.8 ml with 1× TES buffer. Make a CsCl step gradient in two SW55 tubes: layer 2 ml of CsCl solution (use TES buffer as the solute) with a density of 1.85 (1.1673 g CsCl/ml), 0.9 ml of sample, and on the top 2 ml of CsCl with a density of 1.55 (0.7421 g CsCl/ml) in each tube. Centrifuge (Beckman SW55 rotor, 35,000 rpm, 24 hr, 20° C).
4. Fractionate the gradients and analyze the fractions for PRD1 DNA in a 0.6% agarose gel (in 50 mM Tris-HCl, pH 8.6, 50 mM H$_3$BO$_3$, 2.5 mM EDTA). Pool the fractions containing the DNA and dialyze against TE buffer (10 mM Tris-HCl, pH 7.4, 1 mM MgCl$_2$). Precipitate the DNA with ethanol if necessary. DNA can also be stored in CsCl at 4° C, and when needed, dialyzed against water in a small scale over a 0.025-μm membrane filter (Millipore) at room temperature for 30 min.

Sodium Dodecyl Sulfate–Phenol Method to Isolate Genomic DNA with or without Covalently Linked Terminal Protein

1. Adjust the protein concentration of the purified virus preparation (in 10 mM Tris-HCl, pH 7.4) to 0.5 mg/ml with the same buffer.
2. Add SDS to a final concentration of 2% (w/v) and, if removal of the terminal protein is desired, also proteinase K (stock 20 mg/ml in water) to a final concentration of 100 μg/ml or predigested (1 hr at 37° C) Pronase (stock 20 mg/ml in 10 mM Tris-HCl, pH 7.8, 10 mM NaCl) to a final concentration of 0.5 mg/ml. Incubate at 37° C for 45–60 min so that the preparation becomes viscous owing to DNA release.

3. Extract several times (three to five) using phenol saturated with 0.1 M Tris-HCl, pH 8.0. Do not vortex. If protease is not used, the terminal protein stays linked to the DNA and most of the material is found in the phenol–water interface. In this case, carefully collect both the aqueous phase and the interface. Finally, extract six times with ether.

4. Add $\frac{1}{10}$ volume of 3 M NaCl and recover the DNA by precipitating it with 2 volumes of cold ($-20°C$) ethanol. Centrifuge immediately (14,000 rpm, 15 min, 4°C), wash the pellet with cold 75% (v/v) ethanol, and dry under vacuum. Dissolve the pellet in TE (10 mM Tris-HCl, pH 7.4, 1 mM MgCl$_2$). Dispense in desired aliquots, precipitate with ethanol, and store dried in $-20°C$.

The purity and concentration of the isolated DNA is determined by measuring the absorbances at 260 and 280 nm. The presence of residual proteins and lipids can be estimated from the A_{260}/A_{280} ratio, which is typically between 1.86 and 1.90 with purified protein-free PRD1 DNA. If the ratio is within these limits, the concentration can be calculated from the A_{260} value (an A_{260} value of 1 equals 50 μg/ml of dsDNA). We have typically obtained 110–140 μg DNA from 1 mg purified virus, the theoretical amount being about 150 μg. The yield of DNA containing the terminal protein is much lower.

The quality of the DNA is analyzed in a 0.6% agarose gel by slightly overloading the gel in order to see possible degradation of the DNA. The biological activity of the DNA containing the terminal protein can be assayed by following its ability to initiate DNA replication. This can be done *in vivo* by transferring the DNA into the host bacteria by electroporation (6) and counting the plaques formed by the progeny phages, or *in vitro* using purified replication components (purified PRD1 DNA polymerase and terminal protein) (13). The DNA containing the terminal protein, purified with SDS–phenol extraction method, is active in the initiation reaction and can be transferred by electroporation into host cells with high frequency.

Bacteriophage ϕ6

The enveloped dsRNA bacteriophage ϕ6 infects plant pathogenic *Pseudomonas syringae* cells. The virion has an outer diameter of about 86 nm. The P3 protein spikes, needed for absorption to the host cell, extend 7–8 nm from the virus surface. A schematic picture of the ϕ6 structure is presented in Fig. 2. The membrane envelope is 6 nm thick and surrounds the nucleocapsid (NC) (22). The envelope contains several integral membrane proteins. The NC has an outer shell protein P8 which surrounds the RNA polymerase complex composed of four proteins (P1, P2, P4, and P7) (23). This complex has the enzymatic machinery for the packaging of the ssRNA genomic segments, for the synthesis of the corresponding ($-$) strands to form dsRNA, and for the ($+$) strand synthesis using the dsRNA as a template (24). In the

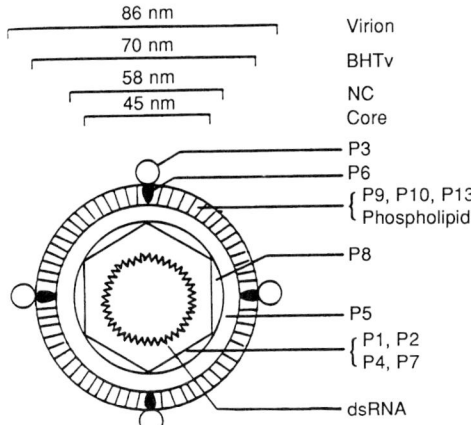

FIG. 2   Schematic presentation of phage φ6 particle. The BHTv refers to a spikeless particle. The locations of φ6 proteins (P1–P13) are indicated. The genome consists of three dsRNA segments L, M, and S. [From T. Li, D. H. Bamford, J. K. H. Bamford, and G. J. Thomas, Jr., *J. Mol. Biol.* **230**, 461 (1993).]

polymerase complex, P1 forms a dodecahedral framework with which the other proteins are associated. Protein P2 contains the active site of the RNA polymerase, and P4 is an NTPase needed for genome packaging. Protein P7 is also associated with the genome packaging although no specific activity has been assigned. The genome consists of three linear dsRNA segments, L (6374 bp), M (4061 bp), and S (2948 bp), which reside inside the dodecahedral polymerase particle.

The receptor for the phage is a pilus which functions as a pathogenesis factor for adherence to the target cells (25). After initial adsorption to the receptor, pilus retraction brings the virus particle into contact with the host cell outer membrane (OM). The virus membrane fuses with the host OM and the NC-associated lytic enzyme digests a local opening to the peptidoglucan layer (26, 27). The NC passes through the opening and interacts with the cytoplasmic membrane (CM). An endocytosis-like event brings the NC into the host cell cytoplasm where the particle is activated (28, 29). The virion-associated RNA polymerase produces ssRNA molecules of positive polarity. The translation of the l segment results in the assembly of empty polymerase complex particles (procapsids) composed of proteins P1, P2, P4, and P7 (30). These particles package all three (+) strand genomic segments and synthesize the corresponding (−) strands inside the particles (31, 31a, 32). After this, the newly synthesized particles produce (+) strands for late gene expression. In the late stage of infection, protein P8 covers the filled procapsid to form the NC. The envelope is obtained from the host plasma membrane where the viral membrane proteins have been synthesized. The enveloped particles reside in the cell cytoplasm before the cell lysis.

In this section, we present the methods for purifying $\phi6$ particles as well as the chemical disruption system used for dissociation of the virus particle to subviral components. In addition, a system for production of the polymerase complex from an expression plasmid system is presented. Because $\phi6$ is a dsRNA virus and thus has a virion-associated RNA-dependent RNA polymerase, the methods for assaying different polymerase activities are also included. These assays measure the functional integrity of the purified subassemblies and range from an RNA packaging reaction to a method where viable virus particles can be produced by an *in vitro* reassembly procedure, thus giving an ultimate test for the quality of different assembly intermediates.

Growth and Purification of $\phi6$

Wild-type $\phi6$ bacteriophages are propagated in a liquid culture of the host bacterium *P. syringae* (2). After cell lysis, the liberated progeny virus particles are concentrated from the lysate with PEG 6000 and NaCl precipitation. The virions are further purified by rate zonal and differential centrifugations.

1. Prepare at least 10 semiconfluent LB plates (500–1000 plaques/plate) of $\phi6$ by plating appropriate dilutions into top agar with the overnight grown *Pseudomonas syringae* pv. *phaseolicola* HB10Y. Incubate the plates overnight at 23°C. Collect the phage-containing top agar, add 3–5 ml LB per collected plate, and incubate the suspension with aeration at 23°C for 4 hr. Remove the cell debris and the top agar by centrifugation (Sorvall SS-34 rotor, 8000 rpm, 15 min, 4°C) and assay the phage-containing supernatant for infectivity. Store the phage stock at 4°C.
2. Set up a 0.2-liter overnight culture of HB10Y in LB at 28°C and incubate with aeration. Dilute it next morning to approximately 1×10^8 cells/ml (3 liter total volume) and grow up at 28°C for the first 2 hr and then continue to grow at 23°C until the cell density is about 5×10^8 cells/ml (takes about 30 min).
3. Infect the cells with the $\phi6$ stock in LB (prepared as in step 1), using an MOI of 10 to 12, and continue the growth at 23°C with aeration. Follow the infection by measuring the turbidity (A_{540}) of the infected culture. Slow down the shaking when the cells start to lyse (about 90 min postinfection).
4. After cell lysis, remove the cell debris by centrifugation (Sorvall GS-3 rotor, 8000 rpm, 15 min, 4°C). Add solid PEG 6000 and NaCl to obtain final concentrations of 9% (w/v) and 0.5 M, respectively. Dissolve the compounds with magnetic stirring at 4°C. Collect the phage precipitate by centrifugation (Sorvall GS-3 rotor, 8000 rpm, 20 min, 4°C) and resuspend the pellets in 27 ml of 20 mM Tris-HCl, pH 7.4.

5. Remove large aggregates from the concentrated phage preparation by centrifugation (Sorvall SS-34 rotor, 10,000 rpm, 10 min, 4°C). Load 1.5 ml of the cleared supernatant onto a 5–20% (w/v) sucrose gradient in 20 mM Tris-HCl, pH 7.4, and centrifuge (Sorvall AH-629 rotor, 24,000 rpm, 50 min, 15°C). Collect the light-scattering zone and concentrate the phage preparation by differential centrifugation (Sorvall T-647.5, 32,000 rpm, 3 hr, 4°C). Resuspend the phage pellets in 9 ml of 20 mM Tris-HCl, pH 7.4, and assay for infectivity. The specific infectivity of the phage preparation is usually about 4×10^9 pfu/μg protein.

Chemical Disruption of Virion and Isolation of Nucleocapsid, Nucleocapsid Core (Polymerase Complex), and Nucleocapsid Shell Protein P8

Phage φ6 can be biochemically disrupted into subviral particles. The stepwise disassembly scheme is presented in Fig. 3. The hydrophilic virus spike protein P3 can be removed by treating the purified phages with butylated hydroxytoluene (BHT) (33). The viral NC can be isolated from the spikeless virus particle by extraction with a nonionic detergent (Triton X-114), which removes the viral membrane and the membrane-associated proteins (34). The NC can be further dissociated to the NC core and P8 by chelating calcium ions that are important for the stability of the P8 shell on the NC surface (35). Also, the NC shell protein P8 can be purified.

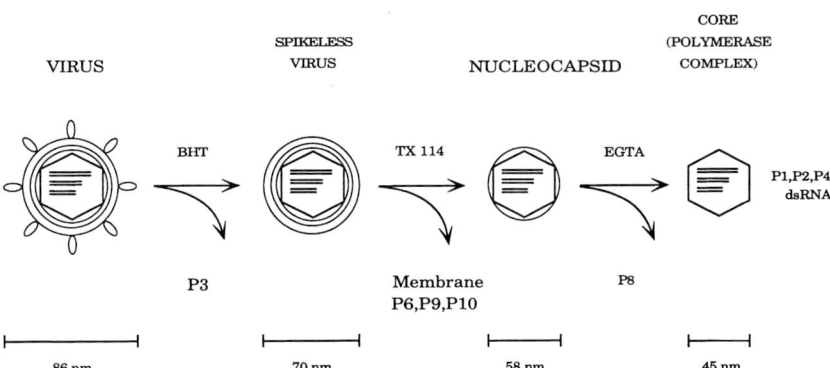

FIG. 3   Chemical disassembly of φ6 virion. The BHT removes the P3 spike protein. Nonionic detergent (Triton X-114, TX 114) is used to dissolve the phospholipids and integral membrane proteins. Removal of Ca^{2+} (with EGTA) releases the NC surface protein P8, yielding the RNA polymerase particle, that is, the NC core. [From T. Li, D. H. Bamford, J. K. H. Bamford, and G. J. Thomas, Jr., *J. Mol. Biol.* **230**, 461 (1993)].

The composition of $4\times$ uncoating mixture is as follows: 80 mM Tris-HCl, pH 7.4, 3 mM MgCl$_2$, 1% (w/v) Triton X-100, 100 mM EGTA, 450 mM NaCl, and 150 mM KCl.

1. Remove the virus spike protein P3 by treating 12 ml of the purified concentration virus preparation with 4 mM BHT at 30°C for 5 min and sediment the P3-free particles through a 20% (w/v) sucrose cushion (in 20 mM Tris-HCl, pH 7.4), using 2 ml virus preparation per 10 ml of cushion, to the bottom of the tube (Sorvall TH-641 rotor, 35,000 rpm, 2 hr 20 min, 10°C). Resuspend the pellets into 12 ml of 20 mM Tris-HCl, pH 7.4.

2. Isolate the viral NCs by extracting the P3-free virus preparation with about 3–5% (w/v) Triton X-114 (precondenced in 10 mM Tris-HCl, pH 8.0, 150 mM NaCl) (36). Incubate the mixture on ice for 5–10 min and apply the sample carefully onto an 8% (w/v) sucrose cushion (in 20 mM Tris-HCl, pH 7.4), using about 5 ml sample per 4 ml of cushion. Incubate at 30°C for 3 min. The elevated temperature causes Triton X-114 phase separation, which is observed as a rapid increase in turbidity. Centrifuge the specimens (a swing-out rotor, 2500 rpm, 10 min, 28°C) to separate the Triton X-114 phase from the water phase containing the NCs. Repeat the procedure once; collect the water phase (NCs). Concentrate the NCs by centrifugation through a 20% (w/v) sucrose cushion in 20 mM Tris-HCl, pH 7.4 (Sorvall T-865 rotor, 40,000 rpm, 3 hr, 10°C), using 5 ml sample on a 20-ml cushion. Resuspend the NC pellet overnight into 2 ml of 20 mM Tris, pH 7.4, on ice. The NC preparation contains approximately 2 mg protein/ml.

3. Mix 250 μl of the $4\times$ uncoating mixture, 200 μl bovine serum albumin (BSA, 5 mg/ml), 10 μl of 100 mM ATP, and the NC preparation diluted to a concentration of 750 μg protein/ml with distilled water. Incubate at 23°C for 1 hr. Sediment 1 ml of the specimen per tube through a 5–20% (w/v) sucrose gradient in 20 mM Tris-HCl, pH 8.0, 1 mM MgCl$_2$, 200 mM NaCl, 1 mg/ml BSA, and 1 mM ATP (the uncoating gradient) to the bottom of the tube (Sorvall TH-641 rotor, 35,000 rpm, 2 hr 45 min, 10°C). Resuspend the NC core (the polymerase complex) overnight in 200 μl/tube of 20 mM Tris-HCl, pH 8.0, 200 mM NaCl, 1 mg/ml BSA, 1 mM ATP on ice and store in aliquots at -80°C.

4. If BSA is omitted from the NC core isolation procedure (see above, step 3), protein P8 can be purified from the top fractions after NC core isolation. In that case, collect and combine the shell protein P8-containing top fractions (about 3 ml/tube). Dialyze the pooled top fractions (P8) at 23°C overnight against 10 mM Tris-HCl, pH 8.0, 0.1 mM EGTA. After dialysis, filter the sample through a 0.22-μm filter and apply it into the Mono Q HR5/5 column (Pharmacia) equilibrated with 10 mM Tris-HCl, pH 8.0, 0.1 mM EGTA. Wash with the same buffer (1.5 ml/min) and follow the absorbance at 280 nm, until the basal level is achieved (about 100 ml buffer). Elute with a 0–1 M NaCl gradient in the same buffer at 0.75 ml/min for 60 min. The major peak, containing protein P8 (as analyzed by SDS-

PAGE), elutes at about 280–290 mM NaCl, and contains about 1.5 mg/ml protein P8. At temperatures below 20°C, the concentrated P8 preparation aggregates and should thus be diluted immediately to about 300 μg/ml in 10 mM Tris-HCl, pH 8.0, before storage at −80°C. The total yield of purified protein P8 is about 5–7 mg if 100 mg of viral protein is used as a starting material.

Isolation of Genomic Double-Stranded RNA from φ6

The three φ6 genomic dsRNA segments can be purified from the nucleocapsid or the NC core particle simply by multiple phenol extractions (37). In this method, the RNA is isolated intact without any enzymatic treatments. It is not necessary to use RNase inhibitors or diethyl pyrocarbonate (DEPC)-treated tubes or solutions, but use an unopened box of Eppendorf tubes and be sure that the phenol is fresh and of RNA quality.

1. Adjust the protein concentration of the NC or the NC core particle preparation to 0.5–1 mg/ml. Extract three times with water or with phenol saturated with 0.1 M Tris-HCl, pH 6.0, and five times with ether.
2. Add $\frac{1}{10}$ volume of 3 M sodium acetate and precipitate the RNA by adding 3 volumes of cold (−20°C) ethanol. Centrifuge immediately (14,000 rpm, 15 min, 4°C). Wash the pellet with cold 75% (v/v) ethanol, dry under vacuum, and resuspend into TE buffer (10 mM Tris-HCl, pH 7.4, 1 mM EDTA). Dispense into aliquots and precipitate the RNA with ethanol. Store dry at −20°C.

The purity of the isolated RNA is analyzed by measuring the absorbances at 260 and 280 nm. The presence of residual proteins and organic solvents can be estimated from the A_{260}/A_{280} ratio, which is typically between 2.10 and 2.16 for φ6 dsRNA. We have estimated the RNA concentration from the A_{260} measurement using the value of single-stranded RNA (an A_{260} value of 1 equals 40 μg/ml of ssRNA). From 1 mg of nucleocapsid we have obtained about 270 μg of RNA, which is near the theoretical value (RNA constitutes about 26% of the nucleocapsid mass). The quality of the dsRNA can also be analyzed in a 0.8% agarose gel by slightly overloading the samples in order to see possible degradation of the three RNA segments.

Production and Purification of Recombinant Empty
Procapsid Particles (Polymerase Complexes)

At present, it is impossible to obtain empty active polymerase complexes by disrupting the virus particles. However, they can be produced in *E. coli* from an expression plasmid that harbors a cDNA copy of the entire large genomic segment containing the genes 1, 2, 4, and 7 (pLM450) (38). The particles self-assemble inside *E. coli* from where they can be isolated and purified by the following procedure.

1. Set up an overnight culture of *E. coli* JM109 (pLM450) cells in 500 ml of LB supplemented with tetracycline (10 μg/ml) at 28°C. Next morning, dilute the culture into the same medium to a density of approximately 1 × 10⁸ cells/ml. Incubate with aeration at 28°C or at room temperature until the cell density reaches 5 × 10⁸ cells/ml. At 28°C, this takes about 2 hr. Induce the expression of the procapsid proteins by adding 500 μl of 1 *M* isopropylthio-β-D-galactoside (IPTG). Incubate with aeration at 28°C for 4–5 hr.
2. Collect the cells by centrifugation (Sorvall GSA or GS-3 rotor, 6000 rpm, 10 min, 4°C). Wash the cells once with 20 m*M* Tris-HCl, pH 8.0, 150 m*M* NaCl. After washing, resuspend the cells into 5 ml of the same buffer. Add the protease inhibitor phenylmethylsulfonyl fluoride (PMSF) to a final concentration of 3 m*M*. Lyse the cells by passing them twice through a precooled (4°C) French pressure cell ($\frac{3}{8}$-inch cell at 20,000 psi).
3. Remove the membrane debris from the disrupted specimen by extracting with 3–5% Triton X-114 in 10 m*M* Tris-HCl, pH 8.0, 150 m*M* NaCl as described above in the NC isolation procedure (step 2). Take the aqueous layer from the top of the sucrose cushion and repeat the procedure.
4. Apply the partially purified procapsid preparation onto 5–20% (w/v) sucrose gradients (in 10 m*M* Tris-HCl, pH 8.0, 150 m*M* NaCl), using approximately 500 μl/ gradient, and centrifuge (Sorvall TH-641 rotor, 27,000 rpm, 1 hr 50 min, 15°C).
5. Collect the light-scattering zone from the middle of the gradient. Measure the protein concentration and analyze the protein composition by SDS-PAGE. The protein concentration is usually 200–400 μg/ml. Divide the preparation into aliquots and store at −80°C.

Plus Strand Synthesis by Nucleocapsid Cores (Polymerase Complex)

The NC core, the viral RNA polymerase complex, is capable of plus strand synthesis when all four nucleoside triphosphates (NTPs) and either manganese or magnesium ions are present (35). In the manganese-containing reaction mixture, ssRNA molecules, corresponding to all of the ϕ6 genomic segments (L, M, and S), are produced. In the magnesium-containing reaction mixture, the plus strand synthesis is regulated so that only M and S are transcribed. The plus strand synthesis can be analyzed by a filter assay of [³²P]UMP incorporation into an acid-insoluble form (34) or by an agarose gel analysis of the RNA products (39, 40).

The composition of 2× transcription reaction mixture is as follows: 2 m*M* each of ATP, CTP, and GTP, 0.2 m*M* UTP, 100 m*M* Tris-HCl, pH 8.9, 12 m*M* MgCl₂ or 4 m*M* MnCl₂, 200 m*M* KCl, 100 m*M* NH₄Cl, and 10 m*M* dithiothreitol (DTT).

1. Dilute the NC core preparation (obtained as described in the section on isolation of φ6 subviral particles, step 3) 1:10 into 20 mM Tris-HCl, pH 8.0, 100 mM NaCl, 1 mg/ml BSA, 1 mM ATP.

2. Dispense 16 μl of 2× transcription mixture into an Eppendorf tube. Add 3 μl of 50% PEG 6000, 1 μl of RNasin (20–40 U/μl; Promega, Madison, WI), 2 μCi of [α-^{32}P]UTP (Amersham PB10203; 3000 Ci/mmol), and 10 μl of the diluted polymerase particle preparation into the tube. The final volume of the reaction is 33 μl.

3. Incubate at 30°C for 1 hr. Add 10 μg carrier tRNA and 1 ml of 10% (w/w) trichloroacetic acid (TCA) and incubate on ice for 30 min.

4. Pour the contents of the tube onto a 25-mm Whatman GF/C (Maidstone, England) filter prewetted with 10% TCA. Rinse the tube with 3 ml of cold 10% TCA. Wash the filter once with cold 10% TCA, five times with 5% TCA, twice with 99% (v/v) ethanol, and once with acetone. Dry the filter and place it in a scintillation vial. Add 4 ml of the scintillation liquid (Optiphase HiSafe III, LKB Wallac, Turkü, Finland) and measure the radioactivity by liquid scintillation counting.

5. If it is desired to analyze the transcription products by agarose gel electrophoresis, after 1 hr of reaction, add 10 μg carrier tRNA and 35 μl of phenol–chloroform–isoamyl alcohol (25:24:1, v/v) and centrifuge (14,000 rpm, 15 min, 23°C). Precipitate the RNA from the water phase by adding 100 μl of 2.5 M ammonium acetate and 2.5 volumes of 99% (v/v) ice-cold ethanol. Keep at −80°C for 20–30 min, centrifuge (14,000 rpm, 15 min, 4°C), wash the pellet with 70% (v/v) ethanol, and dissolve into 15 μl of 1 mM EDTA. Analyze the RNA in a 1% agarose gel containing about 0.25 μg/ml ethidium bromide. The ethidium bromide is necessary to separate the ssRNA segments from the dsRNA. Detect the RNA fluorescence with UV light and dry the gel. Expose on X-ray film to detect the three separated ^{32}P-labeled dsRNA segments and the produced ssRNA plus strands.

ckaging of φ6 ssRNA Segments by Recombinant lymerase Complex Particles (Procapsids)

The empty φ6 procapsid is capable of packaging each of the φ6 ssRNA segments independently (31, 31a). Thus, the packaging reaction can be carried out either with any one or with all of the φ6 ssRNA segments. The energy for the packaging reaction is provided by nucleoside triphosphates that are cleaved to diphosphates. The packaged RNA remains single-stranded but is protected from RNase action. In the following assay for RNA packaging, ^{32}P-labeled RNA is used as a substrate. The unpackaged RNA is hydrolyzed by adding an excess of RNase A. In TCA precipitation, the packaged RNA forms an acid-insoluble fraction which can be quantitated by liquid scintillation counting.

All φ6 ssRNA segments can be produced in the (+) strand synthesis reaction by the isolated polymerase complex in the presence of MnCl$_2$ (see above). The individual ssRNA segments can be separated by a sucrose gradient centrifugation step

(31). Alternatively, (+) strand RNA can be produced from a cDNA copy of the genomic segment that is cloned under the T7 promoter in an RNA synthesis vector (such as Pharmacia pT7T3 19U). The *in vitro* transcription reactions are carried out with [^{32}P]UTP to produce radioactively labeled RNA segments (see the specifications of the supplier of the T7 RNA polymerase).

The packaging reaction is extremely sensitive to the quality of RNA. All reagents, tubes, and pipette tips should be RNase-free. Gloves should be worn to prevent RNase contamination. The standard packaging reaction contains 12.5 μl of 2× buffer, 2.5 μl of 10 mM ATP, 200 ng ssRNA segment(s) [3–5 × 10^5 counts/min (cpm)], 1 μg procapsid (from procapsid purification procedure), and water to 25 μl. The composition of 2× buffer is as follows: 100 mM Tris-HCl, pH 8.9, 10 mM MgCl$_2$, 160 mM ammonium acetate, 4 mM DTT, 0.2 mM EDTA, 20% PEG 4000, and 2U/μl RNasin (Promega). Unlabeled RNA is added to the radioactively labeled RNA to obtain a total of 200 ng of each segment.

1. Start the reaction by adding the procapsids to the reaction mixture. Incubate at 30°C for 60 min.
2. Add 5 μl of RNase A (2 mg/ml); incubate 15 min at 30°C. The high RNase concentration can contaminate the pipette with RNases. Therefore, use either pipette tips that have filters or another pipette when adding the RNase.
3. Add 10 μg carrier tRNA to the side of the tube (not to the reaction mixture, as it contains RNase). Stop the reaction by adding 1 ml of cold 10% TCA. Keep the tubes on ice at least for 30 min.
4. Precipitate the acid-insoluble material onto the GF/C filters (Whatman) as described previously (plus strand synthesis reaction protocol, step 4).

Each experiment should contain two control reactions. In one control, ATP is replaced with a noncleavable analog, β,γ-methyleneadenosine 5'-triphosphate (AMP-PcP), that prevents the packaging reaction and thus gives the value for background precipitation. The other control reaction is a standard reaction except that the RNase treatment step is excluded. This measures the total amount of labeled RNA (in cpm) in the reaction. In a typical reaction, about 20% of the labeled RNA (e.g., 60,000–100,000 cpm, depending on the amount of the label used) is packaged into procapsids. The control reaction with AMP-PcP should contain less than 5000 cpm.

Minus Strand Synthesis of Packaged Polymerase Complex Particles

In the ϕ6 (−) strand synthesis reaction, a complementary strand is synthesized inside the procapsid particle using the packaged (+) strand as a template. The (−) strand synthesis requires the packaging of all three (+)RNA segments (31, 32). In the packaging reaction, the packaging activity was measured as protection of prelabeled RNA from RNases, whereas the measure for (−) strand synthesis is the incorporation of [^{32}P]UMP into dsRNA. Therefore, the packaged RNA in the (−) strand synthesis reaction is unlabeled.

The standard (−) strand synthesis reaction contains 12.5 μl of 2× buffer (same as in the packaging reaction), 2.5 μl of 2 mM each of ATP, CTP, GTP, and UTP, 1.0 μl [α-^{32}P]UTP (Amersham PB10203, 3000 Ci/ml), 500 ng each RNA segment (s, m, and l), 1 μg procapsid (from the procapsid purification procedure), and water to 25 μl.

1. Start the reaction by adding the procapsids to the reaction mixture. Incubate at 30°C for 60 min.
2. Take a 5-μl sample and measure the incorporated radioactivity by precipitating with 10% TCA (see plus strand synthesis reaction protocol, steps 3 and 4).
3. Add 180 μl of water to the rest of the reaction and extract once with 200 μl of phenol–chloroform–isoamyl alcohol (25 : 24 : 1, v/v). Take the aqueous phase and extract once with 200 μl of chloroform–isoamyl alcohol (24 : 1, v/v).
4. Precipitate the RNA by adding $\frac{1}{5}$ volume of 7.5 M ammonium acetate and 2.5 volumes of 94% (v/v) ethanol. Keep at − 20°C for 20 min. Centrifuge (14,000 rpm, 15 min, 4°C), discard the supernatant, dry the RNA pellet, and dissolve it in 10 μl water.

The amount of incorporated radioactivity gives a measure of RNA synthesis activity. Because the procapsid will switch to the (+) strand transcription mode after the (−) strand synthesis is completed, it is important to analyze the reaction products by agarose gel electrophoresis. The strand separation gel system (40), where the (+) and (−) strands are separated from one another, will give an estimate of the (+) and (−) strand synthesis activities. The (+) strand synthesis is inhibited by a low concentration of the NTPs (0.2 mM of each) in the reaction mixture. It can, however, be further inhibited by adding 0.4 mM CaCl$_2$ (final concentration) to the reaction mixture (35).

Formation of Infectious Nucleocapsids in Vitro

Purified φ6 NCs cannot enter the host cells through the OM. The NCs must reach the host periplasmic space to be able to interact with the CM (28). Host cells can be rendered competent for infection with purified NCs by repeated washings with salt and sucrose, and subsequent addition of lysozyme. This leads to the partial removal of the OM and peptidoglucan, permitting the NCs to interact with the CM and initiate a productive infection (41).

The NC shell protein P8 is necessary for the interaction with the CM, and neither the RNA-filled procapsids nor the NC cores are capable of penetrating the CM. However, infectious NCs can be formed by *in vitro* assembly of the purified shell protein P8 onto the dsRNA-filled procapsids (42) or onto the NC core particles (29).

1. Set up an overnight 10-ml culture of MP0.16 in LB and grow at 28°C with aeration. MP0.16 is a receptor-less strain of *P. syringae* (42).
2. Dilute the culture next morning into LB to obtain a 40-ml culture with about 1×10^8 cells/ml. Grow to 5×10^8 cells/ml and then collect the cells by centrifugation (8000 rpm, 2 min, 4°C). Resuspend the cells gently and wash three times with 10 ml of ice-cold 0.5 M NaCl and twice with ice-cold 20% (w/v) sucrose in 30 mM Tris-HCl, pH 7.9. Resuspend the resulting spheroplasts in 1 ml of ice-cold 10 mM potassium phosphate, pH 7.2, 3% (w/v) lactose, 2% (w/v) BSA, and 50% (v/v) LB and keep on ice. The cell density in the spheroplast preparation will be about 1×10^{10} cells/ml. The cells can be stored on ice for 1–2 hr.
3. To assemble the P8 on the RNA-filled procapsids or NC cores, mix 1 μg of filled procapsids or NC cores and 2.2 μg of P8 in 30 μl of 0.5 mM CaCl$_2$, 100 mM KCl, and 100 mM ammonium acetate. Incubate at 23°C for 1 hr.
4. Add 5 μg/ml lysozyme and 20 μl of the above mixture to 110 μl of freshly prepared spheroplasts. Dispense 100 μl of the specimen on a Millipore VSWPO2500 filter placed on an LB plate overlaid with LB top agar containing 3% lactose and 20 mM potassium phosphate, pH 7.2, and incubate at room temperature for 1 hr to infect the spheroplasts.
5. Remove the infected cells from the filter, dilute into 10 mM potassium phosphate, pH 7.2, 3% lactose, 500 mM NaCl, 2% BSA, and assay for plaques using HB10Y as a host. The average yield of the NC infection varies between 10^6 to 10^7 infective centers/ml.

Conclusions

Materials prepared as above have proved to be suitable for physical studies such as cryoelectron microscopy, Raman spectroscopy, and crystallization. Development of the ion-exchange chromatographic methods for virus purification has particularly improved the quality of the material. The different packaging and polymerase assays described for the ϕ6 RNA polymerase complex as well as the *in vitro* assembly methods should help to design comparative studies of viruses belonging to the reovirus family.

References

1. M. M. Bradford, *Anal. Biochem.* **72,** 248 (1976).
2. V. M. Olkkonen and D. H. Bamford, *Virology* **171,** 229 (1989).

3. D. H. Coombs and N. R. Watts, *Anal. Biochem.* **148,** 254 (1985).
4. J. Sambrook, E. F. Fritsch, and T. Maniatis, "Molecular Cloning: A Laboratory Manual," 2nd Ed. Cold Spring Harbor Laboratory, Cold Spring Harbor, New York, 1989.
5. J. K. H. Bamford, C. Luo, J. T. Juuti, V. M. Olkkonen, and D. H. Bamford, *Virology* **197,** 652 (1993).
6. C. Lyra, H. Savilahti, and D. H. Bamford, *Mol. Gen. Genet.* **228,** 65 (1991).
7. D. H. Bamford, L. Rouhiainen, K. Takkinen, and H. Söderlund, *J. Gen. Virol.* **57,** 365 (1981).
8. D. H. Bamford and L. Mindich, *J. Virol.* **44,** 1031 (1982).
9. T. N. Davis, E. D. Muller, and J. E. Cronan, Jr., *Virology* **120,** 287 (1982).
10. J. Caldentey, C. Luo, and D. H. Bamford, *Virology* **194,** 557 (1993).
11. D. H. Bamford, T. McGraw, G. Mackenzie, and L. Mindich, *J. Virol.* **47,** 311 (1983).
12. D. H. Bamford and L. Mindich, *J. Virol.* **50,** 309 (1984).
13. H. Savilahti, J. Caldentey, K. Lundström, J. Syväoja, and D. H. Bamford, *J. Biol. Chem.* **266,** 18737 (1991).
14. J. Caldentey, L. Blanco, D. H. Bamford, and M. Salas, *Nucleic Acids Res.* **21,** 3725 (1993).
15. L. Mindich and D. H. Bamford, *in* "The Bacteriophages" (R. Calendar, ed.), p. 475. Plenum, New York, 1988.
16. J. Cadentey, J. K. H. Bamford, and D. H. Bamford, *J. Struct. Biol.* **104,** 44 (1990).
17. J. K. H. Bamford and D. H. Bamford, *Virology* **177,** 445 (1990).
18. J. K. H. Bamford and D. H. Bamford, *Virology* **181,** 348 (1991).
19. L. Walin, R. Tuma, G. J. Thomas, Jr., and D. H. Bamford, *Virology* **201,** 1 (1994).
20. L. Mindich, D. Bamford, T. McGraw, and G. Mackenzie, *J. Virol.* **44,** 1021 (1982).
21. S. Grimes and D. Anderson, *J. Mol. Biol.* **209,** 91 (1989).
22. J. M. Kenney, J. Hantula, S. D. Fuller, L. Mindich, P. M. Ojala, and D. H. Bamford, *Virology* **190,** 635 (1992).
23. J. L. Van Etten, L. Lane, C. Gonzales, J. Partridge, and A. Vidaver, *J. Virol.* **33,** 769 (1976).
24. P. Gottlieb, J. Strassman, X. Qiao, A. Frucht, and L. Mindich, *J. Bacteriol.* **172,** 5774 (1990).
25. M. Romantschuk and D. H. Bamford, *J. Gen. Virol.* **66,** 2461 (1985).
26. D. H. Bamford, E. T. Palva, and K. Lounatmaa, *J. Gen. Virol.* **32,** 249 (1976).
27. D. H. Bamford, M. Romantschuk, and P. J. Somerharju, *EMBO J.* **6,** 1467 (1987).
28. M. Romantschuk, V. M. Olkkonen, and D. H. Bamford, *EMBO J.* **7,** 1821 (1988).
29. V. M. Olkkonen, P. M. Ojala, and D. H. Bamford, *J. Mol. Biol.* **218,** 569 (1991).
30. D. A. Cuppels, J. L. Van Etten, D. E. Burbank, L. C. Lane, and A. K. Vidaver, *J. Virol.* **35,** 249 (1980).
31. P. Gottlieb, J. Strassman, X. Qiao, M. Frilander, A. Frucht, and L. Mindich, *J. Virol.* **66,** 2611 (1992).
31a. M. Frilander and D. H. Bamford, *J. Mol. Biol.* **246,** in press.
32. M. Frilander, P. Gottlieb, J. Strassman, D. H. Bamford, and L. Mindich, *J. Virol.* **66,** 5013 (1992).
33. D. H. Bamford, *in* "Bacteriophage Assembly" (M. S. du Dow, ed.), p. 477. Alan R. Liss, New York, 1981.
34. P. M. Ojala, J. T. Juuti, and D. H. Bamford, *J. Virol.* **67,** 2879 (1993).
35. P. M. Ojala and D. H. Bamford, *Virololy* **207,** 400 (1995).

36. C. Bordier, *J. Biol. Chem.* **256,** 1604 (1981).
37. J. K. H. Bamford, D. H. Bamford, T. Li, and G. J. Thomas, Jr., *J. Mol. Biol.* **230,** 473 (1993).
38. P. Gottlieb, J. Strassman, D. H. Bamford, and L. Mindich, *J. Virol.* **62,** 181 (1988).
39. M. E. Ewen and H. R. Revel, *Virology* **178,** 509 (1988).
40. N. Pagratis and H. R. Revel, *Virology* **177,** 273 (1990).
41. P. M. Ojala, M. Romantschuk, and D. H. Bamford, *Virology* **178,** 364 (1990).
42. V. M. Olkkonen, P. Gottlieb, J. Strassman, X. Qiao, D. H. Bamford, and L. Mindich, *Proc. Natl. Acad. Sci. U.S.A.* **87,** 9173 (1990).
43. T. Li, D. H. Bamford, J. K. H. Bamford, and G. J. Thomas, Jr., *J. Mol. Biol.* **230,** 461 (1993).

Index